■ Encyclopedia of Human-Animal Relationships

■ Encyclopedia of Human-Animal Relationships

A Global Exploration of Our Connections with Animals

Volume 3: Eth–Liv

Edited by
Marc Bekoff

GREENWOOD PRESS
Westport, Connecticut • London

Library of Congress Cataloging-in-Publication Data

Encyclopedia of human-animal relationships : a global exploration of our connections with animals / edited by Marc Bekoff.

   p. cm.

Includes bibliographical references and index.
ISBN-13: 978-0-313-33487-0 (set : alk. paper)
ISBN-13: 978-0-313-33488-7 (vol 1 : alk. paper)
ISBN-13: 978-0-313-33489-4 (vol 2 : alk. paper)
ISBN-13: 978-0-313-33490-0 (vol 3 : alk. paper)
ISBN-13: 978-0-313-33491-7 (vol 4 : alk. paper)

1. Human-animal relationships—Encyclopedias. I. Bekoff, Marc.
QL85.E53 2007
590—dc22   2007016552

British Library Cataloguing in Publication Data is available.

Library of Congress Catalog Card Number: 2007016552

ISBN–13: 978–0–313–33487–0 (Set)
ISBN–13: 978–0–313–33488–7 (vol. I)
ISBN–13: 978–0–313–33489–4 (vol. II)
ISBN–13: 978–0–313–33490–0 (vol. III)
ISBN–13: 978–0–313–33491–7 (vol. IV)

First published in 2007

Greenwood Press, 88 Post Road West, Westport, CT 06881
An imprint of Greenwood Publishing Group, Inc.
www.greenwood.com

Printed in the United States of America

The paper used in this book complies with the Permanent Paper Standard issued by the National Information Standards Organization (Z39.48-1984).

10 9 8 7 6 5 4 3 2 1

For my parents, who always encouraged the animal in me.

■ Contents

■ Alphabetical List of Entries

Animal Assistance to Humans (*See also* **Animals at Work**)
Animal-Assisted Interventions
Assistance and Therapy Animals
Assistance Dogs
Giant African Pouched Rats Saving Human Lives
Horse-Assisted Therapy: Psychotherapy with Horses
Horse-Assisted Therapy: Reaching Troubled Young Women
India and Animal Therapy
Psychiatric Service Dogs
Service Dogs: A Personal Essay

Animal Rights. *See* **Ethics and Animal Protection**

Animals as Food
Dog Eating in the Philippines
Dogs and Cats as Food in Asia
Dogs in China May Be Food or Friend
Entomophagy (Eating Insects)
Global Diversity and Bushmeat
Veganism

Animals at Work (*See also* **Animal Assistance to Humans**)
Animals in the London Blitz, 1939–1945
Animals in War
Dog and Human Cooperation in Hunting
Stock Dogs and Livestock

Anthropomorphism (*See also* **Human Perceptions of Animals**)
Anthropomorphism
Human Observations of Animals, Subjective vs. Objective
The Myth of "Sexual Cannibalism" by the Praying Mantis

Applied Anthrozoology and Veterinary Practice
Applied Animal Behaviorists
Bustad, Leo Kenneth (1920–1998): Pioneer in Human-Animal Interactions
Veterinarian Training and the Human-Animal Connection
    Sidebar: Stresses at the Vet's Office

Archaeology
Archaeology and Animals

Art
Animals in Art
Animals in Celtic Art
Insects in Art
Rock Art and Shamanism

■ List of Entries that Feature Specific Animals

This list includes those entries that focus on particular animals. Other entries also include information about these animals but not to the extent that the following essays do. Please see the index for all mentions of the large number of different animals that are included in these volumes.

Amphibians

Zoos and Aquariums
Amphibians and Zookeepers

Apes. *See* Great Apes

Bats

Conservation and Environment
India's Bats and Human Attitudes
Living with Animals
Bats and People

Bears

Conservation and Environment
Bears and Humans
Cruelty to Animals
The Bear Bile Industry in China: A Personal Essay

Bees. *See* Honeybees

Birds (*See also* Chickens; Emus; Falcons; Kiwis; Parrots; Penguins; Ravens; Turkeys)

Bonding
Discrimination between Humans by Emus and Rheas
Communication and Language
Birdsong and Human Speech
Interspecies Communication—N'kisi the Parrot: A Personal Essay
Similarities in Vocal Learning between Animals and Humans
Conservation and Environment
Birds and Recreationists
The Passenger Pigeon

Culture, Religion, and Belief Systems
>   *Birds as Symbols in Human Culture*

Ethics and Animal Protection
>   *Raptor Rehabilitation and Behavior after Release*

History
>   *Honeyguides: The Birds that Lead to Beehives* [sidebar to *Harvesting from Honeybees*]

Hunting, Fishing, and Trapping
>   *Falcons, Hawks and Nocturnal Birds of Prey*

Living with Animals
>   *Birds and Problems from Humans*

Music, Dance, and Theater
>   *Interspecies Music*

Rehabilitation of Animals
>   *Wild Bird Rehabilitation—Rare Intimacies: A Personal Essay*

Cats

Animals as Food
>   *Dogs and Cats as Food in Asia*

Bonding
>   *Companion Animals*

Culture, Religion, and Belief Systems
>   *Animal Mummies in Ancient Egypt*
>   *Chinese Youth Attitudes towards Animals*

Literature
>   *Children's Literature: Cats*

Living with Animals
>   *Cat and Human Relationships*

Research
>   *Animal Use by Humans in Research and Experimentation*

Cats, Large (*See also* Tigers)

Enrichment for Animals
>   *Emotional Enrichment of Captive Big Cats*
>   *Do Big Cats Purr?* [sidebar]

Cattle

Animals as Food
>   *Veganism*

Culture, Religion, and Belief Systems
>   *Cattle Mutilation*
>   *India's Holy Cows and India's Food: Issues and Possibilities*
>   *Judaism and Animals*

Ethics and Animal Protection
>   *Compassion in World Farming*

Health
>   *Animal Reservoirs of Human Disease*
>   *Humans, Viruses, and Evolution*

■ Ethics and Animal Protection
Attitudes toward the Treatment of Animals

Human attitudes toward the use of other species are complex, multidimensional, and often inconsistent. Many factors affect our judgments about how animal should be treated. These generally fall into four categories.

1. *Factors related to the species.* How we feel about the use of a species is affected by its characteristics. These include, for example, its size (large animals tend to be valued more), appearance (animals with appealing features such as large eyes have higher status), its perceived intelligence, and similarity to humans (e.g., elevated status is afforded whales and chimpanzees).

2. *Factors related to the role of the species in human life.* Attitudes toward the use of animals are also affected by the species' historic relationships with humans. The feelings that people have about species that are considered pets, such as dogs and cats, are very different from animals such as rats and cockroaches that are almost universally regarded as pests. Attitudes toward some types of animals may have their origins in our distant past. For example, Arne Öhman and Susan Mineka (2003) believe that humans have evolved a special mental mechanism that accounts for the fact that the fear of snakes is a particularly common and easily acquired phobia.

3. *Factors related to culture.* Cultural traditions about the use of animal vary widely. In parts of India, cows and monkey are considered to be sacred. In Korea, some puppies serve as family pets, whereas other are served as the main course of the family dinner. The way a species is viewed within a culture can change rapidly. An example is the attitudes of Americans toward wolves; once regarded as dangerous pests to be shot on sight, the image of the species has undergone a reversal in recent decades, and wolves are now venerated in many circles, though they are still reviled by some ranchers.

4. *Factors related to individual differences.* Many factors are associated with differences of opinion about how animals should be treated. However, most of these variables explain only a small proportion of individual differences in attitudes toward other species. Gender is an exception. Nearly all studies that have examined gender differences in attitudes toward the use of animals have found that women are more concerned with animal welfare than men. Women, for example, are more likely than men to donate money to animal protection organizations, demonstrate for animal rights, and become involved in dog and cat rescue activities. Men, on the other hand, are typically more supportive of animal research than women and are much more likely to engage in recreational activities such as hunting. Further, research on gender differences in attitudes toward animal welfare crosses national boundaries. For example, women have been found to oppose animal research more than men in Japan, Canada, Australia, and across Europe.

Several studies have examined the relationships between personality traits and attitudes toward the treatment of other species. While a few personality traits seem to be associated with pro-animal attitudes (e.g., interpersonal sensitivity), the correlations between most personality scale scores and attitudes toward other species tend to be low. Other variables that are correlated with attitudes toward animal welfare include empathy, ethical idealism, misanthropy, the belief that animals have minds, and exposure to pets at an early age.

Assessing Attitudes toward the Use of Animals

Social scientists have developed sophisticated methods to assess public attitudes in many spheres, including human-animal interactions. Studies of animal-related attitudes generally fall into two categories. In the first category are national surveys conducted by professional polling organizations such as the National Opinion Research Center and the Gallup Organization. These polls are typically commissioned by corporations, nonprofit organizations, or government agencies that seek to gauge public opinion about specific issues. They are typically based on telephone interviews of representative samples of about 1,000 adults and have a margin of error of three or four percentage points. These polls are often narrowly focused on issues of interest to the sponsor, such as attitudes toward the use of animals in medical research, pet ownership, and partici-pation in animal-related recreation such as hunting and fishing. Despite the fact that hundreds of these high-quality polls have included questions related to attitudes about animals, scholars have largely ignored this important source of information. One reason is that the results of these studies are rarely published in academic journals. Fortunately, private polling data related to animal issues are increasingly readily available to scholars and the public at little or no cost through electronic databases.

The second major source of information about attitudes toward the use of other species is university-based research. Many of these studies use college students as participants. Unlike commercial polls, which tend to focus on a few specific issues such as pet ownership or attitudes toward fur, these studies typically use multi-item scales designed to assess factors that mediate individual differences in global support for animal-welfare issues. An example is the twenty-item Animal Attitude Scale, which includes questions related to the use of animals in research, zoos, blood sports, and agriculture. This and similar scales have been used to examine the relationship between attitudes toward animal use and factors such as gender, empathy, personal moral philos-ophy, and personality traits.

Inconsistencies in Animal-Related Attitudes

The responses that people give to opinion surveys sometimes seem inconsistent. Consider, for example, responses to an item in a 1993 national survey of attitudes toward the use of animals. About half of the individuals surveyed agreed with the statement, "Animals are just like people in all important ways." At the same time, however, well over 95 percent of American regularly consumed the flesh of the creatures that they believed were "just like people in all important ways." These types of inconsistencies can become institutionalized and even legislated. Take the legal status of animals under the Animal Welfare Act, the primary federal legislation in the United States that provides protection for animals used in research. While the law pertains to the use of nonhuman mammals in science, Congress has mandated that rats and mice are not considered to be "animals" under the provisions of the Act. Thus, while hamsters are afforded full pro-tection under the law, laboratory mice have no legal status.

There are several possible reasons for the inconsistencies in attitudes toward the use of animals. One is that many—perhaps most—people have given relatively little thought to animal-related issues. When questioned by a pollster, their responses about the treatment of animals may reflect off-the-cuff judgments that are not part of an inte-grated ethical perspective. These superficial opinions are referred to by social psycholo-gists as "non-attitudes" or "vacuous attitudes." Individuals such as animal activists who have given considerable thought to moral issues related to nonhuman species are more likely to exhibit constellations of animal-focused attitudes that reflect a coherent and consistent approach to animal ethics. Some inconsistencies stem from the roles that we assign animals. Thus while mice—historically pests—are not considered animals under the Animal Welfare Act, dogs—historically pets—are given special status; the Act requires that dogs in laboratories be given daily exercise.

Changing Attitudes toward Animals

Attitudes toward some uses of animals have changed substantially in recent decades. Consider support for biomedical research. A 1948 poll by the National Opinion Research Center asked 2,500 Americans, "In general, do you favor or oppose the use of live animals in medical teaching and research?" At that time 84 percent of respondents indicated that they favored animal research. Beginning in 1985, the National Science Board began to regularly gauge support for animal research by asking representative samples of American adults how they felt about the following statement: "Scientists should be allowed to do research that causes pain and injury to animals like dogs and chimpanzees if it produces new information about human health problems." When the question was first asked in 1985, nearly two-thirds of respondents supported animal research. By 1999, the country was evenly split on the issue of research with dogs and chimpanzees. In 2001, for the first time, more people opposed research on dogs and chimpanzees than supported it. Note, however, when mice are substituted for dogs and chimpanzees in the question, the proportion of people who support animal research increases dramatically (67 percent approve and 30 percent disapprove).

While an increasing number of Americans feel that animals deserve serious moral consideration, this does not always translate to changes in behavior. An example is attitudes toward meat. A 2005 telephone survey of about 1,000 American adults conducted by CBS News found that only 2 percent of respondents said that they were presently vegetarians, 6 percent said they had been in the past, and 92 percent had never considered themselves to be a vegetarian. In fact, the proportion of vegetarians may have decreased in recent years; a 1992 Time/CNN poll of 1,400 people found that 7 percent of respondents considered themselves to be vegetarians at the time of the poll.

Theoretical Models of Attitudes toward Animals

Several theoretical models have been developed that provide useful perspectives on human attitudes toward other species. Stephen Kellert (1996) has extensively studied the different patterns of attitudes that individuals exhibit when thinking about nonhuman animals. Kellert believes these attitudes form a constellation of nine "nature values." He refers to these as naturalistic, ecologistic, humanistic, moralistic, scientistic, aesthetic, utilitarian, dominionistics, and negativistic. Kellert has found that these basic values are affected by factors such as age, gender, and nationality. He has also traced changes in attitudes toward animal over historical time.

James Serpell developed a simple yet elegant model of attitudes toward other species. He argues that these attitudes fall along two dimensions. The first dimension is "affect"—the degree that individuals feel positively or negatively about a species. This dimension varies between highly positive (love, sympathy, identification) to highly negative (fear, loathing, misidentification). The second dimension is "utility"—the degree that a type of animal is perceived as beneficial or detrimental to human interests. Serpell points out that attitudes toward members of a single species can vary along these two dimensions. For example, guide dogs for the blind are high on both positive affect and utility. In contrast, in Korea, dogs raised to be eaten exhibit high on utility and neural affect, and stray dogs in Saudi Arabia are considered pests—high negative affect and detrimental utility. Our attitudes toward the use of animals are influenced by the same types of psychological and cultural factors that affect attitudes toward social and political issues generally.

Further Resources

Herzog, H., Rowan, A., & Kossow, D. (2001). Social attitudes and animals. In D. Salem & A. Rowan (Eds.), *The state of the animals 2001* (pp. 311–28). Washington, DC: Humane Society Press.

Kellert, S. (1996). *The value of life: biological diversity and human society*. New York: Island Press.

Öhman, A., & Mineka, S. (2003). The malicious serpent: Snakes as a prototypical stimulus for an evolved module of fear. *Current Directions in Psychological Science, 12*(1), 5–9.

Serpell, J. A. (2004). Factors influencing human attitudes to animals and their welfare. *Animal Welfare, 13,* 145–51.

Harold Herzog

■ Ethics and Animal Protection
Best Friends Animal Society

Best Friends Animal Society is the United States' largest companion animal sanctuary and a worldwide network of people who care about animals. In the 1980s, most major metropolitan humane societies and animal control agencies operated according to standards that had changed little since the 1950s. Prior to this, the battles being fought by the humane community had more to do with how to kill shelter animals more humanely than whether they should be killed at all. No serious consideration was being given to finding alternatives to killing homeless pets as a means of population control until the "no-kill" movement emerged in the 1980s.

Best Friends Animal Sanctuary was founded as a no-kill refuge in 1984 by a group of friends who were concerned about the millions of homeless pets (approximately 17 million per year in 1987) being killed at shelters and humane societies all over the country. Best Friends quickly became one of the flagships of the grassroots no-kill movement that swept the country in the 1990s.

The sanctuary, at the heart of the Golden Circle of southern Utah, is home, on any given day, to about 1,500 dogs, cats, and other animals who come for special care from shelters and rescue groups all over the country. Most of them are soon ready to go to good, new homes. Others, who are older or sicker or who have suffered extra trauma, find a home and a haven at the sanctuary and are given loving care for the rest of their lives or until they, too, are re-homed—however long that may take. Best Friends believes that there is a potential home for every homeless pet. The work of Best Friends is made possible entirely through the donations of members and supporters.

Driven by its grassroots membership, the Best Friends Network enables people worldwide to work together to help animals by setting up geographical communities or communities of common interest. Approximately 25,000 people visit the sanctuary every year, as part of a vacation or to take part in educational programs, internship opportunities, and programs for youth. Not surprisingly, many visitors leave for home with a new pet in tow, or with a plan as to how they can help homeless pets in their own community.

The work of Best Friends is premised on the knowledge that kindness to animals builds a better world for all of us. The no-kill philosophy dictates that as long as an animal has or can have a satisfactory quality of life, they are worth the investment of time, effort, and money.

Special medical and behavioral care for animals at the sanctuary is made possible by donors called the Guardian Angel members of Best Friends. This level of care includes, for example, dedicated areas for seniors and hydrotherapy capabilities for the rehabilitation of dogs recovering from orthopedic surgery or suffering from degenerative conditions such as arthritis. There are also specially designed training facilities for dogs with behavioral problems. For cats, there are special areas and trained caregivers for animals with spinal injuries, feline leukemia virus, and FIV. The sanctuary also provides state and federally licensed wildlife rehabilitation, including the largest raptor flight aviary in the region.

No More Homeless Pets Programs

In the mid-1990s, working with local humane organizations, Best Friends began launching No More Homeless Pets campaigns in several cities and states across the country, helping individual people, humane groups, and entire communities to set up spay/neuter, shelter, foster, and adoption programs in their own neighborhoods, cities, and states. By 2000, the number of animals being killed every year in shelters had dropped to below 5 million. The goal of Best Friends is to reach a time when no companion animals are being killed in shelters for want of a home.

Rescue, Rapid Response, and Model Programs

In 2005, in the wake of Hurricane Katrina, Best Friends established one of the primary animal rescue and recovery shelters in the Gulf Coast region and rescued more than 4,300 displaced pets, which were either reunited with their families, placed in new homes, or put into foster care. Best Friends rescue centers remained open for almost nine months to provide for even the most traumatized of animals.

In another example, in Reno, Nevada, Best Friends came to the rescue of 1,500 rabbits in a backyard hoarding situation, establishing a rescue ranch, bringing in veterinarians and volunteers from around the country to provide medical care and spay/neuter, and then placing the rabbits in good homes.

As with the sanctuary and the No More Homeless Pets campaign, all such efforts are model programs, created by Best Friends and then brought into the mainstream of animal welfare. Present activities include developing model programs and collaborating with individuals and organizations across the country to address the situation of the types of animals who are destroyed in municipal shelters: feral cats, large dogs with behavior issues, older animals, and, increasingly, rabbits.

Further Resources

Best Friends magazine: Published six times a year by the society, *Best Friends* magazine helps to drive the no-kill movement, while at the same time offering good-news and humorous and quirky stories about animals and nature.

Best Friends Sanctuary and Network. http://www.bestfriends.org/

Michael Mountain

■ Ethics and Animal Protection
Biomedicine and Animals

Introduction

The history of the use of animals in medical research goes back a long way. The most famous physician of ancient times, Galena Claudio from Pergamo (129–200), used animals for his research on human anatomy. After the decline of medical science in the Middle Ages and following its rebirth during the Renaissance, from the eighteenth century on the idea that animals could realistically contribute to the study of human medicine was increasingly accepted. The number of experiments involving animals grew exponentially toward the end of the nineteenth century, partly thanks to Claude Bernard, a French scientist who was a great advocate of the use of animals in physiological research. Not only did the number of experiments increase, but the number of different species used in the experiments also increased. As a matter of fact, besides cats and dogs, mice, rats and, later, different species of birds, amphibians, reptiles, and fish entered the research laboratories. In 1940 one million laboratory animals were used in Great Britain alone; in 1960 the number rose to 3.5 million, 90 percent of them being mice and rats. The number of laboratory animals reached a plateau at the end of the 1970s and began to decrease at the beginning of the 1980s, thanks to the discovery of new methodologies that could be used as alternatives to animal research. However, new discoveries from time to time in medicine and biology are likely to generate the need for new animal experiments. Therefore, although the total number of animals used is generally decreasing, fluctuation in the trend has to be expected.

The first legislation concerning animal experimentation was issued in Great Britain in 1876. In the United States a 1985 federal law calling for the humane treatment of animals during experiments and the improvement of their laboratory conditions was adopted. In Europe there exist two important documents regulating the use and protection of animals utilized for biomedical tests. Twenty-six member countries of the Council of Europe proposed the first one in 1985 in Strasbourg. This document is not a law, but is instead an agreement among the participating countries. The second document is European Directive 86/609, a series of legislative norms adopted in 1986 by the Council of the European Community. All members had the duty to translate the Directive into local laws. Directive 86/609 is currently under revision—a very complicated process. Different committees and subcommittees have been created to work on specific parts of the Directive. The effort involves different actors in this process, from protectionist groups to purely scientific associations. Furthermore, individuals of different backgrounds have been contacted to give advice on specific points. The new Directive is scheduled to be ready in 2009 or 2010.

Scientific Aspects

The Darwinian theory of evolution affirms that the morphological and physiological level of similarity between the human species and another species is inversely related to the length of time from the existence of a common ancestor between the two species. Therefore, for example, humans and nonhuman primates are more similar than humans and rodents. This theory overlooks different biological disciplines, including, thanks to the work of Konrad Lorenz and others, the study of behavior. If we adopt this methodological and theoretical framework, we can legitimize, on scientific grounds, the use of animals in experiments that aim to increase the knowledge of the biology of the human species. This knowledge is then instrumental in improving the quality of human life.

Furthermore, the use of animals in experiments is due to ethical considerations, which strongly limit the use of other humans as experimental subjects. Methodological reasons are important as well: a nonhuman animal represents an easier experimental subject, mainly because of the ability of replicating experiments and better control of different variables that can influence the expression of a certain biological phenomenon. Even though the use of animals can be justified in terms of obtaining information useful to improving general human health, not all of the animal experiments are automatically justified and valid. The researcher has to be sure that, given a particular experimental hypothesis, the data obtained are accurate and replicable. At the same time, he has to guarantee that the level of suffering inflicted on experimental animals is as low as possible. The accuracy of the scientific data and the ethical aspects of a particular experiment are two factors to be taken into consideration. The relative importance of these two factors has to be weighed in each particular experiment in the relative scientific, ethical, and social context. In this respect it is useful to make reference to the criteria suggested in 1986 by Patrick Bateson, professor in animal behavior at Cambridge University, for judging the feasibility of a particular animal experiment. Bateson indicated three fundamental factors to be balanced when deciding on a particular experimental protocol: (1) the scientific quality of the proposed project; (2) the predicted value of the project in terms of applicable therapeutic and more general medical aspects; and (3) the potential level of animal suffering involved. Therefore, for example, an experiment that potentially can cause a certain degree of animal suffering can be acceptable only if the results of the project are predicted to be of high importance to better understand and prevent a particular disease.

The choice of a particular animal model depends on the particular problem under analysis. The model has to be valid in the sense that the essential biological characteristics of interest are present and intact. For example, the mollusk *Aplysia* is a very useful model for studies aimed at understanding the molecular aspects involved in animal-learning processes because this invertebrate is characterized by a very simple and accessible nervous system. However, because of its limited behavioral repertoire and the phylogenetic distance from the human species, this animal is not very useful as a model for identifying the neuropsychological aspects related to the display of certain behaviors typical of mammals, such as *Homo sapiens*. Appropriate models to study the biological bases of complex behaviors are rodents, such as mice and rats. Among the advantages of these models are the ability to house them in a relatively easy way, the capacity to easily breed them in captivity, and the availability of inbred strains, that is, animals with known genetic characteristics. This latter factor is of particular importance, because it allows the researcher to study the effect of a particular treatment on animals with the same genetic background, therefore holding constant a series of potentially confounding variables. Furthermore, the use of inbred strains helps to reduce the number of individuals utilized, because there is no need to use large numbers to buffer potential variability.

Different models differ for the level of analysis. We can study the behavior as a whole, or we can look at the neurochemical mechanisms that underlie certain behavioral modifications. Therefore, each model can answer particular questions but, at the same time, can set limits on the research to be carried out. There are limits related to the actual level of knowledge to be gained and ethical limits as well, because each experiment is also characterized by a particular level of invasiveness. In recent years the scientific community has progressively become more aware of the ethical implications of the use of animal as models in biomedical studies. However, the results of a recent questionnaire on the use of alternatives to animal experimentation showed that the choice to use a particular alternative method is based more on the scientific considerations rather than ethical ones. This attitude is not only shown by researchers traditionally habituated to the use of animals, but by researchers active in alternative fields as well.

Ethical Aspects

Researchers are particularly committed to the scientific reliability of animal experimentation. But animal experimentation is not debated only by scientists or professionals—it is also a hot topic in the public sphere. Activists protest against the torture of animals in laboratories and lobby against those who make their business using animals. As we shall see, philosophers also analyzed the issue of the moral acceptability of using some animals as a means for human ends (not just in experiments, but in food production as well). Also, protesters against animal experiments seem to be particularly concerned with the reliability of data obtained from animal models. Actually, in the public debate, one of the most common arguments against the use of animals appeals to the unreliability of such experiments. Experiments on animals cause pain, take their lives, and, most important, do not work. In brief, the standard form of the scientific argument against animal experiments sounds like this: the animal model is too different from the human model to provide valid results. The argument is often backed with historical cases of products or medicine proved safe on animals that turned out to be dangerous for humans (the infamous case of thalidomide is probably the most cited one).

Paradoxically, the scientific reliability of experimentation on animals is not the paramount issue in evaluating its moral acceptability. If reliability were the definitive argument to prove the acceptability of an experiment, then we should conclude that experimentation on humans ought to always be preferred to that on animals. Testing on humans has a success rate much higher than testing on animals. But if we think about our attitude concerning experiments on humans, we realize that reliability is not the ultimate argument to debate its morality. To be ethically sound, each experiment carried out on humans ought to meet some requirement: first of all, the informed consent of the subject of the experiment must be acquired. The same can be said of experiments on animals: there are some moral issues that overwhelm the technical validity of the experiments.

These issues have been the object of analysis of some contemporary moral philosophers. Since the declaration of Jeremy Bentham (1748–1832) in favor of extending the principle of equality to nonhuman animals, many theories have been developed to face the central question of the moral status of animals. In fact, the judgment of the morality of experimentation on animals is directly dependent on the way we believe that animals ought to be treated. And the way animals ought to be treated is deduced from the moral status we afford them—that is, the way we answer the question of whether animals have moral value and what kind of value it is.

As a start, then, the question *if* animals can have moral value must be answered. We can answer such a question in two ways. First, we can say that in practice animals are often treated with moral respect. Think to our pets: we name them, we take care of them, we think they have a character, we try to treat them with dignity, and so on. In brief, humans can have and actually have moral relations with animals. At least some animals (most of our pets) are already part of our moral community and already have moral status. Second, we can reason more theoretically. What are the grounds for human moral status? Religion and philosophy have always stressed the possession by humans of a soul or the capacity to reason or speak. But soul is a matter of faith, and we cannot really prove it. Also, we (rightly) treat with respect humans who lack reason and language, such as infants and severely disabled people. If we look at our moral practice, it seems to be the capacity to suffer and experience pleasure and happiness that are the fundamental criteria to treating people with moral respect. Human suffering calls for relief and human happiness seems to be a good to be promoted.

But why just *human* suffering and pleasure? As we have already seen, Darwinism has changed the way we look at the living world. As the philosophers James Rachels (1941–2003) has brilliantly argued, evolutionism has forever banned the idea that human beings and other living organisms are separated by an ontological gap. Differences are in degree, not in nature. So, if we are interested in the issue of whether an individual ought to be treated with moral respect, we should look just at his or her capacities. Treating the same capacities in different ways because individuals of different species possess them is an infringement of the principle of equality and it is a moral error, which has been named *speciesism* by philosophers such as Peter Singer. Being a speciesist is to be a kind of racist or sexist: it means to discriminate on grounds that are not relevant for morality (such as the color of the skin or the gender).

Those who endorse such arguments and argue for a moral status for nonhuman animals (at least those with some kind of conscience of their suffering and pleasure) look at animal experimentation as a practice that is *prima facie* morally wrong. It causes death and suffering to beings that ought to be treated with respect. Benefits to humans are not a good excuse for this kind of moral wrong, in the same way as they are not regarded as a good excuse for experimentation on humans without their consent. This statement can lead to an *abolitionist* view about animal experimentation: it is a moral crime and it should be banned. Though fascinating, the abolitionist view clashes with reality: experimentation on animals is widespread, and consensus on the moral status of animals is poor in society. More feasible seems to be a *reformist* approach to animal experimentation. According to this view, suffering and death of animals are still regarded as moral wrongs, and we have the duty to minimize them. In general, alternatives to the use of animals ought to be sought with more determination, and the ultimate end should be the abolition of animal use in experimentation (or at least its reduction to the absolute minimum required). In particular, for experiments that are still done, measures to minimize the wrong done to animals have to be taken. First, each experimental protocol ought to be judged in light of its net positive effect: the benefit we expect (in terms of happiness and pleasure) should be significantly greater than the harm we cause (the suffering of animals). Second, animals that are used ought to be treated with the highest possible respect, their suffering should be minimized, and their welfare promoted. This kind of approach seems to be well expressed by the "3Rs" model. The model, elaborated by Russell and Burch (1959), asks scientists to design their experiments following three ethical requirements: (1) If it is feasible, the same goals must be achieved without using animals (Replacement); (2) if animals have to be used, the number of animals ought to be the minimum compatible with scientific soundness (Reduction); and (3) if animals have to be used, their suffering before, during, and after the experiments has to be brought to an absolute minimum and their welfare has to be promoted (Refinement).

Conclusion

The problem of the use of animals in biomedical experimentation looks more like a mosaic than a clear-cut scenario. For example, the use of alternatives to animals has been possible in certain cases; it is being looked for in other cases, and is just not yet possible in others. The answer to these issues cannot be just a straight "yes" or "no". There are pros and cons in a problem characterized by many actors: social, philosophical, and scientific. Totalitarian attitudes are not going to be very useful. What is really needed, instead, is real transparency from the scientific community on the aims and methodologies involved in animal experimentation as well as the absence of prejudices from the people who are dubious on the justifications of such practice. Anybody involved in this scenario has to be aware and feel involved in the different implications of a particular choice.

The European legislation on animal experimentation, as we have said earlier, will change. Why? Much has to do with the changing attitude of humans toward animals in general, therefore influencing the ones used in experimentation. In this context science can act as a powerful modifier of attitudes. For example, recent studies on animal behavior, in particular those related to cognitive ethology, have increased the level of general perception on the similarities between humans and nonhumans. Furthermore, a combination of cognitive ethology and animal welfare research has given ground to the idea that animals are able to experience pain and suffering more than previously thought. The perception of the general public about these issues, and the consequential ethical concerns, can have an influence on the attitude of scientists toward experimental animals. This process is probably more present in the younger generation of researchers, who are more inclined to accept new points of view concerning their work. Scientists are an integral part of the society, and in this view science can change cultural attitudes toward animals in society. However, at the same time, new, more humane, cultural attitudes toward animals in society can change science and the way we use animals in biomedical research. We do not have to forget that the quality of the experimental data go hand in hand with the quality of life of the animals we keep in captivity.

See also

Ethics and Animal Protection—*Attitudes toward the Treatment of Animals* Research

Further Resources

Bateson, P. (1986). When to experiment on animals. *New Scientist* (No. 1496), 30–32.

Davis, H., & Balfour, D. (Eds.). (1992). *The inevitable bond.* Cambridge: Cambridge University Press.

Dawkins, M. S. (1990). From an animal's point of view: Motivation, fitness, and animal welfare. *Behavioral Brain Science, 13,* 1–25.

Hursthouse, R. (2000). *Ethics, humans and other animals.* London: Routledge.

Rachels, J. (1991). *Created from animals. The moral implications of Darwinism.* Oxford: Oxford University Press.

Russell, W. M. S., & Burch, R. L. (1959). *The principles of humane experimental technique* (2nd ed.). London: Methuen.

Singer, P. (1990). *Animal liberation* (2nd ed.). New York: Avon Publishing.

Augusto Vitale and Simone Pollo

■ Ethics and Animal Protection
Compassion in World Farming

Compassion in World Farming (CIWF) is an international organization campaigning solely on farm-animal welfare issues. CIWF has led successful campaigns in the European Union (EU) to ban sow stalls (gestation crates) for pregnant pigs, narrow crates for housing veal calves, and barren battery cages for laying hens. All three systems are now being phased out within the twenty-seven countries of the EU.

CIWF was founded in 1967 and is based in the United Kingdom, with offices and representatives in Europe, Asia, and Africa. CIWF coordinates a European Coalition for Farm Animals and works in partnership with welfare organizations across the world, including in the United States.

CIWF aims to prevent intensive "factory farming" worldwide. In factory farms the animals are often kept in isolation, as with pregnant pigs and veal calves, or crowded together with little space for natural movement, as with battery-caged hens and young pigs being fattened for slaughter. These animals are often bred to grow quickly and to have certain meat characteristics to increase profitability, but these methods have been found to have adverse effects on their welfare. Broiler chickens reared for meat now grow so quickly that their legs often cannot support their heavy bodyweight, causing them to become severely and painfully lame. Most dairy cows have been bred to produce ten times as much milk as their calves would have suckled from them. Lameness and painful mastitis result.

CIWF combines campaigning and raising public awareness to prevent the animal welfare problems that it believes are inherent in intensive farming systems. CIWF lobbies politicians to achieve legal change for farm animals and lobbies the food industry to raise its welfare standards.

CIWF Trust, the charitable wing of the organization, publishes reports and educational materials in a variety of languages; many of the reports are available on its Web site (www.ciwf.org).

Among its various missions is to suggest that people should buy only free-range or organic animal products, as these allow the animals a better quality of life. CIWF encourages people to eat less meat overall and to buy only these more humanely produced products.

The organization is also concerned with the issue of the sustainability of global meat production and consumption. Farm animals, especially those reared indoors in intensive systems, consume huge quantities of cereals and soya. Already most of the world's soya and over a third of cereals are grown not to feed people but to be consumed by farm animals. The return on this investment is poor—a steer roughly produces edible meat that is only one-tenth the weight of the grain that it eats. In addition, there is evidence that intensive farms produce vast quantities of highly polluting slurry effluent, which causes degradation of the environment, affecting soil, water, and air quality.

And there is concern that large-scale business agriculture spreading globally makes it difficult, if not impossible, for small, traditional farms and farmers to keep pace with such operations, and increasing numbers lose their livelihood, causing great hardship.

CIWF engages in lobbying of global governments and international funding agencies to prevent investment in factory farming, with all its adverse effects, and to encourage investment in systems that are environmentally sustainable and where animals are treated humanely.

CIWF's main mission is to achieve a world in which farm animals are recognized as sentient beings with intrinsic value and where farming methods are adapted to meet the needs of the animals.

Further Resources

Compassion in World Farming. http://www.ciwf.org/

Joyce D'Silva

■ Ethics and Animal Protection
Compassionate Shopping

The human-animal interaction that takes place between a shopper and a farm animal may be considered to be distant by most. However, the link is perhaps stronger than one might at first realize. The decisions a shopper makes each time they enter a supermarket do, in effect, cast a vote for the kind of farming system in which animals are kept. As such, each shopper has an indirect hand in the way farm animals are raised and in their welfare. A recent Eurobarometer survey on consumer opinions toward farm animal welfare found that 43 percent of shoppers within the European Union consider animal welfare some or most of the time when making a purchase. As such a large consumer base is concerned with the welfare of the farm animals they are purchasing, more and more retailers and manufacturers are turning to better welfare systems. For example, this is reflected in the recent 2006 decision by U.S. ice-cream giant Ben & Jerry's to source all their eggs from cage-free systems.

There is a vast array of goods on offer to today's shoppers at their local supermarket—all vying with each other to be the cheapest or the tastiest or the most colorful or eye-catchingly packaged. How is a shopper to decide which of the products on display to choose? Of course, the decision will be based both on what they can afford and on their own feelings toward the way different products reach the shelves. In a competitive market, where cost is king, ethics and compassion are often forgotten, but many believe that people should think more carefully about what they buy and eat.

There are many ways for shoppers to buy ethical produce—fair trade, nongenetically modified, and environmentally sustainable products are all widely available. Animal-friendly products have been on the market for some time, particularly with respect to animal testing, but now, more and more consumers are starting to consider welfare-friendly food products to be important. Every year billions of animals are farmed intensively for their milk, their eggs, or their meat. Unfortunately, the majority of these animals will spend their lives in cramped, crowded conditions, often without access to an outside area or natural light. Laying hens, for example, have been genetically selected over generations to produce more and more eggs. Standard laying hens are housed in groups of five or six in battery cages, stacked on top of each other. Inside a battery cage, the hens do not have enough space to stretch their wings, and they do not have access to litter to scratch around and bathe in (for hens, bathing in dusty material is a way of reducing the number of parasites and keeping clean). Within the European Union (EU), barren battery cages are due to be banned from 2012. However, in many countries no legislation exists to ban systems such as this, which have been scientifically shown to lead to animal suffering.

Meanwhile, the broiler chicken, which is raised to produce chicken meat, has been bred to grow incredibly quickly. It takes less than six weeks from hatching for a broiler to reach full adult size and be ready for slaughter. This is almost a third of the time it takes for a laying hen to reach a similar size. During their short lives, thousands of these birds will be crowded together in windowless sheds. Looking into a shed such as this, one is greeted with what looks like a moving white carpet—the birds are so tightly packed that it is often difficult to see the floor. Indeed, these birds may have just a letter-sized area of space each, and as there is currently no law to protect these birds across the EU and in other countries, some birds may have even less. The combination of a fast growth rate and crowded, dark conditions means that the chickens' bones are weak, and because they are unable to move around freely, their leg muscles do not strengthen enough to carry the massive weight of the breast muscle. This is one of the reasons why so many broiler chickens suffer from lameness and other deformities.

Despite being the same species, broiler chickens and laying hens look drastically different, and this is all down to the way that they have each been bred to meet higher production targets and create greater profit margins. This is the fundamental reason why chicken meat is so cheap and why a standard batch of a dozen eggs costs so little—well under 50 pence in the United Kingdom. But it is not just chickens and hens that are subjected to this kind of farming. Dairy cows, pigs, beef cattle, and turkeys have all been bred with productivity in mind and are all commonly reared under intensive conditions. As the dairy cow has been so specifically bred to produce vast quantities of milk, only female offspring are valuable. Many male calves are shot at birth or else are reared for veal. While the EU has banned veal crates, where the calves spend their lives tethered and unable to turn around while existing on a diet lacking sufficient fiber, many other countries still use such practices. Arizona was, in 2006, the first U.S. state to ban veal crates.

The Rise of Compassionate Shopping

Many people choose to shop with compassion by becoming vegan (not eating any animal products) or vegetarian. However there are many people who wish to continue to eat meat who are moving toward preferentially choosing animal products that come from animal welfare–friendly production systems. Additionally, more national and international companies are choosing to use their meat and animal products from better welfare systems. For example, McDonald's in the United Kingdom sources all whole eggs and the majority of their egg products from noncage systems, and Internet giants Google and AOL have recently opted to source all the eggs used in their staff cafeterias from noncage systems. Supermarkets such as Marks and Spencer in the United Kingdom, Albert Heijn in the Netherlands, and Waro, Migros, and Coop in Switzerland sell eggs from free-range systems only.

There are a number of things to look out for when shopping. Assurance schemes, often identified by a logo on the food packaging, aim to guarantee a specific quality of product. However, assurance schemes are interested in different aspects of production, so they are not equivalent. The UK's Red Tractor logo, for instance, confirms compliance with a set of standards that in the main reflect minimum legal regulations for raising that kind of farm animal. The problem is that many farmed animals are not protected by species-specific laws. Where laws do exist, they usually represent only the bare minimum for welfare. Similarly, the UK Lion Quality mark on eggs is primarily a symbol of food safety, not welfare. The Lion Quality symbol may therefore appear on both free-range and "battery" cage–produced eggs and shows that the eggs have been produced by hens that have been vaccinated against salmonella, that the eggs are traceable to their source, and that the "best-before" date stamped on the shell and pack shows that they are fresher than required by law.

The Freedom Food label in the United Kingdom is a Royal Society for the Prevention of Cruelty to Animals (RSPCA)-monitored farm assurance and food labeling scheme that aims to improve farm animal welfare. All farm animals under the Freedom Food scheme must be reared according to RSPCA animal welfare standards. These standards are species-specific and cover each stage of an animal's life, including how it is handled and transported. In the United States, the Humane Farm Animal Care (HFAC) assurance scheme requires that livestock have access to clean and sufficient food and water; that their environment is not dangerous to their health; that they have sufficient protection from weather elements; that they have sufficient space to move naturally; and a number of other features to ensure the safety, health, and comfort of the animal. In addition, the standards require that managers and caretakers are skilled and competent in animal husbandry and welfare and that they have a good working knowledge of their system and the livestock in their care.

The label "organic" means that the product comes from a farming system that does not use chemical fertilizers or pesticides. Most organic animal products come from animals with access to the outdoors, and in the EU, organic hens cannot be reared in barren battery cages. However, in many countries, just calling a product organic does not necessarily mean the animals have had good welfare. Many supermarkets may have their own organic branding labels. The UK Soil Association's Organic Standard represents the highest organic standards.

All eggs sold in UK supermarkets must now be labeled with details of the kind of farm they come from. "Eggs from caged hens" means battery-caged hens' eggs. These birds are intensively farmed in a crowded cage with very little space to flap their wings, perch, or dust bathe. These hens will have no outside access, or see natural light. "Barn hens" are kept in barns with up to nine birds per square meter. They have perches and litter in which to dust bathe and scratch around. "Free range" hens have access to open-air runs during the day. This allows the hens to exercise and explore, flap their wings, and dust bathe and, as such, has the potential to provide much better welfare. "Woodland" eggs come from free-ranging hens that have access to an outside area containing trees and bushes for cover. This encourages the hens to explore more and take advantage of the entire outside area, so is seen as an even better alternative to classic free range hens. "Soil Association Organic" eggs are similar to free range but are kept in smaller groups with more space. Most of their food will have been crops grown without the use of chemical pesticides or fertilizers.

With meat, generally if nothing is mentioned on the label about the kind of farm it comes from, then the farm is likely to conform to basic welfare standards in the country of origin and nothing more. The majority of chicken and pig meat comes from animals raised in factory farms. In the case of both pigs and chicken, Freedom Food free-range and organic chickens will have been raised in higher welfare systems. For pigs, "outdoor bred" means that the pigs were born to mothers kept outside and are likely to have been moved indoors for the fattening period. "Outdoor reared" means that the pigs were born and raised outdoors.

One of the ways people have adapted their interaction with animals and the impact of farming on the environment is to buy and eat less meat. For example, 10 percent of all greenhouse gases come from farming animals. Furthermore, overgrazing by keeping too many animals on too small a piece of land can lead to global desertification, potentially exacerbating global-warming issues. Perhaps even more worrying in this respect is that since the 1960s, approximately 200 million hectares of rainforest have been cleared to make way for cattle grazing and growing crops for use as animal feed. And despite ever-increasing concerns of water scarcity around the globe, we still use 100,000 liters to produce 1 kg of beef. By contrast, to produce 1 kg of potatoes, just 500 liters of water are required. There are currently fifteen farm animals for every human on our planet. All trends suggest that the number of farm animals will continue to increase for the foreseeable future to meet ever greater demands. By choosing to eat less meat, while choosing meat from systems with better welfare conditions, consumers reduce demand and may promote sustainability and a move toward better welfare systems for the animals we keep.

Understanding what different labels or logos mean and what they represent in the farming world is an important step to making informed choices about the food we eat.

Consumers have enormous capacity to create change, simply through what they buy. When shopping compassionately, consumers are actively recognizing their relationship with the animals they eat and are using their wallets to turn their ethical beliefs into a reality for animals farmed for food.

See also

Animals as Food—*Veganism*
Ethics and Animal Protection—*Compassion in World Farming*
Ethics and Animal Protection—*Factory Farm Discourse*

Further Resources

Burgess, K., & Pickett, H. (2006). *Supermarkets and Farm Animal Welfare 'Raising The Standard': Compassion in World Farming Trust Supermarket Survey 2005–2006*. Retrieved March 20, 2007, from http://www.ciwf.org/publications/reports/Supermarkets_Report2005.pdf

European Commission. (2005). *Attitudes of Consumers towards the Welfare of Farmed Animals*. Retrieved March 22, 2007, from http://ec.europa.eu/food/animal/welfare/euro_barometer25_en.pdf

Eyton, A. (1991). *The Kind Food Guide*. London: Penguin Books.

Fox, M. W. (1997). *Eating with conscience: The bioethics of food*. Oregon: NewSage Press.

Lawrence, F. (2004). *Not on the label*. London: Penguin Books.

Mepham, B. (Ed). 1996. *Food ethics*. London: Routledge.

Prince, R. (2006). *The savvy shopper*. London: Fourth Estate.

Robbins, J. (1987). *Diet for a new America*. Walpole, NH: StillPoint.

Turner, J., & D'Silva, J. (Eds.). (2006). *Animals, ethics and trade*. London: Earthscan.

Lisa M. Collins

■ Ethics and Animal Protection
The Enlightenment and Its Views of Animals

The Enlightenment

The Enlightenment was a philosophic movement, anchored in the eighteenth century, that provided the basis for modern Western ideas about natural rights, religious tolerance, and social progress. Over the past two centuries its core principles have become the foundation of many democratic, rights-based political systems, particularly in Western Europe and North America. From a human perspective, the Enlightenment has been a dynamic and life-shaping force. Its ideas connect vitally to the lives of today's citizens of the United States, Canada, Australia, Western Europe, and other democratizing societies across the globe. From the perspective of animals, the implications of Enlightenment thinking are considerably more complex. On the positive side, a number of key Enlightenment thinkers, notably John Locke, Thomas Paine, and Voltaire, expressed concern for animals. They urged humans to treat all living creatures with kindness. On the negative side, the Enlightenment's overall emphasis on reason as a basis and justification for rights has meant that nonhuman animals have generally been excluded from the circle of rights-worthy creatures.

To understand the Enlightenment, it is vital to keep in mind that it was a broad movement that included many different thinkers from Western Europe and a very young United States. At the heart of the movement was the fundamental notion, or recognition, that all humans have the capacity of reason—the ability to think. (Indeed, the imagery of light refers to the "light" of reason and knowledge.) To Enlightenment thinkers, reason was a type of power—the driving engine of all human beings. This might not seem revolutionary to us today, but it was controversial at the time, particularly because of its implications for education, democracy, and human equality. Enlightenment philosophers emphasized the multifaceted abilities of the human mind—that humans have the ability to think logically and analytically, that they can envision creative solutions, and that they can imagine a better future. According to Enlightenment thinkers, education should be encouraged in order for individuals to develop their innate powers. Society would benefit as well. If everyone worked together, better political and economic systems would be developed. This last point is the radical one. Many Enlightenment writers recommended that people should have the ability not only to participate in their own governments but also to express criticisms of all forms of authority and to shape new political systems.

Historically, the net effect of this philosophy on the lives of human beings has been extraordinary and transformational; for animals, ironically, it is much more complicated, as will be discussed below in the focus of the essay. Not only did it supply a framework for supporting such modern practices as public education and for embracing new ideas about freedoms of speech and press, but it also offered a direct and ultimately devastating blow to hereditary monarchies (kings and queens) that had ruled much of Western Europe since the Middle Ages. Enlightenment thinkers such as Thomas Paine and Thomas Jefferson played important roles in the American Revolution. Jefferson's masterpiece, "The Declaration of Independence" (1776), articulated the principles that "all men are created equal" and that government's purpose is to protect the human's rights to life, liberty, and pursuit of happiness; to this day, the U.S. Declaration stands as an exemplar of Enlightenment expression. Jefferson, Paine, and some fellow Americans were involved also in some of the proceedings of the French Revolution of 1789, another landmark revolution in which people challenged the authority of the monarchy. During much of the eighteenth century, France also had been a hub of Enlightenment philosophy, home to such luminaries as Voltaire (1694–1778), who urged people to be tolerant of each other, and Denis Diderot (1713–84), who organized a massive compendium of knowledge, entitled *Encyclopédie*.

Although the Enlightenment ushered in new and liberating possibilities for large numbers of human beings, its effects on the lives of animals are much more complex and fluid. To this day, advocates of animal rights must struggle with the Enlightenment's complicated legacy. Although many of its key thinkers expressed concern for animals, many others did not. Moreover, the movement's focus on everything human—human reason, human rights—has meant that animals were pushed to the margins. For instance, the Enlightenment's focus on reason, meaning specifically human reason or brainpower, is encapsulated in the rallying cry of Immanuel Kant in his famous essay "What is Enlightenment?" (1784). Kant urges the reader to cultivate his or her mind. "*Sapere aude!*" or "Dare to think!" is his main point. To prod the reader, Kant compares unthinking individuals to immature people, to children, and to cowards, and at one point, he briefly invokes the image of tethered animals (Kant 2006). Although Kant's writing is focused on humans, not animals, he clearly has little regard for non-human creatures. In a lecture to a group of students, Kant bluntly summed up his views when he commented that "Animals are not self-conscious and are there merely as a means to an end. That end is man" (quoted in Singer 1991).

Philosophical Views of Animals

On the positive side, a number of Enlightenment thinkers shared a humanitarian outlook that did include a genuine consideration of animals and their needs. Expressions of a profound humane sensibility can be found in the works of such figures as John Locke (1632–1704), Thomas Paine (1737–1809), and Voltaire (1694–1778), as well as in names lesser known. The focus of these individuals' writings was not animal rights per se. However, they expressed a deep sense of empathy, or feeling, with animals, and they were dismayed by human cruelty to helpless animals. Enlightenment thinkers such as Paine and Voltaire believed that acts of brutality were despicable and immoral because they signaled an ugly departure from the peaceful and ethical ways of reason. To be a rational, enlightened individual meant, by definition, that one refrained from destructive passions of anger and violence. Paine expressed this viewpoint in a passage from his book titled *Age of Reason* (1794). His comment, frequently repeated nowadays in humane literature, was that "Everything of persecution and revenge between man and man, and everything of cruelty to animals is a violation of moral duty. . . .The only idea we can have of serving God is that of contributing to the happiness of the living creation God has made."

To a large degree, Paine was echoing the views of earlier English Enlightenment philosopher John Locke. Locke, best-known for his political philosophy of social contract, argued in *Thoughts Concerning Education* (1692) that parents should raise their children in an atmosphere of moderation, charity, and reason, and that adults should resist inflicting harsh punishments on children. Although Locke's intention was to instill in children a capacity for rational living, his childrearing recommendations underscored both the importance of human compassion and the importance of acknowledging animals' feelings and needs. Locke's suggestion was that "Children should from the beginning be bred up in an abhorrence of killing and tormenting any living creature. . . and indeed, I think people should be accustomed, from their cradle, to be tender [kind] to all sensible creatures."

Other examples of the humane impulse included David Hume's (1711–76) advocacy of kindness to animals and Voltaire's protests against the practice of vivisection. In his condemnation of vivisection (the practice of operating on living animals), it is interesting to note that Voltaire, who is perhaps most famous as a critic of man's inhumanity to one another, directly took to task the mechanists of his day, those individuals who contended that animals were unfeeling machines that could be experimented on for purposes of scientific "discovery." Voltaire noted the irony, praising dogs as creatures that surpassed many humans in such fundamental virtues as loyalty and friendship.

Animal Feeling as Justification for Animal Rights

Most significantly, there were several eighteenth-century figures who highlighted the importance of compassion as a moral trait while also grappling directly with key questions about the status of animals. For those thinkers, rights were not so much a matter of "rights to" but of "rights from": the right to be free from pain, abuse, imprisonment, and other human-inflicted cruelties. Of greatest importance was England's Jeremy Bentham. Bentham was a utilitarian philosopher, not an Enlightenment philosopher, and indeed he was critical of many Enlightenment ideas. Yet he was an individual of the same time period, and, like his peers, he struggled with similar philosophical challenges. Bentham clearly was aware that human reason was valued highly by his Enlightenment contemporaries, but he realized also that some humans were less than reasonable when it came to their callous and cruel treatment of helpless animals. Bentham explored the issue of the human-animal relationship in his work. In a manner that was both pointed and provocative, he wrote, "The question is not, can they [animals] reason? Nor, can they talk? But can they suffer? Why should the law refuse its protection to any sensitive being? The time will come when humanity will expand its mantle to everything which breathes."

Bentham's ideas, articulated in his book *Introduction to the Principles of Morals and Legislation* (1780), provided a thoughtful alternative to the predominating Enlightenment paradigm that emphasized reason as a basis for rights. Although Bentham's work has inspired and bolstered the efforts of humane advocates and animal welfare proponents over the past 200 years or so, the stark reality is that sentience—feeling, sensation, capacity to suffer—has yet to become the determining factor in the granting of rights in modern societies. Even to this day, opponents of animal rights campaigns will sometimes continue to ridicule the concept of animal rights by making jokes about the absurdities of animals in voting booths and college classrooms. These sarcastic jests tend to operate as a diversionary means to distract the public from the gravity of the issues.

Although Bentham was the most innovative pro-animal philosopher of the late eighteenth century (even if he himself was not an Enlightenment thinker), similar ideas about the importance of sentience (feeling) were expressed a few decades earlier by French Enlightenment thinker Jean-Jacques Rousseau (1712–78) and a few decades later by nineteenth-century philosopher Arthur Schopenhauer (1788–1860). While discussing the cruelties perpetrated against animals slaughtered for food, Rousseau commented, "It appears, in fact, that if I am bound to do no injury to my fellow creatures, this is less because they are rational than because they are sentient beings." Schopenhauer mirrored a similar framework in his book titled *On the Basis of Morality* (1841). Stressing the importance of compassion, he suggested that human respect for the needs of animals—namely, their right to be treated humanely—was a matter of great moral significance.

Moving from the late 1700s to the present era, we find that the ideas of the past—of the 1700s—still play a lively role in today's culture. In the opening section of Peter Singer's landmark 1975 book *Animal Liberation,* featured prominently are the ideas of Jeremy Bentham. This is revealing, because Singer's book not only inspired and launched the modern animal rights movement of the past thirty years but it also effectively outlined many of the key historical and philosophical obstacles that face advocates of animal rights. Significantly, many animal rights advocates of the past few decades have taken heed of Singer's discussion of Bentham and have tried to usher in new ways of thinking about animals—namely, as subjects deserving of rights. One approach has been to encourage human empathy for animals and their needs, particularly for those animals who suffer in such unnatural and cruel places as laboratories and factory farms. Also successful has been the endeavor by humane organizations to foster a sense of compassion for companion animals. Over the past few decades, anticruelty laws protecting dogs, cats, and other domestic animals have been strengthened in many areas of the United States and Europe, and contemporary studies suggest that many people in the modern West regard their companion animals as valued members of their families.

Animal Cognitive Abilities as Justification for Animal Rights

Another strategy, aided in large part by the work of pioneering scientists, has been to focus on the distinctive cognitive (thinking) abilities of animals, particularly of dolphins and the great apes. Laying the groundwork for these efforts of the past fifty years was Jane Goodall's groundbreaking research in the 1960s on the tool-making activities of East African chimpanzees. By demonstrating that animals have the ability to fashion tools, she effectively countered the older view of tool-use as a distinctively human capacity, and she prompted scientists to reexamine traditional categorizations of animals and humans. Many scientists since have built on Goodall's base, including Roger and Deborah Fouts and Francine Patterson. Their efforts, independent of each other, to teach the basics of American Sign Language to chimpanzees and gorillas, respectively, have enriched modern understandings about the cognitive abilities of the great apes. One outcome of this research has been the Great Ape Project, which aims to harness modern research to demonstrate that the similarities between humans and the great apes are so compelling, so overwhelming, that basic protections should be accorded to them.

Although it remains to be seen whether the Great Ape Project will accomplish all its goals, the campaign has raised awareness and it highlights a broader objective of the modern animal rights movement—to spotlight the cognitive abilities of certain animals with the hope that modern people will accord those groups meaningful legal protections. This emphasis of cognition stems at least in part from the ideas and values of Enlightenment and its emphasis on reason as a measure for gaining certain rights.

Evaluation of the legacy of the Enlightenment is a complicated venture. In essence, the movement generated several long-term but contradictory impulses with respect to animals and human-animal relationships. On the one hand, many key thinkers expressed concern and compassion for all living beings. Some of its major thinkers were truly "enlightened" individuals—people with great minds and big hearts, and their sense of empathy for animals is inspirational to this day. The ideas of individuals such as Locke, Paine, and Voltaire have played a dynamic role over the past two centuries to foster awareness of society's most vulnerable members, both human and animal alike. Moreover, their message of kindness has underpinned the major humane campaigns of the nineteenth and twentieth centuries, and it is clear that numerous animals, particularly companion animals, have benefited from their insights. However, the Enlightenment's prevailing emphasis on reason—and not sentience—as a basis of valuation for legal rights and protections has proved extremely problematic. To this day, activists must continue to struggle to find ways to gain the public's acceptance for rights for animals. It has been an ongoing challenge.

Further Resources

Cavalieri, P., & Singer, P. (Eds.). (1993). *The great ape project: Equality beyond humanity*. New York: St. Martin's Press.

Fordham University. *Internet Modern History Sourcebook: The Enlightenment*. Retrieved March 22, 2007, from http://www.fordham.edu/halsall/mod/modsbook10.html

Ishay, M. (2004). *The history of human rights: From ancient times to the globalization era*. Berkeley: University of California Press.

Kant, I. "What is enlightenment?" in *Modern History Sourcebook*. Retrieved March 30, 2006, from http://www.fordham.edu/halsall/mod/kant-whatis.html

Locke, J. *Thoughts Concerning Education*. Retrieved March 30, 2006, from http://www.bartleby.com/37/1/12.html

Paine, T. *Age of Reason*. Retrieved March 30, 2006, from http://www.ushistory.org/paine/reason/reason14/htm

Rousseau, J-J. *A Discourse on the Origin and Foundation of the Inequality Among Mankind*. Retrieved March 30, 2006, from http://www.animalrightshistory.org/arh_bibliography/rou_jean-jacques-rousseau.htm

Singer, P. (1990). *Animal liberation: New revised edition*. New York: Avon Books.

Voltaire. *The Philosophical Dictionary*. Retrieved March 30, 2006, from http://history.hanover.edu/texts/voltaire/volindex.html

Yolton, J. W., et al. (Eds.). (1991). *The Blackwell companion to the Enlightenment*. Oxford: Basil Blackwell Ltd.

Christina Stern

■ Ethics and Animal Protection
Environmental Philosophy and Animals

Animals play many important roles in our everyday lives, which leads to a wide variety of human-animal relationships. For example, some animals share our homes as companions, while others are served up for dinner. So, it is not surprising that many scholars within the field of environmental philosophy focus on questions about the moral nature of these many different relationships.

Relationships between humans and animals have intrigued Western philosophers for decades. One of the most frequently and passionately debated questions is whether humans should treat animals with the same moral considerations we apply to other humans. Plato and Aristotle, perhaps two of the most famous of the ancient Greek philosophers, gave much thought to this question and have greatly influenced today's contemporary philosophers working in the field now known as environmental ethics. Many have also learned about moral relationships between humans and animals from non-Western cultures, particularly indigenous and aboriginal societies, who tend to have very different approaches to such relationships than those that have developed in Western, industrialized societies. While there are many different cultural and historical philosophies of the environment, in this essay, we will examine five important categories that have developed in contemporary Western environmental ethics, as they relate to human-animal relationships in modern, industrialized society. They are anthropocentrism; ecocentrism; biocentrism; environmental etiquette; and ecofeminism.

Perspectives in Environmental Philosophy

Modern environmentalism gives us many examples of environmental ethics. Organizations such as Greenpeace and the World Wildlife Fund (WWF) promote the idea that nature should be protected from harm. However, often the philosophies underlying such beliefs go unexamined. There are many assumptions of what is of value and to whom involved, which may not be apparent to the general public. This is where the academic study of philosophy has much to offer, to help us understand how and why moral arguments are made. For instance, a common argument about the importance of forest conservation is that human beings depend on forests to provide oxygen for us to breathe, medicines for our illnesses, lumber to build homes with, places for recreation, and so on. Because the forests are valuable to us in these vital ways, it is morally wrong to destroy them. This kind of reasoning appeals to a particular way of looking at the issue of forest conservation and to specific conceptions of what a forest is in relation to human beings (in this case, a forest is seen as a complex set of natural resources that are meant to be utilized by human beings). In other words, this argument presents a particular kind of ethical point of view, or *perspective*.

Anthropocentrism

While each individual has a relatively unique point of view, we can group perspectives into larger categories based upon what they have in common. For example, a *human-centered perspective,* called *anthropocentric,* looks at the issue from the point of view that only human beings deserve moral consideration. It is called human-centered because it means that humans are the center of moral value. This value judgment corresponds with the assumption that the natural world exists for the use and welfare of human beings and that humans are separate from and superior to natural ecosystems.

Therefore, a common argument made about human-animal relationships from this perspective is that although human beings have no moral obligation toward animals, we should treat animals with kindness because it is related to how we treat other human beings. Immanuel Kant is recognized for championing this idea. He argued that violence toward animals could lead to violence against people, so such actions should not be allowed, as they would compromise our moral duty to other human beings. This argument is based on the premise that it is immoral to violate human rights, such as the basic human right that no person should be subjected to torture or cruel treatment.

Kant also distinguished between relationships that treat people as a *means* to something from relationships that treat people as an *end* in themselves. For example, the bus driver serves as a means (driving the bus) to an end (getting people where they need to go). This relationship does not present any moral challenge as such, but Kant insisted that people never be treated as *only* a means. To return to our example, the bus driver should always be treated with moral consideration by all of the passengers because bus drivers are human beings and thus they are ends in themselves as well as a means of transportation.

This concept of means and ends relates to the notion of *moral value* in environmental ethics, in which someone or something that acts as a means has *instrumental value* and someone or something that is an end in itself has *intrinsic value*. From a human-centered perspective, only human beings can have intrinsic value and animals can only ever have instrumental value. That means that animals are valuable to us because they can serve as "instruments" for us. For instance, many humans value dairy cows because they are instruments that provide milk. Returning to our original example of forest conservation, we can see that it too presents an anthropocentric perspective, which appeals to the instrumental value of the forest (providing oxygen, medicine, lumber, and so on).

Ecocentrism and Deep Ecology

In direct contrast to this anthropocentric perspective is one called *ecocentric,* or ecology-centered, perspective, in which nature exists for itself, regardless of any instrumental purposes it may fulfill for human beings. Thus, nature has intrinsic moral value. This perspective assumes that humans are an integral part of natural ecosystems and are not morally superior to the trees, rivers, forests, and animals of the ecology. This ecocentric philosophy is also known as *deep ecology*. So, to look at the issue of forest conservation from an ecocentric, or deep ecological, perspective, human beings have a moral responsibility to protect the forest, which includes all of the plants, animals, and natural elements that it consists of. The forest, as a functioning ecosystem, has intrinsic value unto itself, regardless of any instrumental purposes it may fulfill for human beings.

Deep ecology has inspired many to think about human relationships with animals, and nature in general, in a different way. For American conservationist and philosopher Aldo Leopold, it inspired a *land ethic,* in which "a thing is right when it tends to preserve the integrity, stability and beauty of the biotic community. It is wrong when it tends otherwise." As an avid outdoorsman, Leopold gradually learned to see the importance of every individual part of the whole ecosystem. For example, as a young National Parks warden, he engaged in a wolf eradication program because of the belief that fewer large predators in an ecosystem would lead to more herbivores, such as deer, and thus provide better recreation in the form of hunting. However, once the population of wolves had been decimated, the population of deer became so large that the deer eventually destroyed the entire forest with their grazing and then began to die of starvation. Leopold's land ethic thus asserts that human beings should act morally to preserve the whole ecosystem, which requires that all the beings and natural processes that make it up are considered important to the whole. This is still very different from giving moral consideration to individual animals. For instance, according to an ecocentric perspective, it would be morally appropriate to kill the deer to bring their population down to a size that does not threaten the whole forest ecosystem.

Biocentrism and Animal Rights

While deep ecology does extend ethical consideration beyond a human-center of moral value to include whole natural ecosystems, it remains distinct from yet another major perspective in environmental ethics known as *biocentric,* or life-centered. This view assumes that all living beings exist for themselves and have intrinsic moral value regardless of any utility they provide for humans or other biological functions in nature. Like ecocentrism, biocentrism assumes that human beings are an equal and integral part of nature; however, unlike ecocentrism, biocentrism assumes that individuals have moral value that should not be compromised in favor of the whole ecosystem. Thus, it would not be morally acceptable to cull the individual deer even if the population poses a threat to the functioning of the forest.

Perhaps the best known argument made from this biocentric perspective is the one for *animal rights*. A philosopher by the name of Peter Singer is thought to be the most outspoken on this issue. He bases his argument on the *utilitarian* premise that an action is morally right when it provides the greatest good for the greatest number. In this moral equation, pleasure should be maximized and pain should be minimized. By this argument, it is immoral for a human being to cause pain and suffering. Thus, he claims that individual animals that can feel pain and have the capacity to suffer should have the same moral rights as human beings. Putting this perspective into practice has many dramatic implications for certain human-animals relationships. Meat eating, for example, comes into question. Many arguments within animal rights call for vegetarian or vegan diets, because meat eating requires the death of animals and, particularly in modern industrial factory farming, can be the cause of much animal pain and suffering.

As with Kant's argument above, this argument for animal rights is also based on the moral inviolability of human rights. Singer asserts that the moral imperatives of human rights should be extended to animals. In other words, humans should think of animals as metaphorical human citizens and treat them with the same kinds of moral considerations as other human beings. This extension of moral consideration is often dependent upon animals meeting the criteria necessary for honorary human status. John Benson describes this as granting animals a "moral passport." For example, in order to obtain this passport, an animal must be proven to possess a characteristic such as sentience (capability of feeling). So, a marine biologist could claim that a bottlenose dolphin has sentience and, therefore, should be given a moral passport so that the dolphin is treated with the same moral considerations as human beings.

Environmental Etiquette and Embodied Ethics

While many philosophers promote a biocentric view, others are critical of it and claim that there are fundamental weaknesses in the argument. One is that by trying to position animals as moral equals because they have traits that are valued in human beings, the argument fails to challenge the assumed moral superiority of human beings in the first place. Another weakness is that the moral passport depends on scientific knowledge of animals to provide facts about them. Three main problems arise here with respect to granting moral consideration to animals. The first is that not everyone agrees on the facts. The second is that even when there is agreement on the facts, the criteria necessary for the moral passport itself can be changed. For example, scientists and philosophers might agree that dolphins are sentient, but that this characteristic is no longer good enough to satisfy the moral criteria, which now must be an ability to use tools.

The third is that some believe that conventional Western scientific knowledge is not well suited to providing the kind of information that is most relevant to human-animal relationships. Philosophers with this perspective contend that the whole argument is backward: that human beings need to give moral consideration to animals first and then build relationships with animals and gain knowledge about them. This is called an *environmental etiquette,* so named by Jim Cheney and Anthony Weston. The implication here is that the way knowledge is made about animals already has moral dimensions. To illustrate, they describe the actions of a man named Jim Nollman, who studies killer whales (orcas) using his own methods of interspecies communication. He goes out on the ocean in a kayak and takes musical instruments with him. He plays the instruments to invite killer whales to come and interact with him and then waits for them to approach him. If the killer whales choose not to come near him or interact, he leaves them alone and goes back to the shore.

Ecofeminism, Ethics of Care, and Interspecies Ethics

An important part of this environmental etiquette is for human beings to practice being *attentive* to animals in order to learn about them and to develop relationships with individual animals. This attentiveness involves paying close attention to animals over a relatively long period of time. Taking relationships into account is a primary characteristic of yet another ethical perspective known as *ecofeminism*. Ecofeminism is short for ecological feminism, which sees a connection between dominant Western society's treatment of animals (and of nature in general) and of women. For example, there is a prevailing assumption in Western culture that men are superior to women, which relates to the assumption that humans are superior to animals. The practical consequences of these assumptions are varied and complex, but the underlying connection is that both women and animals are seen and treated as moral inferiors.

As with a biocentric perspective, ecofeminism assumes that human beings are not separate from, nor superior to, nature and that all individual living beings have intrinsic moral value. The main difference between them is that ecofeminism also recognizes intimate connections between the domination and control of nature, women, and non-Western groups of people by a *patriarchal* society. This means that Western society is, historically and presently, dominantly influenced by a small powerful group typically made up of Caucasian males. This dominant patriarchy expresses ideals that represent *dualisms,* or hierarchical divisions of value, such as humans are separate from and superior to animals. (If you recall from earlier, this dualism is a fundamental assumption in anthropocentrism.) Ecofeminism also assumes that personal actions and beliefs have political implications and therefore insists on justice for any oppressed groups and individuals, which includes animals as well as people.

Ecofeminism again sets itself apart from other ethical perspectives by refusing to dictate moral principles that are abstract and meant to have universal application. Instead, ecofeminism promotes an *ethics of care,* which is based upon specific relationships and is dependent on the unique context within which an issue arises. Returning to the issue of meat eating, for example, from a biocentric perspective one might argue for animal right's by setting a universal moral principle that to kill a sentient being is morally wrong. This principle is meant to apply to all people, everywhere, all of the time. However, it becomes very suspect when we consider peoples who live in the arctic, who hunt and eat caribou, seals, and whales. By the universal principle, an Inuit person would be acting immorally. It does not take into account the deeply spiritual and cultural connections between the people and the animals, the historical ties and contemporary context of endangered cultural practices and beliefs, nor can it recognize the geographical setting and the physical reality of survival in that specific place. In this instance, the abstract, universal approach only seems to work for an economically privileged, geographically free sector of individuals characterized by a distinct lack of cultural, spiritual, and ecological ties to place (in its fullest sense) and independent of subsistence concerns.

An ethics of care is not only sensitive to a situation or spatial context, however; it is also concerned with time. Developing a moral relationship involves paying attention to an other for relatively long periods of time. This is particularly vital to developing moral relationships with animals, which corresponds with the way we learn about other animals. Thus, making knowledge about others has ethical implications. According to feminist philosophy and ecofeminist ethics, respect for the individual being, whether human or animal, who is the subject of study should be given up front and learning should occur through careful attentiveness. Applying this practice of etiquette directly to human-animal relationships results in an *interspecies ethics,* which means it is an ethics that exists "between different species." So, for example, it means that there should be an ethics of moral consideration between the human species and other animal species such as orcas. It is an ethics recently described by an Australian feminist environmental philosopher named Val Plumwood, who states that human beings need to be open to communicating with animals so that we may learn how to best interact with them and develop moral relationships with them. Plumwood describes environmental ethics as a process in which etiquette is performed according to specific contexts, which lies in direct contrast to the nature of traditional Western ethics as a set of abstract principles with universal application. It is an embodied performance wherein posture, gesture, tone, and action are expressions of ethics.

In other words, it is an ethics that is expressed through the way we interact and communicate with others using our bodies. To illustrate, imagine walking through a park on a crisp fall day. We suddenly see a squirrel on the path negotiating a pinecone that has fallen from the tree above. We have many options for how we proceed to interact with this individual squirrel. We can run excitedly toward the squirrel with our arms flailing and screaming. We can quietly and calmly walk toward the squirrel. We can squat down where we stand and wait to see if the squirrel approaches us. We can make soft chirruping and clicking noises to see if we catch the squirrel's interest. It is very likely that our first two options would result in the quick retreat of the squirrel up the closest tree and that engaging in our last options might invite the squirrel to come closer and allow us to observe each other for a short while. Over time, if we are attentive to the way individual squirrels respond to us, we may learn how best to interact with each other. What is key to the ethical nature of such embodied interactions, according to ecofeminism and practices of embodied etiquette, is that we give the squirrel the *choice* to interact with us or not. It is imperative that we attempt to learn about squirrels, and all other animals, on their own terms, not just according to ours.

Conclusion

There are many different views in environmental philosophy regarding moral relationships between humans and animals. Each philosophical perspective carries with it certain practical implications for the treatment of animals and the kinds of relationships we develop with them. As such, they raise ethical questions, including: "Is it morally right to use cows as instruments for making milk?" and "What are ethical ways of learning about whales?" The answers vary depending on which ethical perspective it is based on. There are still many disagreements, and these issues continue to be debated in environmental philosophy. Yet, however diverse the arguments are, one thing can be agreed upon: our relationships with animals are very important to us. Otherwise, why would we spend so much time and effort arguing about it?

See also

Conservation and Environment
Ecofeminism
Ecotourism

Further Resources

Armstrong, S. J., & Botzler, R. G. (Eds.). (1993). *Environmental ethics: Divergence and convergence.* New York: McGraw Hill.

Benson, J. (2000). *Environmental ethics.* London: Routledge.

Cheney, J., & Weston, A. (1999). Environmental ethics as environmental etiquette: Toward an ethics-based epistemology. *Environmental Ethics, 21,* 115–34.

Donovan, J., & Adams, C. (Eds.). (1996). *Beyond animal rights: A feminist caring ethic for the treatment of animals.* New York: The Continuum Publishing.

Leopold, A. (1949). *A Sand County almanac, and sketches here and there.* New York: Oxford University Press.

Plumwood, V. (1993). *Feminism and the mastery of nature.* New York: Routledge.

———. (2002). *Environmental culture: The ecological crisis of reason.* New York: Routledge.

Singer, P. (1975). *Animal liberation.* New York: RandomHouse.

Weston, A. (1994). *Back to earth: Tomorrow's environmentalism.* Philadelphia: Temple University Press.

Traci Warkentin

■ Ethics and Animal Protection
Factory Farm Industry Discourse

Historically, people have used other animals as resources. From hunters in the New World tracking the gigantic woolly mammoth, to Cheyenne tribes hunting buffalo, to European settlers raising cows, pigs, and chickens, humans have viewed nonhuman animals as a natural resource for their use and consumption. The various discourses that surround this practice differ, depending, in part, upon how the humans in question conceive of their relationship to nature and, consequently, to the animals they view as natural resources. For instance, the hunting practices and ritualistic use and sacrifice of revered and respected animals are aspects of a well-worn story about the spirituality of Native American cultures. Another familiar narrative concerns family farmers who settled the plains and raised domesticated animals on their small farms to use and consume for subsistence purposes. Still another account recalls Inuits who—by eating, wearing, using as tools, and so on— ecologically made the most of all parts of the seals they hunted, in order to survive the harsh Arctic climate.

At least one common thread weaves through this patchwork of tales. In these three narratives, the human animals are directly connected to the land and to the nonhuman animals they use and consume—they live with, hunt alongside, know, understand, and respect both the land and the animals they exploit. While it is important to avoid the romantic tendency of overlooking these cultures' negative marks upon the environment, it is also necessary to note that current discourses surrounding the use and consumption of other animals are unlike those in any other previous narrative. This chapter surveys the enormous power of the discursive practices constructed to support the factory farm industry and discusses how those practices contribute to the ways Americans conceive of their relationships with nonhuman animals confined on factory farms. In particular, this chapter focuses on two common and codependent corporate discourses: (1) the widespread use of *doublespeak* to describe particular processes internal to the industry, and (2) the discursive creation of "speaking" animals in advertisements to sell the products of those industrial processes to the public.

Theoretical and Historical Considerations

It is important to begin this chapter with several theoretical and historical considerations. First, *discourse,* as philosopher Michel Foucault (1978, 1981) understood it, represents the production of knowledge and power through language, and *discursive practices* are those institutional formations (or *epistemes*) within which meanings of and between contradictory discourses are constructed. This meaning construction is the way that institutions establish particular orders of truth, or what becomes accepted as commonsense "reality" in a given society. *Doublespeak* is a style of discourse that—although descriptive—is intentionally misleading by being ambiguous or disingenuous. As professor of English Richard M. Coe (1998) describes it, "doublespeak techniques include the abuse of euphemism, nominalization, abstraction, presupposition, jargon, titles, and metaphor and other tropes as well as inflated language, gobbledygook, symmetrizing, stipulative definition, and ambiguity (weasel words)" (p. 193).

It is also important to note that *nature* is understood in this chapter as being as much a cultural construction as it is something "out there" that we discover and use. As environmental historian William Cronen (1996) puts it, nature is a profoundly human construction. This is not to say that the nonhuman world is some-how unreal or a mere figment of our imaginations—far from it. But the way we describe and understand that world is so entangled with our own values and assumptions that the two can never be fully separated. (p. 25)

Nonhuman animals, then, as a part of nature, are included in this cultural construction. The way we think about and value those animals is less a matter of what and how they *are* than it is about how *we construct them to be.* That is to say, even though they exist as beings in the world with their own purposes and value, it is through the lenses of human purposes and values that we construct a relationship with them. Understanding how particular dis-courses have historically constructed how we view our relationship with other animals helps us to recognize how specific forms of that relationship are presently possible.

Geography and environmental studies scholars Lisa M. Benton and John Rennie Short (1999) suggest that the roots of nonhuman animals (nature) being understood as a commodity can be traced back to a New World "technological metadiscourse" (p. 2) that constructed nature as a resource for human purposes and that continues to influence current discourses. The technological metadiscourse (also influenced by colonial dis-courses) emphasizes how technology mediates between humans and their environment and makes possible a consumerist approach to natural resource exploitation. As a result of this emphasis, human progress is understood as an ability to control, subdue, and improve nature in order to increase its use-value to humans.

The commodification of nature intensified as settlers moved westward, hunting buf-faloes to near extension and installing domesticated "farm" animals in their place. Although "farm" animals have been raised since Roman times in confined quarters to restrict their movement and, in turn, to fatten them, the lineage of current factory farm practices is relatively unique. In the 1930s and 1940s, late-stage capitalism, with its emphasis on technological innovation and efficiency, enabled animal "farming" to become an industry by developing highly sophisticated and automated practices. By the 1950s, the cattle industry, in its packing practices, had developed an assembly-line production mode that made the process more efficient and extended that mode of production to slaughter-ing practices. Contemporary practices are a direct offshoot of these historical stages, with at least one significant difference: the scale at which "factory farms" (or high-intensity farms) operate has increased tremendously in recent years.

The discourses that surround the current factory farm industry can be understood through the lenses of two helpful metaphors. As William Cronen (1996) discussed them, "Nature as Commodity" and "Nature as Virtual Reality" discourses assist in humans' con-struction of nonhuman animals and help us to develop our relationship with them. Put differently, "Nature as Commodity" and "Nature as Virtual Reality" are particular discursive manifestations of an overarching technological metadiscourse that influences how we think about and relate to nature and, by extension, other animals. The factory farm industry's internal use of doublespeak is a compelling example of a "Nature as Commodity" discourse, and it is discussed in the next section. The industry's marketing, particularly its "Happy Cows" advertising campaign, is an exemplar of a "Nature as Virtual Reality" discourse, and it is discussed in the fourth section.

Animal as Unit/Nature as Commodity: Internal Doublespeak

As Richard Coe pointed out in the previous section, doublespeak is an intentional linguistic practice that can serve to confuse, hide, erase, or deceive about meaning, and it is used in various contexts. Professor of English William Lutz (1987) offers several examples of doublespeak from a military context:

It's not a Titan II nuclear-armed, intercontinental ballistic missile with a warhead 630 times more powerful than the atomic bomb dropped on Hiroshima, it is just a *very large, potentially disruptive re-entry system,* so don't worry about the threat of nuclear destruction. It is not a neutron bomb but a *radiation enhancement device,* so don't worry about escalating the arms race. It is not an invasion, but a *rescue mission,* or a *predawn vertical insertion,* so don't worry about any violations of United States or international law. (pp. 385–86)

In a business context, closing an automobile plant in which eight thousand people lose their jobs can be called "a volume-related production schedule adjustment," or the company could be said to have "initiated a career alternative enhancement program" (Lutz 1996, p. 12). Put simply, when doublespeak is employed to deceive or confuse, it is usually because whoever is doing the deceiving or confusing wants to hide the negative impacts (or potential negative impacts) of actions.

In the factory farm industry, doublespeak is employed to describe a variety of different practices. Joan Dunayer, author of *Animal Equality: Language and Liberation,* offers a wide-ranging, contemporary collection of examples of doublespeak used by various human institutions to hide the treatment of nonhuman animals. For instance, rather than calling the factory farm industry a *livestock industry,* the National Cattlemen's Association (NCA) has urged the use of *animal agriculture* instead. Whereas *industry* could set off environmental warning bells, *agriculture* is generally understood as a safe, natural practice. Moreover, instead of *factory or high-intensity farm,* the NCA has recommended using the euphemism *family farm,* and some of the largest companies in the world use that designation.

The terms *beef, veal, pork,* and *poultry* are common euphemisms employed by the industry and by consumers to designate parts from cows, calves, pigs, and birds such as chickens and turkeys. The use of these euphemisms disguises the fact that the body parts purchased and consumed are the objectified remains of former animal subjects. It is a familiar example of nature (in this case, nonhuman animals) as commodities. Government agencies also employ doublespeak in this commodification process. For instance, the Internal Revenue Service designates "farm" animals as *inventory* and *items that are financed, capitalized, depreciated, invoiced, leased, and liquidated.* The USDA also commodifies animals, calling them *grain- and roughage-consuming animal units.* These "units" are made up of *veal calves, dairy cows, beef cows, feeder pigs, breeder pigs, layers,* and *broilers.* Discursively understood as objects intended to produce sellable products, nonhuman animals on factory farms are commonly handled as objects.

A current factory farming textbook, *Intensive Pig Production,* labels sows' restricted living quarters *individual accommodations* and *modern maternity units.* Joan Dunayer's (2001) description of a typical housing situation for pregnant sows seems less innocuous:

Sows endure each pregnancy isolated in a stall with iron bars and a concrete floor. The stall is so narrow that a sow has room only to stand up and lie down. If the stall is open in the rear, she also is chained to the floor by a neck collar or body harness. . . . Typically, a sow reacts to tethering with violent escape attempts lasting up to several hours: repeatedly she yanks on her chain; screaming, she twists and thrashes; sometimes she crashes against the stall's bars before collapsing. After her futile struggle to break free, she lies motionless, groaning and whimpering, her snout reaching outward beneath the bars. Finally, she sits for long periods with her head drooping and her eyes vacant or closed. (p. 126)

Piglets, if born too small, are called *runts* and are *euthanized.* Euthanization involves holding a piglet by his or her back legs and slamming his or her head against the floor.

Often, the piglets are still conscious after the first attempt to kill them. Calling this practice *euthanasia* is doublespeak, because it hides the violence and suffering inflicted on these animals behind a euphemism that portrays the practice as humane or sterile.

Forced molting is a practice to which hens are subjected so that they will lay more eggs. Hens are caged in total darkness and denied food for at least ten days (and sometimes water for one to two days). According to the industry *Commercial Chicken Production Manual,* this practice can increase productivity and profits, and doublespeak is used to describe it. This practice is called *induced molting* (hiding the violence of force), *fasting* (as though it were voluntary), or *recycling* (which, while mimicking environmental discourse, only recognizes the additional egg-laying cycle while completely erasing the bird).

The production of *veal* also carries with it the use of doublespeak. Housed in narrow crates, the calves are partitioned off from other calves, frustrating their attempts to socialize—to lick or nuzzle each other. They are fed only an iron-deficient formula and no roughage (to keep their muscle tissue white), and the formula is mostly water, powdered milk, and fat. These calves, because of this diet, frequently have chronic diarrhea (which is often fatal) and suffer from ulcers, heat stress, and bloat. The housing situation, according to the industry, is for the calves' own protection: it keeps them from "running around a little bit too much" (Dunayer 2001, p. 132). Calling the conditions merely unaesthetic, industry insiders suggest that each calf has its own *private stall* that *features partitions for privacy*. And, referring to the calves' restricted diet, the industry designates the calves as *milk-fed, special-fed,* or *fancy-fed*.

For factory farmers, "farm" animals are merchandise—they are objects or things that are bought, raised, and sold in the marketplace to consumers. This animal-human relationship is what results when, as William Cronen noted, discourse is shaped by the metaphor "Nature as Commodity." This animal-human relationship (or at least a positive image of it) also becomes a valuable commodity when the discourse is public. That is, when it comes to marketing, positive images of animals' health and welfare under the care of humans become just as saleable as the products procured from "farm" animals. On the one hand, internal discourse hides the actual conditions in which "farm" animals live; and, on the other hand, industry marketing creates an alternative reality to replace that actuality. In the next section, we look at an example of industry discourse—advertising—that reflects William Cronen's "Nature as Virtual Reality" metaphor.

Animal as Speaking Subject/Nature as Virtual Reality: External Marketing

William Cronen's metaphor, "Nature as Virtual Reality," is useful in helping us understand how humans' technological creation of alternative natures can both erase humans' experience of the actuality of nature and, at the same time, replace that actuality with a virtual reality created by humans. For instance, computer simulations of animals or technologically saturated natural environments like Sea World come so "close to constructing an alternative reality" that, even though we might conceive of animals and natural environments as *the real* realities, "it is increasingly possible to inhabit a cultural space whose analogues in nature seem ever more tenuous" (Cronen 1996, pp. 43–45). Put simply, what technology mimics, it also tends to erase and replace.

In the case of factory farm industry marketing, technologically created speaking "animals" in technologically created environments attempt to create a relationship with human consumers by demonstrating the pleasantness of their living conditions. In the process, this industry-created virtual reality erases the actuality of factory farm conditions by replacing them with a pleasant virtual reality. This discourse functions on at least two important levels in constructing an animal-human relationship. The first function is an obvious one: to sell product. The second function of this discourse is a bit more complicated: nonhuman "farm" animals are replaced by "animal" subjects who are functionally identical to human subjects and with whom we can emotionally connect; at the same time, however, the discourse nourishes our consumerist relationship to them as objects for our use.

The first function of speaking "animals" in advertisements is an evident one: to sell animal products to human consumers. An effective approach to marketing and advertising is the use of a spokesperson who, through his or her particular appeal, can lend power to messages intended to be persuasive. Whether it is Charlie Tuna or the Foster Farms chickens, the talking spokes-"persons" for the industry are usually persuasive, in part, because they appeal to humans' emotions. A speaking "animal's" appeal functions as a way to connect to human consumers and create an animal-human relationship that will increase sales. The use of speaking animals selling themselves is largely taken for granted and rarely questioned at the level of ethics. On December 11, 2002, however, the animal advocacy organization People for the Ethical Treatment of Animals (PETA) filed suit against the California Milk Advisory Board, claiming that its "Happy Cows" advertising campaign is fraudulent and unethical.

"Happy Cows" is a popular advertising campaign that features television, radio, and billboard ads, along with merchandise for adults (e.g., calendars and T-shirts) and children (e.g., coloring books and board games). These advertisements and other marketing tools create an alternative reality to the actuality of factory farm conditions. The advertisements feature healthy, clever, and funny animals enjoying their easy lives and happily consenting to "contributing" their share to the "family" business. For instance, in one television spot, an old-fashioned wooden barn lies in the middle of an empty, green field, with two cows inside on a bed of hay, awakening to the sound of a rooster. One yawns and sighs while the other says, "Morning. So what do you think, then . . . get an early start on that alfalfa on the back forty?" The other responds, "What's the hurry, hit the snooze." As the rooster is kicked out of the barn, the following words appear in the foreground, and a voice-over is heard speaking them: "Great Cheese comes from Happy Cows. Happy Cows come from California. Real California cheese. It's the cheese" (quoted from the PETA lawsuit, 2002).

Another television ad plays with the stereotypical California obsession with health and fitness by featuring two bulls commenting on a cow's healthy appearance and lifestyle: In another grass-covered valley, one bull asks the other, "You're from back East, right?" "Huh?" asks the other. "Yeah, trust me on this one, man. The babes out here are different." The Easterner asks, "How come?" "I don't know. All the sunshine, I guess. Clean air, good food, something. They just really take care of themselves." Just then, a dairy cow walks by, and as they spot her they say, "Whoa." "Oh, yeah—hey, you work out?" As the shot widens to show more of the valley sparsely populated with grazing cows throughout, the same catchphrase is repeated: "Great Cheese comes from Happy Cows. Happy Cows come from California. Real California cheese. It's the cheese" (quoted from the PETA Federal Trade Commission complaint document, 2002).

What PETA argued in its lawsuit is that these depictions are fraudulent and unethical because they do not reflect the actual conditions of animals on factory farms. Cows are usually raised in extremely crowded conditions on high-intensity drylots where they often stand in an amalgamation of urine- and feces-saturated dirt (or mud, when it rains). They rarely graze on grass, and on the biggest California farms, cows never graze but are fed highly concentrated feed to which antibiotics are added. Cows are not raised together in families; female cows are usually kept pregnant so that they can be milked constantly. They live short lives (five to six years on average, while the natural lifespan of a cow is twenty-five to thirty years), and the cows are not in control of when they eat, when they sleep, or whether or not they work. Although this is the actuality of factory-farmed animals, most consumers are unaware of those conditions. Instead, the advertising campaign creates an alternative reality and the possibility of an animal-human relationship that may be more comfortable for human consumers to emotionally manage.

A second, related function of advertisements that feature "speaking" animals is reminiscent of an earlier cartoon featuring a virtual animal. The principles involved in creating a spokes-"person" that emotionally connects with humans in order to create a comfortable consumerist relationship is similar to an invented creature called a *schmoo:*

> The schmoo is a fictitious being invented by American cartoonist Al Capp. The schmoo delights in being eaten by humans, and indeed it desires nothing more. We might . . . imagine real animals that, were they able to communicate with us, would express the same desire. (Fox 1999, 164–65)

As professor of philosophy Michael A. Fox reasons in his book *Deep Vegetarianism,* this virtual reality raises significant ethical questions about a number of issues, especially if the virtual were to become actual. One of those is the possibility that genetic engineering could create "animals" that, like machines, are without consciousness, are unable to suffer, and whose limbs regenerate. This possibility, if it were to become an actuality, could create animals with no reasons for existing except those selected by humans. Like other technology, this technology would radically change humans' relationship with other animals. Advertisements create a comfortable human consumerist relationship with other animals while, at the same time, creating the conditions for the possibility that, when it is possible to produce animal-machines for consumption, we will be discursively conditioned to be comfortable doing so.

The discursive function virtual animals serve blurs the line between humans and other animals while, at the same time, encouraging a human behavior that implicitly reinforces the line. A commonly accepted distinction between humans and other animals is humans' "exceptional" ability to speak. Since Rene Descartes, humans have emphasized numerous traits as superior to other animals—the ability to consciously consider ourselves and to enjoy an emotional sense, and the capacity to possess and use language—all of which other animals allegedly do not enjoy. The virtual creation of a speaking "animal" subject that is rational, feeling, and self-conscious blurs the distinction between humans and other animals. We trust these speaking subjects because they use words to communicate with us, and emotionally they seem just like us. Of course, we don't believe that there are actual animals that exist like the ones in animatronic or animated television commercials (though it would be an interesting future research project to learn how children view these images). However, without an actual connection to the factory farm animals, all we have is a connection to a simulation of them.

However, these blurred boundaries also, simultaneously, serve to tacitly reinforce the boundaries between human and nonhuman animals. By "giving" virtual animals the abilities to speak, reason, and feel while, *at the same time,* taking for granted that their only actual value to humans is found in their consumable products, the boundary between humans and other animals is once again distinct. Put simply, factory farm industry discourses like the "Happy Cows" campaign have created speaking "subjects" that are functionally no more than objects, in the end. Ultimately, we are left with a paradox: nonhuman animals finally can speak to and reason with humans, but only from the perspective of an industry selling them to a consuming public.

Conclusion

Current factory farm industry discursive practices in the form of (internal) industry doublespeak have been examined alongside an exemplar of (external) industry advertising in order to illustrate how these intertwined discourses help create, sustain, and promote particular taken-for-granted ideas about animal-human relationships. When nonhuman animals are viewed by humans as resources, languaged as commodities, and handled as objects, advertising campaigns that function to supplant that actuality with a virtual reality support the conditions for the former. And, when virtual reality becomes the primary source of knowledge about the actuality of factory farm animals, the erasure of that actuality supports the virtual.

At the same time, researchers have suggested various ways to engage and understand doublespeak and advertising discourses. Joan Dunayer suggests that relanguaging objectified animals as subjects is one place to start. She also proposes relanguaging sterile-sounding practices related to humans' treatment of nonhuman animals in order to reflect their actuality—in graphic detail, if necessary. Dunayer offers a resource of linguistic options in the form of "Style Guidelines" for writers who are interested in reporting inhumane conditions. Finally, she provides a "Thesaurus of Alternatives to Speciesist Terms" that can be used by those who would prefer to remove the speciesist assumptions embedded in many words and phrases. At the very least, Dunayer maintains that changing the way we talk about other animals may begin to help us reconceive animal-human relationships in more humane ways.

Richard Coe and William Lutz suggest that critical-reading strategies be taught earlier and more often in the public schools, so that children can learn to recognize and understand doublespeak, and media literacy in connection to advertising ought to also be part of this approach. Both also recommend that adults would also benefit from learning about media literacy and critical evaluation of messages that may employ doublespeak.

Some researchers disagree that language-based changes are the best way to improve animal-human relationships. Professor of English, David L. Clark (1997), writes in this respect:

> Language is the implacable human standard against which the animal is measured and always found wanting; but what if the 'animal' were to become the site of an excess against which one might measure the prescriptive, exclusionary force of the *logos*, the ways in which the truth of the rational word muffles, strangles and finally silences the animal? (p. 191)

William Cronen (1996) suggests, instead, that humans need to recognize and honor nature and other animals as "Other," especially when it comes to use:

> Learning to honor the wild—learning to remember and acknowledge the autonomy of the other—means striving for critical self-consciousness in all of our actions. It means that deep reflection and respect must accompany each act of use, and means too that we must always consider the possibility of non-use. (p. 89)

Choosing nonuse (that is, not buying and consuming products associated with factory farming) is another option for those who are so inclined. Ultimately, each researcher would agree that being informed and understanding practices that could damage animal-human relationships is an important place to begin if we are to truly honor that relationship.

See also

Literature—*Human Communication's Effects on Relationships with Other Animals*

Further References

Benton, L. M., & Short, J. R. (1999). *Environmental discourse and practice.* Malden, MA: Blackwell Publishers.

Clark, D. L. (1997). On being "the last Kantian in Nazi Germany": Dwelling with animals after Levinas. In J. Ham & M. Senior (Eds.), *Becoming beast: Discourses of animality from the middle ages to the present* (pp. 165–98). New York: Routledge.

Coe, R. M. (1998). Public doublespeak, critical reading, and verbal action. *Journal of Adolescent & Adult Literacy, 42,* 192–95.

Cronen, W. (1996). Introduction: In search of nature. In W. Cronen (Ed.), *Uncommon ground: Rethinking the human place in nature.* New York: W. W. Norton and Company.

Dunayer, J. (2001). *Animal equality: Language and liberation.* Derwood, MD: Ryce Publishing.

Foucault, M. (1978). *The history of sexuality: Volume 1: An introduction.* New York: Pantheon.

———. (1981). The order of discourse. In R. Young (Ed.), *Untying the text: A Poststructuralist reader.* London: Routledge.

Fox, M. A. (1999). *Deep vegetarianism.* Philadelphia, PA: Temple University Press.

Lutz, W. (1987). Language, appearance, and reality: Doublespeak in *1984. ETC: A Review of General Semantics, 54,* 383–91.

———. (1996). *The new doublespeak.* New York: HarperCollins Publishers.

People for the Ethical Treatment of Animals. (2002, April). Federal Trade Commission complaint document. Available at http://www.peta.org/feat/caldairy/complaint2.html

———. (2002, December). California Dairy Board suit document. Available at: http://www.peta.org/feat/caldairy/newsuit.html

Cathy B. Glenn

■ Ethics and Animal Protection
Fashion and Animal-Friendly Trends

Fashion has long been considered a visual expression of individuality or of people's relationship with the world around them. Likewise, the use of animal products in fashion has also reflected people's relationship to nonhuman animals throughout history. Indeed, humankind's use and consumption of animals has been mirrored in the fashion industry, which has long used animal products, including fur, feathers, bone, teeth, and leather. However, as humankind's relationship with animals has evolved, and animal suffering and sentience have become more widely accepted, "animal-friendly" or "cruelty-free" alternatives have been increasingly utilized by some designers.

The beginnings of animal-friendly fashion in Western society date back as early as the aesthetic dress movement in the late 1800s. Largely characterized by simple lines, beautiful fabrics, and loose, flowing gowns, the movement also rejected the use of many animal products popular at the time, such as bird plumes, and even whole birds, on hats (Kastner, 1994). Many "Aesthetes," as they were called, were also vegetarians, and the movement has been linked with the beginnings of the vegetarian movement.

With the passing of nearly two centuries, the animal protection and rights movement has evolved into a lifestyle-based movement and has given rise to many companies founded on these principles. Some cosmetic companies, for instance, started labeling their products as "cruelty-free" and "not tested on animals" and giving consumers the ability to choose animal-free products. There has been controversy surrounding this labeling, however, because the U.S. government does not define these terms and has not set any standards to accompany them (Massachusetts Society for the Prevention of Cruelty to Animals, 2006).

Largely in response to these discrepancies, in 1996 the Coalition for Consumer Information on Cosmetics (CCIC) developed an international, non–animal testing standard, the Corporate Standard of Compassion for Animals. The CCIC has produced the "Compassionate Shopping Guide," a small, free foldout for consumers, featuring U.S. and Canadian companies that do not test "finished products, ingredients or formulations on animals" and that have met their rigorous standards (Coalition for Consumer Information on Cosmetics, 2004).

Clothing and accessory companies have also begun to offer vegan and cruelty-free products, which are products made without the use of animal products. These have gained popularity as new materials and techniques continue to develop. Companies such as Fabulous Furs, Moo Shoes, Via Vegan, and Chrome Bags have been very successful in marketing clothing, shoes, bags, wallets, and belts. Some designers have also successfully marketed such products simply as mainstream products, without a "vegan" label, as a way to attract a wider audience. Interested consumers have little trouble finding these products. A quick Internet search for vegan clothing or products will turn up multiple links featuring animal-free and vegan products. Shopping guides and additional resources do exist, however, including the "Cruelty-Free Pocket Shopping Guide" distributed by People for the Ethical Treatment of Animals (PETA).

Many well-known, mainstream designers have also recently launched lines featuring clothing and accessories made without the use of animal products. Synthetic materials are now available, such as high-quality faux leather—including vinyl, Naugahyde, stretch vinyl, pleather, ultraleather, and PVC—and can be found in different styles of animal skins and prints, such as alligator or python, and in varying levels of suppleness and strength. Oftentimes, these materials are cheaper to use than their traditional animal counterparts, and companies are able to offer stylish alternatives at half the price. The new faux leathers on the market can be deceptively "real" in their appearance and feel. Although the majority of the companies featuring such products may not adhere to any specific animal-welfare ethic, they do offer consumers an animal-free alternative. Nine West, Liz Claiborne, and Kenneth Cole, among others, are examples of companies that have launched affordable, chic, and urban faux leather lines.

For the most part, consumers can determine themselves whether a product is made without the use of animal products. Simply reading labels will often reveal whether a product is made from an animal. Additionally, most companies now have a Web site or, at the very least, a customer service number, and further information on products can be easily obtained.

Whether large companies are motivated by cheaper production costs or consumer demand, the end result is the availability of a variety of animal-friendly fashion products. In addition, the proliferation of products made without the use of animals by both small and large companies reflects humankind's increasing recognition of animal sentience. Although the fashion industry will invariably make use of animal products, perhaps in perpetuity, alternatives continue to be developed.

Further Resources

Caring Consumer: http://www.caringconsumer.com/c-free.asp

The Coalition for Consumer Information on Cosmetics. Compassionate Shopping Guide. March 28, 2006, http://www.leapingbunny.org/pdf/ccicguide_full.pdf

Kastner, J. (1994). Long before furs, it was feathers that stirred reformist ire. *Smithsonian, 25*(4), 97–104.

Massachusetts Society for Prevention of Cruelty to Animals. Cruelty-free labeling: what does it mean? March 28, 2006, http://www.mspca.org/site/PageServer?pagename=advo_Cruelty_Free_Labeling

Newkirk, I. (2005). *Making kind choices: Everyday ways to enhance your life through earth- and animal-friendly living.* New York: St Martin's Griffin.

People for the Ethical Treatment of Animals. PETA's Shopping Guide to Compassionate Clothing. March 28, 2006, http://www.peta.org/living/clothingguide.asp

Josephine Martell

■ Ethics and Animal Protection
Goodall, Jane

Valerie Jane Morris-Goodall was born in London on April 3, 1934, to Mortimer and Margaret Myfanwe (Vanne) Morris-Goodall. Her father was a telephone-cable-testing engineer who, in his spare time, pursued his real passion: racing cars. Her mother worked as a secretary. After they married in 1933, she became a housewife and companion to her husband, then a mother to Valerie Jane and a second daughter, Judy, born four years after the first. In the autumn of 1939, as World War II began, the family settled temporarily with Mortimer's mother and step-father in the countryside, surrounded by farm animals: horses, cows, geese, and chickens. The child's future career as an animal behaviorist was foreshadowed one afternoon when she disappeared, returning home at dusk with bits of hay in her clothes. After her worried mother inquired where she had been, the child explained that she had wondered how hens laid eggs. Where was an opening big enough for an egg to pass through? She hid for five hours underneath the hay inside a hen house until a hen demonstrated.

Mortimer joined the army, and Vanne and the two girls moved to the seacoast-resort town of Bournemouth to live with her widowed mother, "Danny," who became the matriarch in a household of all women. Mortimer remained on active duty until 1951; the marriage ended in divorce.

Growing up in Bournemouth, young Valerie Jane—or V.J., as she was called—read, with great enthusiasm, books that expressed her own feelings of deep kinship with animals and nature, including the Dr. Dolittle and Tarzan series. She would settle into a perch of one of her favorite trees and read, and thus it was high in a leafy world that she first dreamed of being Tarzan among the wild animals in Africa. Her nonreading life developed some of the practical interests and skills that would enable her to follow that dream. She cared for pets: caterpillars, a canary, cats, guinea pigs, snails, and tortoises. She took riding lessons at a local stable, earning tuition by shoveling manure. She took nature walks on the nearby cliffs and ravines, accompanied by a dog named Rusty—a mixed-breed spaniel, shiny black with a white blaze on his chest, owned by the people who managed a nearby hotel. Rusty, Goodall liked to say in later years, was her first instructor in animal behavior, and he taught her that animals have emotions, personalities, and minds.

By the end of adolescence, V. J. chose to be known as Jane. Along with that name change came the more challenging task of finding a job. Starting in May of 1953, she began learning shorthand, typing, and bookkeeping at a secretarial school in London. She emerged eleven months later with skills sufficient to land a job as secretary at the Oxford University Registry, starting in the summer of 1954. A year later, she returned to London to work in a commercial film studio. In the spring of 1956, however, Goodall received a letter from a former school friend, Marie Claude (Clo) Mange, whose father had purchased a farm in Kenya. Clo Mange invited her friend to Kenya, where she could stay at the farm—and so Jane Goodall quit her job in London and returned home to Bournemouth, where she lived with her family rent-free, worked as a waitress in a local hotel, and saved all her wages and tips. She left England in March, 1957, on the *Kenya Castle,* sailing three weeks around the southern tip of Africa, disembarking in Mombasa, and reaching Nairobi by train on Sunday morning, April 3, coincidentally her twenty-third birthday.

Because she needed a job, Goodall soon moved from the Mange farm into town and began working for an engineering firm. At the same time, her interest in animals led her to phone Louis S. B. Leakey, paleoanthropologist and curator of the Coryndon Museum in Nairobi, and, upon his invitation, she visited the museum. At the end of that visit, he offered her a job as his secretary. Soon after, he also offered her the chance to accompany himself and his wife, Mary, on their usual summer archaeological expedition to Olduvai Gorge in eastern Tanganyika.

Louis Leakey was a brilliant iconoclast who believed that humans evolved in Africa. His life's mission was to sustain that belief through excavating early human and homonid artifacts and bones in East Africa. He was also interested in tracing the evolution of human behavior, and thus he began considering expeditions to study the behavior of humanity's closet living relatives, the African great apes (chimpanzees gorillas, and bonobos), on the theory that any behaviors humans and their closest relatives held in common would likely have been characteristic of their shared ancestors. Near the end of that summer at Olduvai, sitting with young Jane Goodall before a campfire one evening, Leakey talked of his plan to send someone out to watch wild chimpanzees inhabiting a remote part of Tanganyika territory at the edge of Lake Tanganyika. After they returned to Nairobi that September, he offered her the opportunity to be that person.

Two major problems stood in the way. The first was getting permission from colonial authorities to camp in the area. As Leakey had been informed by the British authorities for that part of Tanganyika, no European woman was allowed to go "into the bush" without a companion. Goodall's mother soon offered to accompany her, so that restriction was satisfied. The other problem was financial. No ordinary funding source for scientific research would risk a penny on such a wild scheme, and so Leakey turned to a non-ordinary source, American industrialist Leighton Wilkie. In February 1959, Wilkie agreed to provide $3,000 to underwrite Leakey's "Chimpanzee Project" for four months, long enough, so Leakey had argued, to allow his two "research workers"—his secretary and her mother—to gather all the necessary data for his own research.

On July 14, 1960, Goodall, her mother, and a cook, Dominic Bandola, arrived by boat at the Gombe Stream Chimpanzee Reserve. The mother began collecting plant and insect specimens and running a simple medical clinic for local villagers; the daughter worked on the chimpanzees. Her intuitive approach was to habituate the wild apes to her presence—and thus, rather than trying to watch them while hidden inside blinds or to sneak up on them, she approached them openly, careful not to alarm them with unnaturally colored clothes, too-sudden movements, and too close an approach. For the first few months, she was seldom able to approach closely enough for detailed observations, however, and the summer of 1960 was an exhausting and discouraging time—more so near the end of August, as both women came down with malaria.

Although most experts presumed wild chimpanzees were vegetarians, no one actually knew. On October 30, Goodall saw a chimpanzee eating meat. Less than a week after that, she saw a behavior even more startling. While walking past a clearing in the forest, she noticed a dark object—a chimpanzee, she soon realized—climbing on a termite mound and using a long thin stick to probe inside. She confirmed, on a second day, that the chimps were fashioning sticks to the right length and size and then using them as probes, fishing into termite-mound entrance holes for reflexively clinging termites.

Leighton Wilkie had financed a project that was supposed to last around four months, but, during that very brief period, Leakey's former secretary had established herself and shown her courage. She demonstrated that her method, approaching the wild apes openly, was appropriate. And her two breakthrough discoveries, meat eating and tool use, inspired the wealthy American organization, the National Geographic Society, to take over funding and turn the research into a long-term project. At the same time, Leakey realized that his "research worker" would require formal scientific training and, ideally, a PhD. With National Geographic funding, then, Goodall continued to study the Gombe chimpanzees during 1961—and at year's end, she began advanced work in ethology at Cambridge University. She received a doctorate in 1966.

The National Geographic Society would support Goodall's research for a full decade. In providing that support, though, the society had acquired rights to any images and articles on the project for possible publication in its magazine, and by late August 1962, a Dutch photographer, Hugo van Lawick, arrived and began taking pictures for such an eventuality. By August 1963, the magazine had released to 3 million subscribers her article—with his photographs—"My Life among Wild Chimpanzees." Hugo Van Lawick, meanwhile, had returned to Gombe in 1963 to take motion-picture footage and continue with still photography for possible future articles; in the spring of 1964, he and Goodall were married.

During the next few years, Jane Goodall managed the Gombe research with the help of assistants and managers and a growing staff of Tanzanians. She also invited to Gombe a number of scientists and science students, who jointly maintained a continuous record of chimpanzee observations and were, by the early 1970s, gathering information on a half-dozen aspects of chimp behavior while studying the baboons and other monkeys, as well as the snakes, birds, fish, and vegetation.

In 1971, Jane Goodall was appointed a visiting professor at Stanford University. The benefits of this new association included foundation support and, starting in 1972, an infusion of serious and talented undergraduates coming from Stanford to support the research. Around the same time, she was appointed professor of zoology at the University of Dar es Salaam, and thus she also began traveling to eastern Tanzania while a few of her Dar students were, like the Stanford students, coming to Gombe. By 1975, the Gombe Stream Research Center was sponsoring the research of two post-doctoral scientists and sixteen British and North American doctorate candidates, along with a continuous stream of undergraduates from the United States and Tanzania; they all were supported, in turn, by over a dozen members of the permanent Tanzanian field staff—well-trained specialists in observing and recording the behavior of chimps and baboons. During the summer of 1974, meanwhile, Goodall and Hugo van Lawick were divorced. Within several months, she had married Derek Bryceson, director of Tanzania National Parks, and they prepared to spend a blissful married life together, commuting between his house in Dar es Salaam and her cabin at Gombe.

The golden dream ended on May 19, 1975, when a boatload of forty armed thugs from Zaire arrived at midnight and seized four students, holding them for money, guns, and the release of political prisoners from Tanzanian jails. Fortunately, the kidnapped students survived their ordeal; after a ransom was paid, the last of the students was released, physically unharmed, in mid-July.

One consequence of the kidnapping was the removal of all foreign researchers, since the kidnappers had targeted only whites. By then, however, the Tanzanian field staff, trained in scientific research methods and having assisted the research for the previous six years, was entirely prepared to take over the long-term monitoring and record-keeping. In fact, the next few years were an extraordinary time in terms of the chimp research. By the time of the kidnapping, the original study community was already separated into two groups inhabiting contiguous territories. Chimpanzee territoriality is based upon communities of related and socially bonded males, females who emigrate from one community to another at early adolescence, and borders patrolled by gangs of males and opportunistically expanded by excursions that can turn lethal. In short, the chimpanzees at Gombe had split into two communities antagonistically engaged in a male-driven activity that provocatively resembled, at a primitive economic level, human warfare. One community finally annihilated the other.

While the Tanzanian field staff ran the Gombe research, Goodall lived at her home (shared with Derek Bryceson) in Dar es Salaam. She monitored events with daily radio contact and monthly visits to Gombe, and she catalogued and analyzed written daily reports. She also began writing a grand scientific monograph that would summarize what she and her colleagues had learned about chimpanzees since 1960. Bryceson died of cancer in 1980; during the years following, Goodall continued with the monograph, which was at last published by Harvard University Press in 1986 as *The Chimpanzees of Gombe*.

To celebrate the release of that encyclopedic work, the Chicago Academy of Sciences organized an international conference of chimpanzee and pygmy chimpanzee (bonobo) researchers for late November 1986. For Jane Goodall, the conference was a relevation, a profoundly disturbing jolt. Researcher after researcher described the crisis before them: rapid human population growth, unsustainable economic exploitation, a loss of chimpanzee habitat nearly everywhere, the sudden decline in ape populations from hunting for meat, and threats additionally posed by a trade in live animals. At the same time, in the wealthy Western world, chimpanzees were abused as pets, trained with beatings for the entertainment industry, and held in appalling conditions by the laboratory-research industry.

By then, Western scientists had returned to Gombe, and so, once again, the generalized data-gathering was supplemented by specialized research. Indeed, active research at Gombe continues on a number of fronts today. Before the 1986 Chicago conference, Goodall would have happily remained among the scientists continuing to follow the chimpanzees at Gombe; she was planning to write volume two of her great scientific monograph. Following the painful revelations of the conference, however, she ended her scientific career and turned to her new life as an activist on behalf of the beings she had known and come to love.

In partnership with other field scientists and non-invasive laboratory researchers, she moved to improve the situation for chimpanzees in captivity. She could rely on the institutional support of her own organization, the Jane Goodall Institute (JGI), which formed alliances with similar groups. In 1984, she had founded ChimpanZoo to improve conditions for chimpanzees in U.S. zoos; now she planned to help chimpanzees in U.S. labs. She wrote papers, visited laboratories, and organized two major conferences on the practical issues of improving laboratory conditions. She lobbied the United States Congress, hoping to increase legal minimum cage sizes for chimpanzees and mandate other changes that could improve the psychological well-being of captive primates.

In Africa, Goodall became involved in creating sanctuaries for orphaned chimpanzees. Primarily, these orphans were babies whose mothers had been shot as meat; some of the surviving babies were kept as village pets, and others were sold to outsiders and became part of an illegal trade. Sanctuaries would enable African governments to enforce their laws against the trade by confiscating babies. By the end of the decade, JGI was running chimpanzee sanctuaries in Kenya and Uganda and, in partnership with the oil company Conoco, a major sanctuary in Congo-Brazzaville.

In response to severe deforestation outside the boundaries of Gombe National Park, Goodall and JGI enlisted local villages in an agroforestry project to reduce erosion and provide trees for fuel and building poles, thus reducing pressures on trees inside the park. This project, known as TACARE (Lake Tanganyika Catchment Reforestation and Education), was started in 1994. By 2004, thirty-two area villages had chosen the TACARE option and planted over a million trees. TACARE by then also offered a hybridization project for high-yield oil palm seedlings; training in sustainable agriculture for other cash crops; savings and micro-credit programs; and village development funds for infrastructure.

During this same period, Goodall began her educational mission, first with a traveling "Chimpanzee Museum" touring African cities, providing basic information about the nature and needs of chimpanzees. In 1991, she formed an organization called Roots & Shoots to promote the environmental education of children and young people—in Africa first and then around the world. By 2003, Roots & Shoots included around six thousand youth groups located in over sixty countries.

Dr. Jane Goodall is, at age 72, author or coauthor of over a dozen books for adults, translated into nearly fifty different languages. She is author or coauthor of some seventy scientific articles and nine children's books and is the subject of a dozen films. For her pioneering scientific accomplishments and her work as an activist, she has been awarded two

More than forty years after her first visit to Gombe, Jane Goodall continues her lifelong commitment to study the endangered chimpanzees. ©Michael Nichols/National Geographic.

dozen honorary academic degrees and received countless other honors. She was elected an Honorary Foreign Member of the American Academy of Arts and Sciences in 1972 and an Honorary Fellow of the Royal Anthropological Institute in 1991. She was appointed Commander of the British Empire in 1995, named a United Nations Messenger of Peace in 2002, dubbed a Dame of the British Empire in 2003, and inducted into the French Legion of Honor in 2006. She remains vigorous and active, traveling three hundred days of the year, speaking, lobbying, and promoting her special vision of a world in which the barriers between people and animals are breached and finally dissolved.

Further Resources

Goodall, J. (2000a). *Reason for hope: A spiritual journey*. New York: Warner Books.

————. (2000b). *Africa in my blood: An autobiography in letters*. Boston: Houghton Mifflin.

————. (2001). *Beyond innocence: An autobiography in letters, the later years*. Boston: Houghton Mifflin.

Greene, M. (2005). *Jane Goodall: A biography*. Westport, CT: Greenwood Press.

Peterson, D. (2006). *Jane Goodall: The woman who redefined Man*. Boston: Houghton Mifflin.

Dale Peterson

The Significance of Jane Goodall's Chimpanzee Studies

Dale Peterson

When, in November 1960, twenty-six-year-old Jane Goodall observed that chimpanzees were fashioning and then using tools, she began a revolution in the understanding of our nearest living relatives. Since many people then believed that only humans made and used tools—indeed, that such an activity might prove to be the best defining feature of *Homo sapiens*—she also began a revolution in our understanding of ourselves. In this particular feature, at least, we were no longer so nakedly separate from the rest of the natural world.

And yet, that first discovery was only the start, the flimsiest first result of research that still continues. Yes, she discovered and documented that chimpanzees eat meat and make and use tools, but from her earliest moments in the field, Jane Goodall also showed other primate watchers how to watch primates. She demonstrated that it was possible to live and walk freely among wild chimpanzees, and that a person—even a bare-legged, pony-tailed, young, and female one—could, over time and with great effort, construct a critical understanding of what wild apes do, how they live, and who they are. She was among the first to promote the idea that chimpanzees and other large animals can have personalities, and her example as a scientist helped to make that once-radical perspective commonplace. She introduced the idea of chimpanzees as actors in their own drama, as individuals operating with complex intent, and thus she helped establish the scientific appreciation of animal will and intelligence. She propagated the idea that wild chimpanzees have emotions similar to those of humans, and that their emotional inheritance helps account for the many ways in which chimpanzee behavior echoes and elucidates human behavior. And she was among the earliest proponents of the idea that chimpanzee behavior exists in a cultural context, that some patterns vary from region to region, community to community, and are passed down through the generations by learning and even active teaching.

Jane Goodall practiced and taught a science that combined the cold purity of traditional European ethology with her own warm embrace of intuitive and ethical ways of thinking, and her work remains absolutely central, and seminal, to the larger achievement of twentieth- and twenty-first-century primatology. Through her remarkable personal generosity, her opening up of a place and a style and a method, she also contributed powerfully to the community of her scientific peers, while she simultaneously helped launch a significant portion of primatology's next generation. In short, through her work and example, Jane Goodall helped foster the birth of a new science. Because this new science sought a better understanding of humanity's closest living relatives, it simultaneously brought a better understanding of humanity itself.

A Declaration on Great Apes

We demand the extension of the community of equals to include all great apes: human beings, chimpanzees, gorillas and orangutans.

"The community of equals" is the moral community within which we accept certain basic moral principles or rights as governing our relations with each other and enforceable at law. Among these principles or rights are the following:

1. The Right to Life
   The lives of members of the community of equals are to be protected. Members of the community of equals may not be killed except in very strictly defined circumstances, for example, self-defence.

2. The Protection of Individual Liberty
   Members of the community of equals are not to be arbitrarily deprived of their liberty; if they should be imprisoned without due legal process, they have the right to immediate release. The detention of those who have not been convicted of any crime, or of those who are not criminally liable, should be allowed only where it can be shown to be for their own good, or necessary to protect the public from a member of the community who would clearly be a danger to others if at liberty. In such cases, members of the community of equals must have the right to appeal, either directly or, if they lack the relevant capacity, through an advocate, to a judicial tribunal.

3. The Prohibition of Torture
   The deliberate infliction of severe pain on a member of the community of equals, either wantonly or for an alleged benefit to others, is regarded as torture, and is wrong.

Source: Great Ape Project, http://www.greatapeproject.org/declaration.php

Ethics and Animal Protection
The Great Ape Project

The Great Ape Project aims for granting basic moral and legal rights to nonhuman great apes—chimpanzees, bonobos, gorillas, and orangutans. The Great Ape Project has placed the idea of basic great ape rights on the international agenda. It acts as a catalyst, and, since its instigation, many other organizations strive for the recognition of great ape rights as well. This has resulted in increased attention for the interests of great apes.

Some countries have imposed a ban on invasive biomedical research with great apes, and the United States—where most research with great apes occurs—has adopted a policy of not killing so-called surplus great apes.

The Great Ape Project was launched in London on June 14, 1993, by Peter Singer, philosopher at Princeton University, and Paola Cavalieri, editor of the Italian journal *Etica & Animali*. On that day they presented the book *The Great Ape Project: Equality beyond Humanity*, which contains contributions from more than thirty subscribers to "A Declaration on Great Apes" (see box below).

This declaration demands the extension of the moral "community of equals" to include all human and nonhuman great apes. Like us, nonhuman great apes are "intelligent beings with a rich and varied social and emotional life." Therefore, it is argued, we should consider them as our moral equals—we ought to respect their basic interests in the same way as we respect similar interests of humans. The protection of these interests needs to be assured through the endorsement of three basic rights, namely (i) the Right to Life, (ii) the Protection of Individual Liberty, and (iii) the Prohibition of Torture. The Great Ape Project demands that great apes will be protected in the wild and no longer exploited for biomedical research and human entertainment. Among the earliest supporters of the Great Ape Project are zoologists/primatologists Marc Bekoff, Richard Dawkins, Roger and Deborah Fouts, Biruté Galdikas, Jane Goodall, Adriaan Kortlandt, Lyn Miles, Toshisada Nishida, and Francine Patterson, and philosophers Dale Jamieson, James Rachels, Tom Regan, Bernard Rollin, and Steve Sapontzis.

Why this focus on great apes? There appear to be three major reasons for doing so: humans' close relationship with nonhuman great apes, the latter's complex mental lives, and the expectation that the cost to stop their exploitation is relatively limited and thus quite feasible. Though the Great Ape Project directs its attention to great apes, many of its contributors see this as a first step in the process of extending the community of equals. Indeed, many of them are prominent advocates for other animals as well.

Humans are closely related to the nonhuman great apes—we actually *are* great apes. Chimpanzees are our closest living relatives, and we shared a common ancestor with them until as recently as five to six million years ago. Around that time, one lineage led to modern humans, whereas another resulted in modern chimpanzees. Due to this recent split, it should come as no surprise that humans and chimpanzees share a lot of their genetic information. Some twenty years ago, molecular biologists concluded from experiments that the DNA of chimpanzees and humans is 98.4 percent identical. (DNA, or *deoxyribonucleic acid*, carries the genetic information that is passed on from parents to their offspring.) Moreover, humans and chimpanzees turned out to be more closely related to one another than each is to its next closest relative, the gorilla. A recent comparison of the human and chimpanzee genomes, which was published in 2005 in the science magazine *Nature*, confirms the small differences in their DNAs. Due to the close relationship between humans and other great apes, it has been suggested by taxonomist Morris Goodman and his colleagues that all great apes should share the same family of the Hominidae, and that chimpanzees should be added to the genus *Homo*, which currently includes only humans (or *Homo sapiens*).

This close genetic relationship stands in sharp contrast with the traditional view of "clever humans" versus "dumb beasts." It suggests that nonhuman beings have rich mental lives as well. The nonhuman great apes have been extensively studied, and these studies confirm that they have complex mental capacities. Their emotional behavior is very similar to ours—they experience emotions such as pain, joy, sadness, anger, fear, boredom, disgust, jealousy, love, and happiness. They develop strong social bonds that may last an entire lifetime (about sixty years). Broken bonds may result in depression and dying from "a broken heart." They lead complex social lives and have been called "political" animals—they may develop strategic coalitions in order to strengthen their leadership positions. Great apes can be very selfish, but they also show empathy and altruism—they may risk their own lives to save (unrelated) others. They use and make tools in the wild, such as modifying branches for termite fishing or using stone hammers and anvils to crack nuts. Wild populations turn out to have developed differing behavioral strategies that are passed on from one generation to the next—they have traditions; great apes are cultural beings. They have complex mental maps of their environments and knowledge of seasonal variations—they anticipate when particular trees in the rain forest will carry ripe fruits. Cognitive experiments indicate that they can solve complex problems through insight, and they are very creative in finding ways to escape in captivity. They recognize themselves in mirrors, which is an indication of self-consciousness. They can memorize numbers and learn basic mathematical operations, such as simple addition. Language experiments point out that they are able to communicate through symbols. Given these rich mental capacities, the Great Ape Project argues that we should recognize nonhuman great apes as persons.

The founders of the Great Ape Project see their efforts for great ape rights as an important first step toward ending the exploitation of other animals as well. Stopping the use of great apes in captivity is held to be more feasible than terminating the exploitation of chickens, pigs, calves, and other animals in animal factories—due to differences in demands of consumer attitudes and the economic interests involved. What has been the influence of the movement initiated by the Great Ape Project, and what challenges does this movement still face?

Chimpanzees are the preferred great ape model for invasive biomedical research, though their use is declining. Very few countries use chimpanzees for biomedical research, and several countries have explicitly forbidden the use of nonhuman great apes for invasive biomedical research. Austria, The Netherlands, New Zealand, Sweden, and the United Kingdom have officially banned the use of nonhuman great apes in biomedical research for moral reasons. Gabon, Japan, Liberia, and the United States are the only countries that continue using chimpanzees for biomedical research and testing. The majority of these chimpanzees are located in the United States, namely some 1,300 individuals. In 2001, the Great Ape Project Census was started, to bring the individual apes in U.S. laboratories (and elsewhere in captivity) into the picture. Organizations such as the Chimpanzee Collaboratory (which counts the Jane Goodall Institute, the Great Ape Project, and the Animal Legal Defense Fund among its members), the Humane Society of the United States, and the New England Anti-Vivisection Society call for the U.S. government to end biomedical research on chimpanzees. The U.S. government agreed that chimpanzees no longer used for research should be retired—not killed. To arrange their retirement, President Bill Clinton signed the CHIMP Act into law in 2000. However, this act is an amended version of the original and allows for the recalling of retired chimpanzees for further biomedical research if deemed necessary.

Biomedical researchers predict that the number of chimpanzees in U.S. laboratories will decrease to approximately 800 around 2010, due to a breeding moratorium for the 850 chimpanzees owned or supported by the NIH (National Institutes of Health), retirement in sanctuaries, and mortality rates. This moratorium was imposed in response to the breeding program developed during the years of the AIDS panic. A breeding project resulted in some 400 baby chimpanzees in U.S. laboratories between 1986 and 1996. However, only one chimpanzee from the at least 150 infected with HIV developed (and died from) AIDS. In 2006, the New England Anti-Vivisection Society launched the campaign "Project R&R: Release and Restitution for Chimpanzees in U.S. laboratories." With the support of chimpanzee advocates like Jane Goodall and Roger Fouts, Project R&R calls for the permanent termination of federal funding for the breeding of chimpanzees in U.S. laboratories. Biomedical researchers oppose this and seek to lift the breeding moratorium.

Considerable evidence exists of the cruel methods routinely applied to train and control chimpanzees and orangutans for the entertainment industry—including performances in circuses and on movie sets. Young apes are taken away from their mothers at an early age and confined in small, monotonous cages. Trainers rely on intimidation and fear to control these curious and active individuals. According to a report by the Chimpanzee Collaboratory, "some trainers pummel chimpanzees with their own fists, beat them with hammers, metal rods, and mop handles. Electric devices also may be used to shock them into submission." The Chimpanzee Collaboratory organizes letter campaigns to protest against the use of great apes by companies and producers.

The Great Ape Project states that great apes should not be held captive to serve human purposes. They should be returned to the wild or, when their return to the wild is not feasible, placed in spacious sanctuaries where they can live their lives relatively free from human interference and exploitation. In 2001, the book *Great Apes & Humans: The Ethics of Coexistence* was published by some people who work in zoos, in response to the Great Ape Project. In this book, zoo advocates point out that many zoos undertake considerable efforts to enrich the lives of nonhuman great apes in captivity. However, the bottom line of the Great Ape Project's position is that a life in captivity cannot be as rich as one in the wild. Captivity allows for far less autonomy or freedom of choice in comparison to a life in the wild. A major difference between zoos and sanctuaries is that zoos want to continue the keeping of nonhuman great apes in captivity and therefore organize breeding programs, whereas sanctuaries usually do not allow the breeding of young ones.

Moreover, many members of the movement for ape rights distrust zoos, because zoo personnel often believe it is acceptable to kill healthy animals that are no longer considered useful for captive breeding programs. Donald Lindburg, the leader of the giant panda team at the San Diego Zoo and long-time zoo administrator and researcher, advocates retiring "surplus" animals rather than killing them, the latter being the position taken by many zoo professionals. In 1991, Robert Lacy (Brookfield Zoo and chairman of the Conservation Breeding Specialist Group) argued in favor of "managerial euthanasia" and suggested that resources going to surplus, hybrid orangutans—progeny from Bornean and Sumatran orangutans—should be made available to breed "purebred" orangutans. However, nowadays, most zoos seem to make an exception for nonhuman great apes. In a response to the Great Ape Project, Michael Hutchins of the Association of Zoos and Aquariums (AZA) and his colleagues commented in 2001: "As great ape zoo populations mature, the question arises of what to do with older, postreproductive individuals. Animal rights proponents argue that zoos have a responsibility to care for captive-bred animals from 'the cradle to the grave.' In the case of great apes, we agree. Despite arguments to the contrary . . . and the fact that it is legal, euthanasia of healthy great apes is not generally accepted in the professional zoo community as an option for controlling populations." One is left wondering what the general zoo policy would have been without the growing influence of the movement for great ape rights.

A tremendous challenge for those who defend the interests of great apes is to deal with the enormous threats faced by the remaining great apes in the wild. No viable populations may remain within the next two decades. Major threats are the logging of forests, hunting for meat (the bushmeat crisis), and diseases such as Ebola. The United Nations has launched the Great Apes Survival Project (GRASP) "to lift the threat of imminent extinction faced by gorillas, chimpanzees, bonobos, and orangutans." GRASP brings together organizations such as the Born Free Foundation, Conservation International, the Dian Fossey Gorilla Fund, the Jane Goodall Institute, the Wildlife Conservation Society, and the WWF, in an alliance that cooperates with various governments to turn the tide. Conservation organizations refer, in particular, to the importance of conserving species, such as their ecological role and their aesthetic, scientific, and economic value. Organizations such as the Great Ape Project add a special dimension by stating that each great ape is a valuable individual whose interests need to be protected, in the first place, *because* of this individual.

The Great Ape Project strives for the passing of a declaration of great ape rights by the United Nations, similar to declarations for children and the disabled. In order to change the legal status of great apes, the Great Ape Project and the Animal Legal Defense Fund have launched the Great Ape Legal Project. It is argued that great apes should be respected as legal persons, whereas courts currently view them as human property. A powerful case for legal great ape rights has been made by Steven Wise (Vermont Law School) in his book, *Rattling the Cage: Toward Legal Rights for Animals* (2000), and Harvard Law School hosted, in 2002, a symposium on legal rights for great apes, sponsored by the Chimpanzee Collaboratory's Legal Committee. The transcript of this discussion with prominent lawyers and scientists was published by *Animal Law* in 2003.

See also

Law—*The Legal Status of Companion Animals*
Law—*Public Policy and Animals*

Further Resources

Beck, B. B., Stoinski, T. S., Hutchins, M., Maple, T. L., Norton, B., Rowan, A., Stevens, E. F., & Arluke, A. (Eds.). (2001). *Great apes & humans: The ethics of coexistence.* Washington and London: Smithsonian Institution Press.

Cavalieri, P. (Ed.). (1996). *Etica & Animali, 8,* 1–177. [Special issue devoted to The Great Ape Project.]

Cavalieri, P., & Singer, P. (Eds.). (1993). *The Great Ape Project: Equality beyond humanity.* London: Fourth Estate.

The evolving legal status of chimpanzees. (2003). *Animal Law, 9.* Portland: Lewis & Clark Law School.

Goodman, M., Porter, C. A., Czelusniak, J., Page, S. L., Schneider, H., Shoshani, J., Gunnell, G., & Groves, C.P. (1998). Toward a phylogenetic classification of primates based on DNA evidence complemented by fossil evidence. *Molecular Phylogenetics and Evolution, 9*(3), 585–98.

The Great Ape Project Census: Recognition for the uncounted. (2003). Portland: Great Ape Project (GAP) Books.

Hutchins, M,, Smith, B,, Fulk, R,, Perkins, L,, Reinartz, G,, & Wharton, D. (2001). Rights or welfare: A response to the Great Ape Project. In B. B. Beck, T. Stoinski, M. Hutchins, T. L. Maple, B. Norton, A. Rowan, E. F. Stevens, & A. Arluke (Eds.), *Great apes and humans: Ethics of coexistence* (pp. 329–66). Washington, DC: Smithsonian Institution Press.

Lacy, R. (1991). Zoos and the surplus problem: An alternative solution. *Zoo Biology, 10,* 293–97.

Lindburg, D. G. (1991). Zoos and the "surplus" animal problem. *Zoo Biology, 10,* 1–2.

Nature, 437(7055). (2005, September 1). [Issue with various articles on chimpanzees, on the occasion of mapping the chimpanzee genome.]

Peterson, D. (2003). *Eating apes.* Berkeley, Los Angeles, and London: University of California Press. (With an afterword and photographs by Karl Ammann.)

Peterson, D., & Goodall, J. (1993). *Visions of Caliban: On chimpanzees and people.* Boston and New York: Houghton Mifflin Company.

Rowan, A. N. (Ed.). (1994). *Wildlife conservation, zoos and animal protection: A strategic analysis.* A workshop held at the White Oak Conservation Center, Yulee, Florida, April 21–24.

Serving a life sentence for your viewing pleasure! The case for ending the use of great apes in film and television. (2003). Washington, DC: The Chimpanzee Collaboratory. [Available at http://www.chimpcollaboratory.org/news/movie.asp]

Stallwood, K. W. (2001). A conversation with Peter Singer. In K. W. Stallwood (Ed.), *Speaking out for animals: True stories about real people who rescue animals* (pp. 34–48). New York: Lantern Books.

Tomasello, M. & Call, J. (1997). *Primate cognition.* New York and Oxford: Oxford University Press.

Wise, S. M. (2000). *Rattling the cage: Toward legal rights for animals.* Cambridge, MA: Perseus Books.

Koen Margodt

■ Ethics and Animal Protection
Hoarding Animals

The Hoarding of Animals Research Consortium (HARC) defines an *animal hoarder* as someone who has "accumulated a large number of animals, which has overwhelmed that person's ability to provide even minimal standards of nutrition, sanitation, and veterinary care; failed to acknowledge the deteriorating condition of the animals (including disease, starvation, and even death) and the household environment (severe overcrowding, very unsanitary conditions); and failed to recognize the negative effect of the collection on his or her own health and well-being, and on that of other household members" (2000). In a typical case, a person is discovered living in squalid conditions with dozens to over a hundred animals—both dead and alive.

Animal hoarding, sometimes referred to as "collecting," is a problem that probably exists in every community. Based on the estimated national animal-shelter population of 6 million, there are about 1,200 to 1,600 cases per year, and based on the human population served, 600 to 2,000 cases per year in the United States.(About 60 percent of these cases involve multiple repeated investigations of the same individual). Cats and dogs are the most commonly hoarded species, but wildlife, dangerous exotic animals, and farm animals have also been involved—even in urban situations. Such animals are frequently ill and malnourished to the point of starvation. Floors may buckle from being soaked with urine and feces, and the air may be difficult for investigators to breathe without protective apparatus. Although the stereotypical profile of a hoarder is an older, single female (living alone and known as the neighborhood "cat lady"), in reality, this behavior seems to cross all demographic and socioeconomic boundaries. As hoarders tend to be very secretive, many can lead a double life with a successful professional career; hoarding behavior has been discovered among doctors, nurses, public officials, college professors, and veterinarians, as well as among a broad spectrum of socioeconomically disadvantaged individuals.

Cases come to the attention of authorities primarily by complaints from neighbors, and unsanitary conditions are the primary problem reported. Sanitary conditions often deteriorate to the extent that public health authorities condemn dwellings as unfit for human habitation. By the time these situations have deteriorated to the point they cannot be ignored, expenses for veterinary care and housing of animals, litigation, and clean-up or demolition of premises can run into the tens of thousands of dollars. Unfortunately, because of ill health, contagious diseases, and the large numbers involved, euthanasia is often the only option for many of the animals rescued from such situations.

For most hoarders, living spaces are often compromised to the extent that they no longer serve the function for which they were intended. Appliances and basic utilities(heat, plumbing, and electricity) are frequently inoperative. Household functioning is often so impaired that both food preparation and maintaining basic sanitation are impossible. However, few hoarders seem to meet the criteria for mental incompetence or immediate danger to the community, so options for intervention are limited. In many animal-hoarding situations, other family members—such as minor children, dependent elderly persons, or disabled adults—are present and are also victims of this behavior. Serious unmet human health needs are commonly observed, and the conditions often meet the criteria for adult self-neglect, child neglect, or elder abuse. From a community health perspective, the clutter can pose a fire hazard. In some cases, fireplaces and kerosene heaters are used for heat. Rodent and insect infestations, as well as odors, can create a neighborhood nuisance. These are important public-health aspects of animal hoarding that go largely unrecognized and which may provide avenues for intervention. In one hoarding case, the air ammonia level after the house had been ventilated by the fire department was 152 ppm; the National Institute for Occupational Safety and Health lists 300 ppm as a concentration immediately dangerous to life or health, and 25 ppm as the maximum average occupational exposure during the workday.

Animal hoarding cases tend to overlap or fall between the jurisdictional cracks of numerous state and local government agencies and departments (e.g., mental health, public health, zoning, building safety, animal control, aging, sanitation, fish and wildlife, child welfare), so it is the rule rather than the exception that they are procedurally cumbersome, time consuming, and costly to resolve. Although common sense suggests that the accumulation of large numbers of animals in homes can have important public health implications—including placing neighborhoods at risk, due to unsanitary living conditions; facilitating the spread of zoonotic diseases; and endangering the health of vulnerable household members, particularly children or dependent elderly—the potential for these consequences in animal hoarding cases is not widely appreciated by government agencies. As a result, systematic procedures for resolving these cases are lacking, as are effective preventive strategies. Difficult issues of personal freedom, lifestyle choice, mental competency, and private property rights further confound resolution of these cases. In one famous case, a woman living in a school bus with 115 dogs had been investigated in several jurisdictions in four states. In each case, she had, essentially, been given a tank of gas and told to get out of town. When finally prosecuted in Oregon, she went through three prosecutors and six judges, finally serving as her own attorney. Her trial lasted five weeks.

The paradox that confounds resolution of these cases is that hoarders often profess great love for animals. A hoarder may claim to be a pet rescuer or a "no-kill" shelter attempting to help unwanted pets, and some maybe professional or hobby animal breeders. All too frequently, these excuses maybe used as effective ploys for the media or as defenses in court. Despite these claims of professionalism and good intentions, hoarders are, by definition, oblivious to the extreme suffering—obvious to the casual observer—of their animals.

When animals are rescued from these situations, euthanasia is often the only practical recourse, because of the extent of the animals' illnesses, poor condition, and lack of socialization. These are cases that no one wants to handle, and it is not uncommon for them to be sidestepped until the situation has deteriorated to the point it cannot be ignored. By that time, expenses for veterinary care can run into the tens of thousands of dollars. Procrastination may also increase the likelihood that the resolution will garner media attention. Greater recognition of this syndrome, as well as improved understanding of standards for responsible animal sheltering and rescue, is needed.

Animal hoarding is not yet recognized as indicative of any specific psychological disorder; however, evidence from case reports indicates that, eventually, many hoarders are placed under guardianship or other supervised living situations, suggesting the incapacity to make rational decisions and manage their affairs. This may well indicate a strong mental-health component. Perhaps the most prominent psychological feature of these individuals is that pets (and other possessions) become central to the hoarder's core identity. The hoarder develops a strong need for control, and just the thought of losing an animal can produce an intense, grief-like reaction. Preliminary research suggests that hoarders grew up in chaotic households with inconsistent parenting, in which animals may have been the only stable feature.

Several possible psychological models exist for animal hoarding. One argument is that a focal delusional disorder could be present, since claims that animals are healthy and well cared for—in the face of clear evidence to the contrary—and delusional levels of paranoia about officials are consistent with a belief system that is out of touch with reality. Similarities have been noted between hoarders and substance abusers, others with impulse control problems, or compulsive gamblers. An attachment disorder could also be present, such that relationships with animals are preferred because they are safer and less threatening than relationships with people. Perhaps the most parsimonious explanation is obsessive-compulsive disorder (OCD). Hoarding of inanimate objects is seen in a variety of psychological disorders, but it is most commonly seen in OCD. Two to 3 percent of the human population suffers from OCD, and 15 to 30 percent of those have hoarding as a primary symptom, although it is unknown what proportion of these individuals hoard animals.

It is ironic that, while the medico-legal framework for intervention to help people in these situations seems to be inadequate, there are comparatively effective and easily implemented laws in place to allow for the rescue of animal victims. Every state has statutes that mandate that caretakers provide animals with sufficient food and water, a sanitary environment, and necessary veterinary care in case of illness or injury. Therefore, hoarding cases are often initially investigated and handled by representatives of the local animal shelter, humane society, or other animal protection group. In cases where an animal protection organization does not have jurisdiction, local police officers or municipal animal control officers may be the initial agents to investigate a case. Because of the severity of animal suffering and the need for expediency, a common scenario is for the animals to be removed, for their own protection, through use of a search warrant, with the hoarder subsequently prosecuted under state anticruelty laws. Unless relinquishment can be negotiated, the animals must be held as evidence until the case is concluded.

In some jurisdictions, violations of animal-cruelty statutes may be summary offenses, prosecuted by local humane agents or animal control officers in front of a magistrate, whereas in others, they maybe misdemeanors—or, in some cases, felonies—requiring prosecution by the district attorney's office. Penalties in the event of a guilty finding can range from a nominal fine to forfeiture of the animals and jail time. Some states mandate psychological counseling of offenders, whereas others make it an option for the court. Some state statutes provide for the recovery of the costs of boarding and medical care for animals in cruelty cases.

Occasionally, there may be prohibitions on future pet ownership or limitations imposed on the number of animals, along with a requirement of periodic monitoring of the situation by authorities. Supervised probation has been recommended over court probation as a better way to ensure compliance. Prohibitions against future pet ownership are effective only to the extent that monitoring is practical. Recidivism is rapid and may be almost universal. In a typical example, two fifty-year-old women and their seventy-three-year-old mother were discovered living with eighty-two live cats—and a hundred and eight dead ones. They fled from the investigation, rented a new apartment nearby, and had seven cats and a dog two days later. Even when monitoring is practical, hoarders can escape enforcement by moving to a new jurisdiction, often only across town or county lines.

Some communities attempt to either prevent or remedy hoarding situations by passing ordinances that limit the number of pets a person can own. There is no data to indicate whether these measures are effective, but what is known is that they are wildly unpopular and difficult to enforce, and they are likely to be opposed by a broad coalition of pet fanciers, breeders, rescue groups, and animal protection organizations. This is a harsh—and probably ineffective—remedy that needlessly penalizes responsible pet owners.

The worst situations may be avoided through regulations that stipulate housing densities, sanitation requirements, and veterinary care, as well as provide for regular inspections of licensed facilities. For example, Colorado has developed licensing requirements and comprehensive standards for the operation of an animal shelter or pet-rescue organization. Such criteria also could help the media and the public, as well as the courts, distinguish between legitimate sheltering efforts and hoarding.

See also

Cruelty to Animals

Further Resources

Arluke, A., et al. (2002).Press reports of animal hoarding. *Society and Animals,10*, 1–23.

HARC. (2000, April). People who hoard animals. *Psychiatric Times, 17*, 25–29.

Patronek, G. (1999). Hoarding of animals: An under-recognized public health problem in a difficult to study population. *Public Health Reports,114*, 82–87.

Vaca-Guzman, M., & Arluke, A. (2005). Normalizing passive cruelty: The excuses and justifications of animal hoarders. *Anthrozoos, 18*, 338–57.

Worth, C., & Beck, A. (1981). Multiple ownership of animals in New York City. *Transactions and Studies of the College of Physicians of Philadelphia, 3*, 280–300.

Arnold Arluke and Gary Patronek

■ Ethics and Animal Protection
Horse Rescue

Horse rescue is necessitated for several reasons: (1) to prevent "unwanted," injured, or aged horses from being brought to slaughter, (2) to protect horses from deliberate cruelty and neglect, and (3) to preserve horses that have incapacitated owners or are being neglected due to having fallen through the cracks of human care-giving. Horse rescue is of paramount interest because of existing slaughter plants for horses in the United States for human consumption abroad. Because horses are viewed as companion animals and work animals in the United States, particularly in racing or equestrian sports, it comes as a surprise that horses have been slaughtered in meat-packing plants. There are only three meat-packing plants in the United States, located in Texas and Illinois. The meat is sold for consumption in Europe and Japan, where there is a tradition of horse-meat consumption.

In the United States, horse slaughter has often been referred to as "being sent to the glue factory," for production of items not fit for human consumption. Horses are not culturally regarded as food in this country. Because horses are companions and even treated by some as pets, Americans are often horrified when they find that up to 50,000 horses a year are slaughtered like farm animals that are raised for meat consumption. The second shock is that the horses most in need of rescue are not mistreated by individuals, but are used as commodities in, or impacting on, a production process; they are wild or work animals such as thoroughbreds, standard bred trotters and pacers, and mares used for the production of estrogen for women's hormone-replacement therapy. Ranchers protesting mustang herds and horse-sporting enter-tainment industries produce the greatest number of so-called unwanted animals. Nibert (2002) argues that economic interests drive the exploitation and mistreatment of animals. He argues that the oppression of humans is linked to the oppression of animals through a process of social devaluation. The privileged take advantage of both minorities and animals through individual and group economic interest. Horse rescue is performed primarily to meet the demands of the beef industry and deal with ani-mals cast off by the horse industry. Media attention has provoked three waves of media attention to the need for horse rescue.

The first wave of media attention and public reaction promoting rescue occurred in response to the killing of mustangs, the wild horses of the western United States descended from mounts lost by Spanish colonizers. At one time numbering in the millions, fewer than 10 to 20 thousand remain, as a result of killing by ranchers. At one time, mustangs were gunned down from airplanes, because ranchers did not want them com-peting with cattle for grazing. In 1950, Velma Brom Johnston, known as "Wild Horse Annie," witnessed a group of injured, captured mustangs being trucked to a slaughter-house and decided to take on the Bureau of Land Management. Hope Ryden, author of *America's Last Wild Horses,* documented the abuse, which included results such as feed-ing canned horse meat to pets. WHOA (Wild Horse Organized Assistance) undertook surveillance and research, passing information to media. In 1971, Congress passed the recently renewed Wild Free Roaming Horse and Burro Act and established protective reserves to preserve the mustang as a symbol of the West. The International Society for the Protection of Mustangs and Burros and the Bureau of Land Management operate a wild mustang adoption program for when herd size is controlled. At present, Front Range Rescue (www.frontrangeequinerescue.org/) operates a sanctuary for wild-horse adoption and is contesting further reductions of wild herds proposed by the Bureau of Land Management. They were also involved in the production of *Cloud: Wild Horse of the Rockies,* which documents a herd they are trying to protect from culling, shown by the Public Broadcasting System on the Nature series.

Currently, the most newsworthy form of horse rescue in the United States involves thoroughbred race horses. This charitable activity has been spurred by the slaughter of famous race horses abroad (Warner, 2004). The death of a famous horse by maltreatment stirs public reaction and brings attention to horse slaughter for consumption abroad.

The second wave of media attention and public mobilization for horse rescue occurred in 1997, when the stallion Exceller was slaughtered in Sweden. Exceller's death prompted the creation of the Exceller Fund (www.excellerfund.org/) for the retirement of thoroughbred horses. Exceller won seven top-stakes races and defeated the Triple Crown winners Seattle Slew and Affirmed in the Jockey Club Gold Cup. Racing through age six, Exceller earned $1,654,003 in races involving horses of top class. As a superior racehorse, Exceller entered stallion service at a breeding fee of $50,000, but his foals did not reproduce his degree of ability on the racetrack, and his fee declined to $2,500 over time. In 1991, a Swedish breeder purchased the horse for breeding purposes. Exceller's new owner, Gott Ostlund, declared bankruptcy and slaughtered the horse, despite pleas to let him live in retirement by the farm owner caring for him, Ann Pagmar. Ostlund stated to the *Daily Racing Form* that the horse was killed "because he was very old." According to Mullany (1997), Pagmar stated

> He was in super condition. The owner never visited to see for himself. It didn't seem to matter. He told me to take Exceller to the slaughterhouse, and I walked him over myself.
>
> I made an appointment because I wanted to get it over with quick, but they were very busy when we got there and we had to wait. Exceller knew what was going on; he didn't want to be there.

The meanness and total lack of necessity of his fate led to the establishment of The Exceller Fund, which is a nonprofit under the umbrella of the Thoroughbred Charities of America. Horses are placed in adoption categories which include: (1) riding potential; (2) "pasture pals" (horses that are old or have infirmities which make them unsuitable for riding); and (3) in rehabilitation. One of the horses helped by the Exceller Fund is named Black Beauty, reminiscent of Anna Sewell's classic story of animal maltreatment.

The third wave of media attention to slaughter, which has produced the most significant legal action, occurred in 2003, when champion thoroughbred Ferdinand was slaughtered in Japan. Ferdinand won the most famous race in the United States—the Kentucky Derby—and the Breeder's Cup Classic, a highly prestigious event critical in evaluating thoroughbred horses for championships (Bayer, 2003). After a highly successful year at four years of age, he was named champion older horse and "Horse of the Year." Like Exceller, Ferdinand had not passed on his superior racing ability as a stallion and was exported for a second try at stud duty. Japan has limited facilities for horses, and Ferdinand was first sent to a riding school and then to a Japanese slaughterhouse to be processed for consumption. His death raised a severe outcry in the United States and prompted the creation of the Horse Slaughter Prevention Bill in Congress.

Since the time of Ferdinand's passing, much more effort has been put into persuading thoroughbred breeders, owners, and trainers to take more interest in the fate of horses that give lesser racing performances and are of no use for breeding, such as gelded male horses and those considered to have lesser pedigrees. This has been motivated by the negative publicity generated by celebrity race-horse slaughter and the interest and affection for horses shown in the United States. At present, thoroughbred-horse rescue organizations are located in: Alabama (1); Arizona (1); Arkansas (1); California (12); Colorado (1); Connecticut (1); Delaware (2); District of Columbia (1); Florida (8); Georgia (4); Illinois (2); Indiana (3); Iowa (1); Kansas (1); Kentucky (12); Louisiana (3); Maryland (8); Massachusetts (4); Michigan (4); Minnesota (1); Missouri (1); Montana (2); Nebraska (1); Nevada (1); New Hampshire (2); New Jersey (3); New Mexico (2); New York (7); North Carolina (2); Ohio (5); Oklahoma (3); Oregon (1); Pennsylvania (12); South Carolina (3); South Dakota (1); Tennessee (1); Texas (5); Vermont (2); Virginia (4); Washington (8); West Virginia (3); and Wisconsin (2).

The Thoroughbred Retirement Foundation (TRF), located in Kentucky, is the most well-known of the horse-rescue organizations. TRF has established a horse rehabilitation and training center at the Blackburn Correctional Complex in Lexington, Kentucky; the Marion County Correctional Center in Ocala, Florida; and the Wakehill Correctional Facility in Wakehill, New York, all near major thoroughbred-breeding centers. The TRF has biographies of better-known rescued thoroughbreds and photos and information about horses up for adoption. TRF can be accessed at www.trfinc.org/. Many horses achieve useful "second careers" as sports and pleasure animals. Certain race horses, however, have been so badly and repeatedly injured during their racing careers that they become "pasture pals," providing companionship for other horses—herd animals that do not easily adjust to being alone in a pasture.

One of the long-standing organizations involved in horse rescue in the United States is the American Society for the Prevention of Cruelty to Animals (ASPCA). Their site can be accessed at www.aspca.org/. This organization, as well as others involved in horse rescue, supports the Horse Slaughter Ban Act, which has met real opposition only from those members of Congress in states with horse-slaughter plants. Their adoption program emphasizes PMU (pregnant mare urine) horses. These animals were forced to stand in stalls, connected to catheters that caught their urine for the estrogen used to manufacture Premarin and other women's hormone-replacement drugs. The drug industry has adopted a Code of Practice for better treatment of these horses, but their foals are still sold and often slaughtered. Health concerns related to use of hormone replacement therapy has led to a decline in use, which has produced "unemployed' mares in need of adoption.

First-person accounts of horse rescues have been written by Richards (2006), Bowles (2003), and Caulkett (2003). Although these books provide soft (anecdotal) evidence of the difference human caring can make in the lives of horses, the connection the authors make to the suffering and the threat of being unneeded, for horses that do not know why they are being mistreated and/or disposed of, sends an important social message to those who care about animals or are willing to learn. *Second Chances* (Caulkett, 2003) details many animal-control cases. *Chosen by a Horse* is the story of an abused woman who rescues a standard bred harness-racing mare with a foal (Richards, 2006). *The Horses of Proud Spirit* (Bowles, 2003) gives many anecdotal examples of human-animal intersubjectivity and social connection (see Irvine, 2004). Much of what is described is in the eye of the human beholders, as they imagine the horses' inner states and potentials. Bowles (2003, p. 21) writes about an excitable thoroughbred horse that she bought to rescue from poor treatment and was advised was above her skill level to handle:

> I felt something deep in my soul for this horse, and I believed with all my heart that her problems stemmed from the fact that I didn't know how to communicate with her, coupled with the life she had led. She had been shuffled around from owner to owner, never forming a bond. I could see in her eyes a sensitivity and intelligence that had been stepped on and snuffed out, probably her entire life. For this inner light to glow with all the radiance I knew she possessed, she needed to trust. And I would do what was necessary to learn to communicate with her and show her that she could trust me.

Bowles learns to communicate by observing how horses communicate with each other, and a social bond is formed with the horse. Richards (2006) rescued a malnourished horse with a history of injury, and she expresses the sense of social connection she feels as the horse regains health:

> [Some] would say that I was anthropomorphizing, but I thought the mare was grateful. In fact, I thought most rescued animals exhibited signs of gratitude, an awareness of having been saved from suffering or death, and their gratitude was expressed in particularly open affection.
>
> Lay Me Down [horse's name] expressed affection by sighing. I saw it as an expression of relief, a letting go of all the tension she'd carried in that big body for such a long-time, the horse equivalent of "Phew, I made it."
>
> She sighed a lot. She sighed when I poured the bran mash into her feed bin. She sighed when I put her blankets on at night, and she sighed in the morning when I took

them off. She sighed at her hay, she sighed when I brushed her, she sighed when I kissed the end of her nose. She sighed at the vet: great big sighs, big enough to spray me with snot sometimes; loud, wet affectionate sighs. I loved her sighs. Sometimes I sighed back. I couldn't help it. I wanted her to know that I felt the same way. I was relieved too. "Phew, we've made it," I sighed. We were both safe.

Massive efforts at horse rescue continue to be necessary in the United States. Horse slaughter was temporarily halted by a provision of the 2006 Agriculture Appropriations Act, which does not permit federal inspectors to be paid by taxpayers to inspect horse meat. Gary Francione (2000) indicates that there is a conflict in treating animals as property and simultaneously representing them as subject to moral concerns. Humane-treatment laws tend to reduce suffering, not eliminate it. He argues that animals cannot be treated as commodities, and that they should be given a status equivalent to humans—who cannot legally be treated as property. The principle of equal considera-tion would accord animals the right not to be treated as "thing." The suggestion has been made that industrially used horses should be given retirement and a chance at a second career, just as a human worker. Until that happens, horse rescue and the loving care given to renew animals will be a larger undertaking than the American public under-stands it to be.

Further Resources

American Society for the Prevention of Cruelty to Animals. http://www.aspca.org/site/Page Server?pagename=pro_equinecruelty

Bayer, B. (2003). Roses to ruin. *The Blood Horse, CXXIX(30)*, 3918–23.

Bowles, M. S. (2003). *The Horses of proud spirit.* Sarasota, FL: Pineapple Press.

Caulkett, L. M. (2003). *Second chances: Amazing horse rescues.* Boca Raton FL: Universal Publishers.

Communication Alliance to Network Thoroughbred Ex-Racehorses (CANTER). http://www.canterusa.org/

The Exceller Fund. http://www.trfinc.org/

Francione, G.(2000). *Introduction to animal rights: Your child or the dog.* Philadelphia, PA: Temple University Press.

Front Range Equine Rescue. http://www.frontrangeequinerescue.org/

Irvine, L. (2004). *If you tame me: Understanding our connections with animals.* Philadelphia, PA: Temple University Press.

Kathrens, G. (Director). *Cloud: Wild horse of the Rockies.* A coproduction of Taurus Productions and Thirteen/WNET New York. http://www.pbs.org/wnet/nature/cloud/

Mullaney, M. (1997, July 20). *Unique superstar Exceller met tragic end.* Retrieved June 16, 2006, from http://www.excellerfund.org/

Nibert, D. (2002). *Animal rights, human rights: Entanglements of oppression and liberation.* New York: Rowan and Littlefield.

Re-Run/Adopt a Thoroughbred, Inc. http://www.rerun.org/

Richards, S. (2006). *Chosen by a horse.* New York: Soho Press.

Ryden, H. (1970). *America's last wild horses.* New York: E. P. Dutton & Company.

Tranquility Farm. http://www.tranquilityfarmequestrian.com/

Warner, J. (2004). *Animal celebrity and animal rights reform: Superstar racehorse Ferdinand's slaughter as an impetus to end horse killing for meat in the U.S.* Paper presented at the American Sociological Association meeting in San Francisco, CA.

Judith Ann Warner

■ Ethics and Animal Protection
Human-Animal Kinship for Animal Rightists and "Deep Greens"

Representations and treatment of animals in two movements are important: the animal welfare/rights/liberation movement—the animal movement for short—and the deep green/deep ecology movement. How do each of these movements come to terms with— or fail to come to terms with—the natural continuity existing between animals and humans?

Reducing to Individual Sentience

Members of the animal movement tend to focus on animal individuals as sentient beings and on our ethics vis-à-vis these beings. The domain for animal defenders is *that* nature that has evolved individual and sentient, *that* nature which can feel pain, pleasure, and fear (Singer, 1990).

Because many animal advocates (short for: the members of the animal movement) live in urban areas—are city dwellers (Francione, 1996; Montgomery, 2000)—the animals they encounter tend to be those we have incorporated into our work and living places, such as production animals in factory farms, animals used as organic instruments in laboratories, and companion animals. That is, urban people encounter animals that are either domesticated or been made to live (and die) in human-manufactured habitats (Sabloff, 2001).

Having said this, animal advocates also focus on hunted animals, and this concerns wild rather than domesticated animals. Recreational hunting has a long history, especially in North America (Cartmill, 1993; Flynn, 2002).

The animal movement's focus on *sentience* stems from the understanding that there is continuity between the human and animal condition. Human sentience has ethical significance. It is at the root of the condemnation of oppression, torture, and genocide. Human-animal continuity implies the acknowledgement that many animals have bodies and nervous systems that resemble ours. If well-being is important to humans, it must be important to animals also. Not only do many animals have bodies like ours, their subjectivity—their mind and their emotional life—also bears resemblance to us. Like us, animals are in Tom Regan's terms "subject-of-a-life" (Regan, 1983). Human-animal continuity in body and mind calls for parallel continuity in ethics, such that ethical obligations vis-à-vis animals cannot be radically different from those vis-à-vis humans.

However, many people in the animal movement tend to be almost indifferent to all nature other than animal nature. Supposedly nonsentient living nature, such as plants and trees, is generally not taken into consideration. Neither are nonliving, inorganic natural entities, such as rocks, rivers, or even ecosystems. In themselves, these parts of nature are not sentient and individually they cannot suffer, so the animal movement often overlooks or dismisses them (Hay, 2002).

The animal movement is highly critical of the traditional cartesian notion of "animal-machine" and constitutes the most important group worldwide to condemn factory farming. But it seems to have no objection against similar things done to plants (Dunayer, 2001). A concept such as "plant-machine" and the intensive vegetable and plant farming that is currently taking place do not raise the same eyebrows. The movement's critique of objectification and exploitation seems to rest solely on the aforementioned definition of sentience. The objectification—including things such as genetic manipulation—of the rest of nature goes largely unnoticed or is dismissed.

By concentrating on sentient beings, animal advocates abstract from the environmental context of animal existence. Many animal activists have no conception of how animals, even as individuals, are integrated into other nature. One sometimes encounters a certain uneasiness among members of this movement about nature's meat-eaters—as though the eating of animals by other animals were something that ideally should not exist. Some animal rightists and liberationists tell me that, were it possible, they would like to "phase out" predator-prey relationships or at least liberate (save) the prey animal from the equation (personal communication in several countries).

Another example of refusing to accept animal meat-eating as a zoological necessity is the tendency among vegetarian/vegan animal advocates to turn their carnivorous companion animals into vegetarians as well by feeding them plant-derived food often accompanied by special dietary supplements. Admittedly in North America standard pet food is hardly ever fresh and tends to come out of a packet or tin, unlike Europe, where one can get fresh and increasingly organic free-range meat for one's companion animals at the local butcher. Although many of these people do acknowledge that their animal's body may not "be built" for vegetarian or vegan food, it is apparently no problem for them that the necessary daily intake of supplements will make that animal totally dependent on the health industry. Inadvertently these people are turning animals into duplicates of themselves: modern consumers of the manufactured products of an industrial age. The animals' lives are humanized and *colonized*—their alienation taken to another extreme. Is this about protecting companion animals from nonethical food or about imposing human ethics on the animal other? Incidentally, much plant-based and processed food happens to be the end-product of unsustainable monocultures—to which many animal habitats have had to give way—and has been put on the market by the same globalized and diversified agro-industrial complex that also produces standard pet foods (Noske, 1997).

Many animal advocates thus seem to have trouble accepting nature as an interdependent system where everything has its place, function, and appropriate physical organization. Organic beings took a long time evolving in relation to each other and to nonliving inorganic nature. Nature is a community where every living thing lives off everything else (food, even vegan food, is living nature in a killed state), and in the zoological realm this means that both plant eating and meat eating have their respective *raisons d'être*. Predation is neither a negligible anomaly nor an ethical deficiency in the ecosystem (Plumwood, 1999).

There seems to be a lack of environmental awareness and environmental critique among many animal advocates. Urbanization, technological optimism, and the modern urbanocentric mindset (Lemaire, 2002) are often taken for granted. I have met animal rightists, themselves living in high-rise blocks in a North American city, who feel they should persuade Inuit people in the continent's north to move down south. The argument offered is that by abandoning the frozen lands their ancestors lived on for so many generations these Inuit could take up a more moral lifestyle vis-à-vis animals and become vegetarians (which at present they cannot be for the simple reason that where they are living hardly anything grows).

There are also reports of animal shelters whose managers on principle do not give companion animals to people with a garden, for fear that by going outdoors such animals could escape and come to harm. Accidental death in traffic is seen as infinitely more horrific than a lifelong existence indoors.

Many members of the animal movement seem to move surrounded by machines in an entirely humanized, electronic techno-world and tend to treat this circumstance simply as a given. The hegemony of the car in modern society, for example, hardly seems cause for concern to them. However, apart from everything else that the car represents, this type of private transport results in numerous animal deaths. According to Wildcare, a wildlife rehabilitation center in Toronto, most injured and orphaned animals brought in are victims of auto transport and to a lesser extent car attacks (personal Communication, Canadian wildlife rehabilitator Csilla Darvasi; for the United States see Braunstein, 1998, and Hogwood & Trocmé, 2003). Whereas cars are causing direct death or injury, habitat destruction connected with automobility and road building causes extensive indirect death and even extinction. Members of the animal movement often show no awareness of the violence involved in bulldozing an acre of land or building a road. One doesn't see much blood, but it causes whole communities of animals and plants to perish (Livingston, 1994).

In sum, the animal movement tends to portray animals as though they were isolated, city-dwelling consumer-citizens, living entirely outside of any ecological context. Such a view amounts to a form of reductionism: *individualistic reductionism*.

Reducing to an Ecosystem

Animals, for people in the deep green/deep ecology movement, are first and foremost wild animals, in other words, fauna living in the wild. It is not sentience or cruelty issues that are central here; it is nature, naturalness, and environment (Baird Callicott, 1989). Incidentally, the word *environment* itself is a very problematic term: it literally means that which surrounds us. By definition it is not "us ourselves." In the term *environment* the separation between ourselves and nature is already final (Noske, 1997).

Deep ecologists tend to come down hard on anything that is no longer considered "environment," no longer pristine or positively contributing to the ecosystem. Feral animals and domesticated animals are not popular in these circles. Central concepts are nature, species, and biodiversity (Low, 2001). Only those animals that are still part of a given ecosystem really count for this movement. Animals are approached as representatives of their species. They are almost equated with their species or with the ecosystem of which they are part. The animal as individual is often downplayed.

Feral animals seem to be getting the worst of both worlds: they are neither an interesting species nor individuals worthy of somebody's moral concern (Rolls, 1969; Soulé/Lease, 1995; Reads, 2003). If anything, they are seen as vermin. It goes without saying that as species they do pose a threat to the natural ecosystems. Rats, cats, rabbits, dogs, foxes, horses, donkeys, pigs, goats, and water buffaloes—animals intentionally or unintentionally brought into the Australian or American continent (by humans)—are threatening local biodiversity. These feral animals can and do destroy the balance in naturally evolved communities. The predators among them sometimes totally wipe out indigenous species whose members have no natural defense against these "foreigners." Herbivorous feral animals can totally devastate habitats that native animals are dependent on (Reads, 2003). (Unfortunately, such ecological hazards are sometimes belittled or downplayed by the animal movement.)

Deep green–leaning people perceive feral animals as members of unwanted species and advocate their destruction, often by very inhumane means. Until recently the National Parks and Wildlife Service in Australia was in the habit of shooting brumbies (feral horses) from the air, thereby indiscriminately massacring herds and disrupting whole horse societies and families. In the north of the continent, water buffaloes are being run down by four-wheel-drives equipped with huge "roo bars." Rabbits are purposely being targeted with introduced deadly diseases, often by means of specially infected fleas, which are then released into their burrows (Reads, 2003). Foxes and feral cats and dogs are being killed by means of poison baits. From the literature on human poisoning (Bell, 2001), and from quite recent cases of food poisoning in China (newspaper reports September 2002) we know what horrendous suffering is involved in death by poisoning. It can't be all that different for animals. Among deep greens, however, the suffering of feral and farm animals hardly counts.

Sentience in the deep ecology/deep green discourse is often treated as some sort of by-product of animal life. So is individuality. The natural capacity of sentience is never included in any notion of environment, ecology, or nature.

Some deep greens/deep ecologists, such as Aldo Leopold, Gary Snyder, and Paul Shepard (cf. Leopold, 1949; Shepard, 1996), endorse modern recreational hunting as a way to be at one with nature. Not many deep greens are taking a critical position on hunting except when it involves endangered species. The issue tends to revolve around numbers rather than the preciousness of individual lives. Neither do deep greens tend to take a critical stance on animal experimentation. After all, professional ecologists and conservation biologists often conduct experiments themselves.

Mostly, experimenters are using individuals of numerically strong species or species especially bred for the purpose, such as white mice and rats. In the eyes of deep greens and deep ecologists, these are no longer "nature," and so their well-being is low on their priority list.

Deep greens/deep ecologists have been known to argue that hunting is part of human nature when it was still in tune with other nature. They usually point toward hunter/gatherer societies. Hunting is natural, they say. (Hunting would indeed be natural if human hunters would kill their prey with their teeth or nails, but they happen to use artifacts such as high-tech hunting or fishing equipment, which makes hunting "cultural" rather than natural.) In deep ecology circles the hunting of animals is felt to be more natural than having animals for companions, which is often seen as degenerate. However, the roots of the phenomenon of companion animals go as far back as hunting. All societies from Paleolithic times onward have been known to keep animals as pets or companions. It occurs in all societies, in all periods of history, and in all economic classes (Serpell, 1986). It may not exactly be "human nature," but apparently many people have felt the need for a face-to-face or touch-to-touch relationship with individuals of another species (Lévi-Strauss, 1973; Tuan, 1984). So much for the "unnaturalness" of companion animals.

Because deep greens do not have much time for domesticated animal nature, they tend to be rather uninformed and unconcerned about what happens to animals in factory farms and laboratories. During various ecotours in the Australian outback it strikes me time and again how no effort whatsoever is made to avoid serving factory-farmed meat to the participants of such a tour. When queried on the issue, the often ecologically astute tour guides tend to demonstrate an entirely value-free and neutral attitude to where the tour food was coming from. Deep greens/deep ecologists might disapprove of factory farming because of its unsustainability and its polluting effect on the nature outside, but not because of the things done to natural beings inside. Production and companion animals simply do not figure as "green" (Noske, 1994).

In sum, the deep green/deep ecology movement tends to equate animals with their species. Equating animals with their species or with their ecosystem amounts to another form of reductionism: *ecosystemic reductionism*.

Disembodied Empathy versus Embodied Antipathy

Both movements are potentially united in their struggle against anthropocentrism: the idea of humanity as the measure of all things. But apart from this there seem to be few platforms where the two groups actually meet: only during some international campaigns, such as the ones against seal hunting and whaling. The first time a group such as Green-peace showed any concern for individual animal welfare was when many years ago in Canada three whales got stuck in the ice. The International Fund for Animal Welfare, though essentially an animal welfare organization, does from time to time put forward arguments to do with habitat destruction and extinction of endangered species.

Strangely enough—because one would expect it the other way round—it is the animal movement rather than the deep ecology movement that invokes animal-human continuity as a line of reasoning for considering animals as individuals. On the other hand, many animal advocates are themselves almost the embodiment of human-animal *discontinuity*. As mentioned before, in this movement there hardly exists any critique of the way present-day technology is alienating humans from their "animalness." This issue is tackled by the deep green/deep ecology movement rather than by the animal lobby.

Again consider the car issue. For all other species, bodily movement is first and foremost organic movement: it involves muscle power, fatigue, or sweat. But for modern humans, bodily movement is more and more being replaced by mechanization and computerizing. They let machines do the moving for them, and as a result they are becoming more and more *unanimallike*. Hardly anybody in animal advocacy circles looks upon this as something problematic that could stand in the way of the natural human condition, that is, our physical animalness. For them this issue appears to have nothing to do with human-animal continuity. But continuity is not just about the "humanlikeness" of animals but also about the "animallikeness" of humans. There is an existential and crucial connectedness between the two. In circles of the animal lobby, however, human-animal continuity remains largely an abstract moral principle that is hardly "lived" in reality. One could perhaps say that this attitude is characterized by *disembodied empathy*: the empathy is real, but its material basis forgotten.

The deep green/deep ecology movement, by contrast, does appreciate the wonders of nature, is conscious of animal-human continuity, and denounces various technologies (including the car) as alienating and harmful to nature. But there exists a strange contradiction here, too. Though in deep green circles it is acknowledged that modern human practices have been extremely exploitative of nature and the wild, this does not seem to have induced much sympathy for exploited animals. Animal victims, be they domesticated or feral, are blamed for their own predicament and in some cases for posing an active threat to what is perceived as real nature.

Although the deep greens, in contrast to their city-based counterparts in the animal movement, are more likely to opt for a natural lifestyle and to be more mindful of a shared animal-human past, this doesn't translate into sympathy with animals that have fallen by the wayside. This attitude could be characterized as *embodied antipathy*. Human-animal continuity is lived and "realized," but instead of empathy it is often accompanied by a disdain for those beings that no longer lead natural lives in the appropriate ecosystem. Denatured though such beings may be, they nevertheless are still close enough to nature to possess the *natural* capacity for suffering, whether it be pain, boredom, listlessness, social and ecological deprivation, or agonizing death.

Another contradiction is apparent here as well. In regions such as North America and Australia the green focus is strong and, as mentioned before, is often expressed by advocating harsh measures against the exotic and the feral (Aslin & Bennett, 2000; Reads, 2003). One wonders what self-image underlies such attitudes. Is this a curious case of human foreigners (in the ecological sense) condemning animal foreigners? Would such people advocate the eradication of themselves, members of a group of exotic white invaders whose adverse impact on the local ecosystem has been well documented? Would they be in favor of curbing all—nonaboriginal—human lives and births, not to mention more drastic measures? If the answer is negative, how can such measures be justified with regard to animals? Downplaying animal sentience and animal cruelty issues while at the same time upholding human sentience arguments endorses ethical discontinuity between humans and animals, albeit perhaps unintentionally.

The recent developments in animal biotechnology are going to be a test case for both movements. (Incidentally, the two movements have so far not been all that interested in each other's literature.) Some animal welfarists have claimed that genetic engineering may enable us to design animal species that are fully adapted to factory farming conditions (Rollin, 1995). Others, among them veterinarians, are toying with possibilities of cloning and engineering "more suitable" and "made-to-measure" transgenic companion animals (Quain, 2002). For deep greens the issue of genetic engineering highlights pressing dilemmas with regard to species integrity (Birke & Michael, 1998).

How will the animal movement react? And will the deep ecology movement tackle the issue at all? Admittedly, the deep green/deep ecology movement concerns itself with species but only with species in the wild. Deep greens may be worried about what will happen if transgenic populations come into contact with naturally evolved wild ones. How will that affect the community of species? Most genetic engineering is done to already domesticated species, the ones the green movement isn't interested in. But recently there have been calls by green-leaning scientists to bring back extinct wild species, such as the Tasmanian tiger (thylacine), by way of genetic engineering.

Common Ground?

How we are to navigate between individualized ethics and ecosystemic reductionism? The animal lobby bestows on the sentient in nature a status of individual humanness: It asks how animals are part of human society and ethics. The movement could perhaps bridge the gap that separates it from deep ecology by overcoming its exclusive focus on sentience. It could extend its compassionate ethics so as to include the nonsentient and even the inorganic. The tricky part would be how to include the whole earth without simultaneously humanizing and colonizing it. Moreover, there always will be clashes of interest between animals and animals, animals and plants, individuals and species, and the organic and inorganic.

If compassionate society is about extending ethics as far as we can, deep ecology is not. It is about compliance with and obedience to nature's measure, nature's rhythm, and nature's limitations (Livingston, 1994). It concerns compliance with a nature that includes things such as mortality, predator-prey relationships, the "previousness" of species, imperfect bodies, and our own finiteness. Instead of asking how animals are part of ethics, deep ecology asks how animals *and* humans are part of nature.

Consider Val Plumwood's musings about "Being Prey." In 1985 this vegetarian ecophilosopher barely survived a crocodile attack in Kakadu National Park in Australia's Northern Territory. Thereby she came face to face with her own *edibility*. It made her realize that not only had she a body but also, like all animals, she *was* a body: She was (potential) meat for another animal to devour. The experience has forced her to rethink the ethics/ecology dualism. It is good to focus on large predators such as crocodiles, bears, and sharks—those that can take a human life—Plumwood states, because these animals present a test for us (also for the two movements). Are we prepared to share and coexist with the free, wild, and mortally dangerous otherness of the earth without colonizing it into a form that eliminates all friction, challenge, or consequence? Predator populations test our recognition of our human existence in mutual, ecological terms, seeing ourselves as part of the food chain: eaten as well as eater (Plumwood, 1999).

The two viewpoints—societal ethics and compliance with nature—at times seem incompatible. It is a difficult dilemma. Mary Midgley (1983) and J. Baird Callicott (in Hargrove, 1992) tried to solve it by arguing that wild animals deserve our protection as part of the ecosystem and that domesticated animals are entitled to our care because they are part of a mixed human-animal community and we have ethical obligations to *all* the individuals of such a community. The problem is that this arrangement would not cover all animals. Feral animals and exotics belong neither to the first group (the original ecosystem) nor to the second (the mixed domestic community). The reason commonly given for persecuting and eradicating these animals is precisely that they do *not* seem to belong to any community. "Pests" are neither interesting as species nor as individuals, and this turns them into outlaws.

Nevertheless, all of us, animals as well as humans, somehow exist in nature and also in society (or at least in a human-defined nation-state). Each and every one of us is a sentient individual, a species member as well as a "place" in the world. In this world nature and society intersect. It is all there is; nobody and nothing exists outside either.

The animal lobby needs to realize the importance of wildness, the relative "otherness" of nonhumans, and what Livingston has called the "previousness" of species. It should guard against an ethical colonization and humanization of nature. The deep ecology movement will need to pay more heed to matters of sentience, cruelty, and suffering in the way it conceives of and treats individual animal beings, including those that objectively do damage to other nature. Many feral species did not choose to live where they are now living. Humanity took them there.

To really do justice to animal-human continuity, we must ask ourselves what it is we (should) do with nature but also how we ourselves are "of nature." According to Plumwood (1999), we cannot in a neocartesian way divide the world into two separate domains: an ethical, human realm and an animal, ecological realm. Everyone and everything exists in both. All food is souls, she says—and ultimately all souls are food.

Further Resources

Aslin, H. J., & Bennett, D. H. (2000). Wildlife and world views: Australian attitudes toward wildlife. *Human Dimensions of Wildlife, 5*(2), 15–35.

Baird Callicott, J. (1989). *In defense of the land ethic: Essays in environmental philosophy.* Albany: State University of New York Press.

———. (1992). Animal liberation and environmental ethics: Back together again. In E. C. Hargrove (Ed.). *The animal rights/environmental ethics debate: The environmental perspective* (pp. 249–61). Albany: State University of New York Press.

Bell, G. (2001). *The poison principle.* Sydney: Picador Pan MacMillan Australia.

Birke, L., & Michael, M. (1998). The heart of the matter: Animal bodies, ethics and species boundaries. *Society & Animals, 6*(3), 245–62.

Braunstein, M. M. (1998). Roadkill: Driving animals to their graves. *Animal Issues, 29*(3).

Cartmill, M. (1993). *A view to a death in the morning: Hunting and nature through history.* Cambridge: Harvard University Press.

Dunayer, J. (2001). *Animal equality: Language and liberation.* Derwood: Ryce Publishing.

Flynn, C. P. (2002). Hunting and illegal violence against humans and other animals: exploring the relationship. *Society & Animals, 10*(2), 137–54.

Francione, G. L. (1996). *Rain without thunder: The ideology of the animal rights movement.* Philadelphia: Temple University Press.

Hay, P. (2002). *Main currents in Western environmental thought.* Sydney: University of New South Wales Press.

Hogwood, S., & Trocmé, M. (2003). The impact of highways on wildlife and the environment: A review of recent progress in reducing roadkill. In D. J. Salem & A. N. Rowan (Eds.). *The state of animals, part II* (pp. 137–48). Washington: The Humane Society of the United States.

Lemaire, T. (2002). *Met open zinnen: Natuur, landschap, aarde.* Amsterdam: Ambo.

Leopold, A. (1949). *A sand county almanac.* New York: Oxford University Press.

Lévi-Strauss, C. (1984). *Tristes tropiques.* Harmondsworth: Penguin Books.

Livingston, J. A.(1994). *Rogue primate: An exploration of human domestication.* Toronto: Key Porter.

Low, T. (1999). *Feral future.* Ringwood: Penguin Books Australia.

Midgley, M. (1983). *Animals and why they matter.* Harmondsworth: Penguin Books.

Montgomery, C. (2000). *Blood relations: Animals, humans, and politics.* Toronto: Between the Lines.

Noske, B. (1994). Animals and the green movement: A view from the Netherlands. *Capitalism, Nature, Socialism, A Journal of Socialist Ecology,* 5(4), 85–94.

———. (1997). *Beyond boundaries: Humans and animals.* Montreal: Black Rose Books.

Plumwood, V. (1999). Being prey. In D. Rothenberg & M. Ulvaeus (Eds.). *The new earth reader: The best of Terra Nova* (pp. 76–92). Cambridge: MIT Press.

Quain, A. (2002). Improving their bodies, improving our bodies. *Artlink, Contemporary Art Quarterly* (Theme issue—The improved body: Animals & humans), 22(1), 33–37.

Reads, J. L. (2003). *Red sand, green heart: Ecological adventures in the outback.* South Melbourne: Lothian Books.

Regan, T. (1983). *The case for animal rights.* Berkeley: University of California Press.

Rollin, B. (1995). *The Frankenstein syndrome: Ethical and social issues in the genetic engineering of animals.* Cambridge: Cambridge University Press.

Rolls, E. (1969). *They all ran wild.* Sydney: Angus & Robertson.

Sabloff, A. (2001). *Reordering the natural world: Humans and animals in the city.* Toronto: University of Toronto Press.

Serpell, J. (1986). *In the company of animals: A study in human-animal relationships.* Oxford: Basil Blackwell.

Shepard, P. (1996). *The others: How animals made us human.* Washington, DC: Island Press.

Singer, P. (1990). *Animal liberation* (2nd ed.). London: Jonathan Cape.

Soulé, M. E., & Lease, G. (Eds.) (1995). *Reinventing nature? Responses to postmodern deconstruction.* Washington, DC: Island Press.

Tuan, Y. (1984). *Dominance and affection: The making of pets.* New Haven and London: Yale University Press.

Barbara Noske

■ Ethics and Animal Protection
Human-Animal Support Services

Human-Animal Support Services (HASS) are programs that help keep people with a disabling or terminal illness with their current animal companions in a mutually beneficial relationship. HASS services include financial, emotional, and practical assistance to the disabled pet owner. HASS agencies can be independent, nonprofit organizations, or they can be programs under the umbrella of other organizations, such as humane associations or societies, veterinary hospitals, schools, or associations, AIDS organizations, or other similar human or animal community-service organizations. As one client of such an organization has noted, "The help given is wonderful. I am disabled mentally, physically, and emotionally, this food and vet assistance helps me get by on a very limited fixed income. I can keep my little family member (and me) in a healthy state of well-being."

Disabling illnesses may have a tremendous impact on daily living, as well as on one's quality of life. For some, this can include a decrease or an end to social opportunities and relationships and can lead to experiencing isolation, depression, and physical pain. Yet animals can provide wonderful, nonjudgmental companionship. Although companion and therapy animals can play an important role for people with chronic illnesses, these people may need assistance in caring for their companion animals.

History

Pets Are Wonderful Support (PAWS), in San Francisco, is an example of a human-animal support service that was born out of the AIDS epidemic in the mid-1980s to address the detrimental effects of living with a disabling illness and trying to maintain a companion animal. People living with AIDS (PWAs) were finding that the devastating effects of AIDS made it increasingly difficult for them to feed and care for their pets. Not only were there stressful financial burdens with regard to no longer being able to work, but there were incredible physical constraints as well. Many lacked family support, and, often, their social support systems were no longer functioning in ways to help them with pet care. Furthermore, the fear of zoonoses—diseases that can be transferred from animals to humans—led many doctors to advise their PWA clients to give up their animals.

No single group in San Francisco was prepared to cope with this particular dilemma. However, a group of San Francisco AIDS Foundation Food bank volunteers, many living with AIDS themselves, grew increasingly aware that some of the food bank clients were neglecting their own nutrition by feeding their rations to their animal companions. Witnessing people sacrificing their own care for that of their animals was a powerful testament to the power of the human-animal bond. People's pets were often their sole source of companionship, affection, and unconditional love; their animals were the reason they got up in the morning. Realizing that being able to maintain one's pets in the face of a disabling disease was a serious quality-of-life issue, the volunteers remedied this situation by getting the food bank to carry pet food and pet-related products.

While the financial needs associated with pet food were obvious, it soon became apparent that pet owners required other services, including veterinary care, assistance with dog-walking, boarding services if the person became hospitalized, and adoption services if the person died. In 1986, a group of people, including founding veterinarian Ken Gorczyca, developed Pets Are Wonderful Support, nicknamed "PAWS," as an official project of the San Francisco AIDS Foundation. In recognition of the importance of the human-animal bond, the organization sought to fill in the gaps between other AIDS services and animal-related organizations to address the particular problems and questions faced by immuno-suppressed pet owners. In comparison to traditional Animal-Assisted Therapy programs, in which animals are brought into the hospital or home for short periods, the PAWS services allowed, in essence, 24-hour therapy for an individual's own companion animals. PAWS evolved to keep the "family" together.

PAWS became an independent, volunteer-based nonprofit organization in 1987. The main client services evolved to include financial, emotional, and practical assistance, as part of their CASS (Companion Animal Support Services) Program. These include veterinary care, a pet food bank, foster care and adoption planning, in-home pet care, and pet transportation. Housing rights advocacy and zoonoses education are part of their Education and Client Advocacy Program. Volunteers are the backbone of PAWS, providing direct services to clients, and office help, event staffing, and educational outreach to the community.

PAWS's purpose has been to deliver support services to keep PWAs and their animal companions together, in mutually healthy environments, for as long as possible. Over the past two decades, similar organizations and programs in other communities have developed worldwide to fill local needs. While PAWS has often served as a model for HASS development, each organization/community is diverse, and the services offered vary among the different programs, depending on the local funding, volunteer base, community need, and fiscal support received. Some HASS organizations, including PAWS, have expanded their services to include other disabled or elderly populations as well as PWAs. In 2002, PAWS expanded its services to qualifying low-income people with disabling conditions other than AIDS in the PAWS Expansion Program (PEP). PAWS additionally works with Veterinary Street Outreach Services (Vet SOS) to serve homeless pet owners and is currently developing a program to provide support for low-income seniors with pets.

Services and Programs

Veterinary Care

The Veterinary Care Program is one of the most important functions of the PAWS organization (and others like it), because keeping a pet healthy is important in keeping the human companion healthy. The program provides an annual physical examination and vaccinations to each client's animal and also allows for the client to ask questions about zoonoses. The annual exam, vaccines, and advice are provided free of charge by Pets Unlimited, a local veterinary hospital and shelter. To help defray the costs of emergency or other essential medical treatments, PAWS additionally offers an annual veterinary fund to each client. Local veterinary hospitals further help by offering discounts to PAWS clients. If the client is unable to take his or her pet to the veterinarian, PAWS will provide transportation volunteers to take the pet. PAWS requires that all pets in its program be spayed or neutered and helps pay for these surgeries for new clients.

Pet Food Bank

The Pet Food Bank, the first service provided by PAWS, is the cornerstone of the organization. The PAWS Food Bank provides monthly allotments of pet food, litter, flea treatments, and pet accessories to clients. The food bank is open every weekend and may be accessed once a month by PAWS clients; over 700 cans and 900 pounds of dry food are distributed each week. Delivery is provided for homebound clients. Volunteers that have gone through orientation help staff the food bank and make the weekly deliveries. Products are donated by pet food manufacturers and distributors, local supermarkets, and individuals. Veterinary clinics also offer discounts on diets for pets with special dietary needs. Additional food is purchased with funds raised by the PAWS development staff and volunteers.

Foster Care

Many times, a family member or friend will take care of the person's companion animals if a PAWS client goes into the hospital or is unable to care for the animal. However, early in the AIDS epidemic, several pets were found in people's homes, left unattended when a pet owner went unexpectedly into the hospital. Throughout the years, there have been many instances in which individuals refused necessary hospital care until they were absolutely sure that their animals would receive proper care. PAWS Foster Care Program was created to fill this need.

Adoption

Many clients are fearful of what will happen with their pet if they die. When a client registers with PAWS, they are encouraged to make an adoption plan, including a living will. The living will addresses the client's fears by designing a plan of action to secure his pet a good home if and when the person cannot provide it himself. Individuals are asked to notify PAWS of the future adopting "parent," to ensure that the pet will be transferred to the appropriate person in the event of the pet guardian's death. If an individual is isolated and has no family or friends willing to adopt his companion animal, PAWS helps to identify possible alternative sources for adoption of their pets.

In-Home Animal Care Services

Due to the nature of illness, the health of some clients may have deteriorated to the point where they are unable to provide total care for their animals. For example, many clients face decreased-mobility issues and need to access dog-walking assistance, often on a daily basis. PAWS volunteers provide necessary in-home care for the needs of the animals. These services include litter-box cleaning, dog walking, aquarium and aviary cleaning, administration of a pet's medications, pet grooming, and flea control. Dog walking is the most popular PAWS volunteer activity. The combination of providing a vital service to a disabled individual while regularly connecting with a loving animal makes it a unique volunteer opportunity. PAWS also works with groomers in San Francisco to provide free yearly grooming for clients' animals. These grooming appointments are not for aesthetic purposes, but for maintaining the health of the animal.

Pet-Associated Zoonoses Education: Safe Pet Guidelines

While evidence supports the fact that most companion animals pose minimal risk for transmitting zoonotic diseases (Centers for Disease Control 2005), those caring for individuals with compromised immune systems need to understand those minimal health risks. At the height of the AIDS epidemic, there was a severe lack of information about how to protect oneself from zoonoses, and many doctors encouraged their patients to relinquish their animals. Veterinarians, questioning the value of separating animal companions from the people who seemed to need the companionship the most, took the lead in educating the public and health care fields about zoonotic risk for PWAs. These efforts eventually led to the publication of the PAWS' Safe Pet Guidelines in 1988. These were the first published guidelines to explain how to minimize zoonotic risks and support the importance of the human-animal bond for people with HIV/AIDS. In the 1990s, many other veterinary organizations and schools, humane societies, and the Centers for Disease Control (CDC) published guidelines modeled after those published by PAWS.

Even today, there is still confusion and misinformation about pet-associated zoonoses, and continued education is essential. All PAWS clients receive a copy of the

Safe Pet Guidelines brochures, which are currently geared toward any individual that is immunocompromised. PAWS provides updated guidelines to physicians, veterinarians, other health care workers, and the public. The most current copy of the Safe Pet Guidelines can be found on the PAWS Web site (www.pawssf.org/).

PAWS Externship Program

The development of the PAWS Veterinary (1998) and Public Health (2003) extern programs allowed interested veterinary and master's of public health students the opportunity to get experience in the human-animal-bond field, as well as the nonprofit sector. In turn, the externs provide important information to the community and anyone working in the human-or animal-health fields. Externs have assisted with and helped produce various projects.

National Conference Series

In 2000, PAWS led a national effort to produce a conference, "The Healing Power of the Human-Animal Bond: Lessons Learned from the AIDS Epidemic." As a follow-up and in partnership with PAWS Los Angeles, PAWS presented "The Healing Power of the Human-Animal Bond: Companion Animals and Society" June 2–4, 2005. This conference program explored the social, psychological, and physical benefits that animals provide people living with HIV/AIDS and other disabilities, and the rights and roles of service animals. Presented information included: data supporting positive benefits to both renters and landlords for allowing pets in buildings, housing law regarding companion and service animals, and roles of psychiatric service dogs. Conference proceedings can be requested from the PAWS Web site (www.pawssf.org/conference/2005/).

Client Advocacy Program

Many disabled, low-income San Franciscans with service animals (including emotional-support animals) face housing problems, including eviction and lack of reasonable accommodation. The Client Advocacy Program provides consultation, direct advocacy, and access to pro bono legal assistance to low-income, disabled Bay Area residents who are having housing-related difficulties because they have a service animal. The two primary issues that this program deals with are (1) requesting a reasonable accommodation for an individual to bring an animal into housing and (2) assistance for individuals being threatened with the loss of housing due to their companion and/or service animal. The program also educates service organizations and community leaders about housing and service animals.

Emotional Support

Perhaps one of the most important functions of PAWS is the provision of caring, emotional support. Both staff and volunteers often provide friendship and guidance to clients in the most difficult of times (such as the loss of a beloved companion animal). Oftentimes, the client relishes the opportunity simply to share a heartwarming story about her animal to someone who understands their intense bond; the clients know that their connections to their animals will never be belittled or dismissed and will always be taken seriously.

Assistance with Starting a New Human-Animal Support System Organization

A great deal of planning is required to start a human-animal support service program. Each community typically has resources to help developing nonprofits. Additionally, existing human-animal service organizations may offer assistance for the initiative. PAWS is deeply committed to assisting either individuals or organizations with the creation of HASS-oriented services and offers a "start-up" packet for individuals wanting a blueprint (www.pawssf.org/startapaws.htm); other HASS organizations are always good resources as well. This packet includes information about how to start a board and how to obtain non-profit status, in addition to PAWS policies and intake forms. Whether starting a new organization from scratch or adding on to an existing program, the task can be daunting. PAWS recognizes that new HASS organizations only serve to further public awareness as to the importance of the human-animal bond and, as such, aims to provide as much assistance as possible.

Services and Programs Provided by Other Organizations

Personal Pet Visitation

PAWS Houston, currently with Methodist and Christus St. Joseph's Hospital, provides a program that enables hospital patients in intensive care to receive visits from their own animal companions on a case-by-case basis. Houston arranges and coordinates the hospital administration and permission logistics and necessary pet transportation for these visits. Many medical professionals are trying to introduce similar programs into other hospitals, because of the tremendous research supporting the positive health benefits of having one's pet close by. This program can also be used to help people in hospices that can no longer keep their pets with them.

Since the mid-1980s, the HASS/PAWS movement has provided health education, service, and emotional support to marginalized members of the community. Each community has its own resources and needs, and each has produced its own unique response. Various examples of programs that have evolved nationwide include independent nonprofits as well as programs of existing animal- or human-service organizations. There is no one way for a HASS organization to exist; what matters is that the community is being served. The following list demonstrates different HASS programs with regards to infrastructure:

1. Independent not-for-profit programs: PAWS in San Francisco, PAWS/LA in Los Angeles, and PETS, DC in Washington DC.
2. Collaboration with local AIDS organization: PAWS NY Capitol Region program with The Albany Damien Center in Albany, NY, and Pet Project at the Monterey County AIDS Foundation.
3. Collaboration with local animal shelters: Phinney's Friends with MSPCA in Boston and the SHARE Program with Marin Humane Society in Novato, CA.
4. Collaboration with veterinary schools and veterinary hospitals: Philly PAWS.
5. Entirely volunteer not-for-profit: Pets Are Loving Support, Guerneville, CA.

Further Resources

Pets Are Wonderful Support! (PAWS). http://www.pawssf.org [See this Web site for current listings of HASS programs.]

Andrea Brooks

■ Ethics and Animal Protection
Interest Conflicts between Animals and Humans

Interest conflicts between humans and other animals are at the heart of practical animal ethics. They are crucial from the point of view of animal-human relations because they materialize many of the practical issues involved in such relations and offer new perspectives for investigating the relations. However, theories concerning value do not as such specify how we should solve conflicts of interests. In order to rectify this, many models for solving interest conflicts between humans and other animals have been offered, of which five are presented here.

Before going further, one option has to be mentioned. It is commonly stated that interest conflicts are to be solved by favoring humans, for only humans have individual or inherent value. This option is commonly used for justifying various instrumental practices using animals, such as meat or fur industries. However, it remains lacking. As theories in animal ethics point out, a categorical value difference between humans and other animals can be criticized with well-supported arguments, and the same arguments offer justification for maintaining that animals also have individual value. Although the topic is complex, it will here be assumed that animals do indeed have individual value. However, giving animals such value does not by itself point out how they are to be treated in all circumstances. What remains to be solved is what to do when two individuals of the same value are in a conflict of interests.

Rights

The rights model, as advocated for instance by Tom Regan, maintains that interest conflicts are to be approached through the concept of "rights." Rights denote what can and cannot be done to other beings and as such offer guidance in interest conflicts. For instance, if an animal has a right for life, killing him or her becomes a moral wrong in most instances. Thus, if the conflict of interest involves the possibility of killing an animal, it should be resolved by refraining from the act of killing.

However, there are differences of opinion as it comes to the role of rights. According to some, they offer a solution to interest conflicts, and according to others, they merely point out such conflicts. Hence, rights can be seen either as a solution to conflicts or as a concept by which to bring forward those conflicts. If the latter approach is accepted, rights are not a sufficient method for solving interest conflicts—they can be used in the political and moral rhetoric but are not an all-decisive political or moral tool. This becomes clearer when considering interest conflicts between many similar rights (or rights holders). For instance, if the right to live of two different beings is in conflict, we need other methods in order to find a solution. Moreover, rights do not offer guidance in all situations. We can have interest conflicts where rights are not the primary concern or where rights are bypassed by other considerations (such situations include choosing on the basis of the number of beings concerned). Hence, it can be argued that there are situations where, for instance, utilitarian considerations may be of more significance than rights.

Some, such as Peter Singer and Mary Midgley, argue "rights" to be a rhetorical tool rather than a serious moral concept. According to this view, rights remain vague in content (what exactly does "a right" include and imply?), are based on weak or imprecise theoretical grounds (rights are argued to be based on overt objectivism and intuitionism), and have suffered inflation (people argue for various rights in various contexts, such as a right to have Mondays off from work). They are thus argued to be unnecessary: We can solve interest conflicts without resorting to rights.

However, the model can also be defended. The theoretical elements are not by any necessity based on overt objectivism or intuitionism, as various theories concerning animal rights exemplify. Rights can also be constructed in a precise manner (vagueness is not a necessity), and they can be offered as a serious justification in order to avoid inflation (hence, inflation is not a necessity either). Thus, we can construct rights on a clear theoretical foundation that avoids intuitionism and offers precision and justification. Furthermore, although they do not offer guidance in all situations, they can be argued to be an important moral concept that is needed in interest conflicts. According to many, we cannot talk of the interests of individuals and conflicts of those interests without reference to rights. The role of rights as "problem pointers" cannot be underestimated, and they have played a significant role in various moral efforts, including animal advocacy (as the animal rights movement exemplifies). Hence, rights remain important, even if they do not offer solutions to all interest conflicts.

Interests

The interest model, as advocated by Peter Singer, concentrates on the nature of the interests concerned in each situation. We are to evaluate and compare the interests involved. A distinction of secondary and primary interests has been suggested by many (such as Donal VanDeVeer). Primary interests refer to interests that are crucial from the point of view of basic well-being (they include, for instance, the interest to get nourishment), and secondary interests refer to interests that may be positive but not crucial (such as the interest to eat dark chocolate). We are to prioritize primary interests irrelevant of species. Hence, for instance, fur farming and cosmetic testing would be morally unjustified.

One issue with the model is the comparison of interests: how are we to compare the significance of interests such as the interest to avoid pain and the interest to retain one's freedom? Moreover, categorization offers a difficulty: how are we to categorize different stages of pain or limitation on freedom into "secondary" and "primary"? The issue becomes even more problematic when taking into account the subjective and cultural dimensions involved: different matters have differing importance depending on the individual and his or her culture.

Fortunately, these problems can be offered some solution. Although comparisons and categorizations are difficult, it has been maintained that they are still important. Many attributes (how difficult is "difficult"?) are not specified but still carry a clear content (x can be difficult, even if "difficult" is not specified). The same applies to interests: Referring to "primary interests" is meaningful, even if it cannot be categorized neatly. We need categories and comparisons, despite their problematic nature, in order to make sense of the surrounding world. Thus, comparisons and categorizations involving interests can be argued to be needed—without them, we could not talk of interests in any general sense, and the whole concept would lose its meaning. Also, subjective viewpoints do not necessarily lead to subjectivism. Although subjective viewpoints are relevant when evaluating interests, they are not the only relevant consideration. The same applies to cultural influences: Although they affect our interests, they are not the main concern. It is possible to give room for subjective viewpoints while also striving to create a neutral enough criterion to enable interest comparison between different individuals. Again, this can be argued to be a necessity if evaluations between different individuals and their interests are to be made.

One concern has been that placing emphasis on interests is unfruitful in relation to animals, as we cannot know their subjective interests in any precise manner. However, although common, this problem too can be answered. Study of animal minds and behavior offers insights into their species specific capacities and the ensuing needs. Hence, animal interests do not remain entirely unknown. It has been claimed that animals, in this regard, are not categorically different from humans (whose subjective interests and experiences we can never fully know). Understanding the interests of animals may be more demanding, but it is by no means impossible.

However, the model also faces other difficulties. Firstly, what are we to do in case the interests are similar? Secondly, interests are not the only relevant aspect in interest conflicts. Matters such as responsibility have been argued to also play a part—for example, the interest of an attacker is not of equal concern as the interest of the attacked.

Mental Capacities

The model emphasizing mental capacities maintains, as presented, for example, by Raymond Frey, that cognitive abilities affect the outcome of interest conflicts. In the case where the interests involved are equal, we are to look at the mental capacities of those involved and favor the one with more complex capacities (or "mental richness"). Hence, we would favor humans instead of animals in the case of animal experimentation, if the humans whom the experiment benefits have more complex capacities than the animals used. The idea is often exemplified by the "lifeboat example," in which we are supposed to save a human instead of an animal from an imminent death (the example will not be discussed further here, for it has been criticized for extremity and for being based on competition and black-and-white understanding of morality: in actuality interest conflicts rarely are this extreme, do not include two parties in a competitive either-or situation, and allow for more than two alternative solutions).

However, the relevance of mental abilities has been questioned. Although they can play a part, it would be simplistic to state that they are always of primary concern in interest conflicts. The argument claims that we cannot favor a human instead of an animal in a situation that involves great pain just because the human has great skills in mathematics—the abilities here would not be relevant to the conflict at hand. Hence, general abilities cannot be used as a justification, and a link between a specific ability and the interest involved has to be clarified. This becomes clearer when using the same example in relation to humans. There are great cognitive differences between human individuals, but this is not seen to justify overriding the interests of those with lower capacities (as Dale Jamieson states, we cannot eat "idiot-burgers"). Hence, the "argument from marginal cases" poses a problem.

VanDeVeer has suggested a "threshold" after which all beings are equal and which would include all human beings. However, such a threshold would by necessity also include many animals. Moreover, it would be difficult to point out the relevant level of mental capacity required in order to be considered equal: What (and why) would be sufficient, and would this limit not ultimately also end up leaving many human beings outside its scope? This problem touches the model in general: To place a threshold at a certain level of intelligence seems arbitrary.

Frey has accepted the argument from marginal cases. According to him, not all human beings are equal in interest conflicts, and skilled animals could, if one was to remain consistent, even be favored at the expense of less able humans. This answer has been claimed to not suffice. Not only does it not answer the questions concerning the relevance and relevant level of mental ability, it also goes against fundamental intuitions concerning whether we may, for instance, use mentally handicapped humans in experimentation.

Also, the comparison between different species is argued to be difficult. What would be considered a mental "richness": great visual abilities, sense of smell, complex coordination, or the capacity to compose symphonies? Ultimately, the model faces the danger of being anthropocentric in that specifically human capacities are favored as the criterion for capacities in general.

Special Relations

The special relations model claims that humans are to favor other humans on the basis of biological kinship or emotive attachment. Baird Callicott has argued that members of a species are morally entitled to favor their own kind, and Mary Midgley has claimed that there is a natural preference for members of one's own species.

This idea faces the "naturalistic fallacy," which correlates how things *are* with how they *ought to be*. Having an inclination to favor *x* does not by itself mean that one should favor *x*. This becomes clearer when considering that humans have many inclinations that concern other humans—we tend to favor our own kinds in various forms. Still, these inclinations are invariably seen as not moral but immoral. What is needed is a reason to make the inclination into a moral norm, and without such a reason, favoritism becomes morally dubious. Thus, the question is, why ought a certain similarity and a tendency to favor be the basis of moral choice? In cases where species is seen as the only acceptable form of "kinship," circularity also poses a problem because species is given special importance, which again is used as a reason as to why species has special importance (species should be the criterion for moral favoritism; moral favoritism proves that species is a morally relevant factor). Moreover, it has to be asked why not draw the line on primates, mammals, and so on—why is favoritism placed precisely at species, even though we have tendencies to like animals resembling us more than animals alien to us?

Regarding attachment, Peter Wenz has argued that the level of interaction between humans differs from that between humans and other animals and as such can give basis for prioritization. Others, such as Midgley and Deborah Slicer, have pointed out that personal elements such as attachment should be given room in animal ethics, and they have maintained that we should not strive for pure neutral equality that abandons all emotion. In regard to interaction, it remains unclear why the level of it should matter. Again, interactions between humans vary greatly—there are real differences in both our actual and potential interactions because we tend to interact with only those closest to us, and we often feel incapable of interaction with those from different backgrounds. Placing emphasis on the capacity for interaction (rather than actual interaction) rests on obscure ground, for attachment is born only from actual interaction—thus, even if we theoretically could be said to be able to interact with all humans, the required attachment would not follow without concreteness.

Also, the level of interaction between humans and other animals can be considered to be deep in many instances. Thus, it has been claimed that we indeed can have required levels of interaction also with other animals, especially as we learn more about their cognitive skills and forms of communication. Moreover, for Wenz, in relation to humans, one-sided interaction can lead to special duties (as in the case of third world countries and the impact Western countries have had on them), and this could also be applied to other animals, on whom humans have a great impact through practices such as agriculture. In fact, this would lead to greater duties toward animals in comparison to humans, as Wenz maintains that those harmed should be offered extra protection. It also has been claimed that attachment is a personal sentiment instead of a general, objective attitude and that it thus does not follow species lines or apply to abstract matters such as entire species (we have to personally know those toward whom we can feel attachment). Moreover, the relevance of personal attachment in interest conflicts remains limited, mostly because many conflicts between humans and other animals do not include individuals directly known to us. Therefore, attachment can be morally relevant, but it has been argued to not offer general guidance on how to solve interest conflicts.

Contextuality

The contextual model favors pluralism and practicality. It argues that, as practical situations consist of various elements, an approach to interest conflicts should take into account various manners of finding a solution. There is not one principle that should be followed at all times but rather many principles, out of which we are to choose on the basis of the context itself. This approach coincides with the so-called "particularism," which argues that particularities of each moral context should be examined when making decisions.

There are various versions of pluralism. Peter Wenz has made a differentiation into minimal, moderate, and extreme versions. The first one is found in most moral theories, and the claim is that the context should be given room after giving priority to theory or a basic principle thereof (thus, although we are to follow one theory and one principle, context should be recognized in a secondary sense). The second one claims that context and theory or principle should be given equal standing, and the third one states that no theory or a fixed set of principles are needed at all because context is to be given full priority. Out of these, the first one is argued to give insufficient room for the context, and the latter is claimed to lead to relativism, inside which there would be only contingent and arbitrary guidelines as to how to solve interest conflicts. Hence, the moderate form is often advocated as the most plausible option (although, for instance, Bryan Norton's convergence theory emphasizes the importance of taking into account various often-conflicting principles and slides away from the moderate form).

If moderate pluralism is seen as the basis for the contextual model, both primary principles and the context need to be given room. We need a theory of value, which points out what types of beings are to be given consideration, and basic principles, which dictate how those beings are to be treated. From these, more specific practical principles can be formulated to suit the differing contexts, and the particularities of each context guide our choices between these principles. If a standard animal ethics theory is to be followed, the theory of value would define that all beings capable of having experiences (i.e., beings with phenomenal consciousness) have individual value and thus are objects of equal consideration. Basic principles would include, for instance, respect for the welfare of individuals and equality between different individuals. From these, more specific principles can be formulated, and the context would partly determine which principle we are to follow. These principles would include some of those mentioned earlier: rights guard the value of individuals, primary interests are to be given prima facie priority, and personal attachment can influence our priorities. For example, in a situation where it is the nature of interests that is of most crucial importance, the primacy of interests is to be given special consideration; when, on the other hand, the situation clearly involves rights, these are to be given emphasis, and so forth. It is important to note that, if formulated on the basis of a standard animal ethics theory, the model concentrates on principles that require that animals are given equal consideration. Thus, for instance, principles that underline the instrumental value of animals would not be adopted.

This type of an approach requires moral deliberation and can be criticized for demanding too much of moral agents. However, the requirement of deliberation can also be seen to be a positive element because moral decisions are not always straightforward. The model also has the benefit of giving specific attention to the animals' point of view. As argued, for instance by Martha Nussbaum, we are to identify with the specific viewpoints of different animals, taking into account their history, needs, and so on. As such, the model gives adequate room for the viewpoints of all those involved and also enhances human-animal relations.

Further Resources

Frey, R. G. (1996). Medicine, animal experimentation, and the moral problem of unfortunate humans. *Social Philosophy and Policy, 13*(2).

Midgley, M. (1983). *Animals and why they matter*. Athens: University of Georgia Press.

Norton, B. (1995). Caring for nature: A broad look at animal stewardship. In B. Norton, M. Hutchins, E. Stevens, & T. Maple (Eds.). *Ethics on the ark: Zoos, animal welfare, and wildlife conservation*. London: Smithsonian Institution Press.

Nussbaum, M. (2001). *The upheavals of thought: The intelligence of emotions*. Cambridge: Cambridge University Press.

Pluhar, E. (1995). *Beyond prejudice. The moral significance of human and nonhuman animals*. London: Duke University Press.

Regan, T. (1983). *The case for animal rights*. Berkeley: University of California Press.

Singer, P. (1993). *Practical ethics* (2nd ed.). Cambridge: Cambridge University Press.

Slicer, D. (1991). Your daughter or your dog? A feminist assessment of the animal research issue. *Hypatia, 6*(1).

VanDeVeer, D. (1979). Interspecific justice. *Inquiry, 22*(1).

Wenz, P. (1988). *Environmental justice*. Albany: State University of New York Press.

———. (1993). Minimal, moderate, and extreme moral pluralism. *Environmental Ethics, 15*(1).

Elisa Aaltola

■ Ethics and Animal Protection
Israel and Animal Welfare

Although enormous progress has been made in the Western world in raising awareness about the human-animal bond and its importance to human and nonhuman species alike, in other countries, this work has just begun. When Concern for Helping Animals in Israel (CHAI) was founded in 1984, animal advocacy in Israel barely existed. There was no Animal Protection Law, no veterinary school, and only two very small animal shelters able to do little to promote spaying and neutering. Animal overpopulation control consisted exclusively of mass poisoning cats and dogs using slow-acting, painful poisons such as strychnine and alpha chlorolose. Abused work animals were a common sight, and humane education was unknown.

For more than two decades, CHAI's desire to raise consciousness in teachers, veterinarians, and government officials, as well as in the general public, about the need to help animals has motivated its efforts and projects. CHAI'S mission is to prevent and relieve animal suffering in Israel and to elevate consciousness about animals through education. Its projects foster empathy, respect, and responsibility toward all living beings and inspire and empower Jews, Arabs, and Christians alike to recognize the interconnectedness of all living beings and to make compassionate choices for the good of all.

Over its first two decades, CHAI participated in the process of drafting Israel's first Animal Protection Law; provided funds, veterinary supplies, and equipment, including the first animal ambulance, to help start shelters in areas where there were none and to assist existing shelters; promoted spaying and neutering and sent the first mobile spay/neuter clinic in the Middle East to Israel; successfully pressed the Veterinary Services to switch to the use of the humane oral rabies vaccine to replace mass strychnine poisonings; cosponsored educational projects, including a Jewish/Arab program and national and international educational conferences with Israel's Ministry of Education on topics such as the connection between violence toward people and animals and integrating humane education in the classroom; co-sponsored, with Israel's Ministries of Agriculture, Health, and the Environment, training in animal shelter management and humane overpopulation control for municipal and shelter vets; and successfully campaigned to end various cruelties, including the Army's use of dogs as live bombs.

Today, CHAI works through its sister charity in Israel, Hakol Chai (everything lives), which was founded in 2001. To prevent and reduce the overpopulation that results in so much suffering, CHAI/Hakol Chai's state-of-the-art mobile spay/neuter clinic provides low-cost operations and education on responsible animal care throughout the country. During the evacuation of settlements in Gaza and the West Bank, the clinic's professional veterinary staff and volunteers played a major role in rescuing and adopting companion and farm animals abandoned in the territories.

The organization has also rescued and rehabilitated abused horses, and actively promotes legislation to prevent their abuse, and is raising funds to construct a horse and

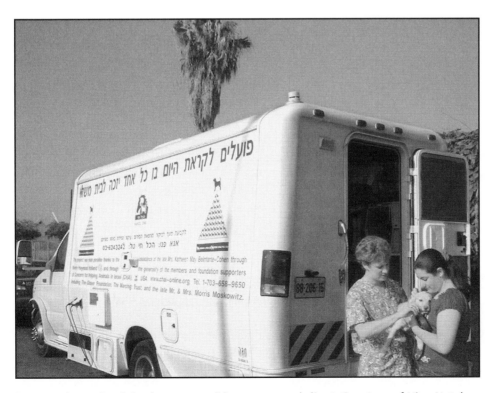

CHAI employee Sarah Levine meets with an unnamed client. Courtesy of Nina Natelson.

Dramatic change is well illustrated before Tikvah was rescued by CHAI, and after being in their care. Courtesy of Nina Natelson.

donkey sanctuary. CHAI's Alternatives Fund offers grants to promote alternatives to the use of animals in laboratories.

All CHAI/Hakol Chai's projects have an educational component because the organization believes that planting seeds of respect, empathy, and responsibility in future generations is essential for positive change. CHAI/Hakol Chai created educational materials and videos for secular as well as Jewish schools and provides education on animal-related issues in schools and community centers. The organization believes that only when the importance of the human-animal bond is understood worldwide will all living beings share a compassionate planet.

Further Resources

Concern for Helping Animals in Israel. http://www.chai-online.org/

Nina Natelson

■ Ethics and Animal Protection
People for the Ethical Treatment of Animals

Widely known for controversial tactics and for holding the view that all animals deserve respect and consideration, People for the Ethical Treatment of Animals (PETA), a non-profit animal protection organization based in Norfolk, Virginia, has brought animal rights issues, especially in the United States, into the public conscience and has attempted to reshape the way the world views animal agriculture, fur, animal experimentation, circuses, zoos, and aquariums. PETA operates under the basic principle that animals are not for humans to eat, wear, experiment on, or use for entertainment.

Founded by Ingrid E. Newkirk and Alex Pacheco in a basement apartment in Washington, DC, on August 14, 1980, PETA has since become the largest animal rights organization in the world, with more than 800,000 members and supporters, and with affiliates in the United Kingdom, Germany, The Netherlands, India, and Southeast Asia.

PETA often uses stunts and undercover investigations to be seen and heard on behalf of nonhuman animals. Since its inception, PETA has shown animal abuse in laboratories, leading to cancelled funding, closed facilities, and hundreds of U.S. Department of Agriculture charges; closed the largest horse-slaughtering operation in North America; convinced dozens of designers to stop using fur; cleaned up substandard animal shelters; helped schools find alternatives to dissection; and provided information on vegetarianism, companion animal care, and other issues.

In 1981, PETA uncovered the abuse of animals at the Institute for Behavioral Research in Silver Spring, MD, launching the Silver Spring monkeys case. PETA's findings led to the first arrest and conviction of an animal experimenter in the United States on charges of cruelty to animals, the first confiscation for animal abuse in laboratories, and the first U.S. Supreme Court victory for animals in laboratories.

PETA's work also led to the closure of a Department of Defense underground "wound laboratory," and successfully campaigned to end General Motors' crash tests on animals. PETA helped pass legislation to create a spay/neuter clinic in Washington, DC, and prompted the first anti-animal cruelty law in Taiwan.

In 1994, California became the first state to file criminal charges against a furrier at V-R Chinchillas after PETA investigators filmed the furrier electrocuting chinchillas by clipping wires to the animals' genitals, a killing method decried by the American Veterinary Medical Association. Other undercover investigations have included Belcross Farm, a pig-breeding facility in North Carolina, which led to the first ever felony indictments for animal cruelty by farm workers. PETA investigators also caught employees at Seaboard Farms, Inc., North America's third largest pork producer, on video, throwing, kicking, bludgeoning, and slamming pigs against concrete floors. As a result of PETA's investigation, the former manager of Seaboard Farms pleaded guilty to three counts of felony cruelty to animals, the first time in U.S. history that a farmer had pleaded guilty to felony cruelty for injuring and killing animals raised for food.

PETA has worked to convince McDonald's, Burger King, and Wendy's to improve the treatment of animals killed for their restaurants and worked with Safeway and other retail giants to improve the treatment of animals sold for food. In addition, the PETA Web site offers GoVeg.com, which allows people to request free vegetarian starter kits filled with recipes and tips on making the switch to a plant-based diet.

PETA has rescued animals, including abused elephants Bunny, Sissy, and Helen, from the entertainment industry and retired them to sanctuaries to live their final years in peace. PETA has rescued other animals from decrepit roadside exhibits and stopped the construction of a dolphin tank. Because of PETA's efforts, some cities and towns across the United States, from Takoma Park, Maryland, to Burlington, Vermont, have passed ordinances banning or restricting animal acts and exhibits, and animal-free circuses such as Cirque du Soleil and the Moscow State Circus are becoming increasingly popular.

Cosmetics companies, such as Avon, Revlon, and Estée Lauder, have signed PETA's Statement of Assurance not to test their products on animals, and hundreds of health charities, such as Easter Seals, Helen Keller Worldwide, Children's Burn Foundation, the

National Alliance of Breast Cancer Charities, and the Spinal Cord Injury Network International, have promised that they will not fund animal experiments.

PETA routinely works with sheriff's offices and prosecutors to stop the abuse of domestic animals and representatives have appeared on television shows such as *Dateline* to discuss the fact that most pet shop puppies come from deplorable puppy mills. PETA operates a mobile spay/neuter clinic and has sterilized thousands of animals belonging to low-income families. In the winter, PETA builds and delivers free doghouses for animals forced to live outdoors.

Although not everyone agrees with PETA that animals are not for humans to use for food, clothing, experiments, or entertainment, the organization has stated that it urges people to find out what happens to animals in these industries so that their choices will be fully informed.

Further Resources

Guillermo, K. S. (1993). *Monkey business: The disturbing case that launched the animal rights movement*. Washington, DC: National Press.

Newkirk, I. E. (2005). *Making kind choices: Everyday ways to enhance your life through earth- and animal-friendly living*. New York: St. Martin's Press.

People for the Ethical Treatment of Animals. *PETA factsheet, PETA's history: Compassion in action*. http://www.peta.org/mc/factsheet_display.asp?ID=3D107

Heather Moore

■ Ethics and Animal Protection
Political Action Committees (PACs) for Animal Issues

From the earliest history of the United States, money has played a role in politics, in particular in political campaigns. Vote buying in one way or another was not unusual. For example, it is said that George Washington bought alcohol for all 391 voters in his 1757 election to the Virginia House of Burgesses. The first legislative efforts to control campaign finances occurred in 1867: the Naval Appropriations Bill prohibited the solicitation of political contributions from Navy yard workers. This bill was followed in 1873 by the Pendleton Act in which the prohibition was broadened so that contributions from all civil service workers were prohibited. Other bills concerning financial disclosure or limitations and prohibitions on contributions to federal campaigns included the Tillman Act in 1907 and its amendments, the Federal Corrupt Policies Act in 1925, an amendment to the Hatch Act of 1939, the Smith-Connally Act of 1943 that prohibited labor unions from contributing funds to Federal candidates, and the Taft-Hartley Act of 1947. Many of the provisions in these acts were either ignored or simply not enforced.

The first Political Action Committee (PAC) was funded in 1943 by the Congress of Industrial Organizations (CIO). This first PAC was named the Political Action Committee, a name that soon became generic, applying to all such organizations. This PAC was formed so that political contributions from individuals could be raised and put into a fund that could provide campaign contributions in accord with the aims and goals of the organization that sponsored it. In this case the union wanted to support President Roosevelt's 1944 election. The formation of the PAC allowed the labor unions to make campaign contributions despite the fact that Congress had outlawed contributions from the treasury of a labor union. This was one of only a few PACs at that time

A major change occurred with the passage of The Federal Election Campaign Act in 1971 and its amendments in 1974. These acts initiated increased disclosure and limitations on Federal campaign contributions and at the same time established the Federal Elections Commission (FEC) with the power to enforce the various provisions of these bills. One of the provisions of this law provided the legal framework for corporations and unions to establish separate segregated funds, consisting of voluntary contributions, that could be used to further Federal elections. Such funding had been prohibited some years earlier by the Tillman Act and the Taft-Hartley Act. The new law provided an exception to the previous prohibition while at the same time it limited campaign contributions. Individuals were limited to $1,000 for each candidate, whereas the PAC was allowed $5,000.

In 1975, shortly after the amendments to the Federal Election Campaign Act had been passed, Sun Oil asked the FEC for an advisory opinion on its plans for political activity, and in particular for the use of corporate funds in administering its PAC and soliciting funds. The FEC published its advisory opinion that same year. It stated that "it is unlawful for any national bank, or any organization . . . to make a contribution or expenditure in connection with any election to any political office, or in connection with any primary election . . . 'the phrase contribution or expenditure' . . . shall not include the establishment, administration, and solicitation of contributions to a separate segregated fund to be utilized for political purposes by a corporation or labor organization." The opinion further stipulated "That it shall be unlawful for such a fund to make a contribution or expenditure by utilizing money or anything of value secured by physical force, job discrimination, financial reprisals, or by monies obtained in any commercial transaction." This opinion allowed Sun Oil to use its general treasury funds to establish its PAC, to administer it, and to cover the costs of solicitation of funds. Although it is required that the funds of the PAC are separate and segregated, Sun Oil was allowed to direct the funds of its PAC and to make any political contribution that it wanted. Although this opinion referred to corporate PACs, it was seen to apply broadly. Two years later, in 1976, some of the provisions of the 1974 amendments to the Federal Election Campaign Act were challenged (*Buckley v. Valeo*), but the limits on contributions were upheld.

At least partly as a result of these changes in the early seventies, the number of PACs suddenly increased. There were 608 registered PACS in 1974 and more than 4,009 in 1984. The number of PACs stabilized at that time as there were only 4,027 in the beginning of 2003. Although the number of PACs has remained more or less the same, their donations have shown a steady increase: from $15 million in 1974 to $220 million in 1998.

Federal law limits the amount of money—known as hard money—that an individual can donate that goes into the so-called Federal account of a PAC. "Soft money" falls outside these restrictions. This money is not regulated by Federal laws since it is not used as a direct contribution to a Federal candidate. Unlimited amounts of soft money can be donated to state or national parties, used for voter registration, used for what has become known as "issue ads," and used in general, for information about causes supported by the PAC.

There are various kinds of PACs, but they are generally considered as either connected or unconnected. Connected PACs are linked to a particular organization, such as a corporation, a union, or a membership organization. Such PACs enjoy the advantage that many of their expenses can be paid for by the parent organization, but they are limited in that they can solicit funds only from those associated with the parent organization.

Connected PACs include those of the National Rifle Association, the National Organization for Women, and the American Medical Association. Unconnected PACs, as the name suggests, are independent or not attached to any other organization and, as such, they can solicit funds from anyone. Emily's List is an example of an unconnected PAC.

PACs are sometimes condemned as a corrupt influence. It is said that PACs buy access and unduly influence elected officials, and even elections, thus minimizing the interests of the individual. This criticism primarily deals with soft money. For example, in introducing the 527 Reform Act of 2005, Senator John McCain complained about the "illegal" and "blatant attempt to influence the outcome of last year's (2004) Presidential election by 527 groups." He asserted that "almost half of the financing for 527 groups in the 2004 elections came from a relatively small number of very wealthy individuals who made huge soft money contributions. . . . This included ten donors that gave at least $4 million each to 527s involved in the 2004 elections and two donors who contributed over $20 million."

On the other hand, the American Civil Liberties Union (ACLU) states that "Political involvement is a bedrock of successful democracies." According to the ACLU, the 527 Reform Act introduced by McCain would "curtail the free speech rights of millions of Americans" in that "the proposed restrictions on fundraising would prevent many political organizations from engaging in important activities encouraging participation in the election process." The ACLU also opposes the 527 Reform Act because it would "restrict the freedom to donate anonymously" and because it "would shift the balance of power toward corporations and wealthy individuals" because the latter could avoid the limits established by the 527 Reform Act "by forming their own organizations or purchasing advertisements on their own." Thus, the ACLU holds that this bill would shift the balance of power "toward larger donors and regular people would be pushed out of the political process."

We are thus confronted with the ironic situation in which Senator McCain, supporting a Reform Act, and the ACLU, opposing it, both claim that they are trying to protect the average person against the ultra wealthy.

The question arises whether such criticism would apply to Humane USA, the nation's first national animal PAC, founded in 1999 by Linda Nealon and Wayne Pacelle, the present CEO and President of the Humane Society of the United States (HSUS). McCain's objections, however, refer to 527 organizations such as MoveOn.com rather than PACs such as Humane USA. In particular, the current condemnation concerns the vast sums of "soft" money 527 organizations raise in their attempts to influence an election. The Web site of Humane USA explicitly states that it does not oppose efforts to regulate "soft" money.

The amount of money Humane USA, the largest animal PAC in the United States, spends for its lobbying and electioneering activities is minuscule, which does not mean that its effects are minuscule. Indeed, it is interesting to speculate exactly how it has managed to be so successful on such a small budget.

Humane USA identifies itself as the "political arm of the animal protection movement." It "has been formed by leaders of major animal protection organizations, including The Humane Society of the United States, The Fund for Animals, Farm Sanctuary, ASPCA, Doris Day Animal League, Animal Welfare Institute, The Ark Trust, Animal Rights Foundation of Florida, and others. Its board of directors, advisory board, and advisors are top grassroots and national animal protection leaders." The groups mentioned are charitable organizations and as such by law are not permitted to donate money to support the political activities of Humane USA, although individuals belonging to these organizations can donate their time or money.

The main task of Humane USA is to involve people in the political process in order to organize an identifiable voting block that can vote for progressive laws for animals. Humane USA has a "single focus" and as such it will support any candidate, regardless of party affiliation, who supports its goals. The one issue that Humane USA considers is animal well-being. Other social issues fall outside the concerns of the PAC.

To achieve these aims, Humane USA, under the guidance of Mike Markarian, seeks to educate people about animal issues so they can vote effectively. The Humane USA Web site explains the various animal bills introduced in each session of Congress, listing which legislators have supported them. A Scorecard, prepared by HSUS, reveals how each senator and representative in Congress voted on each of the various animal bills. Humane USA makes endorsements, raises money to donate to candidates on both the state and federal level who are friendly to animals, and also organizes activists to work on political campaigns.

The emergence of Humane USA demonstrates the newfound commitment to cooperation and political activity for the humane movement. Note that in identifying itself Humane USA uses neither the words "animal rights" nor "animal welfare" but rather a neutral term to describe itself as an animal protection political action committee. No doubt this is also a political statement not to identify itself with either view and an effort to unite all of those concerned with the animal well-being.

The emphasis on political activity is a sign of the maturity of the animal movement. In the past many of the activities of animal activists involved conferences, letter writing campaigns, and demonstrations. There is now an increasing realization that these efforts, although important, by themselves will not bring about the desired changes. The existence of Humane USA and a number of smaller state organizations devoted to political activity, such as the League of Humane Voters and its chapters in New York, Animal Legislative Action Network in California, and the Animal Protection PAC in New Jersey, reveal a general and growing acknowledgment of the importance of political activity.

Animal PACs are unique. Most, if not all, other PACs are funded by people whose self-interest is involved. Labor PACs, for instance, may lobby for better wages or better working conditions, gay and lesbian PACs for certain legal rights, and hemophiliac PACs for more research on hemophilia. If animal PACs are successful, there is no direct or concrete benefit for animal activists; their only reward is the idea that some of the pain and suffering has been eradicated from the world.

See also

Ethics and Animal Protection—*Political Rights of Animals*

Priscilla N. Cohn

■ Ethics and Animal Protection
Political Rights of Animals

Although nonhuman animals are the objects of legislation governing their welfare, they seem *prima facie* to lack political subjectivity, which is to say that they do not seem to be agents who can represent themselves politically. Thus, it would seem that humans must speak on behalf of nonhuman animals, representing them in the exclusively human political domain.

This exclusion of nonhuman animals from the political sphere was, of course, classically signaled by Aristotle when he defined man as the *zoon politikon,* the political animal, therefore implying that other animals are not political, which is to say, cannot play a part in the life of the *polis,* the city, the basic unit of Greek civil life. Now, it was not only nonhuman animals who were in principle excluded by Aristotle from political life, but also the mass of humans who were non-Greeks, and even the vast majority of Greek humans who were female, slaves, and/or children. It may thus be argued that the exclusion of nonhuman animals from political participation might be ended, just as women and the common people have acquired political suffrage in democratic societies.

Certainly something like this claim seems to be true for animal *rights:* at first, we had "the rights of man," the "human rights" (extended not only to "men" but also to women, children, and what were once regarded as inferior races), and now "animal rights," which have actually been enshrined in law. The rights of animals indeed give them a form of political subjectivity under the law.

Types of Political Rights for Animals

Although he is not interested in animal "rights" so much as animal *liberation,* Peter Singer is one among the many who have argued that there is an historical progression at work here. The key concept to this view of the political status of animals is Richard D. Ryder's concept of *speciesism,* adopted most famously by Singer. This concept implies the condemnation of the exclusion of animals from political consideration because of their species, just as racism has excluded some humans on the basis of a subcategorization of human animals.

The anti-speciesist animal liberationists do not, however, argue for the extension of full political rights to nonhuman animals. It seems that there is still a level that nonhuman animals cannot attain, namely, participation in political decisions. Certainly they do not have the rights of suffrage, the right to vote or participate otherwise in political processes, even though these processes produce legislatures that claim the rights to legislate about animals. In this, they are in a similar position to human children. Both groups are held to lack sufficient rationality to determine their own futures, as was once held to be the case also for slaves and women, and hence are barred from playing a formal role in the political process.

However, this lack of a *de jure* role in political processes does not mean that nonhuman animals and infant humans are not *de facto* political agents. It is clear that human children in fact have nontrivial political influence, both through influencing their parents and other enfranchised humans and by influencing things more directly, carrying out small acts of resistance, organizing politically within schools, and so on.

Some of the more explicitly "political" actions of human children, such as joining political youth groups or participating in school governance, are of course not undertaken by nonhuman animals. However, animals *are* able to undertake actions that have political import.

Peter Singer's views have tended to contradict the prevalent view of nonhuman animals as passive political objects by claiming that animals have desires (or rather *preferences*) that they manifest that can readily be discerned. Thus, Singer argues that when an animal tries to escape captivity it is expressing its preference not to be captive in a readily discernible way. In this Singer accords more importance to the agency of animals than does Ryder, whose *pianism* emphasizes the capacity for suffering of animals as the source of our ethical obligations to them.

Although animals resist our control enough to show that they do not *want* to be controlled, their lack of political subjectivity in the full sense possessed by human animals seems to be confirmed by their inability to resist effectively on the human level. Animals' resistance is such that they are readily contained by now-perfected measures. Through changing the animals themselves (through selective breeding) and building environments, fences, cattle prods, cages, and so on, "domesticated" animals are now controlled to the extent that their resistance, although still commonplace and obvious, is apparently neutralized. Unlike humans, nonhuman animals in such situations seem incapable of, for example, secretly organizing to stage an uprising against their captivity.

This inability is a major cause of a certain contemptuousness from the traditional left against the placement of animal liberation on the same plane as the liberation of humanity, seeing political struggle as being an exclusively human affair. They are therefore out of sympathy with talk about how animals are "exploited" in much the same way as human workers, despite the fact that animals are often exploited in the same facilities as humans and by the same people.

However, the fact that animals are incapable of political organization in a narrow sense does not in fact mean that their resistance has been entirely negated. Just as African elephants can be understood as actually farming grass on the African savannah by their habit of uprooting trees, animals may be understood to have political agency via the actual political effects of their actions, specifically, the ability of animals through their expressions of anguish during human maltreatment to affect humans into acting to protect them. Such animal actions may certainly be seen as an essential cause of the discourse of animal rights and animal liberation itself.

Certainly, this manipulation by animals of humans is like the manipulation of adults by children, which is to say that it is naïve, lacking in cynicism. Indeed, part of the reason why humans are moved to help nonhuman animals is their very "innocence."

Everyday Power Relations

On a macro-political level it does seem that animals do not have political subjectivity, because they cannot participate in government, but on a micro-political level, a level of analysis most prominently put forward by French philosopher Michel Foucault, the level of everyday power relations, this is by no means obvious. In a household, for example, animals seem quite capable of exercising power, defined by Foucault as the ability to act on the actions of others. For example, a cat is quite capable of behaving in such a way as to *purposefully* get its owners to give it food, in much the same way as a human can to another human. Pets and other animals are able to enter into power relations with human beings in which they entice, seduce, or threaten humans or are in turn cajoled or seduced by humans. It would even be possible to extrapolate from this micro-political subjectivity to a macro-political influence: Pets owe their very survival to an ability to bond with and command the loyalty of owners, which will in turn lead to owners taking relevant political action to represent the demands of pets.

Recent "poststructuralist" thinking about political agency tends to turn the tables on traditional thinking about political subjectivity in a different way. Thinkers such as Singer merely see an existing progressive trend broadening in the future to include respect for animals in addition to the rights of man. Poststructuralist and structuralist approaches to subjectivity, on the other hand, are suspicious of such a progressive view of history, and rather see the free man as the latest constructed reality, which is to say that they see subjectivity itself as something socially and historically constructed rather than as a fact about human nature.

Rather than raising the possibility of promoting animals to the level of political subjects then, poststructuralism means viewing such a promotion as a change in the nature of society and of political subjectivity itself. Structuralism and poststructuralism represent a self-consciously anti-humanist turn in the history of thought. Although this it is not one that generally speaks about nonhuman animals, it is nevertheless also by this token an anti-anthropocentric turn. The more political of the structuralist and poststructuralist critics of the humanist notion of subjectivity, primarily Foucault, view the domain of the political in ways that allow us to understand animals' involvement in political structures and hence their relation to political subjectivity in such a way as to make them continuous with human political subjectivity. Of course, it nevertheless appears to be the case that animals' political capacities are inherently and permanently limited to a level lower than that of most adult humans. As Singer has frequently pointed out, however, there are adult human adults who are handicapped and therefore have similarly limited political capacities.

See also

Ethics and Animal Protection—*Political Action Committees (PACs) for Animal Issues*

Further Resources

Aristotle. (1995). *Politics.* Oxford: Oxford University Press.
Carruthers, P. (1998). Animal subjectivity. *PSYCHE,* 4(3). Available online at http://psyche.cs.monash.edu.au/v4/psyche-4-03-carruthers.html
Garner, R. (2004). *Animals, politics and morality* (2nd ed.). Manchester: Manchester University Press.
Singer, P. (1975). *Animal liberation: A new ethics for our treatment of animals.* New York: Random House.
Williams, A. (2004). Disciplining animals: Sentience, production, and critique. *International Journal of Sociology and Social Policy,* 24(9), 45–57.

Mark G. E. Kelly

■ Ethics and Animal Protection
Practical Ethics and Human-Animal Relations

Human-animal studies raises complex and often controversial questions about the ethics of humanity's relationship with other animals. These questions have implications for humanity's interaction with wild, companion, farm, and research animals, as well as for how and why we study human-animal relations. One manner of addressing such issues is through practical ethics. Practical ethics generates a situated moral understanding that is well suited to grappling with the complexity and diversity of our responsibilities in a more-than-human world.

Ethics

Definitions of ethics can differ vastly. Most of these differences are rooted in attempts to explain ethics in terms of something else. For example, various academics have tried to associate ethical concerns with personal preferences, emotional responses, religious beliefs, social expectations, and genetic determinism. Personality, empathy, spirituality, social custom, and science may all enrich ethics at various points and times. Yet, we should be careful not to let this obscure the meaning and importance of ethics itself.

To discover the meaning of ethics, we can look to Socrates, a Greek philosopher whose definition of ethics has been at the core of ethical thought for several thousand years. Socrates saw himself as a gadfly and a midwife. As a gadfly, he pushed people to think harder. As a midwife, he helped them develop their thoughts to a higher level of rigor. For him and his followers, ethics was (and is) about "how we ought to live" (from Plato's *Republic,* Book 2, 312d). What this brief statement means is this: ethics is about the moral values that inform (or should inform) our lives. When we engage in ethics, we are not only exploring our ideas about what is good, right, just, and valuable; we are also articulating principles of conduct based on these ideas. Overall, ethics helps us formulate rules of thumb that provide guidance as we strive for what the ancient Greeks termed *eudaimonia*—what we now refer to as *flourishing*.

To help us flourish both as individuals and a community, ethical dialogue has two interrelated functions—one of critique, and the other of vision. As part of the critique, we examine what promotes or detracts from the well-being of ourselves and others. In so doing, we identify how our worldviews, social institutions, decisions, and actions affect our lives. As part of the vision, we consider how we might improve our individual and collective lives by proactively pressing for positive changes in states of affairs that are either wrong or in need of improvement. Because these functions are connected, ethics is not a static ideology of "right versus wrong." Instead, it is a living tradition of thought that, in light of reason and evidence, is continually revising and renewing itself.

Ethics is also a form of power. It is a not a physical power like military force—rather, it has the power of ideas. Ethics can reveal the moral issues at the heart of a situation. Once a problem is made visible through ethical reflection, it can then guide our responses in trying to resolve that problem. It also is an indispensable means of holding people and social systems accountable. Think of what it means to call someone a liar. If the claim is accurate, and the lie has injured people, then an ethical judgment about the intentions behind the lie—as well as the actions because of the lie—has a moral power that is difficult to deny.

The power of ethical ideas is thereby indispensable in community life. It is an element of our social customs and laws, as moral norms help justify (and critique) our individual and collective beliefs and behavior. It is also the inspiration for social movements seeking animal and environmental protection, as well as human rights and social justice. What is accepted or legal is not necessarily ethical, and social norms and laws that were once accepted have now been rejected (e.g., slavery) or are under attack (e.g., speciesism).

Finally, ethics is not only for human beings. People may be the only creatures on Earth who have abstract systems of thought called ethics. In this sense, ethics is an artifact of human culture. This does not mean our ethical considerations must exclude other creatures. The moral community is a mixed one, populated by humans and other animals, all of whom share an intrinsic value and moral standing alongside the rest of nature. In addition, individuals and groups, ecosystems, and societies represent different foci and scales of ethical reason. People, animals, and nature all have a well-being that ethics helps us appreciate and protect.

Theoretically Rich and Empirically Situated

The world's moral complexity and the kind of ethical reasoning necessary to grapple with it was no secret to Socrates. He practiced a form of moral reasoning that was fully engaged with the empirical world, and one that differs markedly from the standard ways in which ethics is often practiced today.

In the standard model of ethics, the right answer is determined ahead of time and derived without the benefit of what we might learn from experience. This is sometimes called *theoretical ethics*. The answers from theoretical ethics are then applied to concrete cases in a top-down, linear, and deductive manner. This is what is meant by *applied ethics*.

Practical ethics proceeds differently. Instead of determining what the right answer must be ahead of time, practical ethics seeks out the best answer by integrating what we learn from a concrete case about a moral problem and the conceptual insights that help us best understand and resolve that moral problem. It is for this reason that practical ethics rejects easy division between theoretical versus applied ethics. Rather, it seeks a situated moral understanding—an ethics that is simultaneously conceptually rich and situated in real life. Practical ethics looks to diverse moral principles, rooted in the empirical reality of cases, to triangulate on the reasons and resolutions to our moral concerns. Several features of practical ethics should be emphasized here.

Pluralism. For the practical ethicist, moral concepts are plural and complementary. The more concepts we have, the deeper our reservoir of potential insights. Thus, the practical ethicist is not precommitted to a single concept that she uses over and over in all situations. She is free to choose from a constellation of concepts. Ideally, her choice reflects those concepts that are most useful in resolving a moral problem. Moral concepts that are commonly used in practical ethics are recognized by such terms as *good, right, fair, just,* and *value.*

Triangulation. Ethical concepts cannot be applied by rote, like a grid of latitude and longitude from which we read off the correct moral "position." Rather, moral understanding is akin to triangulating on the best ethical position. When triangulating over land or sea, one needs several reference points to properly plot one's position. These reference points may be stars (e.g., Polaris the North Star) or landmarks (e.g., mountain peaks). The same is true in ethics, where the reference points are well-developed moral concepts. To triangulate, one needs a plurality of these concepts to find one's way in the moral landscape.

Principles and Maxims. Moral concepts can be used as either principles or maxims. A *principle* is a moral concept used to clarify our thinking. It provides guidance to our reasoning about how we ought to live. The concepts of human rights or the intrinsic value of animals are examples of such principles. A *maxim* is a moral concept used to clarify our actions. Maxims provide more focused guidance than principles, and are especially directed toward what we *ought* to do—that is, what actions we should undertake. The golden rule ("treat others as you would want to be treated") is an example of a maxim. Overall, principles justify the use of certain maxims that guide our action, while maxims align our actions with broader moral concepts.

Rules of Thumb. Moral concepts are not rigid or absolute laws. They are rules of thumb that help us locate better from worse ways of thinking and acting. Both principles and maxims actively and dynamically reveal the ethical issues at stake, and provide guidance on what we ought to do about them. They do not, however, make moral decisions for us. Rather, they are the tools through which we exercise moral judgment.

Praxis. The term *praxis* refers to putting theory into action. Praxis is not a one-way relation where one deductively reasons from theory to action. It is a two-way relation where theory and action are reciprocally informing. In practical ethics, the principles and maxims we use to reveal ethical issues and guide our subsequent actions are selected in light of the case at hand. It is a form of practical reasoning where theory and reality is not disengaged from each other, and ethics can be situated in the world.

Context. Concrete moral problems are situated in space, time, nature, and culture. All ethical issues, therefore, have a geographical, historical, environmental, and cultural context. The stock of moral concepts in use and the actions that a moral agent can take are enabled and constrained by the context in which one operates. These are the sites and situations in which moral problems, the controversies that swirl around them, and their possible resolutions exist. In this sense, questions about both ethical theory and practice are eventually always situated in the world.

Judgment. The proper matching of principles, maxims, and cases takes experience and skill, a feature that practical ethicists refer to as *judgment.* Having good judgment means one can correctly match the most appropriate moral concepts to the case at hand. This is best done when we balance the facts on the ground with our best ethical understanding and chart a course of action from there. Ethical reasoning is always a matter of making our best interpretation of events.

Truth. From the standpoint of practical ethics, there is rarely a single, indisputable judgment that is right (or wrong). Reasonable people will differ on what the best concepts (principles and/or maxims) for understanding a particular case might be. They may also differ on what a reasonable course of action might entail. Recognizing that absolute truth (veracity) is rarely possible, practical ethics seeks the best account of truth that is possible (verisimilitude).

Situated. The recognition that absolute moral truth is very difficult to come by is not a reason to endorse ethical relativism. With its emphasis on praxis and context, practical ethics is not only situated in the world, but it takes the creative middle ground and situates itself between absolutist and relativist interpretations of ethics. It does so believing we can distinguish better from worse moral reasoning or courses of action. We do so in light of the evidence at hand, and the rigor of our thinking. Akin to the evolution of scientific knowledge, both reason and evidence are tested through dialogue with one another and in light of our experience over time. By fusing our conceptual horizons of moral understanding, this allows each of us to come to a deeper and better understanding of a moral problem and its possible resolution(s).

Examples of Principles and Maxims

Below are a few examples of principles, maxims, and how they might apply to concrete cases. A complete discussion of the history, philosophy, and use of practical ethics is beyond the scope of this entry. Nonetheless, these examples should give a flavor of the practical ethics approach. These examples are not meant to be conclusive. Rather, they are suggestive of how a practical ethics about nonhuman animals works. Moreover, there is not a hard and fast line to be drawn between using a moral concept as a principle or as a maxim. In general, principles are more abstract and thought-oriented, while maxims are more concrete and action-oriented.

Principles (Guidelines for Thought)

Geocentrism. We should acknowledge the moral value and standing of people, animals, and nature. This principle values animals and their habitats, while encouraging recognition of humanity's membership in a wider moral community. Geocentrism incorporates the insights of anthropocentrism (the moral value of people and their communities), biocentrism (the moral value of individual people and animals) and ecocentrism (the moral value of biodiversity and ecosystems). Significantly, it helps us sidestep the pitfalls and arguments between the latter three viewpoints.

Equal Consideration. We should give equal consideration to the well-being of people, animals, and nature. This is an adaptation of Peter Singer's principle by the same name. This principle helps us actualize geocentrism by identifying and balancing our responsibilities to people, animals, and their mutual habitats. Note that equal consideration does not imply equal treatment. When creatures differ in their capacities and modes of life (e.g., people, foxes, voles), then equal consideration requires appropriate differences of treatment.

Hard Cases. When faced with a situation pitting humans against animals, first solve the underlying problem, then look for alternatives—and, as a last resort, choose a geographic compromise that protects the entire community's well-being. This principle helps us think through the complications raised when we give equal consideration to the well-being of human and nonhuman others. Our universal need for geographic "space" (habitat, resources, etc.) makes win/lose conflicts a fact of life. We should first seek to resolve the underlying conflict and prevent its recurrence. If we cannot do this, then we should seek alternative modalities that protect the well-being of people and animals, both as individuals and as populations. If this is not possible (and sometimes it is not), then we need to optimize our land-use and planning, so that some kinds of human life can flourish in one area, while other kinds of nonhuman life can flourish in another area.

Moral Carrying Capacity. People should live within an overall carrying capacity that protects the well-being of nonhuman individuals, biodiversity, and landscapes. This principle is crucial, as it helps us avoid the hard cases mentioned above. While technology and social organization may mitigate the upper limit on the earth's carrying capacity for humans, there is a definite and negative impact of societal growth and consumption on the nonhuman world. Humans must take responsibility for limiting their use of the earth's carrying capacity. This is nowhere more true than in urban areas where, with globalization, the entire world has become a hinterland—a pool of resources for urban life. Part of this will involve making geographic compromises by protecting adequate habitat for nonhuman life both within and outside of the urban environment.

Precaution. The idea behind precaution is similar to the medical principle, "First do no harm." The concept is a principle for dealing with the uncertainty that pervades questions of both ethics and science. Precaution states that a lack of certainty is not an excuse for actions that may create harm or are irreversible. In the face of uncertainty, precautions should be taken to minimize the risks to people, animals, and the rest of nature. Of central importance here is that the burden of justification for actions causing suffering or harm lies with the advocate(s) of an action. One has no inviolable right to engage in activities with risk of harm (e.g., polluting a water source) simply because the range and extent of that harm is not yet well documented.

Maxims (Guidelines for Action)

Integrity. We should endeavor to respect the psychological, physical, and social integrity of wild and domestic animals by minimizing stress, using noninvasive and non-lethal techniques in cases of conflict, and avoiding the disruption of social organization and ecological relationships.

Graduated Response. In cases of human-animal conflict, there are a continuum of responses, from nondestructive and nonlethal to destructive and lethal. We should seek to resolve a problem with nondestructive and nonlethal responses first. Where one starts on this continuum depends on the severity of the problem.

Harm-Benefit Ratios. During the design phase of research, policy, or management strategies regarding nonhuman animals, we should calculate harm-benefit ratios for each action. Such ratios help us explore whether the probable benefits to science, society, or nature can outweigh the foreseeable harms to wildlife and its habitats.

Mutual Benefits. Whenever possible, we should adopt those actions that provide mutual benefits for people, animals, and nature. Vague assertions about human benefits or risks to public health are rarely sufficient reasons to sacrifice the well-being of animals. This is a more positive and proactive principle than the harm-benefit ratios mentioned above.

Reduction, Refinement, Replacement (The Three Rs). When using invasive or harmful procedures in the laboratory or in the field, we should practice "the three Rs": *reduction* of our number of actions, *refinements* in our technique, and *replacement* with noninvasive and nonharmful procedures. This principle promotes best practices by mandating that we: (1) impact the fewest number of animals, (2) cause the least amount of physical or psychological harm, (3) minimize the harassment of animals or populations, and (4) maximize the number of alternatives to the direct use or control of other creatures.

End Points. Invasive or harmful actions should specify humane "end points," so that if an action proves harmful, we know when to stop. When an action based on a policy or management strategy is proving harmful, it should have a predefined endpoint. After the action is brought to a halt, the situation should be reassessed to produce a better course of action.

Conclusion

In modern times, variations on the practical approach to ethics have been advocated by Anna Peterson, Anthony Weston, Arne Naess, Georg Hans Gadamer, Jurgen Habermas, Dale Jamieson, and Stephen Toulmin. It was Mary Midgley, however, who set the tone early on in animal ethics—as well as in human-animal studies (a discipline which might also be dated to her early publications). In her book *Animals and Why They Matter* (1984), Midgley carefully explores the dominant theories of animal ethics. She does so as part of an appreciative critique, seeking out conceptual insights while noting shortcomings when a theory or concept is misapplied. She does not ask her readers to choose another theory, per se, but to appreciate and carefully use the full range of concepts that are made available through a diversity of theories. In other words, she asks that we generate a situated moral understanding, one that takes both moral concepts and the facts on the ground as equally important and mutually informing.

Unfortunately, it is not a simple task to put practical ethics to work. The world is ethically fraught and dynamic. The unparalleled power of humans to do good or ill in the world means that no narrow set of rules can hope to represent our moral concerns, much less give adequate guidance beyond basic prohibitions against egregious wrongdoing. So there is no ethical rule-book we can follow to give us certain answers. No theory, method, or concept on its own is adequate to the task.

Nevertheless, this does not mean we lack the ability to clarify and state our ethical responsibilities. We have access to a constellation of moral concepts that we can use as principles and maxims to triangulate on better (versus worse) accounts of how we ought to live. There may be no God-given, scientifically proven, absolute moral truths to set our sights upon, but we can adjust our moral compass for guidance in how we improve or detract from the well-being of ourselves and others. That is the task of practical ethics.

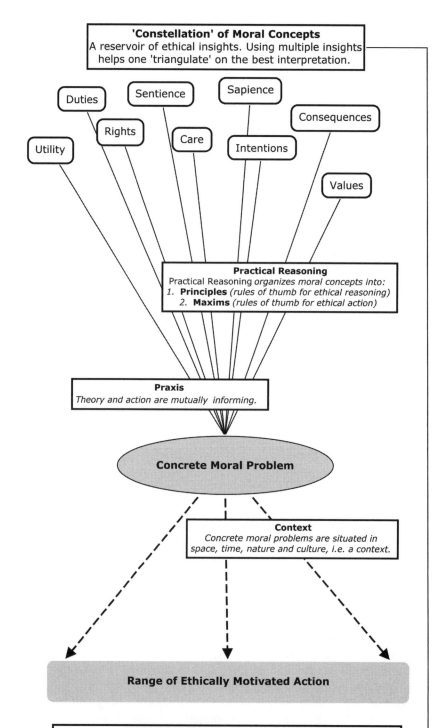

See also

Education—*Human-Animal Studies*
Ethics and Animal Protection—*Wolf Recovery*
Geography—*Animal Geographies*)

Further Resources

Gadamer, H-G. (1993). *Truth and method* (2nd, rev. ed.). New York: Continuum.

Habermas, J. (1993). *Justification and application: Remarks on discourse ethics.* (C. Cronin, Trans.). Cambridge: MIT Press.

Jamieson, D. (2002). *Morality's progress: Essays on humans, other animals and the rest of nature.* New York: Oxford University Press.

Lynn, W. S. (2004). The quality of ethics: Moral causation in the interdisciplinary science of geography. In R. Lee & and D. M. Smith (Eds.), *Geographies and moralities: International perspectives on justice, development and place* (pp. 231–44). London: Routledge.

———. (2006). Between science and ethics: What science and the scientific method can and cannot contribute to conservation and sustainability. In D. Lavigne (Ed.), *Gaining ground: In pursuit of ecological sustainability* (pp. 191–205). Limerick, Ireland: University of Limerick.

Midgley, M. (1998). *Animals and why they matter.* Athens, GA: University of Georgia Press.

———. (2005). *The essential Mary Midgley.* New York: Routledge.

Naess, A. (1989). *Ecology, community and lifestyle: Outline of an ecosophy.* Cambridge: Cambridge University Press.

Peterson, A. L. (2001). *Being human: Ethics, environment and our place in the world.* Berkeley: University of California Press.

Rolston, H., III. (1994). *Conserving natural value.* New York: Columbia University Press.

Singer, P. (1979). *Practical ethics.* Cambridge: Cambridge University Press.

Toulmin, S., & Jonsen, A. R. (1988). *The abuse of casuistry: A history of moral reasoning.* Berkeley: University of California Press.

Weston, A. (2006). *A practical companion to ethics.* New York: Oxford University Press.

William S. Lynn

■

Humans, Animals, and Activism: Rocky Mountain Animal Defense

David Crawford

The impetus in 1994 for creating Rocky Mountain Animal Defense (RMAD, pronounced R-MAD, www.rmad.org) was the following:

- To establish consistency in the local animal advocacy movement
- To create a community in which animal advocates could hone their skills and develop their philosophies
- To develop a unified community voice to speak out on behalf of animals

RMAD's mission has changed throughout the years, but its essence has remained the same: to help eliminate the human-imposed suffering of animals.

(continues)

■

Humans, Animals, and Activism: Rocky Mountain Animal Defense (continued)

In its first decade, RMAD covered the key issues, namely, the use of animals in laboratories and in entertainment, the use of animals for food and for clothing,

the use of animals for companionship, and the persecution of wildlife (including hunting, species reintroductions, and habitat destruction). During this period, the organization's tactics also became increasingly diverse and included the following:

- Working with citizens of Estes Park, Colorado, to pass an initiated ordinance prohibiting the exhibition of animals, which effectively precluded a Plexiglas® zoo from being built at the east entrance to Rocky Mountain National Park
- Initiating a well-publicized boycott of the Nalge-Nunc Corporation (makers of the popular Nalgene water bottle) for the company's unmitigated support of animal research
- Engaging in direct action in opposition to circuses and hunting
- Litigating against the State of Colorado for violating its own constitution in failing to protect wildlife from poisoning
- Educating the public by displaying banners on interstate overpasses, running newspaper advertisements, and delivering presentations in schools and clubs

Today, RMAD continues to support the key issues, but the organization focuses on two programs: prairie wildlife and vegetarianism.

RMAD's prairie wildlife work dates back to 1995, when Nicole Rosmarino (now of Forest Guardians) made the case for adding a prairie dog campaign to RMAD's work. Prairie dogs are a keystone species of prairies in eleven U.S. states, Canada, and Mexico, providing a prey base and creating habitat for numerous species.

Rosmarino argued that the foot-long animal had lost the vast majority—current U.S. Fish and Wildlife Service statistics say 99 percent—of its habitat across its historical range and that the onslaught was continuing. Indeed, prairie dogs continue to be killed because they are perceived as a threat to ranching and agriculture, and their homes are destroyed as development increases in the West. What's more, they are highly susceptible to death from plague, which humans introduced to this continent around 1900.

RMAD's first prairie dog action was a demonstration and civil disobedience against a killing contest near Denver in April 1995. In these killing contests, participants are rewarded for various accomplishments, including most animals wounded (mortally or otherwise). With support from Animal Rights Mobilization, RMAD's action got front-page coverage in Denver's two newspapers and top-story billing on all three network affiliates. In 1999, under pressure from numerous organizations and individuals, Colorado severely restricted contest killing in the state.

Prairie wildlife advocates continue to face many challenges. As the advocacy movement matures, organizations are specializing key activities such as buying out grazing permits, pursuing federal and state protections, and establishing private reserves. With RMAD's support, the Prairie Dog Coalition has emerged, helping support and improve advocacy efforts.

RMAD's prairie wildlife work today focuses on the isolated colonies along Colorado's front range. RMAD works on policy affecting such colonies, technological improvements in colony management, and continuing education and sensitization of the public to the needs and well-being of the animals themselves.

RMAD's vegetarian program focuses on education and community. The organization engages in numerous outreach events each year, maintains a provegetarian Web site, and hosts monthly potlucks.

The organization's principles have through the years converged with the principles of Gandhian nonviolence. Love and compassion are critical operating themes. Collaboration continues to be a fundamental organizational strategy. As RMAD matures, it moves from reaction to pro-action. It establishes better boundaries, recognizing that it cannot be everything to everyone. It attempts to serve as a model of advocacy.

Further Resources

Rocky Mountain Animal Defense. http://www.rmad.org/

■ Ethics and Animal Protection
Raptor Rehabilitation and Behavior after Release

Birds of prey (e.g., hawks and owls) are very vulnerable birds. One reason is that they have been seen as competitors or as threats by humans for a long time, which has led to their persecution. In addition, they occupy the top trophic levels—those devoted to predators and superpredators (i.e., predators of other predators)—and, consequently, their number is limited; moreover, they have the ecological tendency to concentrate, in their bodies, the noxious substances that are accumulating in the lower levels of animals in the trophic chain's intermediate positions, and are subsequently more affected by poisons than other species.

The past century's industrialization and overcrowding greatly worsened the impact of human activity on the welfare of wildlife, revealing a negative scenario. For instance, birds of prey are harmed by a wide range of causes, including persecution by man, hunting and poaching, ingestion of lead shot, collision with vehicles, feeding on prey contaminated with pesticides, and egg stealing. Fortunately, growing concern for wildlife and an understanding of the necessity of preserving natural habitats has followed, and the rehabilitation of birds of prey is a consequence of this concern.

One definition of *raptor rehabilitation* can be "the treatment and care of sick, injured, or orphaned birds until they are returned to the wild." Such a program is usually carried out by individuals or at specific rehabilitation centers (RCs), which usually belong to private institutions or work jointly with university veterinary clinics. In some countries, falconers are actively involved in raptor rehabilitation, because of their knowledge and experience with falconry techniques, which can be successfully applied in such rehabilitation. The most frequent reasons for admission are fractures caused by collisions with vehicles and man-made structures, as well as injuries caused by gunshots. There are many other, less-frequent, causes for admission: emaciation due to starvation (or unknown reasons), poisoning from agricultural pesticides and heavy metals, ocular lesions, electrocution, and parasitic infestations. These factors are not restricted to a particular bird species, and the range of species affected is obviously very wide.

Unfortunately, the trauma or debilitation is often so severe that a large proportion of birds die or must be euthanized soon after admission. Among the survivors, some birds cannot fully recover physically, preventing any possibility of successful release. As a result, usually only 30 to 40 percent of raptors are released after treatment.

There has been controversy about the biological and conservation value of rehabilitation. Skeptics feel that the low outcome does not return the large amount of labor invested, and the low, above percentages offer support to this opinion. There is also doubt about the real biological value of rehabilitation itself and its contribution to conservation of the populations. In contrast, people involved in rehabilitation trust in the value of their efforts—particularly from the biological and ethical points of view. Beyond bird welfare, there is the opportunity to increase our knowledge about morbidity and the improvement of therapy. There are also indirect effects coming from the enhancement of veterinary and biological skills and public education; the latter aspect, in particular, is important when working in parallel with habitat protection and conservation.

The goal of rehabilitation should be to durably restore a bird of prey in the wild population. The contribution to conservation requires that the bird become integrated into the breeding population. From this standpoint, the outcome can only be considered positive if we can assess the reproduction of the released bird. However, in natural populations, not all adults contribute to reproduction; in fact, there is constantly a nonbreeding surplus of birds used to fill in open places. On the other hand, the release of even a restricted number of birds can be a great conservation contribution to many species that are existing as numerically small, localized populations—or those that are at the edge of extinction.

Rehabilitated individuals have to be housed in captivity—and close to humans, even—for a long time. Needless to say, this context is highly stressful for wild birds entering an RC, and this can, therefore, easily affect their behavior. The points that can potentially have adverse effects, both during captivity and after release, on a successful reintroduction will be here briefly considered.

When keeping birds of prey in captivity, every effort must be made to maintain the birds' welfare. Thus, the rehabilitation program should deal with both medical and ethological requirements. Among the many factors of concern specifically involving behavior for maintaining the birds' welfare, the imprinting phenomenon is well-known and typical of very young birds. It rarely affects adults or juveniles in rehabilitation but is of concern to orphaned nestlings. Imprinting on humans is likely reduced at RCs using (cross-) fostering or, at least, rearing nestlings in groups. It may have serious implications for reproductive results in birds, from captive-breeding projects, intended for release. In contrast, habituation to captivity as a whole is dangerous even for adult birds. Although it does not require well-defined sensitive periods, as does imprinting, prolonged close contact with staff must be considered mismanagement, as it can induce some form of socialization or lack of fear, particularly in juveniles. The released bird might not choose the optimal habitat and may stay in areas with human presence. Moreover, the higher the level of socialization to humans, the more frequent is abnormal aggression toward the handler or partners of the same species.

Infantilism—the regression to juvenile behavior patterns in adults—is fairly common in captive raptors. Screaming and mantling (that is, lowering the opened wings as a mantle) are evident examples. Such behavioral regression is likely caused by stress and weight loss. It is possible to record infantilism more often in females than in males, as juvenile behavior patterns are frequent in females' courtship repertoire. Although bird-of-prey species are usually nonsocial ones, social deprivation can affect normal behavior. This occurs when birds—particularly, if young—are kept alone or without visual contact with members of the same species for long time. The effects are proportional to the duration of seclusion and the age of the bird and are more evident in species with long developmental periods. Socially deprived juvenile raptors show behavior disorders and are unable to relate to members of their own species.

Prolonged captivity can reduce release success in some species, because it promotes habituation to humans and reduces flight ability and responsiveness to natural prey. Experienced common kestrels (*Falco tinnunculus*), housed necessarily for more than four months due to several causes, performed less-varied responses to the several types of prey than those in captivity for less than three months. Familiarity with a specific type of food resulted in kestrels performing the same predatory sequence when attacking any prey.

It is common to feed all species with the same type of food, often day-old chicks. The reason is usually the necessity to save both money and staff labor. A varied diet specifically designed for any species is important, particularly in the case of prolonged captivity, and must provide certain constituents at balanced levels. We should consider that birds of prey are adapted to feed on a wide range of prey, and even the most specialized species do remain very opportunistic predators.

Although it is widely recognized that both Falconiformes (kites, eagles, falcons, and allies) and Strigiformes (owls) perceive a strong releasing signal for predation by prey movements, food supply is usually made up of dead items. Young, inexperienced birds of prey may, therefore, have difficulty in developing correct predatory ability. Moreover, prolonged food supply limited to one food source, particularly if "artificial" (such as chicks), might have negative effects, because the raptor may learn to concentrate upon that food only. In common kestrels, even experienced birds may be affected, as has been observed by a weakened response to live prey.

The use of live prey is banned or considered unethical in some countries, although they are accepted as a food source for some species. Animal rights movements concerned with animal welfare oppose the use of animals for research purposes. However, we must also consider and assess all aspects of the captive raptors' biology, and predatory ability is certainly one of the most important. While being concerned with prey welfare, it is also necessary to be concerned with the predator's welfare. There is evidence of behavior alterations among several mammalian predators in zoos, and a bird of prey may be affected from prolonged inability to perform the full predatory sequence in captivity.

Most efforts have been concerned with the captivity phase of rehabilitation, whereas the release phase has attracted less attention. Some studies have evaluated if a bird is ready for release from the anatomical, behavioral, or physiological points of view, but little attention has been paid to the description of general activity and behavior of the birds after release. Most of these studies have dealt with a limited number of orphaned nestlings, mainly eagles and owls. A few studies have, instead, evaluated rehabilitated birds and used large samples, although all focused on survival rate and dispersal distances. The latter parameter is of limited utility, as it cannot ascertain the bird's previous movements, particularly when survival's lasted for months or years. Continuous radio-tracking is very useful for monitoring released birds, but it is of limited use with fast- or long-range-moving birds (or in mountainous areas). This problem can be overcome with GPS or satellite-telemetry technology. However, high costs—combined with the difficulty of receiving the radio signals when the bird is in areas with electromagnetic interference—limit their use, at least as routine application. Nevertheless, although these studies allow the inferring of detailed information about movements— even long-range ones—they do not allow us to learn anything about the behavior of rehabilitated raptors. This is fundamental, particularly in the early period, to assess the birds' adaptability to the wild.

Such types of investigations may be costly and force a reduction in sample size, but they contribute to fill in the gap of our knowledge. One pioneer study involved rehabilitated adult and subadult Eurasian buzzards (*Buteo buteo*) in northern Italy. The radio-tagged buzzards, released in different seasons, acclimated gradually to the new environment and remained in the release area for variable amounts of time. Most of them left the area voluntarily, as they were seen soaring very high and then gliding along a straight path before being tracked for the last time. Interpretation of results of the others was complicated by the fact that conclusions had to be drawn from the loss of signal. This could mean that the buzzards flew out of signal range successfully while not tracked, but it could also mean that the birds were, for instance, electrocuted, and that, at the same time, the transmitter was damaged.

As a whole, the buzzards interacted frequently with resident conspecifics, sometimes even defending a territory adjacent or inside that of a wild buzzard. Although frequently mobbed by corvids, particularly hooded crows (*Corvus corone cornix*), released birds did not move very far. Interestingly, the flight frequency was inversely correlated to the captivity duration. They also showed strong habitat preferences in relation to the season—probably due to prey availability—and showed the skill to hunt several taxa—but not birds, although Eurasian buzzards are known to feed on birds. This, too, may be the result of limited flight performance due to captivity.

Another study investigated the post-release behavior of orphaned northern long-eared owls (*Asio otus*) and tawny owls (*Strix aluco*). The owlets were not injured but simply had been found alone and forwarded to the RC. Moreover, in contrast to the Eurasian buzzards, which had experienced life in the wild, orphaned owlets were released in a completely new (for them) environment. Birds of both species rarely flew immediately after release. They remained in the area for a few days before disappearing. Such a difference from the buzzards is likely based on species adaptation.

It is possible that owls that were no longer detectable after release died suddenly, and there was an interruption of the tag signal. Although hand-reared, the owls were not particularly attracted to locations with human presence. However, long-eared owls,

Falconiforms, *such as this Eurasian buzzard, can be easily radio-tracked using high-frequency tags attached to a couple of tail feathers, without interfering with flying and hunting performance. The tag will drop, together with the feathers, when the bird is moulting. Courtesy of Vittorio Corona.*

in particular, remained close to the release pen for long periods, apparently showing a lack of confidence with the natural environment. Mortality of northern long-eared owls was likely due to starvation and was higher than that of both tawny owls (the one dead bird was likely preyed upon) and Eurasian buzzards (three birds were electrocuted). This likely suggests that released orphaned owls differ mainly in hunting skill, with some species being more able than others to recognize and hunt natural prey, although without previous experience. Further studies are necessary to ascertain breeding success in rehabilitated birds of prey.

Raptor rehabilitation is, then, a rather complex process. It consists of several different phases, all equally important and closely connected. Post-release monitoring is important in judging whether rehabilitation has been successful. This can be achieved only if the bird can join the natural population, possibly as a breeder. The contribution of rehabilitated birds to the natural population is difficult to assess, since even the nonbreeding individual can play a role as nonbreeding surplus. Finally, indirect beneficial effects of rehabilitation (for instance, public education) should also be considered.

Further Resources

Csermely, D. (2000). Behaviour of hand-reared orphaned long-eared owls and tawny owls after release in the wild. *Italian Journal of Zoology, 67,* 57–62.

Csermely, D., & Corona, C. V. (1994). Behavior and activity of rehabilitated common buzzards (*Buteo buteo*) released in northern Italy. *Journal of Raptor Research, 28,* 100–07.

Humphreys, P. N. (1981). The problem of rehabilitation. In J. E. Cooper & A. G. Greenwood (Eds.), *Recent advances in the study of raptor diseases* (p. 165). Keighley: Chiron Publications.

Ingram, K. (1988). Survival and movements of rehabilitated raptors. In R. L. Glinski, B. G. Pendleton, M. B. Moss, M. N. LeFranc Jr., B. A. Millsap, & S. W. Hoffmann (Eds.), *Rehabilitated raptors* (pp. 277–81). Washington, DC: National Wildlife Federation Scientific Technical Series No. 11.

Kirkwood, J. J. (1991). Introduction and rationale for rehabilitation. In London Zoo, The Hawk Trust, & The Hawk Board (Eds.), *Raptor rehabilitation workshop* (pp. 11–18). Newent: The Falconry Centre.

Davide Csermely

■ Ethics and Animal Protection
Regan, Tom, and Singer, Peter

Two philosophers stand out in the field of ethics and animals, one of the most dynamic and divisive fields of philosophical inquiry: Tom Regan and Peter Singer.

Tom Regan

Tom Regan was initially interested in human rights. While exploring civil rights, he discovered the works of a great spiritual teacher from India, Mahatma Gandhi. Gandhi's work and life helped Regan to see that rights and moral duties need not be restricted to human beings, but might well be extended to include other living beings. Regan eventually wrote *The Case for Animal Rights,* an expansive, detailed book that holds the first comprehensive defense of animal rights. Regan's moral theory, "the Rights View," extends basic human rights to nonhuman animals.

Regan states that to have moral rights "is to have a kind of protection we might picture as an invisible No Trespassing sign" (*Empty,* 38). What does it mean to have this protection?

> First, others are not morally free to harm us; to say this is to say that others are not free to take our lives or injure our bodies as they please. Second, others are not morally free to interfere with our free choice; to say this is to say that others are not free to limit our free choice as they please. In both cases, the No Trespassing sign is meant to protect our most important goods (our lives, our bodies, our liberty) by morally limiting the freedom of others. (*Empty Cages: Facing the Challenge of Animal Rights,* 39)

Regan explains that there are many important values we must take into consideration when making moral decisions, but moral rights outweigh all others. Just because Suzie values money does not mean she can sell Lois into slavery, because Lois has a right not to be enslaved. Just because one person likes the taste of flesh does not mean he may eat his neighbor, because his neighbor has a right to life. Moral rights cannot be overridden, even for the benefit of *many* others. Might does not make right, and neither the preferences nor the pleasure of the majority justifies overriding the basic moral rights of the few. Consequently, if someone infringes on the moral rights of other people—or other species—those who object are not asking for privileges or favors; they are simply demanding what is already owed.

Why do certain beings have rights? Regan notes that some creatures have a welfare: they fare well or ill depending on whether or not their interests are benefited or harmed. For example, a nonhuman animal such as a dog or a horse has a welfare—it can be harmed or benefited, depending on what happens to it in the course of its lifetime. It is harmed if it does not have enough to eat, if it is confined in areas that are small or dirty, and if it is slaughtered.

Regan applies the Golden Rule: it is wrong for humans to do to other animals what human beings would not want done to themselves. Even if other animals cannot communicate with words, or use reason to form philosophical arguments, it is wrong to harm them. Other animals have a welfare (they can be harmed), and so they have rights. Moral agents—those of us who are able to make moral decisions—must take into account the rights of these other beings. In the Rights View, moral agents are obligated to consider how actions affect *any* individuals that have a welfare—*any* individual that can be harmed—not just human beings. All animals that have a welfare also have rights, and rights carry obligations for moral agents.

Regan coins the term "subject-of-a-life" to describe individuals with qualities that grant moral rights. A subject-of-a-life has

> beliefs and desires; perception, memory, and a sense of the future, including their own future; an emotional life together with feelings of pleasure and pain; preference- and welfare-interests; the ability to initiate action in pursuit of their desires and goals; a psychophysical identity over time; and an individual welfare in the sense that their experiential life fares well or ill for them. (*The Case for Animal Rights,* 243)

Subjects-of-a-life have inherent value, because they have a welfare; they have moral rights, and we have concurrent moral obligations. Regan's Rights View grants rights and provides moral protection to "normal mammalian animals, aged one year or more" that qualify as subjects-of-a-life (*Case,* 190). This means that cats, sheep, cattle, pigs, and people have rights, if they are "normal" and at least one year of age.

Regan notes that a moral line drawn at one year of age does not establish that living beings outside this category do not and should not have rights, but rather that those within this category clearly should and in fact *do* have rights. In other words, many more animals are likely to have rights, such as eagles and chickens, frogs and lizards, and human beings under one year of age. Regan draws the line at one year simply because there is a need to have a definitive, defensible demarcation of who *clearly* has basic rights.

Inasmuch as human beings have basic rights, Regan concludes that all subjects-of-a-life have basic rights. Furthermore, just as no human being is more or less valuable than another, no subject-of-a-life is more or less valuable than any other. Chris does not have more inherent value than Mick, nor does Mick have more inherent value than Spot or Rover. Regan notes that moral ranking too easily leads to elitist attitudes that result in practices such as segregation—or even slavery. *Every* animal with a welfare, *every* subject-of-a-life, has equal inherent value, inasmuch as every human being has equal inherent value.

Normal mammals aged one year or more, as subjects-of-a-life, have the moral right— or valid claim—to be treated with respect; they have the right not to be harmed (*Case*, 327). They do not need the right to vote or the right to free speech any more than a toddler needs these rights. Yet, inasmuch as they have a welfare, they need and have the right not to be harmed or killed prematurely—just as surely as a toddler has the right not to be harmed or killed prematurely. Subjects-of-a-life can be harmed; moral agents have a moral duty to respect the rights of animals who are subjects-of-a-life. In the Rights View, when we make moral decisions, we must consider *all* entities that have a welfare, and we must assess *each* individual *equitably* in order to honor individual rights and equal inherent value (*Case*, 325). This means that if a moral agent wants to eat meat for lunch, he or she needs to ask if sinking teeth into Bessy respects her right not to be harmed. Do I need to eat Bessy? If I do not need to eat flesh, can I respect her right not to be harmed when I exploit her body and destroy her life for my lunch? Regan offers a clear "no."

Furthermore, Regan asserts that we are obligated, as moral agents, to come to the aid of Bessy when her rights are not respected. In *The Case for Animal Rights,* Regan notes that it is our moral duty to aid subjects-of-a-life whose basic rights are violated (*Case*, 249). Moral agents have a *duty* to assist pigs and cattle, dogs and cats, when others fail to respect their basic rights—when others cause them harm (*Case*, 249). Regan makes it clear that those now engaged in a nonviolent struggle for animal liberation are, like Gandhi, moral exemplars.

Peter Singer

Peter Singer's studies in philosophy focused on *consequentialism,* which led him to question contemporary human relations with other animals. He is the author of *Animal Liberation,* one of the earliest and most widely read books exploring ethics and animals.

Jeremy Bentham, a consequentialist, noted that humans are not the only species to seek pleasure and avoid pain. Rats, pigeons, oxen, and deer also suffer, and they also seek to avoid suffering. Bentham noted that the key moral question does not focus on whether or not a being can reason or talk; the critical moral question is whether or not a being can suffer. Because animals have the capacity to feel pain, Bentham argued that they are morally important.

Consequentialists assess whether or not an act is moral based on outcomes, consequences, or utility; *utilitarianism* is the best-known consequentialist moral theory. *Classical utilitarians* (e.g., John Stuart Mill) measure utility in terms of maximizing pleasure and minimizing pain. Peter Singer represents another form of utilitarianism: *preference utilitarianism.* Singer judges the morality of actions "not by their tendency to maximize pleasure or minimize pain, but by the extent to which they accord with the preferences of any beings affected by the action or its consequences" (*Writings*, 133). Simply put, a person's interests are determined by what they prefer, and a moral act is one that maximizes the satisfaction of interests.

Each individual's interests must be weighed against those of other individuals to determine a morally preferable action. In this process, individual self-interest gives way to collective interests:

> [M]y own interests cannot count for more, simply because they are my own, than the interests of others. In place of my own interests, I now have to take account of the interests of all those affected by my decision. This requires me to weigh up all these interests and adopt the course of action most likely to maximize the interests of those affected. Thus, I must choose the course of action which has the best consequences, on balance, for all affected. (*Practical Ethics*, 12–13)

In Singer's utilitarian theory, a moral act is determined by assessing which action best furthers the interests of all those who will be affected.

Sentience is essential to having preferences (*Practical*, 50). A piece of coal cannot have interests or preferences because a lifeless object cannot suffer. A cow, on the other hand, has a central nervous system and can suffer. Any being that *can* suffer will almost always have an *interest* in *not* suffering, and if we are to bring about the greatest utility, the best consequences for all concerned, we must take into account the preferences of nonhuman animals. To disregard the interests of individuals who don't happen to belong to our species is as irrational as ignoring the interests of people from other races, age groups, or religions. Singer writes, "Pain and suffering are bad and should be prevented or minimized, irrespective of the race, sex, or species of the being that suffers" (*Practical*, 54). He concludes, "[i]f a being suffers, there can be no moral justification for refusing to take that suffering into consideration" (*Practical*, 50).

Singer accepts that no two individuals are equal in all respects, but argues that similar interests ought to be weighed equally. Races, genders, and age groups show different abilities and interests, as do species. Equality is an ethical principle, not a fact of existence (*Practical*, 18). No two individuals are truly equal—no two human beings are equal in mental, physical, or artistic ability—yet, each sentient individual has interests. Moral agents ought to treat equal interests equally.

Because equality is a moral principle and not an actual reality, Singer's theory does not demand equal *treatment* for all sentient creatures; Singer's utilitarian theory requires *equal consideration of interests*. Equal consideration of interest requires that moral agents move beyond personal and group interests to consider the interests of all. Neither race nor species are morally important where suffering is concerned: both a chicken and a human being can suffer. For Singer, the interest an individual takes in not suffering is no more or less important than the similar interest of any other sentient being (*Animal*, 5).

This does not mean that each individual must be given an equal chance to fulfill personal interests. An individual's interest in freedom is more basic than a slave owner's interest in owning a slave. A leghorn chicken's desire not to suffer is more basic and critical than a human being's desire to taste the flesh of a dead leghorn chicken. To avoid significant suffering is more important than gratifying taste buds.

In one of his more controversial statements, Singer explains that people show bias in favor of human beings whenever they do an experiment on nonhuman animals for which they would not use a human being "at an equal or lower level of sentience, awareness, sensitivity, and so on" for the same experiment (*Practical*, 59).

[T]he fact that some people are not members of our race does not entitle us to exploit them, and similarly the fact that some people are less intelligent than others does not mean that their interests may be disregarded. . . . [T]he fact that beings are not members of our species does not entitle us to exploit them, and similarly the fact that other animals are less intelligent than we are does not mean that their interests may be disregarded. (*Practical*, 49)

Species is not relevant with regard to pain and suffering. If it is morally wrong to use sentient human beings who are at an equal intelligence level, whether for scientific experiments, food, or entertainment, then it is morally wrong to use a sentient rat, pig, or monkey. The suffering of one species cannot morally count for less than the similar suffering of another species.

As a preference utilitarian, Peter Singer requires moral agents to minimize suffering and maximize the satisfaction of interests. He notes that *species* is morally irrelevant with regard to pain and suffering. Moral agents are therefore obligated to take into account the suffering of pigs and turkeys in slaughterhouses, of homeless cats and dogs, of rats and birds in scientific experiments, and of dolphins and chimps in circuses. Singer's work, like Regan's, requires moral agents to rethink contemporary exploitation of, and indifference toward, nonhuman animals.

Philosophers Tom Regan and Peter Singer are well known for their important work in the field of animals and ethics. They are also known as activists. Both scholars *live* the high ideals they teach; both Regan and Singer give generously to charities, both are vegans, and both continue to work tirelessly for animal liberation.

Further Resources

Finsen, L., & Finsen, S. (1994). *The animal rights movement in America: From compassion to respect.* New York: Twayne.

Regan, T. (1984). *The case for animal rights.* New York: Routledge.

———. (2004). *Empty cages: Facing the challenge of animal rights.* New York: Rowman & Littlefield.

Singer, P. (1979). *Practical ethics.* Cambridge: Cambridge University Press.

———. (1991). *Animal liberation* (2nd ed.). London: Thorsons.

———. (2000). *Writings on an ethical Life.* New York: Ecco.

Lisa Kemmerer

■ Ethics and Animal Protection
Schweitzer, Albert

Albert Schweitzer (1875–1965), the world-renowned and revered medical doctor who worked for over fifty years in the hospital he founded in Lambaréné, in the Republic of Gabon in equatorial Africa, received the Nobel Peace Prize in 1952 for his humanitarian work. Schweitzer was also an accomplished organist, theologian, and philosopher whose personal belief in a "reverence for life," which must encompass all living things, both informed and gave purpose to his life and service. The ideas behind this philosophic principle were influenced by his observations and experiences with animals during his youth, and they developed more fully when, as a doctor in Africa, he assumed responsibility for the health and welfare of both his human patients and the many animals he encountered.

Early Experiences

Albert Schweitzer grew up in the Alsatian village of Günsbach, now part of France, with pet animals—particularly dogs and cats. Indeed, later in life, Schweitzer quipped that he did not consider it an insult to be compared to a dog (Joy, 1950). He was unusually sensitive to the suffering of these and other animals, especially if he himself caused it, and "he questioned why his own will to live and to be free of pain should ever be in conflict with the same eager will cherished by a deer, a cow, a pig, a dog, a cat, a horse, a mouse, a bird, a fish—perhaps, even an insect." (Free, 1982). He thought that we had ignored animals in our compassion for living things, and he included them in his own prayers as a boy: "Dear God, protect and bless all beings that breathe, keep all evil from them, and let them sleep in peace." (*Memoirs*, 1997)

As a boy he observed how an old horse was beaten on its way to slaughter, later writing: "The sight of a limping old horse being dragged to the slaughterhouse in Colmar by one man while another beat it with a stick haunted me for weeks." (*Memoirs*, 1997) He also felt guilt after disciplining his dog, Phylax.

The most well-known incident of Schweitzer's youth is vividly portrayed in the 1957 Academy Award-winning documentary about his life. In one segment of it, Schweitzer joins a friend to shoot a slingshot at a bird; it is even more meaningful in this reflective description he wrote some years later:

> We approached a leafless tree in which birds, apparently unafraid of us, were singing sweetly in the morning air. Crouching like an Indian hunter, my friend put a pebble in his slingshot and took aim. Obeying his look of command, I did the same with terrible pangs of conscience and vowing to myself to miss. At that very moment the church bells began to ring out into the sunshine, mingling their chimes with the song of the birds. It was the warning bell, half an hour before the main bell ringing. For me, it was a voice from Heaven. I put the slingshot aside, shooed the birds away so that they were safe from my friend, and ran home. Ever since then, when the bells of Passiontide ring out into the sunshine and the naked trees, I remember, deeply moved and grateful, how on that day they rang into my heart the commandment "Thou shalt not kill."
>
> From that day on I have dared to free myself from the fear of men, and when my innermost conviction was at stake, I have considered the opinions of others less important than before. I began to overcome my fear of being laughed at by my classmates. The way in which the commandment not to kill and torture worked on me is the great experience of my childhood and youth. Next to it, all others pale. (*Memoirs of Childhood and Youth*, 1997)

Africa and Reverence for Life

After receiving degrees in philosophy and theology and working as a professor for several years, Schweitzer came to believe that he should devote himself to a life of service to humanity and decided to study medicine. Upon receipt of his medical degree, Schweitzer sought a position with a French medical mission, and, in the spring of 1913, with his wife, Hélène, he sailed for Lambaréné, Gabon, in equatorial Africa, to establish a hospital for the African people of the area. Schweitzer was 38 years old. When he arrived in Africa after the long voyage from France, he was immediately struck by the treatment of domesticated animals in Dakar:

> I have never seen such overworked horses and mules as here. On one occasion when I came upon two natives who were perched on a cart heavily laden with wood which had

stuck in the newly mended street, and with loud shouts were belaboring their poor beast, I simply could not pass by, but compelled them to dismount and to push behind till the three of us got the cart on the move. (Free, 1982)

Lambaréné was teeming with all sorts of animals and insects, and Schweitzer quickly came to admire and respect the creatures who shared his environment. He believed that humans should heed animals and other living things, and that our neglect of them in our ethics was a major gap. His key idea of a "reverence for life" came in a flash of inspiration, as he was traveling on the Ogowe River, having thought long and hard for several days about a moral principle that would be the foundation of ethics:

Late on the third day, at the very moment when, at sunset, we were making our way through a herd of hippopotamuses, there flashed upon my mind, unforeseen and unsought, the phrase "reverence for life." The iron door had yielded. The path in the thicket had become visible. (*Out of my life and thought,* 1998)

Schweitzer defined ethics as "a compulsion to extend to all the will-to-live around me the same reverence for life that I extend to my own. The fundamental principle of morality so necessary for thought is given here. It is good to maintain life and to promote life; it is evil to destroy and to restrict life" (Joy, 1950). "We need a boundless ethics which will include the animals also" is the inscription on the Albert Schweitzer medal given by the Animal Welfare Institute since 1954.

Schweitzer believed we humans should be mindful of our own humble place alongside other animals and living things, and think deeply about the relationship we have with them. He was appalled that certain sports, such as falconry, were becoming popular again in the 1930s, simply as artificial and cruel games for the amusement of humans. He admonished that we should only cause pain and suffering to animals if it were absolutely necessary, and then in full recognition of the pain caused, as this was our moral responsibility to other forms of life.

Yet Schweitzer was not sentimental about life in the forest. He realized that hard decisions had to be made about animals. Still, he saw that it was an awesome responsibility:

From the natives I buy a young fish eagle, which they have caught on a sandbank, in order to rescue it from their cruel hands. But now I must decide whether I shall let it starve, or whether I shall kill a certain number of small fish every day in order to keep it alive. I decide upon the latter course. But every day find it rather hard to sacrifice—upon my own responsibility—one life for another. (Free, 1950)

Observations of Animals in Africa

Animals surrounded Schweitzer's hospital on the banks of the Ogowe River, where many of them came to eat and drink. A number of these animals were partially domesticated by Schweitzer and his hospital staff. These included Sizi, the cat; Thekla, the pig; Parsifal, the pelican; and Léonie and Théodore, the antelopes. Regarding Thekla:

The intelligence and close resemblance of pigs to human beings was not wasted on Dr. Schweitzer. He gave a home to a series of Red River hogs—all called Thekla, for an obscure operatic character. The first Thekla was brought to the hospital compound as a scrawny piglet. Not only did she respond in weight gain to tender loving care, she became something of a "problem child." Mischievous, she seized every opportunity to become a one-pig hurricane: romping about madly, knocking against doctors, nurses, furniture, killing chickens, even snatching food from patients and their families. Threats were made on her life. Penning up did no good. She would dig out. Reluctantly, Dr. Schweitzer sent her to the London Zoo. When he visited her there years later, he called her name and she recognized him. Clearly, the loss of Thekla left an empty space in his heart. As he had to her, he often sang succeeding young Thekla to sleep with the gentle Brahms Lullaby. (Free, 1950)

Parsifal was an orphaned pelican who stayed at the hospital. He guarded the door to Schweitzer's small room at night. Schweitzer called him "Monsieur Le Pélican," and dedicated a book to him. Josephine was a wild boar Schweitzer tried to tame and keep on the grounds of the hospital. He would not believe that she would behave like a boar and kill chickens, and he initially tried to look the other way when this happened. However, when he could not control her repeated killings, Schweitzer butchered Josephine and cured her flesh for eating.

Léonie, an antelope fawn, was brought to Schweitzer after some local hunters had unsuccessfully tried to trap its mother, who had escaped. He named it Léonie and nursed it on a bottle. Léonie and another antelope, Théodore, would accompany him on strolls to the river and lick the salt-infused sweat from his arm.

As he grew older, Schweitzer became increasingly aware of the stewardship of the animal world—even of insects—and he would admonish visitors not to use pesticides, noting that they were dangerous for human health as well. Toward the end of his life, Schweitzer showed reverence for life on a very personal level when he discontinued the eating of meat or fish. Even though he had grown up eating meat and fish, he now believed that there was no distinction among living things:

> In the past we have tried to make a distinction between animals which we acknowledge have some value and others which, having none, can be liquidated when we wish. This standard must be abandoned. Everything that lives has value simply as a living thing, as one of the manifestations of the mystery that is life. (Free, 1950)

Albert Schweitzer stated "the school will be the way" for transforming society toward more compassion—especially for animals. His interactions with and care of the animals around him are today studied in the new interdisciplinary field of anthrozoölogy, which examines the human-animal bond for benefits in mental and physical health for both humans and their companion animals.

Further Resources

Brabazon, J. (2000). *Albert Schweitzer: A biography*. Syracuse, NY: Syracuse University Press.

Fenner, J. (2000). *Schweitzer's prayer for animals*. Retrieved April 2000 from http://home.pcisys.net/~jnf/prayer.html

Free, A. C. (1982). *Animals, nature, and Albert Schweitzer*. Retrieved May 28, 2007, from http://www.awionline.org/schweitzer/as-idx.htm

Joy, C. (Ed.). (1950). *The animal world of Albert Schweitzer*. Hopewell, NJ: Ecco Press.

Schweitzer, A. (1997). *Memoirs of childhood and youth*. (K. Bergel and A. R. Bergel, Trans.). Syracuse, NY: Syracuse University Press.

———. (1998). *Out of my life and thought: An autobiography*. (A. B. Lemke, Trans.). Baltimore and London: Johns Hopkins University Press.

A. G. Rud

■ Ethics and Animal Protection
Tolstoy, Leo

The Russian writer and social activist Leo Tolstoy (1828–1910) is best known for his novels *War and Peace* and *Anna Karenina,* but these writings became frivolous and irrelevant to Tolstoy as he matured. His later works focus on morality rooted in Christian ideals and reveal Tolstoy as a writer committed to social activism. His works offer reasoned spiritual justification for nonresistance, pacifism, and vegetarianism.

In his early life Tolstoy was a hunter and soldier, but as faith gained importance in his life he found it impossible to justify such violence. In his short writing, "The First Step," Tolstoy explains the horrors he witnessed in a slaughterhouse. Witnessing how other animals met their death, he wrote that being a Christian requires personal diligence toward spiritual goals and that such goals naturally lead to abstinence from eating flesh. If an individual "seriously and sincerely seeks a good life," they will abstain from consuming flesh because eating other creatures requires us to kill, which is "contrary to our moral sense" (407–08). Near the end of his life, reflecting on his moral and spiritual choice to stop eating other animals, Tolstoy wrote that twenty years of vegetarianism "never cost [him] the slightest effort or deprivation" (letter to G. P. Degterenko).

Tolstoy's moral commitment to nonviolence was grounded in the gospels; his commitment to nonresistance stemmed from Matthew 5:38–39: "You have heard that it

Leo Tolstoy writing in his study in 1908. Courtesy of the Library of Congress.

was said, 'An eye for an eye and a tooth for a tooth.' But I say to you, Do not resist an evildoer" (NRSV). Reading this, Tolstoy observed that Christians are called to do good by others *always*. "Do not return evil for evil," Tolstoy writes in *My Religion,* "but always do good to all men,—forgive all men" (115).

Tolstoy called Christians to refuse to participate in violence, including both law enforcement and the military because these organizations are designed to cause harm to others. Tolstoy dialogued with soldiers on the streets of Moscow and with police in parks, asking how they, as Christians, could kill other men or arrest poor and homeless people for stealing only what they needed to live. In *The Kingdom of God Is Within You,* Tolstoy encouraged citizens to "refuse to take the oath of allegiance, to pay taxes, to take part in court proceedings, in military service and in duties on which the whole structure of the government is based" (239). Refusal to participate, he argued, would bring inhumane, ungodly governments to a standstill.

Gandhi read Tolstoy's works when he was in Africa, just beginning his own struggles against oppression. These two great leaders exchanged a handful of letters, and Tolstoy thereby helped shape Gandhi's philosophy. Gandhi, in turn, influenced Martin Luther King Jr. Contemporary movements of civil disobedience and nonviolent resistance—those who refuse to comply with or engage in violence against life—owe much to the Russian writer and social activist Leo Tolstoy, including social activists dedicated to animal liberation.

Further Resources

Green, Martin. (1983). *Tolstoy and Gandhi, men of peace.* New York: Basic Books.
Tolstoy, Leo. (1885). *My religion* (Also called *What I Believe*). New York: Thomas Y. Crowell.
———. (1887/2004). *What shall we do then?* (Also translated as *What To Do?* and *What, Then Must We Do?*). Amsterdam: Fredonia Books.
———. (1893/2006). *The kingdom of God is within you.* Lenox, MA: Hard Press.
———. (1904). *The first step.* http://www.animal-rights-library.com/texts_c/tolstoy01.htm or *The complete works of Count Tolstoy,* v. XIX (pp. 367–409). Boston: Colonial.

Lisa Kemmerer

■ Ethics and Animal Protection
Wolf Recovery

Ethics and Wolves

Wolves are not the only examples through which to explore the ethics of humanity's relationship to animals. They have, nonetheless, a special resonance in many human cultures—as beasts of waste and desolation, as vital ecological agents, as creatures exemplifying the best of humanity, as wild beings we can respect in all their familiarity and strangeness. Wolves move people, pro and con, and this opens up possibilities for dialogue about human-animal relations.

From an ecological perspective, wolves are an indicator of landscape health. They are indispensable "top carnivores" that promote the health of ecosystems, as well as a "flagship species" whose cache helps protect or restore other animals and plants that are not so charismatic. Yet the ability of wolves to thrive in wild and humanized landscapes may also be a cogent indicator of our own moral health. If we can learn to live with wolves—large predators require substantial habitat and human tolerance—then we will, per force, have taken significant steps toward living in a sustainable manner. If this was to occur, wolves would be both one instance of and a model for our ability to coexist with a more-than-human world.

The recovery and presence of wolves (and other predators) in a rapidly urbanizing and globalizing world raises old fears and new issues. Learning to share both natural and humanized landscapes with wolves is a difficult personal and cultural shift of perspective for some. It also entails real and unavoidable possibilities for social and political conflict. We should expect such difficulties when we try to optimize the well-being of people and wolves. To help mitigate or resolve such conflict, we need an ethics of wolf recovery. This ethics should not only help reveal the moral issues at stake, but provide guidance on how we ought to live with wolves in a shared landscape.

When speaking about ethics and wolves, one can get caught up in particular ethical theories and what they might say about wolves. This does little to advance our thinking. Ethics is not about rigid rules or dogmatic theories. It is really about the moral values that inform (or should inform) how we ought to live. With this in mind, let us look at several of the topics that have emerged in recent debates over humanity's relationship to wolves.

Ethics, Science, and Public Policy

The first topic is the relationship between ethics, science and public policy. How do we conceive of ethics and the role of ethics in science and public policy?

There are competing definitions of ethics, as well as ideas about how ethics applies to animals like wolves. At root, however, ethics is dialogue about "how we ought to live" (from Plato's *Republic,* Book II, 312d). Moral dialogue, and the insights that come with it, help people envision how we can live so as to improve the well-being of people, animals, and nature. Such dialogue also helps us critique those actions and institutions that detract from our well-being.

Ethics is especially helpful as a moral compass for public policies about wildlife and the environment. One frequently hears that such decisions should be based on science. There is nothing to argue here if this means science is an indispensable element of such decisions. But if it means that science is the only basis for making such decisions, then it is manifestly false and in need of correction. Why? Science can help us make decisions through findings of facts, by outlining possible courses of actions, and by projecting foreseeable outcomes. Yet, science cannot tell us what we ought to do—or, put another way, it cannot tell us what is the *right* thing to do. This is always an ethical decision. However obscured by the disappointments of everyday politics, ethics is at the root of our struggles over articulating, adopting, and implementing public policy. By integrating scientific and ethical information, we are in a better position to "triangulate" on the best public policy.

As we can see, then, our individual and public decisions to live with or without wolves are not best conceived of as a matter of following our own opinions, much less our economic or political interests. Nor are they about balancing the preferences of citizens, pro or con, in a setting where stakeholders bargain from their own self-interest or policy perspectives. Instead, they are ethics-laden deliberations and negotiations about whether to share the landscape with a large predator like the wolf.

Intrinsic Value

The second topic is whether wolves have intrinsic value. The idea behind intrinsic value (sometimes called *inherent value*) is that one has importance or worth in and of oneself, without reference to what one's value is to someone or something else. Do wolves have intrinsic value?

Through the course of history, people have disagreed over the intrinsic value of both people and animals. Racism, sexism, and ethnocentrism are, at their heart, attempts to diminish the intrinsic value of a person or group that is seen as different or "other" than one's own. Speciesism is a related phenomenon and has been an ongoing problem with how some societies relate to wolves. For example, the First Nations of North America by and large respected the intrinsic value of wolves and saw them as fellow residents in a common landscape. In contrast, European colonialists denied intrinsic value to any animal except human beings and saw the wolf as a pest to be eliminated. When someone says a creature is "just an animal," that is a clue as to how they assess the moral value of other beings.

To better understand this issue, we have to first appreciate why people are said to have intrinsic value. Human beings are intelligent and social creatures—we think, feel, and relate. We are aware of our surroundings, as well as our individual "selves." This is why we are termed *Homo sapiens*—the "wise earthly ones." Because of our self-awareness, we have an individual worth independent of the use anyone has for us. This belief in our own intrinsic value is the core reason why we are taught to treat people with respect and why we have developed ethical principles to guide our thought and behavior. Moreover, our well-being can be helped or harmed by others, as well as by social policies. It is no wonder, then, that love, friendship, democracy, and justice are so important. They are interpersonal and institutional ways that help us treat individuals and communities with the respect that moral beings like ourselves deserve.

Some believe that since wolves are not human beings, they cannot have intrinsic value. Instead, they have *extrinsic value* (sometimes called *instrumental value*). To have extrinsic value is to be of use to someone or something else; one's value is extrinsic or outside of oneself. The roles wolves play in ecosystems, or their economic worth (or cost) to human beings, are examples of extrinsic value. From this extrinsic point of view, controversies over human-wolf relations are addressed through a policy process that sees wolves as a "natural resource," available for "sustained harvest," and requiring "rational" wildlife management as driven by "science." This is coded language; it implies that wolves are no different than any other agricultural commodity, or that they are simply functional units of ecosystems. To think otherwise is to be muddled, emotional, and irrational.

Ethically, there is a problem with claims that wolves have only instrumental value. The wisdom of native cultures, cognitive ethology, and common sense tell us that wolves are also intelligent and social creatures. This was so obvious to some early civilizations that they modeled their societies after wolves. Wolves were amongst the first animals to socially interact with human beings and may have been the direct ancestors of the domesticated dog. Think about the extraordinary communication between handlers and their hunting, herding, and working dogs—or between a child and a puppy at play. The special emotional and social bond that has evolved between people and dogs required a species like wolves that, like us, can think, feel, and relate. And, like people, the well-being of wolves can be helped or harmed, most particularly by human actions. From this point of view, wolves are moral beings, too, creatures with their own intrinsic value independent of someone's use or antipathy for them.

Note that wolves are not the same kind of moral beings as people, and we should not try to treat people and wolves in the same way. They have no right to free speech, although that does not mean that there are not right versus wrong ways to treat them. A better way to think about wolves is that they have a mix of intrinsic and extrinsic values, what is termed *co-value*. As individual creatures, they have an inherent value in and of themselves. As members of a pack and predators in an ecological community, they also have an instrumental value. Again, this is not too different from the co-value of humans. We are moral beings who matter in and of ourselves, and we are instrumentally valued (or undervalued) by employers, for example, for whom we are primarily human resources.

What all this means is that wolves likely have some kind of intrinsic value, and their well-being ought to be taken into account as we debate, adopt, and implement public policies about wildlife and land-use. Unfortunately, we rarely give explicit voice to these ideas in public meetings, even though they are one set of background understandings that powerfully inform the debate over wolves. This voice may speak in many tongues, including that of religion, spirituality, psychology, and emotion, all of which may approach the value of wolves from somewhat different directions. The key to understanding this diversity of voices is to look for the underlying meanings and not to bicker over terminology. Overall, whether one thinks wolves are moral beings or not, we must appreciate that this belief is a core element of the wolf debate, and a legitimate reason why people care so much about wolves.

Land-Use

The third topic to have recently emerged from debates about human-wolf relations is land-use. The focal concerns here are whether and how to coexist with wolves in humanized landscapes.

As noted, ethics is a key element in the policy making process. It helps to ensure we are clear about both the facts and values which are the basis of our policy judgments. This is especially important with respect to wolf recovery. The heat of political and cultural conflict over wolves is more intense than that of any other wildlife issue. Without both ethics and science to keep our positions grounded, the debate over wolves can quickly dissolve into unproductive name-calling.

So, what kinds of creatures should be allowed to live in humanized landscapes? Where the basic needs of wolves are concerned, most experts note that wolves inhabit a range of habitats, need neither wilderness nor untouched forests to survive, and frequently live around human communities—including large cities. The ecological needs of wolves are, therefore, not a serious barrier to their flourishing. Rather, wolves will thrive with sufficient prey, habitat, and solitude, as long as we exercise the ethical and social discipline to leave them alone.

This insight becomes more powerful when we understand that our beliefs about intrinsic value influence our willingness to share the landscape with wolves. A case in point is the vilification of wolves in North America. Historically, anti-wolf sentiment took the form of a moral argument against wolves. Wolves were considered villains, varmints, and vermin; criminals preying on blameless deer, cattle, and sheep; the spawn of Satan despoiling the land. We do not call violent criminals and sex offenders "predators" for no reason; it is a carryover from a cultural bias that equates predation with victimization. As a consequence of this reasoning, humans tried to exterminate wolves with a vengeance, and, in many portions of the wolf's range, we succeeded.

Over the last century, scientists, ethicists, and advocates challenged this view of wolves. Scientists argued for the value of predation in the natural world and demystified what it means to live with wolves. Ethicists debunked the idea that wolves are evil, malicious, or immoral. They also clarified the moral reason why we should respect their well-being. Advocates mobilized broad public support for the defense of wolves as an endangered species, which is crucial to conserving biodiversity. Taken together, these efforts have interjected more reasonable ecological, social, and moral criteria into public policy. These efforts have also transformed our relationship with wolves. Where that relationship was formerly one of outright hostility, it is now more diverse, with the largest segment of the public supportive of coexisting with wolves.

With respect to how we can coexist with wolves, there are inevitable conflicts that need to be addressed. Wolves occasionally prey on livestock and companion animals.

Neither of these are substantial problems for our society or our economy. For example, the number of livestock lost to wolf depredation is miniscule when compared to the number of cattle and sheep that die of disease, accidents, weather, and other natural causes. Even so, the overall statistics can mask the substantial economic and/or emotional impact on individuals who lose companion animals or livestock.

In response to these legitimate concerns, three proactive strategies have arisen: education in living with predators, range management practices, and compensation funds for lost livestock and working dogs. The education programs teach people the importance of wolves to a healthy landscape, as well as how to avoid conflict with them in the first place. Wolves are attracted to easy meals, so bringing pets indoors at night and securing our garbage are examples. The education programs may also touch upon best practices for range management. These include predator-proof fencing, temporary pens for livestock birthing, purpose-bred guard dogs, and aversion technologies, such as motion-controlled sound devices. Compensation funds either pay individuals the fair-market replacement cost of livestock or working dogs lost to wolves or subsidize the purchase and use of the best practices mentioned above.

Note, however, that there is also a growing debate over whether a natural background rate of predation should be accepted in landscapes with wolves. National and other levels of government frequently subsidize livestock production, as well as the "wildlife services" that kill wild animals to benefit agricultural interests. Wolves are often killed in an effort to reduce their depredations (however small in number), and little attention is given to proactive measure of conflict mitigation, much less other improvements in herd management that would reduce the overall mortality of livestock. This raises the question of whether citizens should, for moral and economic reasons, be footing the bill to kill wolves and other predators, instead of both requiring and helping agricultural communities transition into more predator-friendly modes of operation.

In addition, there is an increasingly tense debate about the scale and kind of development activities that people should practice in "wolf country." The conversion of wild into agricultural or urban landscapes has been an ongoing feature of human settlements for thousands of years. The scale and impact of this development has picked up rapidly in the last one hundred years, due not only to the explosive growth in human population, but the intensification of resource use. We now refer to this as the problem of *sprawl*. Sprawl is of two sorts—urban and economic. It involves the expansive growth of low-density urban landscapes (think strip malls and track housing) and the development of formerly wild or inaccessible areas for agriculture, forestry, and extractive industries. Sprawl destroys agricultural and wild habitat around cities and towns, fragments wild land into smaller and smaller blocs of less species-rich habitat, and brings people into closer proximity to wild animals (such as wolves). Controlling sprawl requires comprehensive land-use planning at local, regional, and national scales.

Wolf Management

The final topic relating to ethics and wolves is about the practice of wildlife management. This involves questions of how one monitors and intervenes in the lives of wolves, whether for scientific research or for the administration of wildlife policies.

In any discussion of predator management, you are likely to hear quite a bit about "sound science." Sound science is supposed to be the evidentiary, theory-rich baseline for managing wildlife and making public policy. Yet, when science is substituted for ethics, our moral compass fails, and we are likely to be led astray. Wolf management provides a particularly powerful example of the moral controversies that can arise from a seemingly technical subject.

The techniques used to study and manage wolves are frequently intensive and intrusive. Wolves are radio-collared, monitored, tranquilized, assessed, captured, incarcerated, and killed on a regular basis. We still have much to learn about wolves, and there are undoubtedly legitimate scientific reasons to study them using such techniques. Managing wolves in this way may also be required to meet certain goals of wolf recovery. It is, for instance, a necessity in the red wolf recovery program, where monitoring and managing wolf pairings helps prevent hybridization with coyotes. Even so, the use of these techniques is not a sustainable model for long-term recovery. They are expensive propositions in terms of time and labor, and a burden on underfunded and understaffed organizations—as well as an annoyance to individuals and communities. As noted before, with sufficient food and space, wolves will flourish. Over time, they will establish their own population levels and distribution in dynamic relationship to the habitat and other resources they need for survival.

There is another, more insidious reason for conducting intensive wolf management: namely, to appease vested human interests that oppose our coexistence with wolves. This kind of management is not undertaken for the benefit of science, much less for the well-being of wolves. Although sometimes justified as maintaining the "social carrying capacity" of wolves, intensive management in this context involves killing or removing wolves with little attention to other proactive measures for mitigating human-wolf conflicts. This approach is also behind the artificially low population goals in some wolf management plans, the designation of certain wolf populations as expendable, and land-use planning that effectively creates wolf-free zones. Wolf recovery and conservation may be the stated goals. The reality of this type of management is quite different;

Missing from the landscape for more than thirty years, the howl of the Mexican gray wolf can once again be heard in the mountains of the southwestern United States. Courtesy of Shutterstock.

it amounts to an institutionalized system of species cleansing that tries to exclude wolves from the vast majority of the landscape.

Vested interests that distort wolf management are ethically problematic in their own right. Equally disturbing is employing lethal and other blunt-force techniques with little apparent concern for the well-being of individual wolves or their packs. For wolves, the social disruption of intrusive management can be severe. Pups without parents starve or are preyed upon. The loss of adult members that teach younger wolves how to survive in the wild, as well as around humans, can lead to heightened mortality and further conflict with people. Wolf packs that are exterminated are replaced by new packs, which may be even less familiar than their predecessor with how to avoid the danger of particular humans on the landscape. What we have here is the makings of a vicious cycle that, from an ethical point of view, we should try to break.

A growing number of voices are objecting to wolves being isolated in a gulag of isolated habitats, surrounded by exclusion and free-fire zones, and subjected to routine and invasive management. From an ethical perspective, such questions raise the possibility that our current wolf management practices detract from the overall well-being of wolves. For all these reasons, debate over this relatively recent topic of wolf recovery is likely to grow increasingly intense in future years.

See also

Conservation and Environment—*Conflicts between Humans and Wildlife*
Conservation and Environment—*Coyotes, Humans, and Coexistence*
Conservation and Environment—*Wolf and Human Conflicts: A Long, Bad History*
Living with Animals—*Wolf Emotions Observed*

Further Resources

Bekoff, M. (2000). *Strolling with our kin*. New York: Lantern Books.

———. (2006). *Animal passions and beastly virtues: Reflections on redecorating nature*. Philadelphia: Temple University Press.

Coleman, J. (2004). *Vicious: Wolves and men in America*. New Haven: Yale University Press.

Evans, N. (1998). *The loop: A novel*. New York: Dell.

Flader, S. L. (1994). *Thinking like a mountain: Aldo Leopold and the evolution of an ecological attitude toward deer, wolves, and forests*. Madison: University of Wisconsin Press.

Hall, R. L., & Sharp, H. S. (Eds.). (1978). *Wolf and man: Evolution in parallel*. New York: Academic Press.

Jickling, B., & Paquet, P. C. (2005). Wolf stories: Reflections on science, ethics and epistemology. *Environmental Ethics, 27*, 115–35.

Lopez, B. H. (1978). *Of wolves and men*. New York: Scribner's.

Lynn, W. S. (1998). Contested moralities: Animals and moral value in the Dear/Symanski debate. *Ethics, Place and Environment, 1*, 223–42.

———. (2002). Canis lupus cosmopolis: Wolves in a cosmopolitan worldview. *Worldviews, 6*(3), 300–27.

Mech, D., & Boitani, L. (Eds.) (2003). *Wolves: Behavior, ecology, conservation*. Chicago: University of Chicago Press.

Naess, A. (1974). Self-realization in mixed communities of humans, bears, sheep, and wolves. *Inquiry, 22*, 231–41.

Nie, M. A. (2003). *Beyond wolves: The politics of wolf recovery and management*. Minneapolis: University of Minnesota Press.

O'Neil, J. M., & Egan, J. (1992). Men's and women's gender role journeys: Metaphor for healing, transition, and transformation. In B. R. Wainrib (Ed.), *Gender issues across the life cycle* (pp. 107–123). New York: Springer.

Robinson, M. (2005). *Predatory bureaucracy: The extermination of wolves and the transformation of the West*. Boulder: University Press of Colorado.

Sharpe, V. A., et al. (Eds.). (2001). *Wolves and human communities: Biology, politics, and ethics*. Washington, DC: Island Press.

Underwood, P. (1994). Who speaks for wolf: A Native American learning story, *Focus* 4(1), 45–51.

William S. Lynn

■ Euthanasia
Animal Euthanasia

The major differences between veterinary medicine and human medicine are not biological, but ethical and economic; in no way is that more evident than in decisions and policies regarding euthanasia.

"Euthanasia" comes from two Greek words: *eu* (good, well) and *thanatos* (death). Euthanasia is a central concern in human-animal relations, as several million animals are euthanized by people each year in animal shelters, veterinary clinics, and research laboratories. That number reaches the billions when food animals—to whom the word *slaughter* is more often applied than *euthanasia*—are added in.

The definition of euthanasia differs in important ways in veterinary medicine and human medicine. In human medicine, the term is restricted to "mercy killing"—killing a patient where death is a welcome relief from a life that has become too painful or no longer worth living. Not all forms of killing humans deserve the "good death" label of euthanasia: capital punishment, for example, no matter how painlessly performed, is not euthanasia.

Human euthanasia is controversial for many reasons. Critics of legalized human euthanasia—and its close relative, assisted suicide—fear that seriously ill or old people could be coerced into having their lives ended. In that case, the death would not be an act of mercy for the person being killed, but one of convenience or economics for the survivors.

Veterinarians are familiar with the euthanasia ideal of mercy for the suffering patient, as well as with the call to end animals' lives for such reasons as convenience and economics. Veterinarians often euthanize patients with serious or incurable diseases, in cases where death really does seem the animal's best option. However, veterinarians may also be called upon to end the lives of animals who are destructive in the home, are inconvenient or aggressive or simply unwanted; shelter workers are similarly required to end the lives of healthy but unwanted animals. In the middle, between the mercy of incurably suffering animals and the destruction of unwanted animals, are those animals who are suffering but not from incurable conditions; these animals, too, may be put to death if their human decision-makers cannot or will not devote the time and money to their health needs.

Because the motives for killing animals are so broad, the definition of the word *euthanasia* in veterinary medicine is similarly broad. It comprises not just mercy killing of incurably suffering animals, but the killing of healthy animals for owner convenience, for reasons of overpopulation, for behavior problems, or as donors of tissues in research laboratories. What makes euthanasia a "good death," when speaking of animals, is not that it is better than continued life, but that the death is caused without pain or distress to the animal. It is method, not motive, that has traditionally defined animal euthanasia.

How Animals Are Euthanized

Human euthanasia comprises both "active euthanasia" (actions such as drug over-doses that kill patients) and "passive euthanasia" (the withdrawal or withholding of treatments that could sustain life). In veterinary medicine, withholding or withdrawing treatment is not typically referred to as euthanasia. Many veterinarians are distressed when owners choose to let a suffering pet die slowly of disease when fast, painless, active, medical euthanasia is an available option. Thus, "passive euthanasia" is not part of the veterinary ideal of euthanasia.

Not all methods of killing animals can be considered "euthanasia," a truly good death. The American Veterinary Medical Association first published guidelines for animal euthanasia in 1963 and has updated them five times (most recently in 2001). Primary criteria for the evaluation of euthanasia techniques are the physical pain and psycholog-ical distress experienced by the animal. Other criteria include: the emotional effect on humans who are present; the availability of appropriate drugs; and the compatibility with the subsequent examination or use of the animal's body and tissues. The veterinary guidelines only cover methods of euthanasia, not issues of why, when, or whether spe-cific animals should be euthanized. They offer no real guidance for veterinarians on how to advise clients on whether to euthanize an animal.

The preferred method for euthanizing individual dogs or cats has not changed in the forty years in which the AVMA has published its guidance. Then, as now, a rapid injec-tion of an anesthetic overdose (usually a barbiturate, such as sodium pentobarbital) is chosen, because it induces unconsciousness rapidly and painlessly; only once the animal is peacefully anesthetized does the drug go on to stop the breathing and the heart. Some-times, the veterinarian recommends an oral or injected tranquilizer several minutes before the anesthetic overdose, making the process even calmer for animal and human. Often, human caregivers choose to be present during the euthanasia of their companion animal and are relieved to see how a suffering animal can leave the world so peacefully.

The AVMA's guidelines have been updated so many times not because euthanasia of a loved, ill animal has changed, but because other circumstances are more challenging. What is the least painful way to euthanize very large animals such as zoo elephants or stranded whales? How best to euthanize dangerous wildlife? How to process the dozens of animals a busy animal shelter euthanizes every day? How to euthanize laboratory rats and mice in a way that minimizes pain and distress while leaving their tissues suitable for study? The AVMA gathers a panel of veterinarians and scientists to review the avail-able science and update recommendations for how to humanely kill these varied animals.

Making the Euthanasia Decision

One of the hardest decisions for an animal's human companion is when and whether to euthanize an ill or aging animal. How does one tell when it is the right time? This author believes there is no such thing as *the* right time, given the range of factors at play.

The euthanasia decision is only partly a medical decision, but it should certainly be made with a veterinarian's input. The veterinarian can do his or her best to provide a medical diagnosis of the animal's condition. But even a diagnosis of an incurable illness does not mean immediate euthanasia is warranted. A combination of good medical and nursing care may keep animals with certain terminal illnesses comfortable for months or years. Conversely, diagnosis of some treatable injuries and illnesses may still result in the animal's euthanasia. This may be because of the cost of the treatments (since insurance coverage for payment of veterinary bills is not common), the time demands of some treatments, or the significant suffering that an animal would likely go through before starting to feel better.

Veterinarians can help animal guardians predict what the animal will experience with a particular illness. Not all heart diseases, for instance, are equal. Some heart disease may result in sudden death, some in decreased exercise tolerance, some in a distressful inability to breathe comfortably. From the animal's perspective, these are three very different heart conditions. Sudden death is sad, but the animal does not suffer in the months leading up to it. Decreased exercise tolerance means the animal will run and play less, but may be content to limit his or her activities without significant suffering. The inability to breathe comfortably, however, may be severely distressing for weeks or months on end. A veterinarian can help understand not just whether the condition is treatable, but how much suffering it causes.

As with heart disease, so with other life-limiting illnesses. Some cancers may be excruciatingly painful, while others are barely noticed until very advanced. Kidney disease can make animals feel extremely ill, but with dietary management and supplemental fluids, they may remain in relatively good health for several months.

There will be medical uncertainty. Veterinarians can tell a person how the average case progresses, but not how an individual patient will. People want to know, "Is this animal suffering?" Like human patients, animals have better and worse days. Veterinarians can help a person learn how to recognize the major signs of an animal's quality of life: interest in food, ability to eat and drink, and ability to move about. None are particularly mysterious, but they require careful observation.

Rarely, however, is euthanasia solely a medical decision, which is why the decision rests with the animal's caregiver, not the veterinarian. The caregiver must decide how much time, energy, and money she or he can devote to end-of-life care for an animal. But even given infinite resources, she must assess when she considers the animal's life somehow no longer worth living. This includes value assessments of how many bad days—and how many good—will tip the balance toward euthanasia. Moreover, a person's beliefs about the value of life—and the possibility of an afterlife for an animal—will affect the course chosen. One person may feel that a half an hour a day of apparent comfort and happiness means that the animal's life is still worth living. Another may believe that half an hour a day of serious sickness or pain makes that life intolerable. Most will believe somewhere in the middle.

Is there an animal equivalent of assisted suicide? It is impossible to know for sure what the animals are thinking, but it is clear that animals sometimes feel far too sick to eat or drink on their own and that this can lead to their deaths. Most veterinarians will treat this anorexia as a clinical problem that can be managed and treated, just as fever, infection, and broken bones are treated; most do not treat this as the animal's attempt to end his or her own life.

Grieving

Pet guardians often grieve the euthanasia of a loved animal just as we grieve the death of our loved human friends and family. Social workers and therapists are increasingly recognizing this important response to animal death. They work to help people come to terms with this loss, rather than trivializing or ridiculing it. Some books on the topic are listed at the end of this article. In addition, following the lead of the University of California Davis veterinary college, various "pet-loss support hotlines" have arisen, most of them associated with veterinary colleges.

Support during grief for the loss of an animal is important, as many people may find their friends and family do not really understand. For many people, the love and companionship of their animal is a central part of their life, and the loss is devastating. This can be true for adults as well as children, but it may be ridiculed as immature or inappropriate by people who are less animal-focused.

Grief over the euthanasia of a companion animal is complicated by the pet owner's knowledge that she or he made the conscious decision to end the animal's life. This decision is rarely easy, and many people will guiltily second-guess their decision in the following days and months. When is a pet's suffering so severe that death is a welcome relief? What limits can loving pet owners put on the time, money, and effort they will devote to keeping their ailing pet comfortable? When do we give up hope for a cure or for improvement? Not only must the decision-maker come to terms with the fact that she made a decision that may later feel wrong, but she must also decide how to discuss this with others, possibly including small children.

Loving pet guardians are not the only people who may feel grief and distress in connection with animal euthanasia. There are also professionals for whom killing animals is part of a day's work: veterinarians, veterinary technicians, research workers, and animal shelter workers. All participate in animal euthanasia, some as part of the decision-making, others powerless to make the decisions but required to perform the euthanasia procedure. Thus, euthanasia training for shelter workers includes not just technical training, but also seminars on dealing with the tragic irony that responsible animal care sometimes includes killing animals.

Further Resources

American Veterinary Medical Association. (2005). *How do I know it is time? Pet euthanasia* (brochure). American Veterinary Medical Association: Schaumburg, IL. Available online at http://www.avma.org/communications/brochures/euthanasia/pet/pet_euth_brochure.pdf

Beaver, B. V., Reed, W., Leary, S., McKiernan, B., Bain, F., Schultz, R., Bennett, B. T., Pascoe, P., Shull, E., Cork, L. C., Francis-Floyd, R., Amass, K. D., Johnson, R., Schmidt, R. H., Underwood, W., Thornton, G. W., & Kohn, B. (2001). 2000 report of the AVMA panel on euthanasia. *Journal of the American Veterinary Medical Association, 218*(5), 669–96.

Carmack, B. J., (2003). *Grieving the death of a pet.* Minneapolis, MN: Augsburg Books.

Kay, W. J., Cohen, S. P., Fudin, C. E., Kutscher, A. H., Nieburg, H. A., Grey, R. E., & Osman, M. M. (Eds.). (1988). *Euthanasia of the companion animal.* Philadelphia: The Charles Press.

Larry Carbone

■ Evolution
Charles Darwin and the Theory of Evolution

On the Origin of Species by Means of Natural Selection begins in the wonderful prose of nineteenth-century science writing tradition:

> When on board the H.M.S. Beagle, as a naturalist, I was much struck with certain facts in the distribution of the organic beings inhabiting South America, and in the geological relations of the present to the past inhabitants of that continent. These facts . . . seemed to throw some light on the origin of species—that mystery of mysteries. (Darwin 1859, p. 1)

There were those before Darwin who talked of both evolution and natural selection, but it was only with the publication of *The Origin,* however, that a *complete theory* of evolution by natural selection was presented—a theory that was backed up by the enormous amount of evidence laid out by Darwin.

What is most surprising about *The Origin* is the subject material of the first chapter. The opening chapter of Darwin's masterpiece talks at length about pigeon breeding in Victorian England. The reason for this seemingly odd subject matter was that Darwin was using familiarity as a tool to brace the reader for what was to follow. He realized that his readers would feel comfortable with a discussion of pigeon breeding, a very popular pastime in Victorian days. Darwin gambled that if he could convince readers that the process that led to the extraordinary *variants* produced by pigeon breeding was similar to the process that led to variation in nature, his task would be a little simpler. The process leading to pigeon variation is *artificial selection,* while the process leading to the wide variety of traits we see in nature is *natural selection.*

Artificial selection is the process whereby humans systematically breed certain varieties of an organism over others. For thousands of years humans have been shaping animals and plants by this process. Ever since our ancestors *selected* some varieties of wheat, corn, and rice over others, and systematically planted their seeds, we have been involved in artificial selection. The same can be said of our systematic preference for the breeding of certain varieties of dogs over others, as well as hundreds of other examples.

Following Darwin's lead, suppose that we, like Darwin's friends and colleagues in Victorian days, want to produce a variety of pigeon with snow-white plumage. Assuming that plumage coloration is under genetic control—as it is in other species of birds—we begin the process of artificial selection by allowing only those individuals in our population with the whitest plumage to breed. We continue this process generation after generation, sorting the birds based on plumage coloration and allowing the whitest—those that are closest to the variant we want to produce—to breed. As time goes on, we would be producing individuals that were coming closer and closer to snow-white feathered birds, and eventually, we'd recognize that we were as close as we were ever going to get to our idealized snow-white pigeon.

For Darwin, natural selection was a process very similar to that of artificial selection. He noted that

> variations, however slight and from whatever cause proceeding, if they be in any degree profitable to the individuals of a species, in their infinitely complex relations to other organic beings and to their physical conditions, will tend to the preservation of such individuals. . . . The offspring, also will have a better chance of surviving, for of the many individuals of any species which are periodically born, but a small number can survive. I have called this principle, by which each slight variation, if useful, is preserved, by the term natural selection, in order to mark its power to man's power of selection. (1859, p. 55)

The two salient differences between artificial and natural selection lay in (1) the selective agent and (2) what is selected. With respect to the selective agent, under artificial selection humans choose the traits that they wish to modify (usually in a specific direction). In natural selection, one can think of "nature" as the selective agent. With respect to what traits are selected, Darwin noted, "Man can only act on external and visible characters, but Nature . . . cares nothing for appearances, except in so far as they are useful to any being. She can act on every internal organ, on every shade of constitutional differences, on the whole machinery of life" (p. 72).Of course, with modern molecular evolution, we can modify more than outward appearances, but Darwin's description remained accurate for the better part of 100 years after he rendered it.

Darwin's process of natural selection was simple, yet extremely powerful in a world in which all organisms are in a constant struggle for existence. He hypothesized that evolution by natural selection was a slow, gradual process. Darwin argued that there could be little doubt that selection acted on small differences between individuals, and that "Many large groups of facts are intelligible only on the principle that species have evolved by very small steps" (p. 225). As a case in point, Darwin asks his reader to imagine the wolf "which preys on various animals, securing some by craft, some by strength, and some by fleetness" (p. 78).When prey items are scarce, natural selection acts with brute force on such wolf populations. Wolves that possess any traits that even slightly improve their hunting success survive longer and produce more offspring—offspring, in turn, who possess the very traits that benefited their parents in the first place. Repeat the process *generation after generation,* and eventually you get wolves that are very efficient hunters. "Slow though the process of selection may be," noted Darwin, eventually you end up with a wolf better adapted for hunting.

The same sort of argument can be made for endless other cases. Darwin, for example, discussed plants that rely on insects for cross-fertilization. Given that insects often eat most of the plant's pollen, could plant traits that foster insect pollination ever be favored? Darwin believed the answer to this question might very well be "yes": "[A]s pollen is formed for the sole purpose of fertilization, its destruction appears to be a simple loss to the plant; yet if a little pollen were carried at first occasionally and then habitually, by pollen-devouring insects from flower to flower, and a cross thus effected, although nine-tenths of the pollen were destroyed, it might still be a great gain to the plant thus robbed; and the individuals which produced more and more pollen, and had larger anthers, would be selected" (p. 81).

Darwin recognized that natural selection not only operated on morphology (as in the wolf case mentioned), but on *behavior*. If certain behavioral variants had strong, positive effects on longevity and reproductive output, selection would favor those behavioral variants over others. Darwin illustrated this nicely using the egg-laying habits of the cuckoo, a bird notorious for depositing its eggs in the nests of other species. How could such a bizarre behavioral trait evolve? What's in it for the cuckoo?

For Darwin, the potential benefits for the behavior of parasitic egg-laying abounded. Following Darwin, imagine that at the start of this evolutionary process *some* cuckoos occasionally laid *some* of their eggs in the nests of another species. Darwin believed that parasitic egg-layers might profit "by this occasional habit through being enabled to migrate earlier . . . or if the young were made more vigorous by . . . the mistaken instinct of another species than reared by their own mother." With such benefits available, if young cuckoos inherited their mother's tendencies to lay eggs in the nests of others, as Darwin thought them "apt to," then "the strange instinct of the cuckoo has been generated" (p. 190). As with the wolf case, *if* one variant of a trait (slim, sleek wolf morphology or parasitic egg-laying habits) is superior to other variants, and *if* some means exists by which traits are passed from parent to offspring, *then* natural selection will favor *behaviors* that create a better-adapted organism.

Darwin summarized his ideas on evolution and the process of natural selection in what he called "two great laws": (1) unity of type and (2) conditions of existence. By unity of type, Darwin was referring to "that fundamental agreement in structure which we see in organic beings in the same class, and which is quite independent of their habits of life" (p. 186). Unity of type was explained by common descent. Organisms in species that share a common ancestor tend to resemble one another in many respects for the very reason that they share common ancestry.

While there has been much debate in the literature over the extent to which Darwin saw natural selection as *the* force driving evolutionary change, his own writings quite clearly demonstrate the enormous power he attributes to natural selection. In addition to ending his introductory chapter of *The Origin* by claiming, "I am convinced that natural selection has been the most important, but not the exclusive, means of modification" (p. 5), Darwin lays out his position in even more detail for the reader in a later passage:

> It may be metaphorically said that natural selection is acting daily and hourly scrutinizing, throughout the world, the slightest variation; rejecting those that are bad, preserving and adding up all that are good; silently and insensibly working, whenever and wherever opportunity offers, at the improvement of each organic being in relation to its organic and inorganic conditions of life. We see nothing of these slow changes in progress, until the hand of time has marked the lapse of the ages. (p. 73)

This eloquent summary of natural selection is so powerful that it is worth ending by analyzing it, phrase by phrase. Much has been made of the fact that Darwin spoke "metaphorically," as if this weakens his case. But how else could he speak? The process he was describing was essentially new, and so he borrowed language to make the ideas comprehensible. And for Darwin there was no escape from selection, as it is "acting daily and hourly scrutinizing, throughout the world." Selection works twenty-four hours a day, every day, forever and everywhere. Only a force of such magnitude could have shaped all the life that we see around us, and for that matter, all life over the last four billion years. Darwin knew full well that he was laying out a revolutionary new idea, and he wanted no doubts about how powerful a force natural selection truly was.

Natural selection does not wait for large differences to emerge before it kicks into action. Any differences in reproductive success, even "the slightest variation," will be acted on by selection. As such, natural selection acts as an editor with respect to an organism's traits, "rejecting those that are bad, preserving and adding up all that are good," and it does so "silently" and by "insensibly working." As such, natural selection acts to make organisms "better," only in the sense of molding an organism that is better and better adapted to the environment in which it lives (i.e., "at the improve-

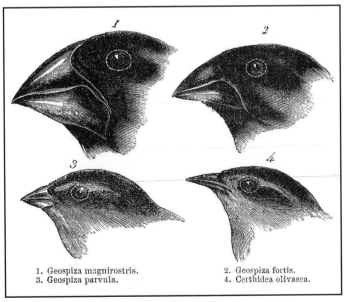

1. Geospiza magnirostris.
2. Geospiza fortis.
3. Geospiza parvula.
4. Certhidea olivasea.

One aspect of Darwin's theory of evolution came from his observation of four species of finches on the Galapagos Islands, showing variation of beak. ©HIP/Art Resource, NY.

ment of each organic being in relation to its organic and inorganic conditions of life"). Lastly, Darwin warned his reader that "We see nothing of these slow changes in progress, until the hand of time has marked the lapse of the ages." Selection often works slowly, but the cumulative effect is extraordinarily powerful—powerful enough, in fact, to shape all the life we see around us.

Further Resources

Darwin, C. (1859). *On the Origin of Species by Means of Natural Selection.* London: John Murray.

Lee Alan Dugatkin

■ Evolution
Nineteenth-Century Naturalists and the Theory of Evolution

In 1859 Charles Darwin published *On the Origin of Species by Means of Natural Selection,* a text with a profound influence on the modern understanding of life. Darwin's book appeared during the booming years of nineteenth-century natural history study. Two primary factors in the development of this field were the scientific revolution of western Europe's enlightenment period and the expansion of western European powers into lands across the globe. The scientific revolution paved the way for the empirical inquiry into the nature of other living organisms while expanded imperial exploration brought the vast richness of the natural world to the attention of European scientists and naturalists. This led to a drive to document flora and fauna across the globe for economic and scientific purposes.

Interest in the exotic brought on by exploration resulted in popular desire for collections of specimens, referred to as "cabinets of wonder," peaking just prior to the nineteenth century. These collections, along with more scientifically motivated efforts, resulted in a haphazard consolidation of animal specimens in Europe, some alive, most, however, dead. Out of this disorganized specimen collecting came the field of comparative anatomy, the root of natural history. Carl Linnaeus and Georges-Louis Leclerc, Comte de Buffon, initiated the first systematic attempt to characterize life on earth through the examination and organization of specimens in these collections—thus establishing a field that characterized animals largely outside of their natural environment. These first systematic classifications set the stage for early nineteenth-century naturalists, including Jean Baptiste Pierre Antoine de Monet de Lamarck and Georges Cuvier, and later Alfred Russel Wallace, Charles Darwin, and Jean-Henri Fabre, to compare the distributions of and relationships between organisms in natural settings, and led to the pioneering work of embryologists such as Ernst Haeckel and illustrators such as John James Audubon.

Both Lamarck and Cuvier sought to establish a universal system by which animals and plants could be studied, characterized, and classified. These extensive taxonomic documentations of life were underpinned by a desire to understand the overarching laws dictating nature's order. Cuvier's study of the internal structures of organisms led to his theory that animals fell into four groups. This idea conflicted with the concept of the great chain of being, the classical idea that all organisms were created on a single continuum from simple to complex, with humans at the pinnacle of the earthly realm. Indeed, many of Cuvier's classifications have held up under modern taxonomic scrutiny. Cuvier maintained that species were fixed rather than mutable, arguing, as did William Paley in *Natural Theology,* that organic structures were too complicated to arise by piecemeal change.

Lamarck agreed with Cuvier that a divinity was involved in the pattern of life but suggested that species were mutable. Other naturalists, such as Geoffrey Saint-Hilaire, proposed that life evolved, but Lamarck was the first to develop a comprehensive theory for how it might happen .Lamarck's theory dictated that species both adapted to their environment through inheritance of acquired characteristics and progressed from less to more organized forms with humans at the pinnacle of organization, as in the great chain of being. The idea of progression articulated a distinction between humans and all other organisms; a distinction Cuvier also believed, although he claimed qualitative rather than progressive differences between humans and nonhuman animals.

In response to the work of early comparative anatomists, many natural historians developed an interest in traveling to remote regions. Generally, their goal was to document flora and fauna in situ and bring back specimens to augment collections. The availability of training and funding offered by the newly bourgeoning natural history museums of Europe and North America helped shift the focus of these voyages from a search for exploitable resources and objects sold for collections to a focus on scientific inquiry. These naturalists, including Darwin, documented organisms in their natural environment and often collaborated with comparative anatomists and other naturalists to identify and characterize specimens. Individuals such as Alexander von Humboldt, who wrote about ecology in the Americas, and Henry Walter Bates, who focused on South America, contributed to the growing science of natural history in the field. Bates's colleague, Alfred Russel Wallace, had a particular impact on the theory of evolution by natural selection.

Although Darwin conceived of the idea of natural selection two decades before Wallace, he and Wallace introduced it with simultaneously published papers presented to the Linnean Society in London in 1858. Wallace's ideas about natural selection focused on change in environment rather than competition among individuals, and therefore, his idea of natural selection was of a much weaker force than was Darwin's. Wallace developed his ideas during exploratory trips to the Amazon and the Malay archipelago, where he documented biogeographical and ecological patterns while reading the same texts that influenced Darwin, Charles Lyell's *Principles of Geology* and Reverend Thomas Malthus's "Essay on the Principle of Population."

Although Wallace and Darwin shared many ideas about the pattern of life on earth, their opinions differed in several areas, especially in the realm of human evolution. Wallace felt that natural selection could not explain the traits of modern humans, and he published a paper titled "The Limits of Natural Selection Applied to Man." This disagreement appears to have stemmed from Wallace's unwillingness to accept similarity between humans and nonhuman animals in cognitive and behavior traits. The implied boundary between humans and nonhuman animals was a stumbling block for many individuals to accept the influence of natural selection. In opposition to those who refused to see humans as on the same behavioral plane as nonhuman animals was Thomas Henry Huxley, an invertebrate morphologist who helped induce the scientific community to accept Darwin's theory. Huxley viewed humans as well as nonhuman animals mechanistically, and he supported Darwin's nontheological explanation of evolutionary change.

The absence of a theological framework limited acceptance of the theory. Those naturalists who presupposed a theological explanation for life and did not accept evolutionary change over time, such as Louis Agassiz, clearly rejected any role for natural selection, but even many that did accept evolutionary change, such as Saint George Mivart and Robert Owen, rejected the theory because of the lack of divine intercession.

Agassiz was one of the most influential North American naturalists of the nineteenth century. He combined a variety of approaches in his search for a divine plan under Cuvier's rubric of four basic animal forms. He recognized that combining traits, such as embryological and anatomical characters, might help in classifying organisms. Noteworthy counterparts of Agassiz, Owen, and Mivart accepted the potential for evolution but rejected Darwin's hypothesis because of the absence of a creator. Owen proposed the hypothesis of a basic blueprint for all vertebrates, and Darwin used his work to develop his theory. Despite this, Owen dismissed the theory because it lacked any theological structure. Similarly, Saint George Mivart believed that in nature was the divine plan, and he too refused to accept natural selection as an agent of evolution. In 1871 Mivart published a collection of arguments against Darwin's theory titled *Genesis of Species*.

Embryological work complemented the methods of field naturalists and comparative anatomists such as Mivart and Wallace, and even augmented it in the case of Agassiz. Two embryologists in particular, Karl Ernst von Baer and Ernst Haeckel impacted the study of natural history. Von Baer, like Owen, emphasized a vertebrate archetype, suggested by the similar forms of many vertebrates during the early stages of development, using Cuvier's proposal of four basic categories of animals. Haeckel followed von Baer in using an embryological approach to study animal taxa, melding the experimental practices of the contemporary physiologists with the theoretical practices of the comparative anatomists and natural historians. He accepted the influence of natural selection on evolution but espoused progressive evolution; an idea at odds with the nonlinear approach proposed by Darwin. Haeckel suggested that ontogeny (development) recapitulates phylogeny (evolutionary pattern over time) and replaced the suggestion by von Baer and others that ontogeny recapitulated progression from lower to higher organisms.

While the study of natural history developed through the approaches of individuals such as Wallace, Owens, and Haeckel, documentation of species developed through the approaches of individuals such as Charles Lucien Bonaparte, Albert Gunther, John Gould, and John James Audubon. Bonaparte studied the birds and Gunther the fish found in the growing collection of specimens in museums across the world; both produced guides useful to naturalists such as Wallace. Though the catalogues did not include much information about the ecology or behavior of the species, and thereby, to some extent, reinforced the objectification of nonhuman animals, they acted as repositories of information useful to natural historians from the nineteenth century onward.

In contrast, the illustrators John Gould and especially John James Audubon attempted to document species in their natural settings by observing focal species in the field and subsequently collecting them for illustrations. Gould, with much help from his wife Elizabeth, produced an extensive series of illustrations of birds in Australia and Great Britain as well as illustrations of mammals. He also influenced the theory of evolution directly by aiding Darwin in identifying the species collected during his journey on the *Beagle*. Darwin was also influenced by Audubon. He attended at least one of Audubon's demonstrations and included references to Audubon's work in his publications. Audubon was probably the most skilled illustrator of birds of the nineteenth century, documenting birds from across the globe including, most especially, North America. He emphasized the importance of observing animals in their habitat and included anecdotes with his illustrations. He also attempted to compose his specimen-based illustrations using behaviors he observed in live individuals. Gould and Audubon reflected differences in the perception of animals during this time. Gould's animals tended to be peaceful nurturing creatures; Audubon's often were aggressive creatures in the midst of hunting or fighting.

While many naturalists of the time, including Audubon, Owens, and Wallace, documented a broad range of species, some naturalists were interested in the individual behavior and ecology of specific species or groups. These individuals, such as Jean-Henri Fabre and Mary Davis Treat, tended to focus on local organisms, about which they compiled detailed observations as well as conducted small experiments, and their work influenced Darwin's theory. The study of behavior at this time, despite the work of skilled naturalists such as Treat and Fabre, was primarily in the form of unsystematic anecdotal descriptions. The attitudes at the time tended to be either overly sentimental or overly exploitative; both viewpoints reflecting an increased distinction between urban and natural environments. George C. Romanes was perhaps the first naturalist, apart from Darwin, to systematically accumulate information in an attempt to demonstrate the continuity of behavior and cognition between humans and nonhuman animals. Romanes's publications reflect much of the attitudes and approaches of individuals who embraced the continuity and evolution of cognition. Unlike Darwin, he tended not to weed out unlikely anecdotes and relied upon his own impressions of emotion that suggested a progression of intellect with humans at the pinnacle.

This tendency toward sentimental objectification in which the animals mirrored human cognition and emotion, and its contrasting tendency, the exploitative objectification expressed in attitudes towards exploration, reflect the social environment within which naturalists of the nineteenth century functioned. All of the naturalists of this time, from the comparative anatomists to the field naturalists, from the embryologists to the illustrators, were formed within and helped form the fundamental shift in the manner in which Westerners viewed nonhuman animals. Darwin's theory of evolution by natural selection had a profound impact by demonstrating the manner in which all life might be related and subject to the same forces. This idea interacted with the increasing industrialization and urbanization of humans that resulted in a more objectified relationship between human and nonhuman animal. It is at this intriguing place in time where the empirical study of animals blossomed amidst an environment of rapidly expanding exploitation and in some cases, decimation, of the very species of interest, and where the way animals are studied today was established.

Further Resources

Bowler, P. (2003). *Evolution: The history of an idea* (3rd ed.). Berkeley: University of California Press:

Darwin, C. (1859). *On the Origin of Species by Means of Natural Selection*. London: John Murray.

Farber, P. L. (2000). *Finding order in nature: The naturalist tradition from Linnaeus to E. O. Wilson*. Baltimore, MD: Johns Hopkins University Press.

University of California Museum of Paleontology. (2006).*The History of Evolutionary Thought: 1800s*. Berkeley, CA. Available at http://evolution.berkeley.edu/evolibrary/article/0_0_0/history_index_02

Jennifer Calkins

■|Extinction of Animals.
See Conservation and Environment

■|Feelings.
See Bonding; Sentience and Cognition

■ Film.
See Media and Film

■ Fishing.
See Hunting, Fishing, and Trapping

■ Geography
Animal Geographies

Geography is uniquely concerned with the relationship between human culture and the natural environment. Animal geography is an important subdiscipline of geography that focuses on human-animal relationships and considers nonhuman animals significant agents in the constitution of space and place. Although always interested in the natural world, geographers had largely overlooked animals or considered them as nondistinct elements of a larger natural world until the emergence of contemporary animal geographies in the 1990s. Before the growing interest in human-animal relations, animals typically appeared in geography as mere biological pieces of a larger ecological system, as instruments for human use, or as forms of symbolic capital. However, the interplay between geography and social theory, cultural studies and environmental ethics in the 1990s led to a rebirth of interest in nonhuman animals, and animal geography flourished as a result.

Jennifer Wolch and Jody Emel brought human-animal interactions to the foreground in the1990s with a thematic issue of *Society and Space* (1995) and the edited book *Animal Geographies* (1998). Chris Philo and Chris Wilbert followed soon after with a collection of essays titled *Animal Spaces, Beastly Places* (2000). Reviving and remaking the face of animal geography, these scholars today recognize animals not as mere biological parts or symbols but as foundational to how we think about ourselves and the world and as intentional beings with inner lives who are worthy of consideration in their own right.

Contemporary animal geographies are about the interplay between animals, culture and society, and the exploration of a broad range of human-animal concerns, including habitat loss and species endangerment, domestication, animal entertainment and display, wildlife restoration, global trade in animal bodies, and many others. Animal geographies are fundamentally geographical because they are essentially about nonhuman animals and their *place* in human society—place meaning both material borders (societal practices that shape the spaces where some animals are welcomed and others are not) and conceptual boundaries that call up questions of human identity and animal subjectivity.

There are three basic themes in contemporary animal geography. The first concerns animals and the making of place (including material boundaries).The second theme involves questions of human identity and animal subjectivity (issues of conceptual boundaries). Finally, animal geographies examine the role of ethics and how we *ought* to treat animals. These organizational themes are not independent of one another, and they frequently overlap and dovetail with concepts such as animal instrumentalism, anthropocentrism, and the human-animal divide. Animal geographers recognize the fluidity of boundaries, emphasizing not only the distinctions but also the connections, overlaps, and similitudes between human and animal worlds.

Material Boundaries: Animals and the Making of Place

Discussions in human geography about the social construction of landscapes have led to the exploration of how animals and their networks leave their imprint on places, regions, and landscapes over time. For example, Kay Anderson considered the human-ordered placement of wild captive animals at zoos. Suggesting that zoos are a means of crafting a human experience of animals, she found that zoo animals have changed the ways humans think of animals—as either worthy of exclusion or inclusion in humans' lives (with carnivorous animals such as lions and bears tending to be at the exclusionary extreme). In addition to tangible places such as zoos, farms, experimental laboratories, and wildlife reserves, animal geographers also consider economic, social, and political spaces such as the worldwide trade of captive wild animals.

Other work in animal geography considers places characterized by the presence or absence of animals and how human-animal interactions create distinctive landscapes. For example, animal geographers have evaluated the impacts of individual and regional land-use practices on wildlife survival in rainforest areas. Furthermore, human-wildlife interactions—which have inevitably increased since the late twentieth century with increasing numbers of people moving into areas inhabited by cougars, coyotes, wolves, bears, and other animals—provide animal geographers with opportunities to consider how those interactions may lead to changes in public attitudes. For instance, predators, which were once admired as symbols of wilderness, have come to be seen as cold-blooded killers in many places. Similarly, animal geographers have considered the reintroduction of predator animals to areas (now) populated by humans, showing how boundary-making policy can also lead to conflict over the proper *place* for animals such as wolves and mountain lions in a human-dominated landscape.

Conceptual Boundaries: Human Identity and Animal Subjectivity

Many scholars contend that the distinctions between nature and society, and between human and animal, are so profoundly ingrained in Western culture that the conceptual separation between them has gone largely unquestioned. More than separating human from animal, animal geographers highlight that humans are often privileged *over* animals, just as society is privileged over nature, the rational over emotional, and the West over the non-West. Some suggest that anthropocentrism (human-centered thinking) is so prevalent that the stark inequality of privilege between humans and animals seems perfectly natural. Those on the bottom of the vertical ranking, because they are deemed inferior, are consequently subjugated and oppressed.

Animal geographer Jody Emel considered how landscapes of the nineteenth-century American West were transformed by a contest for space and dominance over land and investment. She found that wolf killing was not, as some argued, a simple material border dispute in the shape of an economic need to protect livestock. Instead, she compared wolf killing to the ways that Native Americans had been treated, suggesting that wolves were killed to secure land and investment. They were also nearly exterminated for their pelts and as trophies, for sustaining the land for big game animals so that human hunters could kill them, and for the sake of scientific data. Wolf hunting continues today, and Emel suggests that what undercuts the matter is based on *conceptual* boundary issues. Drawing on the work of cultural ecofeminists, she contends that cultural phenomena and economic factors interact with each other in complex ways. And while many of the issues related to wolf eradication have to do with class and economy, they also stem from a dominant construction of masculinity that is based on mastery and control through the hunt.

Animal geography explores questions related to the relationships between human and animal identities and how and why the demarcation between humans and animals shifts over time and space. Calling for a theoretically inclusive approach to thinking about humans and animals, animal geographers today explicitly recognize that both humans and nonhumans are embedded in natural and social relations and networks with others upon whom their social welfare depends. Essentially, animal geographers argue that dualistic, human-centered thinking has led to instrumentalism, exclusion, and exploitation of the nonhuman world. Such views suggest a reconceptualization of what has been called the *human-animal divide*—naturalized thinking that casts humans as vastly different from (and superior to) animals.

It is the hierarchical nature of dualistic thinking—the vertical ranking of humans over animals—that has helped to promote what is frequently seen today as human distance from and control over the sphere of nonhuman animals as the *other*, while minimizing nonhuman animal claims to space and to elements of agency, reason, and ethical consideration. Consider again wolf eradication efforts. Jody Emel explained how, in the human process of constructing the landscape, the wolf came to represent traits that were contrary to the image of masculinity and humanity on the American frontier (morality, progress, and civilization). Instead, wolves represented savagery, lack of mercy, and unfair habits of pack hunting and cowardice, and such representations were what ultimately devastated wolf populations. Not only that, such thinking and practices were compared to the racist and sadistic treatment of people falling below European-American males on the hierarchy of beings.

Ethics, Humans, and Other Animals

Much of the animal geography literature is critical about and concerned (whether explicitly or not) with the normative task of advancing the well-being of animals. Animal geographers who are interested in animal well-being argue that the concept of a human-animal divide should be remade; instead of an oppositional dualism of humans on the one side and animals on the other, animal geographers envisage a complex network of interdependences between humans and animals. Animal geographers also encourage thinking about animal agency and subjectivity, emphasizing that a great many animals have intentions and are communicative subjects with potential viewpoints, desires, and projects of their own.

Jennifer Wolch and other animal geographers criticize the theoretical and methodological impulses that have shaped geography in the past, particularly those that privilege cognition and language as the markers of an exclusively *human* geography while ignoring our ethical kinship with other animals. Instead, many animal geographers think that nonhuman animals are equally important subjects of human geography and conceptualize them as "strange persons," or as marginalized, socially excluded people. Indeed, recent findings on animal thinking, culture, and politics from comparative psychology, primatology, and cognitive ethology have provided extraordinary insight into nonhuman animal consciousness and capacities for complex thought and behavior in many animals. But because animals cannot, as a practical matter, directly challenge human policy decisions (which take place in a thoroughly human venue), animal geographers recognize that they require human representatives to speak and act in their interests. For better or worse, humans are regulators in today's political world, largely deciding whether animals are in (or out of) place.

To advance the plainly normative project of animal geography, many have worked to unveil societal values so that they can better understand how animals have shaped particular *moral landscapes*. David Matless analyzed Broadland's moral geography as a set of ideas about how human life should be lived in relation to given environments. For example, he compared a perspective based on a hunting approach (with all the violence hunting entails) with other, more preservationist approaches. Matless found that such conflicting cultures of human-animal relations resolutely shaped local society-animal relations and ideas about the proper place of animals in that space. Similarly, James Proctor argued that the spotted owl conflict in the Pacific Northwest was part of a larger debate over the moral landscape as revealed in the relationship between people and the forest. On the one hand, environmentalists argued that old growth forests and wildlife predated (and existed without) people; as such, people had an ethical obligation to protect them. On the other hand, pro-timber advocates contended that logging was the best way to manage and sustain the forest (emphasizing that human welfare depended upon such management). Proctor suggested that the fate of the spotted owl became so hotly disputed because it called into play the two conflicting visions of the Pacific Northwest's moral landscape.

On the whole, in their efforts to advance the well-being of both humans and animals, animal geographers argue for the inclusion of animals in the moral community. The practical consequences of such inclusion are considerable, made clear by the spotted owl case: How are we to decide who or what is most important in environmental policy-making, for example? And who, exactly, gets to decide? Especially when human-animal needs clash in a world of finite space, a framework of normative principles suggested by animal geographies—principles inclusive of animal interests and desires—can guide human-animal relations and resolve the moral dilemmas that relate to conflicting wants and needs of both humans and animals.

Merging the spheres of ethics and geography, some animal geographers have developed such frameworks for moral understanding. For example, Bill Lynn's *practical ethics* approach calls for a recognition of the ethical questions that are present in all human and animal geographies, appreciating that moral value is the foundational concept for remapping how humans understand and relate to the animal world. Granting moral inclusion and subjectivity to nonhuman animals is where animal geography largely departs from the theoretical positioning found in contemporary nature/culture debates in geography, which remain largely anthropocentric and ethically agnostic. In sum, animal geography maps the processes involving human-animal relations while remaining concerned about the well-being of humans *and* the other animals who are part of those processes.

See also

Ethics and Animal Protection—*Practical Ethics and Human-Animal Relationships*
Urban Animals

Further Resources

Anderson, K. (1995). Culture and nature at the Adelaide Zoo: At the frontiers of 'human' geography. *Transactions of the Institute of British Geographers, 20*(3), 275–94.

Emel, J. (1998). Are you man enough, big and bad enough? Wolf eradication in the U.S. In J. Wolch & J. Emel (Eds.), *Animal geographies: Place, politics, and identity in the nature-culture borderlands* (pp. 91–118). London: Verso.

Lynn, W. S. (1998). Contested moralities: Animals and moral value in the Dear/Symanski debate. *Ethics, Place and Environment, 1*(2), 223–42.

———. (2002). *Canis lupus cosmopolis:* Wolves in a cosmopolitan worldview. *Worldviews, 6*(3), 300–27.

Philo, C., & Wilbert, C. (2000). *Animal spaces, beastly places.* London: Routledge.

Whatmore, S. (1999). *Hybrid geographies: Natures cultures spaces.* London: Sage.

Wolch, J., & Emel, J. (1995). Bringing the animals back in. *Environment and Planning D—Society and Space, 13*(6), 632–36.

———. (Eds.). (1998). *Animal geographies: Place, politics, and identity in the nature-culture borderlands.* London: Verso.

Kristin L. Stewart

■ Health
Animal Bite Prevention

A videotape program called *Animal Safety Is Fun* is in use in all fifty states and seven countries around the world. It is used by libraries, civic groups, veterinarians, animal health technicians, humane societies, and animal control officers to help improve the human-animal bond. It serves to help children be more safe and confident around animals. The program teaches them to respect the animal's territory and how to avoid antagonizing dogs who may be predisposed to biting.

Overview

American Pet Food Institute surveys report that we live with approximately 60 million dogs (in 50% of our homes) and 75 million cats (in 30% of our homes). The dog population is growing at about 1.5 percent per year and the cat population at about 2 percent per year.

Numerous sources of animal bite statistics are available. However, most authors agree that only half to one-quarter of animal bites are reported to the appropriate authorities because many people who have been bitten do not seek medical attention. Hence, the numbers are likely higher than presented here.

On average, there are 2–3 million dog bites reported annually in the United States. Dog bites represent 75 to 80 percent of all types of animal bites. The latest Humane Society of the United States (HSUS) survey reports 4.7 million dog bites in 1995, with 2.8 million of those bites occurring with children. It is estimated that 30 to 50 percent of bites are provoked by the person bitten, usually over territorial issues. Other reports state that there are 300 to 700 bites per 100,000 people annually. And others claim that 60 percent (1.8 million) of all dog bites (3 million per year) are in children. In national surveys, 47 percent of school-age children report having been bitten at least once. Of those children bitten, 55 percent were boys, and 39 percent were girls. Dog bite injuries are the number-one childhood public health problem reported because they make up more than the total incidence of measles, mumps, and whooping cough.

Children are overrepresented as a special group within dog bite statistics. Five percent of children five to nine years old have been bitten by a dog. That represents more than 30 percent of dog bites in less than 9 percent of the population. Newer studies claim that dogs that the child knows may inflict many dog bites. Dog bites represent 5 percent of all emergency room admissions, and 585,000 dog bite wounds require medical care each year. There are fifteen to twenty dog bite–related fatalities annually in the United States.

Recent studies within the insurance industry report that more than one billion dollars annually is expended for claims against homeowners' policies as a result of dog bite injuries. Some insurance companies either refuse to insure a household that has a dangerous dog breed residing within that home or add on an extra fee for insuring a household with such an animal present.

In 1995, 2,851 letter carriers (about eight per day) were bitten while on their routes. The Humane Society of the United States and the United States Postal Service have declared the second week of June as National Dog Bite Prevention Week. The hope is to draw attention to the problem to encourage preventive measures.

Bites

Anatomic bite-location statistics may provide valuable insight as to where, why, and how dogs bite people. In order of frequency, dogs bite people most often on the right arm, both arms, hands, and legs. Sixty-five percent of all face bites are in children younger than ten years old. Dogs tend to bite bodies and body parts that move frequently and suddenly. It is instinctive for dogs to chase moving objects. Unfortunately, it is instinctive for children to move or run and scream in fear when alarmed by dogs, a combination that results in a potentially volatile situation.

Prevention

With this information, veterinarians and animal support groups have discussed what can be done to minimize or prevent dog bites. Preventive measures fall into two general areas. The first is via animal intervention—"the biter." The second is via human interaction—"the bitee."

"The Biter"

Traditional methods of animal bite prevention include animal control, licensure, leash laws, obedience and socialization classes, and veterinary behavior consultation and treatment. Spaying and neutering decreases the likelihood of dog bites. A spayed or neutered dog is one-third as likely to bite as the unsterilized dog.

The breed of choice to incriminate in dog bites changes with the popularity of breeds. Some municipalities and government agencies discriminate against certain breeds in their laws and ordinances. Others respond legally to the individual dog's behavior (e.g., third reported bite results in specific legal action against the pet's owners). Breeds change, but situations do not. The dog's behavior is usually the result of the owner's shortcomings in the previously described areas. In summary, neuter, train, and socialize!

"The Bitee"

Human intervention to minimize dog bites centers around education. Children and adults should be taught not to provoke animals to bite. They should leave stray dogs alone and be taught to recognize and respect a dog's territory. As a veterinarian, my personal focus is training children to avoid negative interaction with dogs.

Animal Safety Is Fun is a videotape program that teaches children safety around animals. The simple guidelines for children and adults are as follows:

1. Leave stray dogs alone.
2. Report stray dogs' presence to the nearest adult so that animal control authorities can be notified.

3. If approached by a dog, "be a tree." "Being a tree" is a neutral, nonthreatening posture that minimizes a dog's antagonism. It is performed by standing with the feet together and the fists folded under the neck with arms and elbows against the chest. This is a good posture to assume when approached by any dog that could be a potential biter or aggressor. Also, there is no harm in doing this when the dog is friendly. Because eye contact may incite aggression, look straight ahead, avoid eye contact with the dog, and speak in a soothing voice, saying something nice. Children are taught to say, "nice dog," "nice doggie," or "good boy." Wait for the dog to walk away. He will usually get bored and leave. Never run!

4. "Act like a log" if you are already lying down when approached by a potentially dangerous dog or if you are knocked down by a dog. This position is performed by lying face down with the fists folded behind the neck and the forearms covering the ears. The feet and legs are together. It, too, is a nonthreatening, neutral position. Most dogs will get bored and walk away. Children and adults should be advised not to be afraid if the dog sniffs around their body because that can be a part of normal canine socialization and familiarization behavior.

The dog bite problem is largely preventable. The best policy is to educate the public concerning responsible pet ownership and to train children, at an early age, to avoid negative interactions with dogs. Of course, no single program can prevent all dog bites, but this one serves, as do others, to increase public awareness of dog bites and to minimize the severity and number of dog bites that occur.

See also

Health—*Dog Bites and Attack*

Further Resources

Animal Sheltering. (1996, January–February). [Available from the Humane Society of the United States, 2100 L Street, NW, Washington, DC 20037.]

August, J. R. (1988). Dog and cat bites. *Journal of the American Veterinary Medicine Association, 193,* 1394–98.

Beck, A. M., & Jones, B. A. (1985). Unreported dog bites in children. *Public Health Report, 100,* 315–21.

Beck, A. M., & Katcher, A. (1983). *Between pets and people.* New York: Putnam.

Cromwell, M. (Author/Distributor), & Riley, E. (Producer/Videographer). (1996). *Animal Safety Is Fun* [Videotape]. Available from ER Productions, Columbus, OH. http://www.animalsafetyisfun.com.

Gershman, K. A., Sacks, J. J., & Wright, J. C. (1994, June). Which dogs bite? A case controlled study of risk factors. *Pediatrics, 93,* 913–17.

Goldstein, E. J., & Richwald, C. (1987, July). Human and animal bite wounds. *American Family Physician, 36,* 101–9.

Postmaster's Dog Bite Prevention Week Kit. (1996). [Available from Dept. D, 2100 L Street, NW, Washington, DC 20037-1525.]

Sacks, J. J., Lockwood, R., Homreich, J., & Sattin, R.W. (1996, June). Fatal dog attacks, 1989–1994. *Pediatrics, 97,* 891–95.

Veterinary Product News, May–June 1996, 12.

Wishon, P. M., & Huang, A. (1989, May–June). Pet-associated injuries: The trouble with children's best friends. *Children Today, 18,* 24–27.

J. Michael Cornwell

■ Health
Animal Reservoirs of Human Disease

Throughout history, humankind has periodically experienced devastating outbreaks of epidemic disease brought about by the convergence of disease pools. When two previously isolated groups of people, each with its own spectrum of endemic diseases to which the population has developed some resistance, establish contact, one or both of them often fall prey to a lethal plague.

Transmission of disease from animals to humans presents a similar threat. Wild and domestic animals suffer from a variety of viruses, diseases caused by microorganisms, and parasites. In stable natural populations, pathogens and their animal hosts coevolve to the point where the disease is enzootic: always present, but rarely lethal or debilitating to otherwise vigorous individuals. Sometimes, human activity results in the convergence of animal disease pools, as in the introduction of anthrax, hoof and mouth disease, and psittacosis (thought to be a major factor in the extinction of the Passenger Pigeon and Carolina Parakeet) into North America via infected domestic cattle and fowl. The resultant epidemics among wildlife mirror those that decimate indigenous peoples.

Three of the most devastating pandemics in human history—bubonic plague in the fourteenth century, the 1918 influenza epidemic, and the current global AIDS epidemic—can be traced to animal pathogens that crossed the species barrier and established a chain of human-to-human transmission. Literally hundreds of other serious diseases periodically jump from animals to humans, and more are being discovered every year. In the last decade, local outbreaks of yellow fever, bubonic plague, avian influenza, West Nile virus, poliomyelitis, SARS, dengue fever, Ebola hemorrhagic fever, and Marburg virus have demonstrated that animal-to-human transmission of lethal diseases is a major concern in an era when political instability combined with rapid global travel make quarantine measures extremely problematical. The United States Centers for Disease Control estimate that 75 percent of emergent diseases—diseases either newly recognized in humans or increasing in geographic distribution and virulence—are of animal origin.

Certain common indigenous practices, often dismissed as superstition, probably minimize exposure to animal pathogens. Except in emergencies, people rarely hunt carnivorous mammals for food or eat the flesh of animals that have died of natural causes. Traditional cultures avoid the flesh of old or obviously diseased domestic animals. Carrion-eaters are universally shunned. Although modern methods of food preparation have reduced some of the risks associated with eating meat, they have introduced others. Bovine spongiform encephalopathy (mad cow disease) is thought to have arisen when cattle, normally a vegetarian species, were fed the rendered remains of diseased sheep. Mechanical processing of beef allows intestinal bacteria to contaminate ground beef, creating a new route for transmission of enteric diseases.

The probability that an animal pathogen will spread to humans depends on a number of factors, including genetic similarity of host, proximity, frequency of interaction, the sizes of the relative populations, and the general disease susceptibility of the human population. Primates, our closest relatives, harbor the largest number of potential human pathogens, and rodents, especially rats and mice, pose a danger because they are so numerous and ubiquitous. Bats, often overlooked, are now recognized as an important agent of rabies transmission. Diseases spread by migratory birds, notably West Nile virus, equine encephalitis, and avian influenza, cannot be excluded through quarantine.

Human susceptibility is a crucial factor in establishing an epidemic. Although often lethal once transmitted, animal pathogens are rarely highly infective to humans, and have not adapted to spread from one person to another. The fourteenth-century bubonic plague epidemic and the current AIDS epidemic illustrate how human susceptibility transforms a sporadic local threat into a global catastrophe.

Plague originated as a disease of marmots in arid areas of Asia, with limited opportunities for transmission to indigenous nomads. As urban centers grew up along trade routes, the possibility of amplification in a dense population of humans and rats increased, as did the possibility for transmission to remote areas. Plague traveled overland to the Crimea, and thence by ship to Italy in the form of infected humans and rats. Europeans, who had no biological resistance and no cultural methods for combating the disease, were already debilitated by famine. Famine spawned political and social instability, undermining efforts at quarantine. Finally, the disease mutated into a form capable of direct human-to-human transmission. In the words of a contemporary, "a third of the world died."

Plague never became established in Europe. In the nineteenth century, plague emerged from the old reservoir in Asia, caused massive mortality in China and India, and was carried around the globe by infected rats on steamships. There are now permanent loci of infection in North and South America and sub-Saharan Africa. A breakdown in public health in any of these places could result in a pandemic.

Human immunodeficiency virus occurs naturally in several species of nonhuman primates in Africa and probably spread to humans when hunters butchered infected carcasses. Until recently, limited mobility and cultural prohibitions against promiscuity prevented amplification in human populations, and individuals with compromised immune systems quickly succumbed to other illnesses. As these factors diminished, the incidence of HIV increased exponentially, reaching 30 percent in some African countries. The high incidence of HIV in sub-Saharan Africa is of particular concern because the area is also the greatest source of emergent diseases. The immunocompromised individual is readily infected with a novel pathogen through casual contact, and his body, once infected, becomes a fertile breeding ground for human-adapted strains.

Diseases of domestic animals that can spread to humans include anthrax, brucellosis, and toxin-producing strains of *Escherichia coli*. Looking at the historical record of measles epidemics and the high degree of similarity between measles and rinderpest viruses, historical epidemiologists have concluded that measles evolved from rinderpest when humans first domesticated cattle, the original host.

Smallpox, monkeypox, and cowpox are caused by three closely related herpoviruses, all restricted to the Eastern Hemisphere before 1500. In their successful efforts to eliminate smallpox, an exclusively human disease, humans introduced cowpox among domestic cattle and thence into wild populations of hoofed animals in South America. Tropical Africa is experiencing a resurgence of monkeypox now that vaccination is no longer routine there. Nearly asymptomatic in its rodent reservoir hosts, monkeypox is a serious illness in primates. It was recently accidentally introduced into North America via African rodents imported for the pet trade.

Transporting wild-caught animals always creates a risk of disease transmission, especially if the animal will be in close contact with humans or other animal species at its final destination. Quarantines fail to detect asymptomatic infections and those with long incubation periods.

Importing animals for the pet trade or for exhibition is risky, but the most serious threat appears to come from the use of wild-caught primates in biomedical research. People engaged in such research are the first targets. Marburg virus, a lethal hemorrhagic fever, first appeared in technicians working in a German facility extracting human pharmaceuticals from African monkey organs. There have been subsequent outbreaks in

Africa. Philippine macaques at a primate facility in Maryland succumbed to a strain of Ebola virus that was apparently infective but nonvirulent to humans. Using rhesus monkeys to develop a polio vaccine resulted in contamination of the vaccine with simian virus 40, which is highly carcinogenic to rodents. This virus is now endemic in North America and may be implicated in human cancers. Xenotransplantation—transplanting animal organs into human bodies—carries an appalling risk because it places living cells containing an unknown load of asymptomatic viruses into a body whose immune system is deliberately suppressed. Blood transfusions and organ transplants between humans are notorious sources of amplification of blood-borne diseases of animal origin. The biomedical route is essentially the only one by which an exclusively blood-borne animal pathogen can be transmitted to a remote human population.

As settled human communities encroach upon natural ecosystems, more humans are exposed to animal pathogens. Human disruption of the natural environment encourages population explosions among rodents, triggering epizootics in the rodent population. Anthropogenic population cycles in native wild rodents have made Hanta virus an emerging public health threat in the American Southwest. In the Northeast, suburban expansion into woodlands, coupled with management practices encouraging high densities of deer mice and white-tailed deer, resulted in frequent human infection with Lyme disease, a serious tick-borne rickettsial disease enzootic to deer mice. The developed world has made great progress in eliminating water-borne diseases resulting from fecal contamination by humans and domestic animals. Recently, giardiasis, caused by a protozoan enzootic in aquatic rodents, has emerged as a disease of hikers and people in rural areas who drink untreated water.

Arboviruses (arthropod-borne viruses) include a number of widespread dangerous pathogens, notably dengue and yellow fever. Elementary sanitation has nearly eliminated cycles of tick-, louse-, and flea-transmission among humans. Mosquitoes are another matter. Once transmission of the yellow fever virus from its wild primate reservoir has occurred, most urban areas in warm climates support enough mosquitoes to sustain a human epidemic. Fortunately, an effective vaccine exists, enabling health workers to break the human cycle before it has grown exponentially and spread to other areas.

Occasionally, it is feasible to combat a disease in the animal population. Anthrax was eliminated from domestic hoofed animals in western Europe and North America through vaccination, and the mandatory inoculation of domestic dogs against rabies has greatly reduced the incidence of human exposure in areas where this is practiced. In Europe, use of bait containing an oral vaccine has achieved some success in controlling rabies in wild foxes. Such direct human intervention only works if the wild animal reservoir population is small and discrete.

Although modern man possesses vastly more scientific knowledge about the risks and mechanisms of animal-to-human disease transmission than did his ancestors, he seems to have grown cavalier about them, ignoring the medical consequences of rampant development in this and other areas and trusting that science will prove capable of curing whatever ill emerges. The AIDS epidemic demonstrates that the world's human population is probably more vulnerable to animal pathogens today than at any other time in history.

See also

Conservation and Environment—*Conservation Medicine Links Human and Animal Health with the Environment*

Health—*Disease between Animals and Humans (Zoonotic Disease)*

Further Resources

Committee on Occupational Health and Safety. (2003). *Occupational health and safety in the care and use of nonhuman primates.* Washington, DC: National Academies Press.

Gousmit, J. (2004). *Viral fitness. The next SARS and West Nile in the making.* Oxford and New York: Oxford University Press.

Hurst, C. J. (2000). *Viral ecology.* San Diego and London: Academic Press.

Krause, R. M. (1998). *Emerging infections.* Biomedical Research Reports. San Diego and London: Academic Press.

Logan, M. H., & Hunt, E. E., Jr. (Eds.). (1978). *Health and the human condition: Perspectives on medical anthropology.* North Sciutate, MA: Duxbury Press.

Marano, N., & Pappanioanon, M. (2004, December). Historical, new, and re-emerging links between human and animal health [Special issue]. *Emerging Infectious Diseases.* [Available online at http://cdc.gov/ncidod/EID]

Pavlovskii, E. N. (Ed.). (1960). *Natural foci of human infections.* Washington, DC: U.S. Department of Commerce, Office of Technical Services.

Wolfe, N. D., Daszak, P., Marm Kilpatrick, A., & Burke, D. S. (2005, December). Bushmeat hunting, deforestation, and prediction of zoonotic diseases. *Emerging Infectious Diseases.* [Available online at http://cdc.gov/ncidod/EID]

Martha Sherwood

■ Health
Benefits of Animal Contact

Society's recent commitment to companion animals stems not only from our general sensitivity to the welfare of animals, but also from a new appreciation of the role animals play in our lives. In the early 1980s, the term "the human-animal bond" came into use to capture that role; the term was inspired by the respected association found between parents and their offspring—the "mother-infant bond."

For there to be a "bond," the effect on each partner has to be significant and mutual. The association between people and their companion animals is indeed significant and mutual. The human-animal bond involves complex psychological and physiological interactions between people and their pets that have profound influence on human and animal behavior and health.

The American Pet Products Manufacturers Association found that in 2004, more than 63 percent of U.S. households contained companion pet animals, and nearly 45 percent had more than one. More than 39 percent of households had dogs, 34 percent cats, 14 percent fish, 6 percent birds, 4 percent horses, and more than 9 percent had other small mammals or reptiles. The sheer numbers of pet animals is only one facet of the "pet experience." Contact with animals is not rare nor is it random. In the United States, the people who interact with pets tend to be younger than the general population; that is, dogs, cats, and small mammals are far more common in families that have children. For instance, whereas about 20 percent of households without pets include children under eighteen year of age, more than 38 percent of pet-owning households have children; 68 percent of households that have small pet animals also have children. Gail Melson of Purdue University found that animals are very important in child development. All children, urban and rural, are very attached to pets, and the attachment is not related to the child's gender.

Companion animals play a variety of roles for people at different ages, but pets allow people of all ages to be alone without being lonely. Nevertheless, in the past, most studies documented the detrimental effects of animal contact, addressing infectious diseases, diseases from animals (zoonoses), parasites from animals, and traumatic injury from bites and kicks. At the same time, our culture has maintained a commitment to protecting the animals kept for companionship, and there is now much scholarship on the importance of animal companionship for both people and animals.

Animals and Human Health

In 1980, the first epidemiological report documenting the value of pet ownership was published. Ericka Friedmann and Aaron Katcher studied people hospitalized after a heart attack and found that 94 percent of those who happened to own pets were alive after the first year, in comparison with only 72 percent of those who did not own any animal. The ownership of any animal correlated with improved survival, not just dog ownership. Following that study, there was a period of research documenting a variety of effects that animals had on humans. Many books and studies documented that companion animals are important to human health.

The mechanisms by which contact with companion animals influences psychological and physical health may not be surprising and indeed may be predictable.

Edward O. Wilson formulated the "biophilia hypothesis," which postulated that during human evolutionary development, there was natural selection for people who had an improved ability to hunt animals and locate plants good for food. Thus, we humans evolved with the inborn instincts (we were "hard wired") to pay attention to animals and the living aspects of the environment. Because we are a social species, and animals are important to us, we find comfort in their association with us.

A more specific suggestion for the health benefits of animal contact comes from "social support" theory, again because we are social animal species ourselves. Social support is observed from the positive benefits of marriage, having a close friend, belonging to a religious group, or even receiving a telephone call from a family member. Much of the research on human-animal interaction was modeled after what was known about the positive health effects humans had on other humans. James Lynch showed that people who are denied good human contact do not do well. It appears that human companionship affects people in the following beneficial ways: decreases loneliness, stimulates talk, encourages touch and care, focuses attention, stimulates exercise, encourages laughter, and encourages social contact. It is apparent that companion animals create similar effects.

Decreasing loneliness

One way people can be protected from the difficulties of loneliness is animal companionship. Nearly thirty years ago, Roger Mugford conducted one of the first pet intervention studies. Older adults (people sixty-five years of age and older) living alone were given either a plant or a small bird (parakeet); television ownership was also considered as a variable and occurred in all groups. Having a bird appeared to improve morale and was associated with having more visitors; the birds served as a "social lubricant." Over the years, more sophisticated epidemiological studies with older adults found animals to improve psychological well-being. Judith Siegel found that elderly pet owners, especially those with dogs, experienced improved morale and went to their physicians less than those without animals. An Australian national survey also found that dog and cat owners made fewer annual visits to physicians and were less likely to be on heart medications than non-owners. Now, many nursing homes have programs that include animal contact and encourage volunteers and other visitors to bring animals.

Stimulating Talk

Nearly all pet owners talk to their pets, and whereas nearly half of adults confide in their pet, more than 70 percent of adolescents do so. Although people interact with different species of pets in ways related to the specific animal, most people find comfort talking to their animals. Unlike talking to other humans, people experience a decrease of blood pressure talking to pets, indicating that they are more relaxed than when talking to people. The decreased physiological arousal indicated by the reduction of blood pressure is associated with typical changes in facial expression and vocal pattern; the face becomes more relaxed with a decrease in muscle tension, especially around the eyes. When talking to pets, people usually talk more slowly and more softly with a slight increase in pitch, much like trying to soothe an infant. Talking to pets is very common and rewarding to both the person and animal.

Encouraging Touch and Care

Children learn important values and attitudes from animals. For instance, Gail Melson noted that by preschool, children can appropriately appreciate the differences between dogs, cats, puppies, and kittens; they know adult animals are caregivers, not babies. Melson notes that young people learn nurturing from even their transient roles as pet caretakers. And this is especially important for young boys who do not have games or activities that encourage nurturing except for pet keeping. Although boys usually increase their knowledge about animals with age, there is usually a decrease in interest and care for human infants, while their dedication to animal care remains. This may be because pet care is not associated with gender, as with human infants, which is viewed as a "mommy" thing. Boys, in particular, may be introduced to the importance of nurturing with the aid of their pets.

Touch is important for all people in all cultures. Anthropologist Erving Goffman observed that in the United States, women usually use more intimate human touch than men do in public places. In contrast, adult men, at least in public places, exhibit the same signs of intimacy and the same frequency of touch with pets as do women. At all ages, caring for animals is a focus of nurturing and a source of comfort.

Focusing Attention

Under laboratory conditions, people of all ages experienced significant decreases in blood pressure when in the presence of animals, when petting their dog, or even just when watching aquarium fish. People who observed an aquarium were able to undergo dental surgery very much like those who were hypnotized prior to the procedure—that is, the dentist could not distinguish patients who had watched a fish tank prior to coming into the dental chair from those who were hypnotized and told to relax ahead of time. This is why fish tanks are becoming more common in dental waiting areas.

One of the major problems facing those who care for people with Alzheimer's disease is the weight loss often experienced by the patient. The tendency of Alzheimer's patients to wander and their lack of attention mean that they eat less food at meals. Special fish tanks appear to hold Alzheimer's patients' attention and have significantly helped in maintaining patient weight. Nursing homes with these fish tanks additionally found that they could decrease the use of nutritional supplements, which are required when the patients do not complete their meals. The fact that observing fish still holds the attention of even those with advanced Alzheimer's disease demonstrates the innate nature of our bond with the living world. That bond, the biophilia hypothesis, is so much part of our life that it survives dementia. At home, many fish-tank hobbyists report that they watch and care for the animals when they are in need of relaxation.

Stimulating Exercise

Many people report that they use their pets as an excuse to walk, jog, or engage in physical activity. Psychiatrist Michael McCulloch studied patients who had depression after suffering serious physical disabilities; those who had pets all reported that the animals stimulated useful physical activities, which otherwise would not have been part of their lives. Being active is very important in managing depression. Physical activity is important for all people, and many studies have shown that people with dogs walk more often than those without dogs. A dog companion even makes the walk more fun.

Encouraging Laughter

The writer Norman Cousins suffered from a painful inflammatory disease, ankylosing spondylitis. In his 1979 book *Anatomy of an Illness as Perceived by the Patient: Reflections on Healing and Regeneration,* he described the role of laughter in diminishing his pain, in even reducing the inflammatory process that afflicted him. Indeed, laughter is a recognized medical intervention, and animals are a frequent source of humor. In McCulloch's study of psychiatric outpatients, animals helped all the patients laugh and maintain a sense of humor. Animals often permit people to laugh at themselves or their surroundings. Look, for example, at the role animals play in cartoons for people of all ages. Animals in humor, like animals in children's books, have no human gender, ethnicity, or age, and thus, they do not hurt anyone's feelings but can just be enjoyed. We can enjoy animal behavior without any distraction.

Encouraging Social Contact

When people are trying to understand and explain what is happening in a social setting, they tend to view behavior as a particularly significant factor, more than external factors such as the setting or context. This is what psychologists call the "fundamental attribution error" because the attributions to the temperament of the person observed may or may not be accurate. Perhaps because of our own dedication to animals, we attribute positive attributes to people observed in normal settings with animals. Because animals improve how people view other people and even places, animals often are used when politicians are seeking approval or advertisers want to interest people in a product for sale.

Experimental observations of normal and handicapped subjects in public situations suggest that the presence of pet animals improves the social attractiveness of human subjects. People perceive people in the company of animals more positively, with more favorable attributes, than people who are without animals. Randal Lockwood showed this to be true using line drawings of people with and without animals present, and Alan Beck and Aaron Katcher demonstrated the same effect using photographs of real people. In the real world, people with animals experience greater social acceptance.

Many events in people's lives are enhanced by animal companionship. People walking with dogs experience more social contact and longer conversations than people who walk alone, and wheelchair-users are more likely to experience positive social interactions when with a dog.

In a psychiatric facility, inpatients were more comfortable talking and participating in group therapy sessions in the presence of birds then they were in the same room without the animals present. Animal contact has many therapeutic implications, and now animal-assisted therapy programs are common in nursing homes and hospitals, and such therapy is a major area of study in schools training health professionals.

One conclusion of all these observations and studies is that animals benefit human health, and our interaction with companion animals is one of our strategies for survival.

Humans and Animal Health

There is also evidence that humans are important to animals. All domestic species were created by, and are sustained by, people. It is not surprising that social animals succeed in groups, and for all domestic species, that group includes humans. Dogs can understand human actions more than wolves and even primates can. Understanding people appears to be part of the domestication process and part of the dog's strategy for survival.

The interaction between humans and animals has profound impacts on both species. As human physiology and behavior are improved in the presence of animal companions, human companionship appears to cause changes in domestic species. James Lynch demonstrated that petting an owned horse elicited a slowing of heart rate, whereas entering and exiting elicited transient, but marked, increase in heart rate. Dogs too experience a significant decrease in heart rate and blood pressure immediately after a person enters the room or when they are petted.

Phenylethylamine (PEA) is a neuroamine (brain hormone) that promotes energy and elevates mood. Whereas a deficiency in PEA renders the person sluggish and depressed, taking PEA rapidly restores well-being, one of the reasons eating chocolate is so pleasurable. Johannes Odendaal found that for both humans and dogs, the levels of plasma PEA were significantly higher after interacting with each other. Apparently, our bond with dogs is mutually beneficial and desired. At the very least, this means humans are obligated to be good animal caretakers and be committed to humane treatment for their animal companions.

Human and Animal Health Together

Because nonhuman animals are often more sensitive to some of the world's hazards, they have been used as sentinels for humans. Indeed, the proverbial "canary in the mine shaft" was just that. We do not have to place companion animals into the human environment; they are already there. Animals in the pet population, which is monitored by owners and veterinarians, receive detailed physical evaluations comparable to those of their owners. Pets can serve as surveillance for diseases and even bioterrorist attacks. Larry Glickman has demonstrated how epidemiological studies of pet animals with spontaneously occurring disease are an alternative to either laboratory animal or human studies. In comparison with human diseases, animal diseases have a shorter latency after exposure and occur with fewer confounding factors that are associated with the complexity of the human lifestyle—for instance, animals do not smoke or experience workplace pollutants. As an example, in contrast to laboratory experiments, spontaneous tumors in pets reflect natural exposures to a wide variety of environmental carcinogens.

Companion animals are an unrecognized alternative for studying many of the health problems facing people today: another way that the human-animal bond helps our health and that, in turn, we can help our animals. People and their companion animals help each other by improving the health and well-being of each other.

Further Resources

Beck, A. M., & Katcher, A. H. (1996). *Between pets and people: The importance of animal companionship* (rev. ed.). West Lafayette, IN: Purdue University Press.

Fine, A. (Ed.). (2006). *Handbook on animal-assisted therapy.* San Diego: Elsevier.

Hare, B., Brown, M., Williamson, C., & Tomasello, M. (2002). The domestication of social cognition in dogs. *Science, 298*(5598), 1634–36.

Johnson, R. A. (Ed.). (2003). [Special issue devoted to human-animal interaction and wellness.] *American Behavioral Scientist, 47*(1), 1–102.

Katcher, A. H., & Beck, A. M. (Eds.). (1983). *New perspectives on our lives with companion animals.* Philadelphia: University of Pennsylvania Press.

Kellert, S. R., & Wilson, E. O. (Eds.). (1993). *The biophilia hypothesis.* Washington, DC: Island Press.

Melson, G. F. (2001). *Why the wild things are: Animals in the lives of children.* Cambridge: Harvard University Press.

Odendaal, J. S. J. (2002). *Pets and our mental health: The why, the what and the how.* New York: Vantage Press.

Alan M. Beck

■

The Benefits of Petting Dogs

Paul McGreevy

Animals who have evolved to live within groups seem to enjoy being stroked and groomed more than members of solitary species. Most socialized dogs, for example, enjoy being petted. This may reflect the social nature of canidae, but in contexts other than courtship and parenting, dogs exhibit less allogrooming (grooming one another) than cats or horses (Houpt, 1991). This is surprising when one considers the close proximity of canine group members with one another, especially at rest.

People who enjoy the companionship of warm-blooded animals, such as cats and dogs, frequently stroke or pet their animals. This may have benefits for both participants. During positive dog-human interactions, such as gentle scratching of the body and ears, concentrations of enjoyable, naturally occurring chemicals such as β-endorphin, oxytocin, prolactin, and β-phenylethylamine increase in both species (Odendaal & Meintjes, 2003).

Physical contact, such as petting, may also be used by humans as a secondary reinforcer (reward) for desired behavior, as an alternative to offering food. Grooming of horses at the withers (the base of the neck), rather than other areas, such as the forearm, has a calming effect on unsettled recipients and is a technique increasingly used to reward horses. Anecdotal reports suggest that many companion dogs prefer to be groomed in certain areas, such as the front of the chest. Despite this, there seems to be a tendency for humans to stroke the top of the head and down the neck. It is believed that, in some dogs, contact in these areas may result in displays of aggression because dog-to-dog contact with those regions (and especially the shoulders) has been related to attempts to assert rank.

A study involving kenneled greyhounds and guide dogs that were groomed for eight minutes in four different areas of their bodies showed that groomed dogs had significantly lower average heart rates than non-groomed animals (McGreevy et al., 2005). This supported earlier studies in smaller numbers of dogs that suggested that heart-rate deceleration is an important component of cardiac response to petting reward (Kostarczyk & Fonberg, 1982), although the greyhound and guide dog study showed that the area of the body where grooming was conducted was not important. However, if having a reduced heart rate is a sign of reduced stress, we can assume that noninvasive interventions producing this effect are generally enjoyable and therefore rewarding.

Of course, the extent to which dogs are rewarded by physical contact depends on their socialization with humans in general and their relationship with the people who are grooming them in particular. Indeed, acceleration of heart rate has been observed when a dog was petted by an experimenter who had previously punished the dog (Anderson & Gantt, 1966).

The duration of grooming and petting by humans is probably greater than any canine self-grooming bout. The effect of physical contact on heart rate may constitute a reward but only after some time. So the immediate reinforcing effects of physical contact are likely to be secondary to other benefits, such as the proximity of social affiliates (friends).

Further Resources

Anderson, S., & Gantt, W. H. (1966). The effect of person on cardiac and motor responsivity to shock in dogs. *Conditional Reflex, 1*(3), 181–89.

Houpt, K. A. (1991). *Domestic animal behavior for veterinarians and animal scientists* (2nd ed.). Ames: Iowa State University Press.

Kostarczyk, E., & Fonberg, E. (1982). Heart rate mechanisms in instrumental conditioning reinforced by petting in dogs. *Physiology & Behavior, 28*(1), 27–30.

McGreevy, P. D., Righetti, J., & Thomson, P. C. (2005). The reinforcing value of physical contact and the effect on canine heart rate of grooming in different anatomical areas. *Anthrozoös. 2*, 33–37.

Odendaal, J. S. J., & Meintjes, R. A. (2003). Neurophysiological correlates of affiliative behavior between humans and dogs. *The Veterinary Journal, 165*, 296–301.

■ Health
Chagas Disease and Insect Behavior

Chagas disease, or American trypanosomiasis, is named after the Brazilian physician Carlos Chagas (1878–1934), who first described it in 1909. Chagas disease exists only in the Americas, mainly in rural areas of Latin America (Centers for Disease Control, 2007) and is caused by the protozoan parasite *Trypanosoma cruzi*.

Chagas disease represents a major public health problem in Latin America and is associated with poverty. Chagas disease is the fourth leading cause of death in Latin America. *T. cruzi* currently infects 16 to 18 million people, causing severe chronic illness and approximately 43,000 deaths per year. It is sometimes, but relatively rarely, found in the United States.

Transmission of the parasite into a human or animal host is accomplished several ways. One way is through a blood transfusion from contaminated blood supplies. It can also be transferred from an infected mother to her offspring and by a blood-sucking insect popularly known as an assassin bug.

Not all assassin bugs are vectors of Chagas disease, and not all assassin bugs prefer human hosts. The main vectors are *Rhodnius prolixus, Triatoma infestans,* and *T. dimidiata.* In the northern portion of South America, Venezuela, Colombia, Suriname, Ecuador, French Guiana, and Guyana, *R. prolixus* is the main vector. Chagas disease can also be found in the Central American countries of Costa Rica, Guatemala, Belize, Honduras, Nicaragua, Panama, and El Salvador, where the main vector is *T. dimidiata.* In the southern portion of South America, such as in Brazil, Argentina, Chile, Bolivia, Peru, Paraguay, and Uruguay, *T. infestans* is the main vector. Others species of triatomines have been found to be vectors of Chagas disease but are epidemiologically less important, such as *Panstrongylus geniculatus.*

People affected by Chagas disease show two stages of the illness. First, there is an acute stage, which appears shortly after the infection. This stage is characterized by a high fever and inflammation at the site where the parasite entered the body. Particularly baffling is the extraordinarily large amount of inflammation that occurs if the parasite contacts the face. The length of this stage depends on the victim's health, age, and nutritional level.

The second stage of Chagas disease is known as the chronic stage, and its onset may take several years. This stage is also characterized by wide variations. Symptoms include enlarged heart and inflammation of other internal organs such as the intestines, stomach, colon, and esophagus. The peripheral nervous system is also affected. The chronic phase of the disease is irreversible.

Chagas disease may be diagnosed by a blood test. Diagnosis is also based on symptoms but must be confirmed serologically. It is also possible to diagnose Chagas disease using a technique known as xenodiagnostic. Xenodiagnostic involves the patient being bitten by a parasite-free assassin bug. At least seven to ten days later, the feces of the bug is collected and examined under a microscope. If the feces contain *Trypanosoma cruzi,* the diagnosis of Chagas disease is confirmed.

An interesting aspect of the disease is that the effects vary widely from one area to another in Latin America. For example, in Venezuela, Chagas disease affects predominately the heart. In Brazil, the disease affects the internal organs such as the intestines. The reason for this peculiarity is unknown at this time.

Although the parasite causing Chagas disease can enter the body by contaminated blood, the major method of transmission is contact with the feces of the assassin bug. The assassin bug finds its human host by odor and temperature. The insects are especially attracted to the odors of lactic acid, ammonia, and carbon dioxide. They prefer warm temperatures and are attracted to light.

Once the assassin bug finds the human host, its proboscis pierces the skin. Unlike the mosquito, which injects a parasite through the proboscis, the parasite responsible for Chagas disease enters the blood stream through the insect's feces. Defecation can occur while the insect is feeding or later when the insect is resting.

The assassin bug can enter a human dwelling several ways. For example, an adult assassin bug can fly into a home, attracted by the light. These bugs can also be attracted to animals that live near the home, including rats, chickens, pigs, dogs, and armadillos. Their preference, however, is to feed on the blood of humans.

Recently, it has been suggested that assassin bugs can learn to find their human hosts and can learn to change their preferences from an animal host to a human host. Unfortunately, many poor people in Latin America often share their dwellings with farm animals and live in homes made from mud with roofs constructed from palm tree leaves. Palm trees are a favorite environment for several of the vectors. The unsuspecting "campesino" contaminates the home when using palm tree leaves to build the roof of a house.

Those suffering from Chagas disease may be treated. The drugs nifurtimox and benznidazole have been used with some success. These drugs are useful in the first stage but may not be useful for those in the second stage. More recently, a compound called TAK-187 has also been shown to be effective during the first phase of the disease. There are no current vaccines against Chagas disease. Although drug treatments and insect control are useful, Chagas disease can only be eliminated when the living conditions of the poor are improved by collaboration between local communities and government.

Further Resources

Abramson, C. I., Romero, E. S., Frasca, J., Fehr, R., Lizano, E., & Aldana, E. (2005). The psychology of learning: A new approach to study behavior of *Rhodnius prolixus* Stal under laboratory conditions. *Psychological Reports, 97,* 721–31.

Centers for Disease Control. (2007). Chagas disease fact sheet. Retrieved May 28, 2007, from http://www.cdc.gov/ncidod/dpd/parasites/chagasdisease/factsht_chagas_disease.htm

Lent, H., & Wygodzinsky, P. (1979). Revision of the Triatominae (Hemiptera, Reduviidae), and their significance as vectors of Chagas' disease. *Bulletin of the American Museum of Natural History, 163,* 123–520.

Salvatella, R., Franca Rodríguez, M. E., Curto de Casas, S., Barata, J., & Carcavallo, R. (1998). Human habitats, dwellings and peridomiciliary sites. In R. Carcavallo, I. Galíndez-Girón, J. Jurberg, & H. Lent (Eds.), *Atlas of Chagas' disease vectors in the Americas* (pp. 601–19). Rio de Janeiro, Brazil: Foundation Insitute Oswaldo Cruz.

Schmunis, G. A. (1999). Prevention of transfusional *Trypanosoma cruzi* in Latin America. *Memorias do Instituto Oswaldo Cruz, 94,* 93–101.

World Health Organization. (1997). *Chagas disease, progress 1995–1996.* Thirteenth programme report of the UNDP/World Bank/WHO Special Programme for Research and Training in Tropical Diseases (pp. 112–23).

Elis Aldana and Charles I. Abramson

■ Health
Diseases between Animals and Humans (Zoonotic Disease)

Zoonotic diseases (those transmissible between humans and animals) represent one of the most urgent and enduring issues we face in human-animal relationships. We are interrelated within the ecology of disease: human and animal bodies are at once separated and linked by an organic boundary transgressed by any number of disease-causing microorganisms. This fact reflects the long history of the biological relationships between humans and animals, and understanding this history is critical. The usual path for zoonotic diseases to follow begins with birds, reptiles, amphibians, or mammals and ends in human populations (although the reverse occurs as well). Torrey and Yolken (2005) estimate that microbes originating with nonhuman animals in the past and those currently transmissible to humans from nonhuman animals cause approximately three-quarters of all human infections. Examples of the former category include tuberculosis, smallpox (a derivation of cowpox), and measles (related to rinderpest in cattle and canine distemper); diseases currently transmissible include bubonic plague, rabies, anthrax, and various parasites. These infections have caused both devastating epidemics (such as bubonic plague) and the less dramatic endemic diseases (such as tuberculosis and dysentery) that have been responsible for even greater mortality over time.

Social and cultural concerns have combined with biology in three important ways to influence the interactions between zoonotic diseases and human-animal relationships: domestication and the resultant opportunities for microorganisms to colonize new hosts; the development of human beliefs and practices that mediated the spread of disease; and human attempts to understand and control zoonotic diseases. Two particular zoonotic diseases, tuberculosis and anthrax, serve as illustrations.

Domestication and Other Human Practices

Taxonomically speaking, many microorganisms are "animals." If they were writing this essay, they might congratulate themselves on their successful ability to expand their range of hosts by taking advantage of close relationships between nonhuman animals and humans. One such close relationship, domestication, involved humans altering animals' natural attributes by controlling breeding and living conditions. To become domesticated, animals had to spend long periods of time living closely with human populations—good conditions for ambitious microorganisms seeking new hosts. Archaeological evidence points to domestication occurring between about 15000 and 2000 BCE, with dogs, sheep, goats, pigs, and cattle predominating.

These species have been the most likely sources for the majority of zoonotic diseases. This point circumstantially supports the idea that domestication was a primary source of initial disease transmission. The current distribution of zoonotic diseases also reflects a basic assumption about different animals' proximity to humans: our closest companions, dogs and cats, have been the source of almost half of our zoonoses, with most of the rest traced to livestock (who live with or near humans but probably not as closely). Evolutionary proximity also plays a role: nonhuman primates share arboviruses and malaria, among other diseases, with humans. Many of these diseases fall into a category that G. J. Jackson has labeled "heirloom infections." The microorganisms and their hosts evolved together from prehistoric times, with the microorganisms being passed along to each successive species of host. Most of these microorganisms and hosts developed apparently cooperative or tolerant relationships; the pathogens are the exceptions. Nonetheless, they are an enduring legacy of the common evolutionary process between humans and nonhuman animals (especially primates, our nearest relatives).

Human practices and beliefs have also encouraged the propagation of infections from domesticated animals and have ensured the evolutionary persistence of heirloom infections. One of the most obvious examples of this is the migration of human populations. As William McNeill has argued, a "confluence of disease pools" occurred between about 500 BCE and 1200 CE, creating large concentrated human populations in east and south Asia and the Mediterranean, as well as in the Middle East. These populations (disease pools) supported zoonotic "crowd" diseases such as smallpox, measles, and bubonic plague that occasionally exploded into epidemics (especially during wars and other sociopolitical disruptions). During this period, the development of land and sea routes connecting Asia and Europe provided new opportunities for the exchange of microorganisms as well as trade goods between the disease pools. For example, the global spread of bubonic plague may be traced by finding the locations of ancient colonies of rodents that harbored plague and then connecting those locations with distant areas of human outbreaks. This analysis shows that plague arrived in Mediterranean ports on trade ships from either northeastern India or central Africa. Global trade thus established bubonic plague in Europe, spawning a series of devastating outbreaks that spanned the first millennium of the common epoch.

Subsequently, during the age of European imperialism (beginning in the 1400s), diseases of ancient animal origin played a role in major global cultural disruptions. Europeans passed along measles, smallpox, tuberculosis, and their other zoonotic infections to the native human populations of the New World with disastrous results. The Incan and Aztec civilizations disappeared, as did whole villages and populations of North American native peoples. Here, too, the evidence of immunologically "naïve" native populations decimated by European diseases supported the theory of disease acquisition through domestication of animals. Most New World native populations, responding logically to the available climate and resources, had domesticated very few species of animals by 1400. Thus, they had not developed the long shared history with microorganisms, such as measles and smallpox, brought to the New World by the European invaders and their animals. The staggering morbidity and mortality they suffered contributed to Europeans' belief that God blessed Caucasian imperialism, but the more proximate cause lay with the ecology of zoonotic diseases.

Two Case Studies: Tuberculosis and Anthrax

In the modern context, tuberculosis and anthrax provide examples of how human-animal relationships and zoonotic diseases have continued to interact. Both diseases have long evolutionary histories common to humans and animals, and both can be spread by direct transmission between animals and humans. The social and cultural contexts of different times and places have encouraged human interpretations of tuberculosis and anthrax as threats, resulting in human intervention to try to control the diseases. This intervention has often focused on severing the direct link between the animal source and susceptible human populations by changing animal husbandry or food-processing methods. Both diseases have remained problematic despite human efforts to control them: bovine tuberculosis is rampant in some nations even as it has been mainly controlled in others, and anthrax has become a modern biological weapon. These microorganisms and other zoonoses continue to elude complete human control.

Tuberculosis of all types was the greatest killer of humans in the modern era (the late eighteenth century to the present). We now know that there are several strains of the bacteria *Mycobacterium* that cause tuberculosis in humans and animals. *Mycobacterium tuberculosis* causes the pulmonary form most common in human populations, and its cousin M. *tuberculosis bovis* is the most common cause of disease in cattle and goats. The extent to which M. *tuberculosis bovis* causes disease in people has long been contested, but evidence points to its role in causing non-pulmonary disease in children who have ingested infected milk or meat. Childhood tuberculosis was (and is) common wherever milk is not pasteurized. Cases still appeared in the United States up to World War II and in Europe until the 1960s; in many parts of the world, it remains a public health problem even into the twenty-first century.

British and American efforts to control this disease over the past 150 years provide a good illustration of the interaction between human-animal relationships and zoonotic disease. Citizens, physicians, and others concerned with tuberculosis infection of children framed bovine tuberculosis as a crisis of national proportions. Progressive reformers targeted corporate trusts and sought to implement federal regulation of food producers in the United States, a nation trying to come to terms with the moral and practical consequences of industrialization and laissez faire government policies. British reformers, enduring tremendous changes in 1901 with the challenge of the Boer War, the death of Queen Victoria, and the slow dissolution of the Empire, blamed bovine tuberculosis for threatening its young people and decreasing the efficiency of its institutions. Thus, the transmission of tuberculosis from cattle to humans symbolized fears of physical deterioration of the nation's individual and communal bodily integrity, threatening imperial power and the moral qualities of the world's leading democracy.

In both nations, control efforts focused on eradicating the disease in cattle (a goal that has been mostly, but not completely, realized) and, eventually, pasteurizing milk. The fight against milk-borne disease encouraged the removal of dairy animals from human-populated areas, greater regulation and surveillance of dairies, and the production of pooled milk from distant sources. All of these changes created greater distance between the animals producing the food and the humans consuming it. Increasing the distance between food-producing animals and human consumers did not, however, mean the complete eradication of zoonotic tuberculosis. Even in the United States, which has zealously pursued eradication, bovine tuberculosis is still endemic at very low levels, and sporadic outbreaks still occur.

Anthrax, a much-feared disease of agricultural areas at least since Roman times, has also managed to elude human efforts to control it. The microorganism that causes it, *Bacillus anthracis,* persists as spores in infected soil for decades to centuries, depending on conditions. It can infect most mammals, including humans, through the skin, gastrointestinal tract, or lungs. It has flourished in large populations of herbivores (the Texas cattle trails remain contaminated with viable spores). More recently, it has become established in industrial mills that process wool, hair, and hides imported from endemic areas of Asia. Like the case of milk production, humans using these industrial products have distanced themselves from the animals that are the source of meat, wool, and other products. Nonetheless, anthrax has persisted because animals and animal parts from endemic areas continue to travel around the world. Moreover, the hardiness of anthrax's spores enables its use in weapons delivery systems. Anthrax has been used or developed as a weapon (against both livestock and humans) in Mesopotamia, Argentina, Norway, the United States, Great Britain, the Soviet Union, Germany, Japanese-controlled Manchuria, Zimbabwe, and other areas. Despite efforts to control it and break the cycle of animal-human infection, anthrax persists as a threat to human health and security.

Zoonotic diseases (both "new" and reemergent) have made a comeback in the late twentieth and early twenty-first centuries. From relatively obscure origins as a simian virus in the mid-twentieth century, AIDS has ensconced itself as the major destructive force of human populations in sub-Saharan Africa. Poliomyelitis resists World Health Organization (WHO) attempts to eradicate it; strains of tuberculosis have acquired resistance to formerly effective antimicrobial drugs; SARS, initially blamed on an infection in Asian civet cats, spread to North America on passenger airplanes within weeks. Perhaps most alarmingly, the virulent H5N1 avian influenza virus has caused to date over 100 cases of the disease in humans, 50 percent of whom have died. Should the virus acquire the ability to be transmitted directly from human to human, many authorities believe that epidemiological conditions are right for it to cause a human influenza pandemic as destructive as that of 1918–19. From the avian point of view, H5N1 influenza has already proved ruinous: millions of birds have either died of the disease or been slaughtered by officials attempting to control outbreaks.

Some authors have argued that the devastation due to zoonotic diseases is the price humans have paid and are paying for their close associations with domesticated animals (see, for example, Torrey & Yolken, 2005; and Diamond, 1997). Animals provided sources and reservoirs of microorganisms that, through the process of domestication, then adapted themselves to proximal human bodies and populations. Modern efforts to control these diseases have largely focused on breaking the cycle of infection by distancing humans from domesticated animals, usually food-producing animals. Torrey and Yolken also outline the diseases transmissible between companion animals (such as dogs and cats) and humans, advocating the close scrutiny or even removal of companion animals.

However, the analysis in this essay shows that some of these prescriptions for control may be misdirected, based as they are on the theory that severing our close relationships with animals will help to eradicate zoonotic disease threats in human populations. The history of disease ecology in humans and animals makes it clear that we cannot completely protect human populations by distancing ourselves from domesticated animals. We are only beginning to understand the complex ecological relationships between populations of humans, animals, and microorganisms. For example, recent genetic studies have yielded some surprising evidence that points to long-term parasitism of animals and humans by bovine tuberculosis and anthrax bacilli. This evidence supports a conclusion that ancient ancestors of these microorganisms parasitized humans and animals separately, prior to domestication. It stands to reason that we can expect these diseases to continue infecting their respective host populations.

We cannot eradicate zoonotic disease by eradicating close human-animal relationships. This is not to say that we should abandon all efforts to control disease; on the contrary, this analysis should warn us that we need to seek complex, historically informed solutions that address global disease problems.

See also

Conservation and Environment—*Conservation Medicine Links Human and Animal Health with the Environment*
Domestication—*The Domestication Process; The Wild and the Tame*
Health—*Animal Reservoirs of Human Disease*

Further Resources

Diamond, J. (1997). *Guns, germs and steel: The fates of human societies*. New York: Norton.
McNeill, W. H. (1976). *Plagues and peoples*. Garden City, NJ: Anchor Press.
Torrey, E. F., & Yolken, R. H. (2005). *Beasts of the earth: Animals, humans and disease*. New Brunswick, NJ: Rutgers University Press.
Zinsser, H. (1935). *Rats, lice and history*. Boston: Little, Brown.

Susan D. Jones

■ Health
Dog Bites and Attack

As part of our frequent and close interaction with dogs, sometimes dogs bite people. Understanding dog bite can help both dogs and people live together better. Dog bite is not a random event and thus can be studied with a classical epidemiological approach, by asking who, what, where, when, how, why, and why not.

Who Are the Victims, and How Often Are They Victimized?

Animal bite injury, especially from dogs, is very common. This is under-recognized because published frequencies of occurrence rely on bites reported to local health departments. There is general agreement that reported bites are only the proverbial "tip of the iceberg." How many bites are still "under water" is not usually known. A survey study by Alan Beck and Barbara Jones of 3,200 four- to eighteen-year-old children found 45 percent were bitten by dogs during their lifetimes; 15.5 percent had been bitten in 1980—more than thirty-six times the reported rates in the area. More that 20 percent of the children seven to twelve years old were bitten in the study year, an incidence greater than for all childhood diseases combined.

Various studies using reported bite-rate data show that people from ages four to nineteen receive about 20 percent of dog bites. Children under fifteen years old make up only 8 to 10 percent of the population, and thus, dog bite injury occurs disproportionately in children. According the U.S. Centers for Disease Control and Prevention, the bite rate—that is, the number of people bitten in comparison with the population at risk—is more than twice as great for people under twenty years old as it is for those older than twenty. Children under four experience even a higher rate. Generally, males are more frequently bitten than females, but that may be more related, in part, to the environmental setting in which the bites occur rather than gender; boys (and dogs) play more outside. Among adults, people who must routinely approach or enter private property, such as letter carriers and meter readers, are at much greater risk of dog bite.

What Is the Nature of the Biting Animal?

The dog is the most frequently reported biting animal, followed by other humans and then cats. Bites from other pets, wild rodents, and wildlife are much less common. Contrary to the public perception, the owned pet dog leads the pack in bite rate; dogs owned by the family of the victim or the neighbor of the victim do much more biting than non-owned or stray animals. Survey studies of carefully controlled populations or bites occurring on military bases give a better indication of actual rate, and in those instances, non-owned or stray dogs account for less than 10 percent of all bites. One reason for the "over-reporting" of strays is the perception that strays cause more disease, such as rabies. Therefore, people tend to seek medical care and report the bite more frequently when bitten by strays than when bitten by an owned animal. Even when rabies was more common, most people requiring treatment were bitten by pet dogs, not strays.

The lack of good estimates of the populations of specific dog breeds makes it difficult to precisely predict which breeds are more or less likely to bite, but various studies indicate some general groupings. Eskimo dogs, German shepherd dogs, Saint Bernards, and Doberman pinschers appear to be more likely to bite than would be expected relative to their estimated population. In general, larger dogs are involved in reported bites more than smaller ones.

What Is the Place or Environmental Setting of an Injury?

The loose or unsupervised dog accounts for the vast majority of bites. Nearly half of over 2,500 reported dog bites occurred when the dog-bite victim was working or playing on or adjacent to the owner's property. The family setting is the second most common location.

When Do Bite Injuries Occur?

Dog bite is more common when there is a greater frequency of occurrence of human-animal interaction (i.e., during the summer months of May to August), with dog bites peaking in the late afternoon and cat bites earlier in the day. More bites occur during weekends than other days of the week. The age, time, and place distribution of bite reflects the increased interaction of young people encountering loose pet dogs. Hildy Rubin and Alan Beck observed stray dogs, and like most wild animals, they were less aggressive toward people than loose pets, and loose pets become less aggressive the further they are from their home.

How Do Bites Cause Injury?

Bites can crush, puncture, and lacerate tissue, causing damage, inflammation, pain, or blood loss, and can introduce infectious agents. Dog bites range from inconsequential to fatal. Although the vast majority of bites do not leave significant injury, the National Center for Injury Prevention and Control (U.S. Public Health Service) estimates that 26.4 percent of all dog bites to children, and 12.4 percent to adults, require medical attention. One reason that dog bites to younger people more often require medical attention is the location of the bites on the body; the pattern of bites to five- to nine-year-olds is 41.5 percent on the head and neck, in comparison with only 8.9 percent on adults, according to the National Electronic Injury Surveillance System.

The normal bacterial flora of the dog and cat mouths includes a mixture of aerobic and anaerobic infectious agents. Infections occur in about 50 percent of dog and 90 percent of cat bites, mainly from *Pasteurella multocida*. Dogs usually get rabies from wild animals, not other dogs, and the potential for rabies necessitates treatment, which is a major medical cost to society.

The most serious consequence of dog bite is death. Fatal dog attacks occur (1) as a tragic consequence of what would usually be an ordinary bite, (2) as a predatory attack, or (3) as an attack, as if in a dog-dog fight.

In the first instance, the fatality is usually because of the location of the bite, usually the head, and the disparate size or strength of the dog and victim. It usually involves a single family-owned animal encountering a young child or infant. The incident may not be dog aggression, but curiosity or play.

In the second case, the dog or, more often, dogs attack the victim repeatedly, treating the victim more as a prey species than a member of the dog or human pack; flesh consumption is often involved, leading to excessive hemorrhage. Often, the dogs form groups, and environmental stimuli, previous acts of aggression against humans or other animals, and social isolation from humans are frequently involved.

In the third case, the dogs bite with the same pattern, and presumably the same motivation, as they would when attacking another dog, using breed-specific behaviors of bite and hold with little regard to the communications that would inhibit most dog interactions. Pit bull–type dogs are almost exclusively involved in these incidents.

Jeffrey Sacks and collaborators studying the 1979–1998 records of dog bite–related fatalities (DBRF) recognized that not having concise estimates of the population of each dog breed placed some limits on the certainty of the data. "Despite these limitations and concerns, the data indicate that Rottweilers and pit bull–type dogs accounted for 67 percent of human DBRF in the United States between 1997 and 1998. It is extremely unlikely that they accounted for anywhere near 60 percent of dogs in the United States during that same period and, thus, there appears to be a breed-specific problem with fatalities."

Many states and cities have so called "generic" laws that regulate "dangerous" dogs, and others have "breed"-specific laws that manage or prohibit new pit bull–type dogs from being owned within their community.

Why Does Bite Injury Occur?

Dog bite can be a result of aggression, predation, a communication signal gone wrong, or a reaction to fear or pain. Classically, aggression between animals is directed toward the same species to settle a territorial or dominance dispute. Dogs communicate dominance and submission with people as if the person was another dog. This may explain why most dog bites occur on or near the dog owner's property (the dog's territory); people choose not to withdraw, do not have the time to do so, or are just ignorant of the communication being signaled that precedes the bite. This explains the high bite rates for letter carriers and meter readers.

How Can Bite Injury Occur Less Often and with Less Severity?

The best way to lessen the risk of dog-bite injury at home is to have a well-socialized and well-trained animal. The dog should be trained to be submissive to *all* members of the family and be comfortable with other people under all conditions. On the street, never pet a strange dog without taking the time to observe it. Approach slowly, noting the animal's behavior as you approach. Try not to surprise or corner an animal; most animals will avoid you if they know you are coming. If a dog approaches, stand still or approach very slowly, avoiding direct stare. Your lack of concern, passive avoidance, may be all that is necessary for the animal to lose interest in you. While apparently ignoring the animal, keep it in view, stopping your approach at the first sign of aggressive behavior, including such behaviors as ears becoming erect, body stance stiffening while the hackles raise (piloerection), growling or low-pitch barking, and keeping the tail in an upright stiff position. If the dog does not lose interest in you, try talking gently, approaching if it appears more relaxed. You can let the dog lick the back of a closed hand if appropriate. If the dog remains alert or increases aggressive behaviors as you come closer, hold your ground, protect your face by presenting your side, and say "down" or "no" in a firm, confident tone.

Be prepared to "feed" an aggressive dog your jacket, rather than your arm or hand. Often pretending to throw an object will frighten an aggressive dog. If you are surprised by an attacking dog, do not hesitate to climb a tree or jump onto a car hood to avoid the bite and gain your composure. If you are walking with a small child and an aggressive dog approaches, stand between the child and the dog and then assert your voice and posture. Do not attempt to lift the child, which will throw you off balance and make the dangling legs a tempting target for the dog to bite.

Bicyclists are often chased. Although most can outrace a small dog past the animal's territory, one should avoid being knocked over. The rider should stop, dismount, put the bicycle between himself or herself and the dog, and yell, "Go home!" If necessary, the rider should use the horn to break the dog's chase response and scare the dog off. Walkers and joggers in problem areas can carry a repellent spray, air horn, or short wooden stick on which the dog can bite.

Avoiding dogs is often not possible or even desired. Enter a dog's property cautiously. Dogs are often more aggressive when their owner is present, so suggest that the owner assist in the meeting.

In sum, each year, an estimated 1.8 percent of the U.S. population is bitten by dogs, but the bite rate is 150 percent higher for children. The safest and most humane way to reduce bite injury is to train and socialize your dog and commit to keeping your dog on a leash or well supervised when on public property. With a decrease in the bite rate, animals will be more welcomed in the community, animals will be surrendered less often to humane shelters, and all people, especially children, will continue to benefit from the many values of animal companionship.

See also

Health—*Animal Bite Prevention*

Further Resources

Beck, A. M., & Jones, B. (1985). Unreported dog bite in children. *Public Health Reports, 100,* 315–21.

Beck, A. M., & Katcher, A. H. (1996). *Between pets and people: The importance of animal companionship* (rev. ed.). West Lafayette, IN: Purdue University Press.

Beck, A. M., Loring, H., & Lockwood, R. (1975). The ecology of dog bite injury in St. Louis, MO. *Public Health Reports, 90,* 262–69.

Borchelt, P. L, Lockwood, R., Beck, A. M., & Voith, V. L. (1983). Attacks by packs of dogs involving predation on human beings. *Public Health Reports, 98,* 57–66.

Gershman, K. A., Sacks, J. J., & Wright, J. C. (1994). Which dogs bite? A case-control study of risk factors. *Pediatrics, 93*(6), 913–17.

Rubin, H. D., & Beck, A. M. (1982). Ecological behavior of free-ranging urban pet dogs. *Applied Animal Ethology, 8,* 161–68.

Sacks, J. J., Sinclair, L., Gilchrist, J., Golab, G. C., & Lockwood, R. (2000). Breeds of dogs involved in fatal human attacks in the United States between 1979 and 1998. *Journal of the American Veterinary Medical Association, 217*(6), 836–40.

Wright, J. C. (1985). Severe attacks by dogs: Characteristics of the dogs, the victims, and the attack settings. *Public Health Reports, 100,* 55–61.

Alan M. Beck

■ Health
Epidemiology and Human-Animal Interactions

Since the 1980s, various attempts have been made to apply the methods of epidemiology to the field of human-animal interactions. Applying this methodology to their studies, some investigators have contributed substantially to the literature on human-animal interactions. A longitudinal study led by Erica Friedmann, then a graduate student, investigating the relationship between pet ownership and one-year survival of coronary patients is one of these early examples. More recently, the case-control method has been applied to the examination of risk factors for dog bites and a retrospective cohort study has assessed risk factors for unsuccessful dog ownership in Taiwan. Notwithstanding the work that has been accomplished, the application of the epidemiologic method to the study of human-animal relationships and human-animal interactions has not been a standard procedure. Belief in the universal beneficial aspects of human-animal relationships has preceded its evidence, and a systematic approach to the study of human-animal relationships is still lacking.

Epidemiology as stated by Lilienfeld, Stolley, and Lilienfeld "is concerned with the patterns of disease occurrence in human populations and the factors that influence these patterns." More generally, epidemiology is concerned with the determinants and distribution of outcomes (health-related or otherwise) in human populations. Epidemiologic investigation is undergirded by the fundamental assumption that the outcome in question is not randomly distributed in the population, and consequently, it is relevant to the heart of many human-animal interaction questions. Indeed, even the most basic human-animal relationship questions present fertile soil for epidemiologic investigation. One such question might be "Why do (certain) individuals, communities, and populations interact with (certain types of) animals, whereas others do not?"

Study the Distribution of Outcomes

First, descriptive epidemiologic studies may be used to determine the extent of animal-related exposures or outcomes in a community. The term "exposure" refers to a condition, factor, or state whose presence or absence is believed to influence the occurrence of an event or endpoint; an outcome is the event or endpoint being investigated, and it may or may not be health-related. A descriptive study examines the frequency and relative distributions of exposures and outcome in a population. Using probability sampling, surveys can be performed, enabling one to answer questions such as the following: "Among persons of different levels of visual impairment, what percentage of persons of a particular age, gender, ethnicity, socioeconomic circumstance, and/or geographic location use guide dogs?" or "What percentage of persons of a particular age, gender, ethnicity, or nationality have favorable views on pet-facilitated therapy?" This type of information is necessary as a preliminary step to identifying the effects of human-animal interactions among populations of persons. In fact, in a time when limited resources are available for research, it is essential in determining whether there is even need for further study.

Identify Causes and Outcomes Caused

Second, analytic epidemiologic studies may be used to identify the causes of human-animal interactions as well as identify outcomes caused by different types of human-animal interactions. They can be used to identify which aspects of human-animal interactions increase or decrease the risk of experiencing certain outcomes and ultimately to answer such questions as which types of persons benefit or are injured by interactions with animals or whether human-animal interactions of persons with a particular outcome distinguish them from persons without that outcome. An important aspect is the choice of population to be studied. This should be determined by whom the information garnered from the study is aimed at benefiting. The study population should be drawn from the same source population to which the results would be generalized. Thus, if the aim is to make a statement on factors associated with benefiting from pet ownership among white, middle-class, HIV-positive males, then the study sample should consist only of those individuals. Correct application of epidemiologic principles would also dictate that the results not be extrapolated to women, nonwhites, and persons of low socioeconomic circumstances.

Observational studies have much to offer by their strict requirement of a (correctly chosen) comparison (or control) group. Observational studies are nonexperimental epidemiologic studies in which the goal is to measure the association between a given exposure and an outcome. They include cross-sectional, case-control, and cohort studies. A cross-sectional study is designed to measure the association of a given exposure with an outcome. Both exposure and outcome information refer to the same time period, and hence causality cannot usually be inferred. A case-control study is designed to measure the causal association of an exposure with an outcome. Individuals are selected based on their case (with outcome) and control (without outcome) status. The proportions of exposed and nonexposed cases and controls are then used to determine the extent of the exposure-outcome association. A cohort study measures the causal association of an exposure with an outcome. Individuals are selected based on their exposure (exposed and nonexposed) status. The groups are then followed through time, and the incidence of the outcome in both groups is compared. A study is prospective if exposure information is gathered prior to the occurrence of the outcome and retrospective if exposure information is gathered subsequent to the outcome having occurred. Case-control and cohort studies may be either prospective or retrospective.

Using these study designs, one not only can assess whether an animal-related exposure is associated with an outcome in a population of individuals, but also can estimate on average how much more (or less) likely the outcome is to be experienced by an individual given that he or she had previously had some animal-related exposure, as opposed to if the individual had not. For example, in order to ascertain whether interaction with a dog reduces depression, a group of residents (diagnosed with depression) of a nursing home can be observed over time. Some would have weekly experiences of interaction with a dog, and others would not. At the end of the observation period, the proportion of those with less frequent bouts of depression (determined by the case definition) among those who received pet therapy can then be compared to the proportion of those with less frequent bouts of depression among those who did not receive pet therapy. This type of prospective cohort study would be useful in ascertaining whether there is a true causal relationship between weekly interaction with a dog and the resolution of symptoms of depression.

An application of epidemiological principles also requires precise and unambiguous definitions of the outcomes under consideration. This is particularly important because investigators of human-animal interactions are often interested in intangible outcomes that are particularly susceptible to subjective interpretation. For example, if it is to be investigated whether or not the presence of a cat may reduce isolation or loneliness, then isolation needs to be defined in ways that can be objectively measured and that render an isolated person clearly and reliably distinguishable from one who socializes. Similar principles would apply to defining human-animal interaction exposures under consideration. For example "presence" of a cat may be defined by length of ownership, hours per day spent directly interacting with the cat, and the presence of physical contact. Clarity in these definitions is a prerequisite for proper interpretation of study results as well as for comparisons with other studies. Causal diagrams may be created in order to explicitly clarify the hypothesized relationship between the exposure under consideration and the outcome in question. This tool permits the investigators, prior to data collection, to determine what information needs to be gathered in order to investigate the relationship between exposure and outcome. This is necessary because some variables, though not of direct interest, are related to both the exposure of interest and the outcome in ways that obscure the true exposure-outcome relationship. For example, in a study aimed at examining how a person's educational attainment affects the time taken to resolve the loss of a pet dog, both gender and ethnicity would need to be adjusted for analysis because they both independently influence educational status and how a person responds to grief. On the other hand, the purpose of the dog need not be controlled for because it is consequent to the person's educational level. It follows from the use of causal diagrams that data-analytic techniques must be sophisticated enough to handle the interrelationships of extraneous variables affecting the effect of human-animal interaction exposures on an outcome of interest.

Determine Policy Implications

Third, results from carefully conducted epidemiologic studies can serve to provide the basis for the development of policy and the justification for public education regarding human-animal interactions. This is true with regard to areas as diverse as whether puppy socialization classes should be encouraged as a means of avoiding later dog aggression and pet relinquishment and whether the presence of dogs and cats in nursing homes should be encouraged because they positively enhance the well-being of residents. Unfortunately, much of the pioneering work and results of earlier epidemiologic studies that potentially could have an impact on public policy and health education in the field of human-animal interactions have not had that impact. For example, Ory and Goldberg found that pets were associated with negative outcomes for elderly women living in rural settings, but with positive outcomes for women in suburban and urban settings; higher socioeconomic status and attachment to pets were associated with more positive indicators of happiness. In their study, Garrity and colleagues found that no important health correlates for pet ownership were reported overall, though benefits were observed for the subgroup of individuals experiencing bereavement during the first year of loss of a spouse. Raina and colleagues found that among the elderly, non-owners showed more deterioration over one year in their activities of daily living than did pet owners, though owners tended to be younger and more active. One possible explanation for the lack of impact of these results on the scientific community at large is the inconsistency in the use of epidemiologic methodology in the field of human-animal interactions. The consequence is that attempts to replicate (or refute) these results have been inconsistent and rare.

The study of human-animal interactions remains a somewhat new field that continues to define its scope and methods. An early emphasis has been to document the health benefits of pets wherever they exist and to develop standardized techniques for making it easier for people to have contact with companion animals. For this to be justified, the gap between belief in the positive effects of animals in society and evidence of these effects needs to be bridged. Use of epidemiologic methods provides one means whereby this gap may be filled.

Further Resources

Garrity, T. F., Stallones, L., Marx, M. B., & Johnson, T. P. (1989). Pet ownership and attachment as supportive factors in the health of the elderly. *Anthrozoös, 3,* 35–44.

Gordis, L. (2000). *Epidemiology.* Philadelphia: Saunders.

Lilienfeld, D. E., Stolley, P. D., & Lilienfeld, A. (1994). *Foundations of epidemiology.* New York: Oxford University Press.

Ory, M. B., & Goldberg, E. L. (1983). Pet possession and life satisfaction in elderly women. In A. H. Katcher & A. M. Beck (Eds.), *New perspectives on our lives with companion animals* (pp. 303–317). Philadelphia: University of Pennsylvania Press.

———. (1984). An epidemiological study of pet ownership in the community. In R. K. Anderson, B. L. Hart, & L. A. Hart (Eds.), *The pet connection: Its influence on our health and quality of life* (pp. 320–330). Minneapolis: Center to Study Human-Animal Relationships and Environments (CENSHARE), University of Minnesota Press.

Raina, P., Waltner-Toews, D., Bonnett, B., Woodward, C., & Abernathy, T. (1999). Influence of companion animals on the physical and psychological health of older people: An analysis of a one-year longitudinal study. *Journal of the American Geriatric Society, 47,* 323–329.

Lynette Hart and Locksley L. McV. Messam

■ Health
Epizootics: Diseases that Affect Animals

The term *epizootic*—stemming from the Greek meaning "on animals"—refers to infectious diseases that potentially affect large numbers of animals, primarily livestock but also horses and wild animals. As the nonhuman equivalent of epidemics, most epizootics are contagious, spreading either from animal to animal or through the environment. Whereas most epizootics are limited to animals, some might potentially spread to humans, most notably anthrax and rabies. The most common epizootics include cattle plague, hog cholera, foot-and-mouth disease, anthrax, rabies, and more recently, avian flu and mad cow disease, or BSE. Historically, two of the most devastating types of epizootics were the cattle plague and foot-and-mouth disease.

Cattle plague, also known as rinderpest, is a deadly viral disease of cattle that is characterized by fever, diarrhea, and oral erosions. Historical records describing cattle plagues go back to ancient times. Periodic outbreaks of this highly contagious disease, which originated in India, plagued Europe throughout the centuries, causing tremendous agricultural losses. For example, during the eighteenth century, major cattle plagues spread through Europe in 1711, 1715, from 1742 to 1748, and again in the 1770s, devastating the livestock holdings of Germany, Belgium, France, and Italy. This epizootic spread particularly rapidly during times of war, such as the Napoleonic Wars and the Franco-Prussian War, when large cattle herds were moved alongside armies. By the late nineteenth century, the frequency of cattle plague outbreaks diminished in Europe as the conditions of livestock-keeping improved, but the disease continued to exist in parts of Africa, where major outbreaks were recorded into the 1980s. A vaccine against rinderpest was developed in the 1990s, leading to predictions that the disease will be eradicated by 2010.

Foot-and-mouth disease, which is also called hoof-and-mouth disease, is a highly contagious viral disease that primarily afflicts cattle and pigs but that can also befall other cloven-hoofed animals such as deer, goats, and sheep. Humans are affected only very rarely. As the name indicates, this disease, which is characterized by high fever, causes painful blisters inside the mouth and on the feet. Although foot-and-mouth disease is highly contagious, most animals would eventually recover; however, because it leads to severe weight loss, the milk and meat production of infected animals is diminished.

Much like the cattle plague, foot-and-mouth disease has occurred throughout history in most parts of the world, causing tremendous agricultural losses. It was endemic in many areas in Africa, the Americas, Asia, and Europe. Only after World War II did the frequency of outbreaks diminish; however, in 1996 numerous outbreaks occurred in Africa, Asia, and South America. Most recently, the United Kingdom was afflicted by a serious outbreak of foot-and-mouth disease in 2001, which cost close to fourteen billion dollars and led to the culling of over seven million animals. A vaccine was developed in 1981, but its use remains controversial.

In general, epizootics have had a devastating effect on animal populations, agricultural production, and national economies throughout history and especially during times of war, when epizootics often spread particularly quickly. Measures to contain, eradicate, and eventually prevent epizootics include quarantines, vaccinations, legal measures, and oftentimes culling. Even though some animal diseases have been conquered, epizootics continue to be a notable threat, as recent concerns over mad cow disease and avian flu indicate. Moreover, they point to the heightened dangers of epizootics that are potentially transmissible to humans, particularly as the mobility of humans and animals expands alongside the globalization of the world.

See also

Health—*Animal Reservoirs of Human Disease*
Health—*Diseases between Animals and Humans (Zoonotic Disease)*

Further Resources

Swabe, J. (1999). *Animals, disease, and human society: Human-animal relations and the rise of veterinary medicine*. New York and London: Routledge.
Wilkinson, L. (1992). *Animals and disease: An introduction to the history of comparative medicine*. New York: Cambridge University Press.
Zinsser, H. (1935). *Rats, lice and history*. Boston: Little, Brown.

Dorothee Brantz

■ Health
Human Emotional Trauma and Animal Models

Human Emotional Trauma: A Personal Experience

My therapy client came in that day agitated and tense, shaking her head and mutely looking at the floor. As her counselor, I knew she was recovering from a kidnapping and rape incident, but in previous sessions, she had been able to look me in the eye. This woman had been severely brutalized, but had so far dealt with it by using logic and affirmations about not letting the incident get to her. What had changed?

What had changed was that she had been walking down the street a few days ago, and a man wearing a red sweatshirt had walked up behind her. As he passed her, she began to experience shortness of breath, dizziness, tunnel vision, and a powerful urge to run away. As he cluelessly walked off, she collapsed and was only able to get up and walk home after some people came to her aid. The incident shook her up so much that she had difficulty sleeping, eating, or going to work.

As we began to take a look at the current incident, one that was obviously re-traumatizing, we uncovered more detail about the original event—that her attacker had been wearing a red shirt and that he was roughly the same age as the young man who had walked by her. She was also able to recount that the original attack had come from behind her. Though other important details of the two incidents did not line up (the clueless pedestrian was whistling cheerfully, for instance), there was enough of a match with the original attack to trigger an intense body response, one that overwhelmed her.

To again be so out of control over her situation formed the core of the re-traumatizing experience.

Key Features of Trauma

This story, like so many that survivors and their therapists can tell, illustrates some important features of trauma. Loosely defined, trauma is a deeply distressing experience that can involve physical or emotional shock or both and can lead to lasting impairment. The disturbing experience can be physical, such as being in a car accident or being assaulted; more emotional, such as seeing someone being harmed; or even more cognitive, as in being constantly verbally denigrated. In many trauma cases, the event can be distressing on all these levels at once. Most trauma specialists feel that trauma can be divided into two categories: effects that come from single or discrete incidents, such as being attacked, and what is called developmental trauma, which arises from repeated, chronic abuse over time. In this second case, the injury develops over time as a result of relatively low-grade, but damaging, relationships, such as a child being raised in a home where he is constantly subjected to instability, neglect, or hostility.

A key feature of both types of trauma is that they overwhelm the organism's current resources for dealing with perceived threats to its integrity. The person is helpless and not in control of the outcome of the event. Symptoms of unresolved trauma can include dissociation, deficits of learning and memory, exaggerated responses to previously tolerated stressors, physiological and behavioral hyperreactivity (such as intrusive reliving of the traumatic event in flashbacks or nightmares). Unresolved trauma can also involve restrictive symptoms such as emotional constriction, paralysis, social isolation, a sense of estrangement, loss of interest, and detachment.

Another key feature of lasting trauma (often called posttraumatic stress disorder, or PTSD) comes from the fact that all organisms are wired to learn from their experiences in some fashion. It is obviously a good thing to learn and remember the circumstances of threatening events, so that we can avoid repeating them. But in the case of unresolved trauma, the organism has also learned that it is helpless. Bessel Van der Kolk (1987), a trauma researcher and clinician at Harvard, has stated that one of the primary features of trauma is a real or perceived lack of control.

This learned helplessness is not a cognitive belief that the person can articulate—often it is deeply buried underneath conscious processes, back in regions of the brain that govern such matters as heart rate, stress responses, and primary emotions such as fear. These regions of the brain tend to "imprint" events—learn to associate red shirts with extreme danger, for instance—in a way that is quite strong and potentially permanent. And because of this, trauma is not something we can talk ourselves or others out of. In therapy it would do no good for me to remind my client that this pedestrian wished her no harm. Her body is telling her otherwise, and its messages of threat are much stronger, utterly believable, and more urgent.

Animal Models

Both humans and animals can be and are traumatized. In fact, it is hard to find an organism that has not gone through some kind of traumatic event at some point in its life. What has plagued and intrigued researchers and therapists is that some people and animals seem to recover from trauma relatively easily, and sometimes they are rendered incapable of leading a normal life afterward. In humans, there is not always a strong correlation between the severity of a trauma and the individual's capacity to recover from it.

In nonhumans, evidence suggests that wild animals may possess extra resources that at times enable them to recover more quickly and completely from a threat. Domestic animals fall somewhere in between; they recover from traumas more easily than humans but less easily than their wild kin. What is going on here that we can use to get better at helping both humans and animals recover from traumas?

To answer this question, scientists and therapists have begun turning to the study of animal behavior. By understanding the biological and behavioral mechanisms that both humans and animals have in common, especially the ones that involve the nervous system and how it regulates states of consciousness and actions, the hope is that we can design therapies that help trauma be an event that we remember and learn from, but that does not compromise our ability to get on with our lives.

Humans and animals share many states of consciousness. Both relax and rest. Both get more alert when something novel or unusual happens. Both will try to defend themselves when under threat. Social mammals likely feel primary emotions such as fear, anger, and joy. These states are biological necessities and are largely governed by the nervous system. In both humans and animals, the state an organism is in largely determines the actions it will take (if you are nervous, for instance, you are much more likely to be defensive). A successful childhood is thought to be one that allows a human child or a wolf pup to practice regulating these nervous system states and their resulting actions in a relatively safe environment so that the young one grows up able to match its actions appropriately to the current situation. Acting appropriately in the current state of events has tremendous survival value and also engenders a feeling of agency, or a deep knowing that one is basically able to handle life's circumstances.

Ellert Nijenhuis is one of the researchers who have looked into what wild animals tend to do in threatening situations, and he has applied these strategies to humans under threat. He believes that there is a similarity between animals' defensive reactions and dissociation symptoms in humans. He notes that when a prey animal first becomes aware of a predator, it tends to momentarily and involuntarily freeze. This is probably because the animal needs a moment to assess the situation, but it is also because most predators detect moving prey more easily and ignore immobile objects, so the attention of a predator may shift to other moving or noisy stimuli. Also, movement cues such as running away tend to actually stimulate predatory behavior, and freezing reduces that likelihood.

Nijenhuis also notes that when a wild animal is about to be attacked, this freezing tends to be combined with analgesia, or not feeling pain. He feels this is functional in that feeling pain could distract the animal from defending itself. We have all experienced this moment, being slightly surprised to discover that we hurt ourselves in some way only after a scary situation is over.

After the freezing phase, when the predator gets dangerously close, the prey animal tends to dramatically change its behavior and suddenly display an explosive escape response, such as fighting for its life or running away. At this point, it will or will not survive the attack. If it survives, it will not only tend to learn something about how to successfully defend itself in similar situations, but it will also remember that it *can* successfully defend itself. As we have seen before, this second lesson may be crucial to the ability to recover from trauma and deal effectively with future traumas.

If an animal is unsuccessful in fending off an attack, the freezing and analgesia tend to return. It is speculated that the analgesia may blessedly numb the pain of death. When an animal successfully defends itself, pain perception returns afterward and stimulates recuperative behaviors such as tending to wounds and resting. From the perspective of a successful survival, we can see that danger tends to evoke analgesia, and afterward, pain evokes recuperative behavior.

Peter Levine, a medical and biological physicist turned therapist who specializes in trauma treatment, has become very interested in this recuperative phase in wild animals. He has noted that one of the methods an animal might use to recover from immobilization during a traumatic event is quite literally shaking it off. It shivers, wiggles, and otherwise moves around in such a way that it helps itself to come down off the stress. We humans do this routinely after we have experienced a fright—we blow out our breath, shake ourselves, and say things like "Whew! That was scary!" This active means of regulating our state seems to be absent or thwarted in people and animals that get caught up in unresolved trauma. Levine states, "Evidence clearly shows that the ability to go into and come out of this natural response is the key to avoiding the debilitating effects of trauma. It is a gift to us from the wild" (1997, p. 17). He feels that the study of wild animal behavior is essential to the understanding and healing of human trauma because the involuntary and instinctual portions of the human brain are virtually identical to those of other mammals.

Looking more deeply into animal behavior, we can see that part of the puzzle is about how an organism defends itself, and part is about how it recovers from defensive behaviors. The nervous system plays a crucial role in both these functions. In terms of defense, both humans and other animals share a "favored four": flight, fight, freeze, and faint. Flight, or running, works when there is a hope of outrunning the predator. Fight translates to hurting the predator so that continuing the attack is not worth the predator's risk of injury. Freezing works when your predator hunts by sensing movement. If you can become invisible by blending in with the background, you have a better chance of escape. Fainting involves playing dead and works when the predator will not eat things that it thinks are already dead. Flight, fight, and freeze all involve the sympathetic branch of the nervous system, the branch that governs getting excited and agitated. Fainting involves the parasympathetic branch of the nervous system, the one that helps us to relax, rest, and go to sleep.

As noted before, a good childhood allows a youngster to practice these defensive behaviors in a safe environment—preferably in the context of play, where there are rules and limits that keep it fun so that the human or animal learns how to perform each defense strategy well and knows which defense to match to what kind of predator or attack. We do not want to try to run from a mountain lion, for instance; it can actually be more effective in an encounter to get big, be loud, and be willing to fight. Trauma researchers have found that people who tend to become disabled by traumatic events are more likely to have had childhoods where this kind of practice was absent, unsafe, or chaotic. Often, these conditions occur in abusive or neglectful families. My client, for instance, had two older brothers that tormented her constantly, in a mean and demeaning way. The most relevant part of that childhood scenario, however, was that the incidents occurred when the children were unsupervised by the working, single mom. She learned, in a very wordless and primitive way, that she was basically alone and without protection.

A potentially more insidious scenario in childhood abuse is when kids learn that *no* defense works. These children learn that they are helpless and that they can only passively wait for an assault to be over. As with an adult under stress, they may make no effort to control their situation or their state. When that child grows up, he or she is at a much greater risk for being traumatized by events that others can manage and for that trauma to be more severely impairing.

Studies that involved exposing animals to inescapable shocks duplicated these human outcomes. In the name of science, animals were electrocuted under conditions where they could escape if they moved off a metal plate and under conditions where they could not escape. Animals who could not escape the pain began to exhibit the symptoms of PTSD, and these symptoms were enduring.

Nijenhuis put it like this:

> Some types of human trauma may be compared to predator attack, inescapable shock, and animal social conflict. For example, phenomenologically, sexual and physical abuse resemble inescapable shock, especially when occurring to children and involving relatives as perpetrators. Such highly aversive, burdening, and painful stimulation is inescapable in that perpetrators often force passive subjugation to sexual abuse. . . . It also reflects a loss of control and familiarity. . . . [Other researchers] found that only a minority of adult females reporting childhood sexual and physical abuse actively resisted the perpetrator, because such defense was perceived as useless, or as eliciting further attack. Passive defense (dissociation and fantasy) increased with severity of abuse. (2004, pp. 114–15)

Nijenhuis also reported that flight responses were shown by about half of these women, but were restricted to hiding in dark corners or under blankets. The women tended to freeze and stiffen. A large majority experienced analgesia and kinesthetic analgesia (insensitivity to touch). What seems to happen in unresolved trauma is that these responses persist even when the trauma is over, and they become integrated into the personality. In other words, in resolved trauma, the person was fearful during the event, and now she is not, whereas in unresolved trauma, the woman becomes a fearful person no matter what.

Often, PTSD in humans involves chronic pain that can last indefinitely. Wild animals tend to escape this outcome. Chronic animal pain seems to be rare. This may be partly because an animal in chronic pain would not survive very long in the wild, but another factor might be that wild animals are not as burdened with shame, guilt, second-guessing, and other cognitive gymnastics that may contribute to keeping ourselves in pain. In humans, thinking is strongly designed to mix with feeling in order for people to solve problems, learn, and be social. This important capacity may get in our way, however, when it comes to managing and recovering from trauma, especially when, like my client, we try to use thinking to resolve the unthinkable. Human trauma tends to interfere with the integration of experiences in consciousness, autobiographical memory, and identity, leaving the trauma stored in our memory on the level of painful sensations and excruciating emotions. Overthinking can interfere both during and after a traumatic event. Levine notes that "lacking both the swiftness of the impala and the lethal fangs and claws of a stalking cheetah, our human brains often second-guess our ability to take life-preserving action. This uncertainty has made us particularly vulnerable to the powerful effects of trauma . . . when confronted with a life-threatening situation, our rational brains may become confused and over-ride our instinctive impulses" (1997, pp. 18–19). This same process may come into play afterward, interfering with our instinctive animal capacity to deal with our body first by mobilizing it and calming it.

Some researchers and clinicians believe, because of this, that trauma therapy may need to involve a more "bottom up" approach, one that first addresses how a traumatized person feels in his or her body—sensations and feelings. What we may need to do is learn how wild animals shake off after a threatening experience, taking care of their bodies by staying active and alert and relatively calm. Levine says that we need to discharge the pent-up energies left over from an attack so that they do not persist in the body, become fixated in it, and cause the complex symptoms of PTSD. Many therapists now believe this can happen by paying attention to sensations and feelings—really listening to them and allowing them to gradually work their way through the body. In a way, it reassociates us after the dissociation that occurred during the threat. Levine puts it like this:

> The experience of the felt sense gives us a backdrop for reconnecting with the animal in ourselves. Knowing, feeling, and sensing focuses our attention where healing can begin. Nature has not forgotten us, we have forgotten it. A traumatized person's nervous system is not damaged; it is frozen in a kind of suspended animation. Rediscovering the felt sense will bring warmth and vitality to our experiences. This sense is also a gentle, non-threatening way of re-initiating the instinctual processing of energy that was interrupted when the trauma occurred. (1997, pp. 86–87)

Both humans and nonhumans possess what is called an "orienting response." When something novel happens, especially something sudden and unexpected (like a loud sound), humans and other animals will reflexively orient toward the stimulus in a "what's that?" mode. This is a way to quickly assess whether or not something is dangerous and to respond to it effectively.

We have all experienced this orienting response, such as when turning to look at where a sudden noise came from. In those of us who are not caught in a trauma response, we turn quickly and see that it is just something normal, and then we might shake our bodies or take a few breaths to relax. This is a normal way for organisms to shed their preparation for possible defense and get back to a more relaxed state.

In traumatized people, the orienting response tends to be diminished, suppressed, or exaggerated. Often a trauma reaction occurs instead of a normal orienting reflex. This could look like freezing, cringing, feeling faint, losing control over one's body, hitting, screaming, or ducking for cover. And even after the traumatized person can see that she is not in danger, she cannot stop these intense, defensive actions, often for hours afterward. It can be so strong that it overwhelms rational thought—she cannot talk herself down, and other people cannot talk her out of it.

One of the first ways to deal with this specific traumatic reaction is to help her to change what she is orienting to—instead of obsessively and defensively listening for another loud sound, for instance, she can look into the eyes of a friend, hold onto his hands, and listen not to the content of what he is saying, but to the soothing tones of his speech. That way, her body can begin to attune to something that is not traumatic, and she can begin to come back to a calm and alert state. Once she can find and hold onto that calm state, only then will she begin the second phase of therapy, where the trauma is gradually confronted, only this time with the resources needed to deal with it. Another strategy is to consciously practice slowly and deliberately turning toward strange stimuli.

Good psychotherapy will help a traumatized person to develop these and other resources, and over time, a normal orienting response can be reestablished, along with other appropriate behaviors. It is especially important for therapists to understand the biological basis of trauma and treat it as not a dysfunction of thought or belief, but as a disturbance of the body's normal methods of coping with danger. In a sense, trauma therapy for humans needs to respect and address our animal nature, helping us to appreciate that danger and wounding go directly to the parts of ourselves that we share with many animals—the need to defend and the need to recover. Over time, my client learned to listen to her body signals and learned methods such as conscious breathing and staying mobile even when she remembered her traumatic event. This helped her to feel normal again, to be able to tolerate people coming up on her from behind, and to look forward to her future.

Dangerous and traumatic events happen to all humans and animals, all the way from stumbling and falling down to torture and prolonged abuse. Both humans and animals are beautifully designed to cope with trauma, and under optimal circumstances, both humans and animals can react appropriately and get back to normal as soon as possible. But if trauma is so massive that it overwhelms even these innate mechanisms, or if it is so early or chronic that the organism adapts to it and "normalizes" it, then it becomes institutionalized in the body. By studying how our animal brethren resolve these threats, we humans may begin to recover strategies we may have forgotten as we developed our cognitive and language capacities. We can celebrate and come into alignment with our animal natures as well as our human abilities.

Further Resources

Levine, P. (1997). *Waking the tiger: The innate capacity to transform overwhelming experiences.* Berkeley, CA: North Atlantic Books.
Nijenhuis, E. (2004). *Somatoform dissociation: Phenomena, measurement, and theoretical issues.* New York: Norton.
Ogden, P., Minton, K., & Pain, C. (2006). *Trauma and the body: Sensorimotor approach to psychotherapy.* New York: Norton.
Rothschild, B. (2000). *The body remembers: The psychophysiology of trauma and trauma treatment.* New York: Norton.
Van der Kolk, B. (1987). *Psychological trauma.* Boston: Cambridge University Press.

Christine Caldwell

■ Health
Humans, Viruses, and Evolution

Viruses cause both the mildest and the most deadly disorders: colds, cold sores, and warts as well as rabies, sudden acute respiratory syndrome (SARS), and Ebola. Some viruses are longstanding pathogens of mammals and evolved along with us in essentially a "host-pathogen arms race." But most of our infectious diseases, viral or otherwise, were acquired from other animals. Thus, the evolution of viruses that regularly plague humans illustrates two kinds of human-animal relationship:

1. Genetic, ancestral relationships among many human and animal viruses mirror those of the host species they infect, indicating that these viruses and hosts have evolved together.
2. The acquisition of new viruses reflects ecological relationships between humans and other animals, such as predator-prey, competition, domestication, or simply sharing an environment.

Because disease mortality can be a powerful agent of natural selection, the infections that plagued our ancestors probably influenced the course of human evolution.

Some viruses have been with us since time immemorial. The mildest of viral infections are caused by ancient viruses such as herpesviruses (cold sores and chronic fatigue syndrome) and papillomaviruses (warts and other skin conditions). These

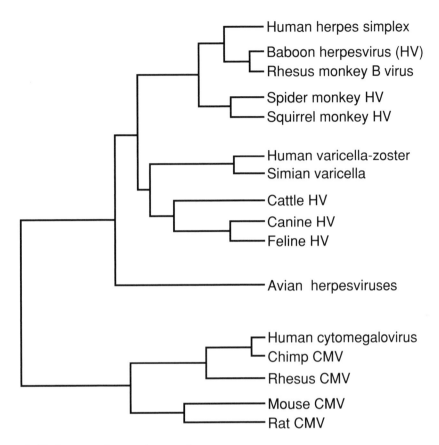

Figure 1: Phylogeny of some herpesviruses.

slowly evolving DNA viruses tend to cause lifelong, chronic, and inapparent infections that produce symptoms only under stress and that are capable of persisting in the small populations found in most wild animals, including our ancestors. When one compares the DNA sequences of this type of virus with related viruses in other animals, one finds that they show the same evolutionary relationships as their host species (Figure 1). That is, the closest relatives of human herpesviruses are those of chimpanzees, then Old World monkeys such as baboons and rhesus macaques, followed by the viruses of South American (e.g., spider and squirrel) monkeys that are less closely related to us than monkeys of Africa and Asia. More distantly related are herpesviruses found in dogs, cats, or cattle, and even farther away are those of birds. "Molecular clock" estimates dating the common ancestors of these viruses based on degree of genetic sequence difference yield similar times to the divergences of the animals they infect, indicating that these viruses have evolved along with their hosts, or have "co-speciated." These viruses, which cause mostly mild and nonlethal conditions, are unlikely to have had much effect on human evolution.

More serious diseases such as measles and influenza are generally newer, and most of them stem from the intense human-animal interactions surrounding domestication and the adoption of an agricultural lifestyle. Living in close contact with herd animals and dogs allowed the exchange of many infectious agents and their adaptation to a new host. This exchange was not only from animal to human; whereas early farmers acquired

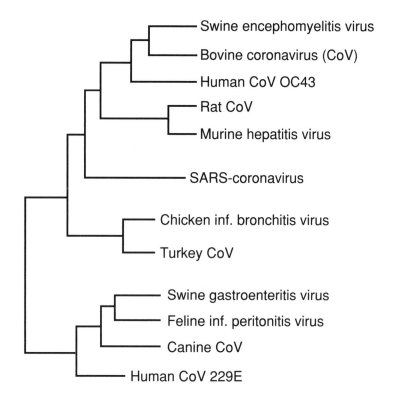

Figure 2: Phylogeny of some coronaviruses.

the ancestors of mumps and measles viruses from their domesticated herds, human cox-sackie B virus is thought to be the ancestor of swine vesicular disease, and the common agents of infant diarrhea (rota- and caliciviruses) have been freely shared among babies, calves, and piglets. Commensal rodents attracted to granaries and houses also contributed their own diseases. The family trees of these viruses show a different pattern from the ancient viruses that co-speciated with us. Here we find human strains closely related to viruses of cattle (measles), pigs (infant diarrhea), or mice (parainfluenza and encephalomyocarditis). Some, such as coronaviruses that cause colds and other respiratory infections, have been transmitted among cats, dogs, pigs, cattle, rats, mice, and people (Figure 2). All of these primarily RNA-based viruses have very high mutation rates, giving them the capacity to evolve quickly. Thus, they easily jump from one species to another and can change and adapt once there to create entirely new diseases. The recent emergence and epidemic of SARS in humans is a case in point. The SARS-coronavirus is now thought to reside in bats and was apparently transmitted to humans via infected civet cats (Normile, 2005).

In a similar fashion, domesticated animals can be the vehicle by which other wild animal infections cause epidemics in humans. Influenza, for example, exists in nature in wild birds but can become a dangerous disease for humans after passing through domesticated poultry and pigs. When pigs are infected with both avian and human strains of the virus at the same time (a common occurrence in parts of the Far East where farming techniques result in pigs and ducks raised together), influenza's segmented viral RNA can reassort to form new hybrid strains. This process, known as "antigenic shift," is believed to have produced the 1918 influenza pandemic as well as the "Asian" and "Hong Kong" flus of 1957 and 1968. The latter two, although serious, were not as deadly as the 1918 flu, probably because of the availability of antibiotics to treat secondary bacterial infections that frequently develop with the disease. Public health officials are currently concerned about the potential for a virulent pandemic originating from reassortment between the avian A(H5N1) and a human type-A influenza strain.

In addition to the most ancient human viruses and those that have recently evolved is a group of agents of in-between age that includes various hepatitis viruses, polio and other picornaviruses, and some retroviruses. These are frequently found in other primates, and some of these may be very old human parasites, but probably not as old as herpes and papilloma. Hepatitis viruses, picornaviruses, and human t-cell lymphotropic virus (HTLV), a retrovirus carried in the cellular DNA of most humans today, became endemic in human ancestors sometime before anatomically modern humans spread across the globe around 40,000 years ago. Some of these may have infected the earliest hominins, but others that share closer relationships with the viruses of monkeys than those of apes (e.g., hepatitis A, polio, and HTLV) were probably acquired some time since our divergence from the human-ape common ancestor. Many are transmitted via contaminated food (including meat) or water and could be acquired just by sharing an environment with their reservoir hosts. Close phylogenetic relationships of some picornaviruses and HTLV-I with similar viruses in baboons are suggestive and may reflect the shared savannah environment of early *Homo* and baboon ancestors.

Our forebears may have acquired many of these viruses when they shifted to eating more meat, a dietary change thought to have been associated with early *Homo*'s movement onto the savannah. No doubt, the earliest hunted or scavenged prey included other primates. In fact, today an important means of transmission of disease to humans is the eating of wild primates as "bushmeat" (Peeters et al., 2002). This is now considered the origin of both human immunodeficiency virus (HIV) and Ebola virus infection in humans. Genetic studies suggest the human lineage lost the ability to synthesize N-glycolylneuraminic acid (Neu5Gc), a cell-surface sialic acid receptor used by several viruses and other pathogens, around two million years ago (Varki, 2001), about the same time the archeological record suggests our ancestors shifted to a more carnivorous diet. The implication is that increased infectious disease load contributed to strong positive selection for a mutation that knocked out the binding site used by a number of infectious agents (including many RNA-based viruses) to initiate infection. Incidentally, in other animals that are able to synthesize it, such as chimpanzees, Neu5Gc is relatively scarce on brain cells, so its loss may have been involved somehow in the evolution of increased brain size in humans.

Today, HTLV, hepatitis, and picornaviruses cause mostly chronic or latent infections, although this may not have been the case in the past. When these viruses were first acquired, they were probably more virulent, and different hominin species may have had different susceptibilities, leading to the possibility of parasite-mediated "apparent competition" whereby disease was a factor contributing to the extinction of some branches of the human family tree. We know the human family tree had many branches that have died out, some of them earlier and some of them late. Genetic studies suggest that HTLV-I was transmitted to humans sometime during the last 50–100,000 years, whereas HTLV-II jumped from bonobos (*Pan paniscus*) to human ancestors about 400,000 years ago. Considering the array of retroviruses found in African monkeys and apes today and the ease with which they cross species barriers, ancient humans may have encountered more than just HTLV. Retroviruses have the ability to insert DNA copies of their RNA genomes into host cell DNA, so they can contribute to the genetic makeup of future generations.

"Endogenous retroviruses" are viral sequences that have become integrated into primate cellular DNA and are the traces of ancient germ-cell infections. These elements have the ability to retrotranspose into different parts of the host genome and affect gene expression directly and thus may have played an active role in the evolution of the human genome (Hughes & Coffin, 2001).

The process of acquiring new viral infections from animals has not stopped. Emerging viruses such as HIV, SARS, Ebola, avian influenza, and others as yet undiscovered have been and will continue to be the source of deadly outbreaks. Modern food-production methods that crowd large numbers of animals together in enclosed spaces increase the probability of disease emergence on a grand scale. Human population growth and the resulting environmental degradation compound the probability of emerging viruses that we will be hard-pressed to contain. Given our large, interbreeding populations and our ability to use technology to shield ourselves from natural selection, some would say that human biological evolution has ended. They do not take into consideration infectious disease, which will likely be the source of selection for genetic resistance factors and strong immune defenses against the emerging diseases of the future.

Further Resources

Garrett, L. (1994). *The coming plague: Newly emerging diseases in a world out of balance*. New York: Farrar, Straus & Giroux.

Hughes, J. F., & Coffin, J. M. (2001). Evidence for genomic rearrangements mediated by human endogenous retroviruses during primate evolution. *Nature Genetics, 29*, 487–89.

Normile, D. (2005). Researchers tie deadly SARS virus to bats. *Science, 309*, 2154–55.

Peeters, M., et al. (2002). Risk to human health from a plethora of simian immunodeficiency viruses in primate bushmeat. *Emerging Infectious Diseases, 8*(5). [Available online at http://www.cdc.gov/ncidod/EID/vol8no5/01-0522.htm]

Peterson, D. (2003). *Eating apes*. Berkeley: University of California Press.

Torrey, E. F., & Yolken, R. H. (2005). *Beasts of the earth: Animals, humans and disease*. New Brunswick, NJ: Rutgers University Press.

Van Blerkom, L. M. (2003). Role of viruses in human evolution. *Yearbook of Physical Anthropology, 46*, 14–46.

Varki, A. (2001). Loss of N-glycolylneuraminic acid in humans: Mechanisms, consequences, and implications for hominid evolution. *Yearbook of Physical Anthropology, 44*, 54–69.

Linda M. Van Blerkom

■ Health
India and Rabies Control

The bond between human and dog had its beginning twelve to fourteen millennia ago somewhere in Eurasia, where a reciprocal relationship between them first emerged. Provided with scraps of food when approaching the early encampments and settlements of man, the wolf soon became a frequent and welcome visitor, warning humans of imminent danger and later assisting them in the hunt for wild animals. Thus began the domestication of the dog and the establishment of a bond between humans and animals that has no equal. Today, humans violate that bond by allowing dogs to breed excessively and then abandoning them in great numbers, thus creating hazards for the dogs themselves as well as a considerable health risk to human society. Frequently, authorities faced with the problems caused by these dogs have destroyed them in the hope of finding a quick solution that turns out to be ineffective. Moreover, by temporarily reducing the population of straying dogs, the authorities improved the chances of survival of the remainder and provided fresh opportunities for newly abandoned dogs. It is now becoming recognized that removal of surplus dogs cannot solve the problem unless combined with other measures such as registration and neutering of dogs and education of the public.

At a certain population density, the birth rate and the death rate become equal, the population comes to an equilibrium, and population growth levels off. This more realistic description of population growth is referred to as logistic growth. The upper limit at which population growth levels off is called the carrying capacity of the environment. Each habitat has a specific carrying capacity for each species. This specific carrying capacity essentially depends on the availability, distribution, and quality of the resources (shelter, food, water) for the species concerned. The density of a population of higher vertebrates (including dogs) is almost always near the carrying capacity of the environment. Any reduction in population density through additional mortality is rapidly compensated by better reproduction and survival. In other words, when dogs are removed, the survivors' life expectancy increases because they have better access to the resources, and there is less competition for resources.

In 1964, appalled by the horrific way the Corporation of Madras was killing street dogs, the Blue Cross of India began to study this issue. The Blue Cross was surprised to learn that the Madras corporation—at 300 years one of the oldest corporations, or governing bodies of municipalities, in the world—started its catch-and-kill program in 1860. Dogs for which complaints were received were often shot on the street, and the complaints generally were about dogs that were biters and, therefore, suspected to be rabid. Section 218 of the Madras City Municipal Corporation Act of 1919 authorized catching and killing any dog on the street that did not have a license tag. S. Theodore Baskaran, the former Post Master General of Tamil Nadu states, "In the early 1970s, the number of stray dogs destroyed by the Corporation was so high that the Central Leather Research Institute, Madras, designed products—such as neckties and wallets—from dog skins." The number of dogs being killed by the corporation continued to rise after this period. So did the number of dogs on the street, and so did the number of cases of human rabies deaths.

The Blue Cross was convinced that if a procedure designed to control or eliminate street dogs had not shown positive results after being implemented for over 100 years, something was wrong. It was also convinced that where a dog had to be killed because it was overly aggressive or suspected to be rabid, the killing must be done in a more humane manner.

In 1964 the Blue Cross proposed a more humane and viable solution to prevent the visible increase in the number of street dogs and the number of cases of human rabies: a sustained catch-and-neuter program coupled with vaccination against rabies. It decided to call the program the Animal Birth Control program, or the ABC program, to show that the control of the street dog population was as easy as ABC.

As could be expected, the Madras corporation's response was to reject the proposal outright. The Blue Cross kept up the pressure on the corporation and began to spay and neuter all street dogs rescued by the Blue Cross. After treatment, the dog would be spayed, vaccinated, and released at the same spot from where it had been picked up. Owners were also encouraged to have their pets spayed and vaccinated free of charge. In the meantime, from an average of less than one dog per day in 1860, the number of dogs killed by the corporation went up to as high as 135 dogs per day in 1995.

In 1990 the World Society for the Protection of Animals (WSPA) and WHO brought out their *Guidelines for Dog Population Management,* followed by WSPA's *Stray Dog Control* guidelines. The report, authored by Dr. K. Bogel, chief veterinarian of the public health unit of the WHO in Switzerland, and John Hoyt says, "All too often, authorities confronted with the problems caused by these dogs have turned to mass destruction in the hope of finding a quick solution, only to discover that the destruction had to continue, year after year with no end in sight." Albert Einstein defined insanity as doing the same thing over and over again, expecting different results. The age-old method of catch-and-kill has not worked and never will.

It was in 1995 that the Blue Cross was finally able to get the Corporation of Madras to agree to try out ABC as an alternative to killing in a part of South Madras. The Blue Cross realized that a citywide ABC program would have been the ideal solution, but the corporation commissioner, Mr. M. Abul Hassan, asked to start the program and then increase its scope. The only assurance he gave the Blue Cross was that he would personally monitor the program and that no dog that had been spayed and vaccinated would be caught. Dogs in the area not covered by the ABC program would continue to be caught and killed by electrocution. The total cost of the program was met by the Blue Cross.

Chennai and Jaipur were the first cities to start sustained ABC-AR (Animal Birth Control-Anti-Rabies; www.bluecross.org.in/adoptathon.html) programs. Within six months, results in the areas covered by the Blue Cross ABC program were promising enough to prompt the corporation to extend the program to the whole of South Madras. By a stroke of luck, Mr. Abul Hassan became the special officer—equal to mayor—of the corporation. The group People for Animals (http://members.petfinder.org/~NJ17/index1.htm) agreed to take up ABC in North Madras, and the corporation converted its electrocution chamber to an ABC center.

Several cities have taken up ABC, but in many cases it has not been a sustained program. In many places where the ABC program was being implemented, local municipalities suddenly ordered the destruction of dogs on a massive scale in a knee-jerk reaction to complaints, and the dogs destroyed were usually the ones that had been spayed and vaccinated at great expense and effort.

A multi-centric study of rabies in India, sponsored by the World Health Organization, for the period 1993 to 2002 showed that the incidence of human rabies cases in India was more or less at a constant level during this period, at about 17,000 cases per year.

The purpose of the ABC program is to bring down the number of street dogs in a humane manner and, more importantly, to bring down the number of cases of rabies. To see whether this has been a success, let us look at the cases of human rabies in Chennai, where dog killing was stopped in 1996 and ABC was implemented. According to the numbers given by the Ministry of Health & Family Welfare Government of Tamil Nadu, and the Corporation of Chennai, numbers of rabies deaths in Chennai have steadily declined from 120 in 1996 to just 5 in 2003. However, the number of rabies deaths in India in cities where there is no ABC program has shown no major change during the same period. The following points must also be kept in mind.

1. The human population of Chennai has increased tremendously in the last few years because of an influx of people from rural areas, as well as because of the expanding city limits.
2. The awareness level regarding rabies has increased thanks to television and radio, and a far higher percentage of cases are reported today than five years ago.

3. Even a few years ago, a village person, currently living in a city, who was not well would often go away to his or her village to seek "traditional" treatment. Many dog-bite victims did so too.

We find a steady decrease in human rabies cases in those places where a proper ABC-AR program is being carried out. In Jaipur, the number of cases of rabies from the walled city where the group Help in Suffering (HIS) is carrying out the spay-and-vaccinate program is currently zero for the third year running.

In Kalimpong, where the program has been carried out by an HIS associate, there had been no reported cases for the last twenty-one months at the time of this writing. In the case of Kalimpong, the anti-rabies program has been much more widespread than the ABC program.

That ABC-AR does indeed work and is the only solution to the street dog issue is beyond doubt. What is now needed is the cooperation of the local municipalities and corporations to implement it properly. The Chennai Corporation has been a trendsetter and has shown the way to other municipalities and local bodies.

Further Resources

Beck, A. M. (1973). *The ecology of stray dogs: A study of free-ranging urban animals.* Baltimore: York Press.

———. (1975). The ecology of feral and free roving dogs in Baltimore. In M. W. Fox (Ed.), *The wild canids* (pp. 380–90). New York: Reinhold.

Bogel, K., & Hoyt, J. A. (1990). *Guidelines for dog population management.* World Health Organization and World Society for the Protection of Animals.

Fox, M. W., Beck, A. M., & Blackman, E. (1975). Behavior and ecology of a small group of urban dogs (*Canis familiaris*). *Applied Animal Ethology, 1,* 119–37.

Sudarshan, M. K. (2003, July). *National Multi Centric Rabies Survey—2003.* Report at the 5th National Conference on Prevention of Rabies at Bhubaneshwar, July 5–6, 2003.

World Health Organization. (2004, May). *Assessing burden of rabies in India.* WHO-sponsored National Multi-Centric Rabies Survey, Final Report.

S. Chinny Krishna

■ Health
Insect Pests in Human Economy and Health

The competition among insects and humans has always been intense. Insects represent approximately 95 to 97 percent of all animal species and compete with us for food, housing, and water. In addition to competing for resources, some insect pests spread disease, whereas others, such as human lice, depend on humans for nourishment.

The branches of science devoted to the economic impact of insects and to the interaction between insects and health are known as economic entomology and medical entomology, respectively. Through research often spanning 100 or more years, these sciences have taken great care to demonstrate that not all insects are pests. Almost all insects are considered beneficial to humans and include such well-known pollinators as the honeybee, bumble bee, and butterfly. Other insects, such as the ladybug and praying mantis, are beneficial to humans because they are predators of insect pests.

Approximately 1 percent of insects are considered pests. An insect pest is any insect that through its behavior injures or spreads disease to humans, animals, crops, and ecosystems. An insect is also considered a pest if its behavior weakens or destroys dwellings and structures. Some well-known insect pests are fire ants, army ants, termites, mosquitoes, boll weevils, and locusts.

The economic impact of insect pests is substantial. Consider, for instance, the desert locust story from the Old Testament. In parts of Africa, the Near East, and Asia, crop damage by locusts is in the billions of dollars and leads directly to almost unimaginable levels of starvation and death. In 2004 an infestation of locusts in China caused over 12 billion dollars in economic loss. It is interesting to note that in the United States, the state bird of Utah is the California seagull because of the role that bird played in saving the 1848 wheat crop from grasshoppers (now known as the "Mormon cricket"), thus playing a vital role in the survival of the Mormons during their first year in Utah.

In addition to locusts, a wide variety of insect pests feed upon fruit, vegetables, and stored grains. Some of the most economically important and dangerous include the cherry fruit fly, the Mediterranean fruit fly, the Colorado potato beetle, the codling moth (which destroys apples), the cotton boll weevil, and the tobacco hornworm. Various species of grain beetles and grain moths reduce the weight and nutritional value of stored grain. Worldwide damage to these crops and grain products is estimated to be in the billions of dollars.

In addition to crops, insect pests reduce the growth rate of trees through defoliation. The gypsy moth, for example, defoliates oak, gray birch, and fruit trees. Some species of bark beetle can destroy a tree by disrupting the transport of water within the tree. Destruction of trees negatively impacts a variety of industries including construction, papermaking, and aesthetics. Moreover, deforestation and defoliation reduce existing home values. The U.S. government spends over 200 million dollars a year to prevent the destruction of forests by insect pests.

Insect pests also attack farm and pet animals. Direct reduction in growth because of blood sucking is insignificant; the greater effect is on the disruption of feeding behavior in the afflicted animal. These insect pests include the stable fly, horn fly, horse and deer flies, face fly, and various species of lice.

The effect of insect pests on human health and welfare is equally devastating. Transmission of microbial diseases and parasites is a major concern of medical entomology. Illness and death caused by insect-transmitted diseases is estimated to be in the high millions. Mosquitoes, for example, can transmit dengue fever, malaria, yellow fever, Japanese encephalitis, and West Nile virus. Malaria causes approximately one million deaths annually and affects over 500 million people worldwide. The U.S. government spends 200 million per year to combat malaria. American trypanosomiasis, which can be contracted through contact with contaminated feces of an infected assassin bug, is the fourth leading cause of death in Latin America, with approximately 43,000 deaths a year. Even the common housefly is potentially dangerous because, although it does not bite, it can carry a variety of bacteria and protozoans that cause cholera and typhoid fever.

In addition to the transmission of disease, some insects are dangerous because their bites or stings can produce an allergic reaction. These insects include wasps, ants, and honeybees. The allergic reaction can be so intense that anaphylactic shock develops, and if the reaction is left untreated, death may result. Annually, there are approximately fifty deaths from insect bites and stings in the United States. Moreover, it is estimated that 6 percent of the United States population has a fear of insects and animals. The intense fear of insects, known as entomophobia, is characterized by cleaning and sterilization rituals that affect the quality of life from those suffering from this irrational fear.

One of the greatest economic effects of insects is the damage caused by termites. It is estimated that over 600,000 homes suffer some form of termite damage in the United States. Annual costs associated with termite damage is in excess of $1.5 billion, and more than 2 million U.S. homes require some form of consistent termite control.

Fear, crop damage, destruction of trees and homes, disruption of animal behavior, disease, and death have spawned one of the largest industries in the world. Pest control is a billion-dollar industry affecting the lives of most humans around the world. In today's society, humans are constantly being exposed to a variety of pesticides and insecticides designed to control insect pests. In response to problems associated with the use of agro-chemicals, alternative forms of pest control are being developed, including genetically engineered fruits, vegetables, and grain, biological control, and the use of integrated pest management strategies.

Further Resources

Alford, D. V. (1999). *Textbook of agricultural entomology*. London: Blackwell.

Mullen, G., & Gurden, L. (Eds.). (2002). *Medical and veterinary entomology*. New York: Academic Press.

Winston, M. L. (1997). *Nature wars: People vs. pests*. Cambridge, MA: Harvard University Press.

Charles I. Abramson and Janko Bozic

■ Health
Parasites and Humans

Whether we like it or not, numerous animals inhabit our bodies and living quarters as parasites. In fact, it has been estimated that up to 10 percent of our dry weight is made up of parasites and other symbionts that rely on us for their livelihood. Although we may be unhappy with this relationship, these animals depend on us for their survival. Some, like the medicinal leech and fly larvae, have been put to use in treating medical problems; consequently, there is a clear benefit to both human and animal. Others, like the tapeworm, follicle mite, flea, tick, and bed bug, appear to benefit at our expense. This essay discusses the idea of parasites and introduces a few of the strangest and commonest with whom we share ourselves.

Parasitism is defined as a type of predation that occurs over a long period of time and usually does not kill the host. It is estimated that nearly 50 percent of animal species are parasitic at some point in their life cycles. Traditionally, it is believed that the parasite benefits at the cost of the host, but in the late 1960s, a new hypothesis was put forth called the "goodness of parasitism," which suggested that the host benefited from certain aspects of hosting a parasite. The hypothesis suggested that parasitism could be considered a mutualistic interaction, where both host and parasite benefit. For instance, in humans, some parasitic worms can be used to fight off certain diseases such as colitis, hay fever, asthma, and other autoimmune diseases. Dust mites, which inhabit our living spaces and sleep with us in bed at night, assist with house cleaning by ingesting the organic trash—for example, skin cells that we slough off every day. But whether these examples are special cases or the rule and whether the benefits outweigh the costs are two unanswered questions.

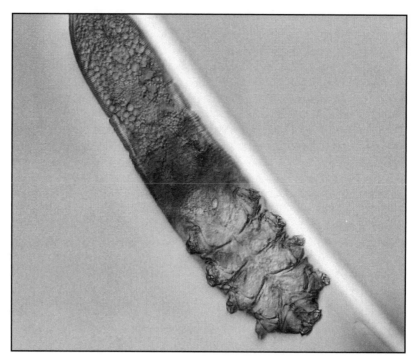

A follicle mite on a strand of hair, magnified 450 times. ©Peter Arnold, Inc./Alamy.

Whatever the case, humans and their parasites have been coevolving for millions of years. As a result, the relationship between human host and parasite is tight, often bizarre, and quite complex. Consider the follicle mite, a small worm-like relative of spiders that inhabits the hair and eyelash follicles of over 98 percent of all living humans. These invertebrates crawl inside a follicle, attach themselves via their legs and mouths to the follicle base and then turn our cells into a brothy soup, which they then suck up. Their feeding is so efficient that they excrete no waste. At night while we sleep, they may leave the follicle and migrate across our skin at a rate of about 1 cm per hour. In healthy people, mites inhabit about every tenth follicle. The mite lays eggs asexually and then mates with her offspring to produce fertilized eggs. The babies wander across our skin to uninhabited follicles. The mites that inhabit hair cells on a mother's breast may leave those hair cells to hitch a ride with a suckling baby, thus beginning the colonization of a new human and ensuring the continuation of the follicle mite species into a new generation of humans. It is estimated that newborns become colonized by these mites within a few hours of birth. Hence, if human babies required no care, and if humans had limited interpersonal contact, these mites would likely perish.

Many parasites have complex life cycles, which involve spending some of their life living inside humans and another part of their lives in our foods (e.g., fruits and meats). For instance, the tapeworm lives inside the human intestine and passes its eggs in fecal material. The eggs hatch, and the larvae are picked up by animals that are often eaten by humans (e.g., insects that live in fruit, fish, cattle, pigs, and so on). The tapeworm larva develops into a cyst in these hosts. If a human then eats an infested food source, the cyst matures, grows a head and neck, latches with its teeth onto the human's intestinal lining, and then begins producing body segments that will mature and ultimately become eggs,

■ Enrichment for Animals

Emotional Enrichment of Captive Big Cats

Louis Dorfman shown engaging in emotional enrichment of an adult male lion. Note the level of contentedness exhibited by this animal.
Courtesy of Scott Coleman.

Behavioral enrichment devices that encourage play are used at International Exotic Feline Sanctuary. For example, pumpkins are a seasonal treat with a strong appeal to tigers and other big cats.
Courtesy of Scott Coleman.

■ Ethics and Animal Protection

Animals in Disasters

Oklahoma City firefighters and animal welfare workers remove Wanette, a Belgian draft horse, from the mud at a pond on a horse farm, 2006. The horse was stuck for more than six hours, and it took rescue workers more than an hour to free the horse.
©AP Photo/The Oklahoman, Chris Landsberger.

Animals in Disasters

Kathlean Baska and Roy Krueger, from the Missouri Boone County Urban Search and Rescue Task Force 1, hold rescued kittens that they found while searching a house in the aftermath of Hurricane Katrina, 2005.

Courtesy of FEMA.

Goodall, Jane

Jane Goodall gives a little kiss to Tess, a female chimpanzee at the Sweetwaters Chimpanzee Sanctuary near Nanyuki, 110 miles north of Nairobi, Kenya, 1997. When young, orphaned chimpanzees first come to the sanctuary, they are given lots of affection to compensate for the loss of their mothers.

©AP Photo/Jean-Marc Bouju.

People for the Ethical Treatment of Animals

A People for the Ethical Treatment of Animals (PETA) campaigner lies in a plastic covered tray, mimicking meat packaging, during a protest in downtown Hamburg, northern Germany.
©AP Photo/The Oklahoman, Chris Landsberger.

■ **Geography**

Animal Geographies

Chinese rescue workers securing a safety rope around a tranquilized wild giant panda in a tree at Dujiangyan city, Sichuan, 2005. The panda led residents of Dujiangyan on a chase after wandering into the town.
©AP Photo/EyePress.

Benefits of Animal Contact

A woman kisses her tame parrot. As social creatures, humans require companionship for their physical and emotional health. Contact with animals provides numerous benefits, including physical touch, encouragement to speak, and an increase in exercise, sociability, and mental focus.

Courtesy of Shutterstock.

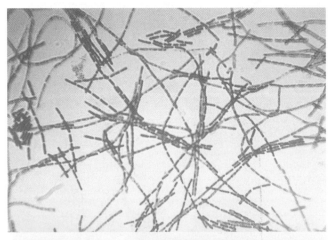

Diseases between Animals and Humans (Zoonotic Disease)

Bacillus anthracis bacteria, better known as anthrax, and tuberculosis have posed a common threat to human and animal populations throughout history. While efforts to limit the spread of the microorganisms, including key changes in the handling of animals and animal products, have been successful in modern times, endemic infections persist. Alarmingly, several countries have recently developed anthrax for use as a biological weapon.

Courtesy of the CDC.

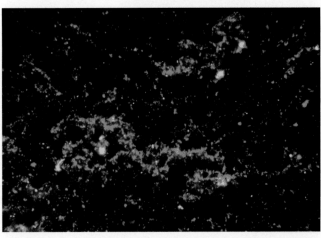

Humans, Viruses, and Evolution

This fifteenth-century anonymous Italian fresco painting shows animals and humans together in farming and in the community. Some diseases, such as measles and influenza, resulted from the intense human-animal interactions surrounding domestication and the adoption of an agricultural lifestyle.

©Scala/Art Resource, NY.

History

Harvesting from Honeybees

A bee flying toward a sunflower. The bee's ability to transform flower nectar into a nonperishable sweetener has attracted the attention of sweet-toothed humans since prehistoric times. Early attempts to retrieve honey often proved fatal to a natural hive, but over the centuries beekeeping has become a domesticated business comparable to farming.

Courtesy of Shutterstock.

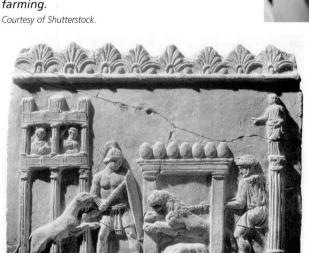

Animals in Ancient Roman Arenas

A deadly contest between lion, lioness, and gladiators is displayed in this terracotta bas-relief from the first to third century.

©Erich Lessing/Art Resource, NY.

History

The Turkey in History

The turkey highlights the growing conflict in Western culture between the age-old presumption that animals exist solely for humans to exploit and the view that nonhuman animals are kin to humans with value and autonomy in their own right.

Courtesy of Shutterstock.

Human Anthropogenic Effects on Animals

Human Anthropogenic Effects on Animals

Marc Bekoff and his good friend, Willie.

Courtesy of Jan Nystrom.

Hunting, Fishing, and Trapping

Falcons, Hawks, and Nocturnal Birds of Prey

A Qatari falconer's bird, a shaheen peregrine, at rest in a tent decorated with Arabian textiles as part an exhibition of traditional life during the Doha Cultural Festival. The practice of using birds of prey for hunting is believed to have originated in the Middle East around 700 BCE, but also gained popularity in East Asian, Mediterranean, and European cultures.

Courtesy of Shutterstock.

Humans as the Hunted

Woodcutters wearing government-issued masks approach a tiger-infested island in the Sunderbans Tiger Reserve. As Royal Bengal tigers rarely attack from the front, the government began issuing masks as a measure to protect those frequenting the island from attack. Between 1975 and 1985, 612 people were killed by tigers in the Sunderbans delta.

© AP Photo/John Moore.

■ Literature

Human Communication's Effects on Relationships with Animals

An animated version of the children's classic Charlotte's Web. *The story features an unlikely relationship between a friendly spider and a young pig trying to escape slaughter.*

Courtesy of Photofest.

▨ Living with Animals

Corals and Humans

An intact reef in Indonesia. Coral exist in a delicate, interdependent relationship with their ocean environment and form the backbone of vibrant underwater ecosystems.

Courtesy of Shutterstock.

Dolphins and Humans: From the Sea to the Swimming Pool

A drive fishery, one of many slaughters of dolphins and other cetaceans in Japan, which still continue.

Courtesy of Toni Frohoff.

thus completing the life cycle. Clearly, parasites such as the tapeworm emphasize the tight and intimate relationship between humans and the lives of their food sources.

As repulsive as these parasite examples may be, many human societies are not, or have not been, so put off by parasites. Up until a few centuries ago, royalty in Europe considered parasites such as lice and fleas to be a common part of life. Even rodents nested in the white wigs of high society! Poets wrote romantically of the insect infestations shared by lovers. And picking parasites off each other or throwing parasites at each other were acceptable behaviors woven into the cultural fabric. Even in modern times among certain societies and cultures, parasites are accepted as a normal part of life.

Currently in the United States, parasitic infections are surprisingly common. It is estimated that one-third of the U.S. population is infected with intestinal parasites, resulting in gastrointestinal symptoms (diarrhea, bloating, cramps, constipation, and other digestive disorders), fatigue, nervous disorders, pain, allergies, nausea, muscle weakness, headaches, fevers, sleeping disorders, and weight changes. Obviously, these symptoms are not always indicative of parasite infection because these symptoms can have non-parasitic causes too.

Many parasites of the world can cause severe diseases and disorders, including bubonic plague, malaria, sleeping sickness, and even blindness. But not all parasitic relationships are harmful. For instance, our skin, mouth, and many internal organs are protected by organisms that help ward off infectious organisms in a kind of turf war. We would not be able to digest many of the foods we eat without the bacterial flora that lives in our intestines. In this sense, each of us is an ecosystem that is protected and even sustained by our parasitic inhabitants.

But even at a more fundamental level, we rely on a historical bonding of parasites. Within each of our trillion or so cells are organelles called mitochondria, which are responsible for turning the oxygen we breathe into energy to keep us alive and allow us to move. There is good evidence that these mitochondria were once free-living cells that were engulfed by another cell. But instead of being digested, the mitochondria formed a symbiotic relationship with their host cell, and this symbiosis allowed multicellular organisms to evolve. Hence, at a cellular level, we are a living manifestation of a successful parasitic relationship between two organisms that have formed a matrimonial relationship that has lasted for hundreds of millions of years.

Finally, there is at least one aspect of parasites related to healthy environments. Scientists have discovered that the healthier an ecosystem, the more parasite species found in the ecosystem. By contrast, polluted ecosystems contain fewer parasitic species, and the few parasitic species they do have occur in large numbers. So when it comes to parasites in the environment, the more species, the healthier.

In summary, we humans are bound to parasitic relationships that have existed since the dawn of multicellular organisms. Parasites have protected us and harmed us, creating a complex relationship that is still poorly understood and underappreciated. It appears that for our environments to be healthy, parasites are here to stay, inhabiting our cells, our organs and bodies, and our food, beds, clothing, homes, and children. We may not like it, but our uneasy alliance with these relatively small organisms made us and keeps us who we are.

Further Resources

Bogitsh, B., Carter, C., & Oeltmann, T. (2005). *Human parasitology* (3rd ed.). New York: Academic Press.

Combes, C. (2001). *Parasitism: The ecology and evolution of intimate interactions.* Chicago: University of Chicago Press.

Conniff, R. (1998, December). Body beasts. *National Geographic.* [Available online at http://www.nationalgeographic.com/ngm/9812/fngm/index.html]

Geoffrey E. Gerstner and Ben Dirlikov

Health
Rabies

Rabies is a disease that has haunted mankind for over two millennia. Roman scholars recorded contraction of the disease from the saliva of rabid dogs, assuming the infectious substance was a poison in the saliva. Nearly half a century earlier, ancient Greeks described dog rabies as "a madness" transmissible through the bite of an affected animal. However, rabies is likely to have been known to man far longer than 2,500 years because the Latin word "rabies" comes from the Sanskrit word "rabhas," which links the disease to ancient civilizations of the Indus Valley at least a thousand years earlier. Of course, rabies is not the result of a poison as conjectured by the ancient Romans. Rabies is the result of a viral infection (like AIDS, the common cold, or flu), but the biological nature of rabies was not understood until the nineteenth century.

Human rabies infection has an incubation period of ten days to more than a year, during which time few symptoms appear. The duration of the incubation period is dependent on where the infection enters the body. Infections initiated in regions closer to the brain (such as the face) hasten the disease, as do infections in regions with more nerve fibers. Typically, the incubation period is from one to two months. Onset of symptoms starts with depression and flu-like symptoms and can include some paralysis near the initial site of infection. These symptoms worsen over the period of approximately a week, leading to uncoordinated movement as paralysis spreads, excessive salivation, and anxiety. Exceedingly painful throat-muscle spasms are triggered by drinking, eating, or even air movement. This causes the unfortunate victim of rabies to develop a strong phobia of water despite severe dehydration (hydrophobia). Coma and death from respiratory failure (paralysis) follow within two weeks of the onset of symptoms of the disease. Once symptoms occur, rabies is fatal, but treatment prior to the onset of symptoms is nearly 100 percent effective in stopping progression of the disease. Thus, early treatment is essential if exposure to rabies is suspected!

Any mammal can be infected with rabies; however, carnivores and omnivores are the reservoir for the disease. Humans contract rabies typically by being bitten by an infected animal. About 35,000 people die worldwide from rabies each year, and over a million people receive rabies treatment worldwide each year for animal bites. Historically, human infection occurred from being bitten by domesticated dogs, pets. This is still true throughout most of the developing world and was the case in the United States until the 1950s, when widespread dog immunization for rabies became prevalent. Rabies in the United States and other industrial nations is now nearly limited to wildlife (90%), but domestic cat cases have increased significantly (Black, p. 724). Primary rabies reservoirs are bat, coyote, fox, raccoon, and skunk populations. However, rabies is known to occur also in rodents and livestock as the result of attacks from predators. Rabies is a reemerging disease in the United States. Raccoon rabies infection has spread from Florida since the 1950s and now occurs throughout the eastern United States. Skunks are the largest reservoir for rabies throughout the middle states and California, and the fox population is an important reservoir in the southwestern states, in Alaska, and in Canada. Infected bats occur throughout the United States. Bat transmission of the disease to humans may occur from respiratory infections contracted from bat-inhabited caves as well as from being bitten by a bat. Recently, rabies has become a severe health problem in China, especially in the southern provinces. As a result, many dogs have been killed. In some places, people are allowed to have one dog and within a certain size limit. Animal rights spokespeople in China have spoken out in favor of vaccination, as opposed to extermination of dogs and other animals.

The rabies virus is an RNA-containing rhabdovirus. That is, the genetic material packaged in the virus's shell is RNA rather than DNA. The genetic material is surrounded by a protective protein coat (Centers for Disease Control, 2005). Treatment for rabies involves thorough cleaning of the wound, application of hyperimmune rabies serum in and around the wound to neutralize virus particles before they reach the nervous system, and injection of a series of vaccines to induce production of antibodies. Once the rabies virus has reached the central nervous system, these measures become largely ineffective because of the blood-brain barrier. Because rabies is caused by a virus rather than a bacterium, antibiotics are not an effective treatment. Vaccinating mammalian pets regularly is the best means of limiting spread of the disease to humans. Bait for foxes and raccoons that contains an oral rabies vaccine appears promising as a control for rabies in wildlife populations in the near future. Not all animals develop the stereotypic "furious rabies." Many animals develop "dumb rabies," in which paralysis predominates. Some animals, such as bats, can remain asymptomatic for prolonged periods and can even become prolonged carriers able to transmit the disease. Thus, care must be taken when any abnormal behavior is observed in domestic or wild mammals.

Further Resources

Black, J. G. (2005). *Microbiology* (6th ed.). Hoboken, NJ: Wiley.

Centers for Disease Control. (2005). http://www.cdc.gov/ncidod/dvrd/rabies/

Halliday, T. (1998). *Pakistan*. Singapore: APA.

Krebs, J. W. (2004). Rabies: Ancient malady, new twists. In D. Schlossberg (Ed.), *Infections of leisure* (3rd ed., pp. 277–300). Washington, DC: ASM Press.

Madigan, M. T., Martinko, J. M., & Parker, J. (2003). *Brock biology of microorganisms* (10th ed.). Upper Saddle River, NJ: Prentice Hall.

Harrington Wells and Charles I. Abramson

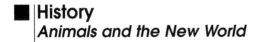

History
Animals and the New World

Introduction

The European colonization of the Americas initiated in 1492 was a meeting not only of human cultures but also of diverse evolutionary histories and nonhuman species. This article focuses on the first 150 years of contact (roughly 1492–1650), while touching on the widening circles that exist beyond the first series of contacts between the Old World and the New. It is somewhat incongruous to employ the phrase "New World" because the lands called by that name had been inhabited for thousands of years. To designate the Americas as a new world—or, as Columbus called it, "another world" (otro mundo)—invokes the perspective of the colonizers. However, the term is also revealing because the European encounter with these unfamiliar lands and their inhabitants expanded perceptions of geography on both sides of the Atlantic. In the wake of such early encounters, both European colonizers and native residents of the Americas scrambled to make sense of the creatures that confronted them, using the only referents they had: the mythologies, theologies, philosophies, folktales, and historical narratives of the worlds they already knew.

Native American Wild Animals

Separated by an ocean, Europe and the Americas possessed distinctive fauna and distinctive human cultures. Soon after European explorers arrived in the New World, they began sending reports that highlighted the contrasts between a largely domesticated Europe and the "Eden" awaiting on the other side of the Atlantic for those bold enough to seize the opportunity. Among the first to make such a claim was Christopher Columbus, whose reflections on his early voyages foreshadow many of the differing interpretations of the New World that followed him. Columbus's journals reveal a man propelled by his faith in providential destiny who viewed himself as a divine instrument and interpreted the New World through the lenses of the biblical scriptures and the works of classical authors. The journals also reveal a concern with justifying the merit of his voyages, particularly to his patrons, King Ferdinand and Queen Isabella of Spain.

Columbus's observations of indigenous peoples and fauna, therefore, enthusiastically lent support to his claims of the Eden-like character of the Caribbean islands, as well as their potential material riches. Wild quadrupeds were scarce on the islands, but the abundance and variety of fish made an impression on Columbus. So did the unfamiliar sea turtle, which, he noted, "looked like a proper pig . . . [though] it was all shell, very hard, and it had no soft place except the neck and eyes." Columbus also puzzled over dogs that did not bark—an observation repeated by other early writers—and expressed excitement about birds of all colors and sizes. In the latter category, parrots were particularly important to him, for according to folk wisdom parrots indicated heat, and heat indicated wealth.

When European sailors and chroniclers could not identify newly encountered animals by reference to familiar Old World species, they often employed mythical classifications, as is evident in the early reports that made mention of mermaids, Cyclopes, phoenixes, men with tails, giants, and other monsters. Europeans also were perplexed by unfamiliar, seemingly hybrid animals such as the armadillo and the iguana. Later historians, such as Gonzalo Fernández de Oviedo, the official "Chronicler of the Indies," outlined basic classificatory systems that first compared the animal in question to Spain's fauna (when this was possible) and then underscored the differences that made the New World animal unique. Oviedo's writing represents a proto-scientific attempt to characterize animals in the New World through scrupulous observation, occasional experimentation, and the corroborating testimony of others, which he took pains to confirm.

In the decades following Columbus's voyages, and in contrast to the limited fauna of the Caribbean islands, the mainland areas of Mesoamerica and South America confronted explorers and conquistadors with a vast diversity of animal species. Visions of an earthly paradise remained influential—so, too, did beliefs in a recapitulation of the Golden Age, which were inspired by the Renaissance humanism of the time. The abundance of wild animals (generally depicted as meek and timorous) was used as evidence to support these various notions of a mythic paradise on earth. Though Christian interpretations, classical texts, and medieval bestiaries served as the primary mode of categorizing the New World's fauna, at least some persons allowed the New World fauna to challenge their preconceived notions. For example, Amerigo Vespucci, the person for whom the Americas are named, expressed doubt over how so many different kinds of animals could have fit into Noah's ark.

Native American Domesticated Animals

By the time Columbus sailed the Atlantic, domesticated animals, including dogs, goats, sheep, cattle, pigs, fowl, cats, and horses, were ubiquitous throughout Europe (as in many African, Asian, and Middle Eastern societies). European domesticated animals were companions and also served various utilitarian purposes, including transportation, food, clothing, aid with hunting, and protection. In contrast, pre-conquest Amerindian cultures had very few domestic animals prior to colonization. One notable exception was dogs, the only domestic animals present among the majority of Native American groups throughout the hemisphere, from Alaska and northern Canada south to Patagonia.

Though the dates and locations for canine domestication remains contested, dogs most likely diverged from their wolf ancestors in Eurasia over 15,000 years ago, long before the first humans crossed the Bering land bridge around 12,000 years ago. It is probable that domesticated dogs participated in the first human settlements in the Americas. The earliest clear archaeological evidence for true domestic dogs in North America dates from around 8000 BCE, and there is evidence of the deliberate burial of dogs in the Illinois River Valley from 6500 BCE. Domestic dogs appear to have been present in almost all indigenous cultures in the Americas except the interior of the Amazonian rainforest.

These indigenous American dogs varied across cultures and ecosystems, from the large, powerful and well-furred Arctic hauling dogs to much smaller and shorter-haired varieties in Mesoamerica. These differences reflected their wide-ranging roles in human cultures. American dogs, like those in Europe, assisted with hunting, herding, and hauling and served as companions and protectors. Unlike European dogs, however, American dogs were used as food sources in some, though not all, cultures. They also played important symbolic roles in many cultures, especially as companions and guides in afterlife journeys. Perhaps in a telling sign of events to come, a Native American dog was likely the first casualty of conquest in North America, killed around 1525 by a Spanish musket ball. Its remains are buried in a Timucua village on St. Simons Island, Georgia.

European Domesticated Animals and the Conquest

Although indigenous American dogs were among the first victims of the conquest, European dogs accompanied the first conquistadors, assisting their masters in the project of colonization. The Spaniards used their dogs, including war-trained mastiffs, to punish and kill human natives. Early colonial testimonies and drawings testify to the terror that these animals inspired.

Europeans brought not only dogs and horses to help in the conquest but also a host of other domestic species. The apparent scarcity of quadrupeds on the Caribbean islands, noted by Columbus and others, was quickly remedied by the importation of pigs and cattle from Spain. As early sixteenth-century chronicler Peter Martyr d'Anghiera noted approvingly, "our cattle are said to be born fatter there [Hispaniola], and turn out much bigger on account of the lush pastures." Martyr also reported similar beneficial results for imported horses, commenting that these animals achieve "perfection" in the New World. With the utilitarian gaze of a soldier bringing wealth to his country, the Spanish conquistador Hernán Cortés noted the abundance of "game and animals and birds" and assessed the valleys of central Mexico as "very well suited for the raising of all sorts of livestock and growing each and every one of the plants of our nation." This emphasis on domestic stock presaged the important changes that such animals would enact upon the New World landscape.

Whether as an aid for conquest (horses and hunting dogs), a source of food (pigs, sheep, and cattle), or carriers of foreign diseases, animals were instrumental in weakening the physical and psychological resistance of America's indigenous inhabitants. The ability of Europeans to establish their presence in the Americas depended in large part upon the relationships they had with nonhuman animals. Domesticated animals, such as horses, provided advantages in military conquest, trade, and speed of movement. European cattle and pigs provided a portable food source and enabled long-term settlements that would have otherwise been more difficult to maintain.

Another advantage that domesticated animals provided Europeans was less visible but perhaps more potent: the close quarters that Europeans had maintained with sheep, pigs, and cattle for centuries bestowed upon them a resistance to pathogens that native inhabitants of the New World did not have. For this reason, biologist Evan Eisenberg wryly remarks that the Age of Exploration was also the Age of Infection. More generally, the transatlantic exchange of peoples and animals brought together continents that had been separated for millennia, eventually disrupting and augmenting ecological relationships on a massive scale.

Two examples illustrate in particularly dramatic fashion how European domesticated animals influenced the course of American history. The first comes from Mesoamerica, where the Aztec (Mexica) empire was rapidly conquered by a small Spanish army led by Hernán Cortés. The descriptions written by Cortés and other Spanish participants, as well as the native accounts that were transcribed by Spanish missionaries, leave no doubt that Spanish horses played a crucial role in the incredible success of his military expedition in Mexico, carried out with vastly inferior numbers (a few hundred men, compared to hundreds of thousands of Aztecs) and completed in two years (1519–21). In addition to the use of gunpowder and cannons, these horses provided both speed of movement, and perhaps more importantly, a psychological advantage. Transcriptions of native accounts recall the armored "deer" and their thunderous hooves that intimidated and mystified the Aztecs. Cortes, always a shrewd tactician, even buried the corpses of horses that were killed in battle to sustain the confusion among the natives over whether the horses were mortal or not.

In the Peruvian highlands, a similar blitzkrieg of the Inca empire took place, conducted by Francisco Pizarro and 168 Spaniard soldiers. Like Cortes' conquest of the Aztecs, mounted cavalry played a decisive role in the Incas' subjugation. In the Andes and on the Peruvian coast, the only large domesticated animal was the llama (alpaca), whose uses were confined to food, transport, and clothing, and thus the Incas had no counterpart for the military might of the horse. Given the sophisticated numbers and statecraft of the Inca empire, one might still consider the dramatic conquest of the Inca an improbable outcome. However, with a clever strategy, superior steel weapons and armor, and the use of blaring trumpets, exploding guns, and horses wearing rattles, Pizarro captured Inca emperor Atahualpa and killed many other unarmed Inca chiefs who attempted to bear his royal platform in a critical encounter at Cajamraca on November 16, 1532. Panicked Indians, running from the slaughter, were cut down by Spanish soldiers on horseback. In subsequent years, the Incas devised methods to ambush Spanish horsemen, but out in open flatlands they were no match against this domestic alliance between the Spanish and their horses.

Despite the importance of these dramatic events, organisms of microscopic proportions may have been most important in the conquest and subjugation of the New World. Since populations in Europe developed concentrated agricultural settlements earlier than American Indians and lived in close proximity to domesticated animal communities, Europeans (and their domesticated animals) hosted a variety of diseases, plagues, and sicknesses. Immunities gained in Europe had no counterpart in the Americas. Infectious diseases, especially smallpox, were the single largest factor in the decline in native populations in America after contact with Europeans, leading to the deaths of up to 95% of the population in some areas. These pathogens, often spreading in advance of the colonizers, killed thousands of native peoples at a time and left vast areas depopulated. The indigenous people that did survive found their resistance to further incursions weakened, their communities fragmented, and their social and religious structures irrevocably transformed.

Animals in the Colonization of North America

As in the southern latitudes, early reports from North America described an abundance of wildlife, particularly of animals that were considered valuable commodities. Early seventeenth-century fishermen reported having to throw Atlantic cod back into the water for lack of space on their boats, and early colonists marveled at sturgeon spawning runs, which so filled the streams that people speculated whether or not they could walk across the backs of fishes to cross from one shore to the other. Bird migrations also awed the Europeans, who told of flocks of passenger pigeons blotting out the light from the sun. Turkeys and deer were found in such great numbers that they could be hunted throughout the year, and colonial promoters were quick to offer visions of ease and plenty to lure prospective colonists.

Several factors lie behind this abundance, including the ways that Native American populations had seasonally managed these environments for centuries. Through systematic and cyclical selective burning, clearing, and planting, Amerindians in southern New England created what historian William Cronon describes as an ecological mosaic, opening up hunting areas that attracted deer, elk, and rabbits, and the predators that fed on them, as well as increasing the rate of nutrient recycling, which enabled wild berries and other harvestable foods to flourish.

The English colonists practiced a different, and ultimately much more damaging, form of husbandry. After 1620, domesticated stock in New England became a significant factor in ecological changes. Initially, pigs and cattle were let loose to roam the countryside and woodlands, creating internal disputes with colonial farmers and external confrontations with neighboring Indian tribes. As the English population increased, wooden fence enclosures proved a necessity for separating crops and livestock, both to protect economic interests and establish property lines that were intended to prohibit native tribes' seasonal use of the land. These fences also encouraged further destruction and displacement of wildlife endemic to the region. For example, though wolf bounties were an early method of clearing the "howling" wilderness (a popular auditory description of the New England forests), land enclosures exacerbated wolves' plight, for as Cronon notes, "Because, unlike Indians, wolves were incapable of distinguishing an owned animal from a wild one, the drawing of new property boundaries on the New England landscape inevitably meant their death."

The colonial fur trade, aided by Native American expertise, also proved devastating to the continent's mammal populations, especially beavers, otters, and moose, and the decline of these animals further altered the New England landscape. The substantial reduction of deer, in particular, not to mention bear, lynx, elk, and other animals, also had the consequence of making once-mobile Native Americans more dependent on agriculture and sedentary settlements.

Conversely, on the Great Plains, imported animals increased Native American mobility. After escaping from Spanish settlements in the early seventeenth century and migrating northward, wild horses were tamed by Native plains tribes, prompting these tribes to leave behind agricultural settlements in favor of more nomadic lifeways. Horses enabled Plains tribes to travel long distances in search of buffalo and improved their efficiency in hunting. Horses also, ironically, allowed some Native North Americans, such as the Apaches, to resist Anglo conquest for much longer than they would have otherwise.

It is worth noting that some European animal "imports" were unintentional. Securing safe passage on colonial ships, smaller animals like the black fly, the cockroach, and the gray rat found new homes on the continent. Thus, by the end of the seventeenth century, portions of the New World were beginning to look very much like the Old.

Cultural Animals

Human beings, as cultural animals dependent on inherited forms of social learning, may need long periods of time to learn of the impacts of their actions and adapt their lifestyles accordingly when confronted with unfamiliar landscapes. In light of this consideration, anthropologist Gary Paul Nabhan rejects the view that pre-Columbian cultures became "instant natives" to their newfound homelands, arguing that it takes time for any culture to become sensitized to ecological constraints. Strong evidence suggests that the arrival of humans who entered North America from Siberia across the Bering Strait (c. 12,000 BCE) contributed to an extinction episode that included the demise of many large animals—a factor that proved critical regarding the lack of large domesticated animals in the Americas. Nabhan argues that the successful adaptations of many contemporary Native American tribes, including their carefully managed environments, is a result of their learning from the mistakes of past episodes of overexploitation—in other words, as a product of time-tested trial and error.

European colonizers of the New World, not knowing the land or its inhabitants, also did not (and in many cases, could not) realize the extensive impacts that their presence and actions would have. In this sense, a "New World" for any human culture may create an illusion of unlimited possibility. Yet any independence felt by humans on unfamiliar soils, and the (temporary) ability to circumvent local resource shortages by extracting these from "new" landscapes, if performed at unsustainable scales, can only eventually lead to limitations and negative feedback from oversimplified landscapes. From a biological perspective, mutualistic relationships between human and nonhuman animals always enact changes, and these episodes of change can be dramatic, as they were in the New World; whether or not these changes are suitable for long-term survival is a matter worthy of reflection.

Further Resources

Benson, E. (1997). *Birds and beasts of ancient Latin America*. Gainesville, FL: University Press of Florida.

Cronon, W. (1983). *Changes in the land: Indians, colonists, and the ecology of New England*. New York: Hill and Wang.

Diamond, J. (1999). *Guns, germs and steel: The fates of human societies*. New York: W.W. Norton and Co.

Gerbi, A. (1985). *Nature in the New World: From Christopher Columbus to Gonzalo Fernandez de Oviedo.* Pittsburgh: University of Pittsburgh Press.

Nabhan, G. (1995). Cultural parallax in viewing North American habitats. In M. Soule & G. Lease (Eds.), *Reinventing nature* (pp. 87–101). Washington, DC: Island Press.

Rivera, L. N. (1992). *A violent evangelism: The political and religious conquest of the Americas.* Louisville, KY: Westminster/John Knox Press.

Sale, K. (1990). *The conquest of paradise: Christopher Columbus and the Columbian Legacy.* New York: Alfred A. Knopf.

Schwartz, M. (1997). *A history of dogs in the early Americas.* New Haven and London: Yale University Press.

Todorov, T. (1984). *The conquest of America: The question of the other* (R. Howard, Trans.). New York: Harper and Row.

Gavin Van Horn and Anna Peterson

■ History
Animals in Ancient Roman Arenas

Humans in every culture have developed spectacles designed to demonstrate our superiority over other animal species. Exhibitions as diverse as lion-taming acts and bull-fights share a common intent: to give clear proof of the ability of humans to confront and defeat a natural world that we perceive to be our adversary. The ancient Romans devised spectacles that included exhibitions of animals killing other animals and humans, and of humans killing animals. The former made it evident that Nature is chaotic, irrational, and violent, whereas the latter showed that we humans are capable of triumphing over savage opponents and thereby creating a secure environment for ourselves. These exhibitions reminded spectators that their membership in the human community provided protection from the brutality of animals. They also advertised the power of the Roman state to crush hostile forces and maintain order.

These spectacles were produced with public funds and government support. For example, at the dedication of the Flavian Amphitheater (later known as the Colosseum) in ancient Rome in 80 CE, 9,000 animals were killed over a period of 100 days. And in 107 CE, the emperor Trajan celebrated his military victory in Dacia (modern Hungary and Romania) with 120 days of spectacles during which 11,000 animals were killed. Although other cultures have certainly found entertainment in events at which animals are tormented and slaughtered, the spectacles of the ancient Roman world are notable both for their scale and for their significant role in the social and political life of the period. At its height in the second century CE, the Roman Empire extended from Britain in the northwest to Syria in the southeast and encompassed people of many different cultures and traditions. Throughout this vast territory, publicly financed spectacles of violence against animals served not only to amuse people but also to gather them together as a community for an event sponsored by officials who were advocates of the Roman state. Even in the most remote regions of the Roman Empire, spectators could rejoice in the assurance that these events offered that the Roman state could keep them safe.

Public displays of animal slaughter had, however, been a familiar aspect of Roman culture long before the city of Rome had procured its far-flung empire. In fact, the origins of these displays can be traced back to activities in agricultural communities, where farmers struggled to eradicate animals that preyed on their livestock or consumed their food plants. At annual festivals designed to celebrate the community's success in producing a secure food supply and thus ensuring its survival, rural residents attended spectacles at which large numbers of pest species, such as foxes and rabbits, were killed. These festivals were celebrated also in urban areas, and although most town-dwellers had no direct experience with the damage done by foxes and rabbits, they understood the symbolic dimensions of the killings. The public destruction of pest species demonstrated that humans were capable of gaining control over their environment and, in particular, that the well-ordered civilization that the Romans had constructed could defeat the forces of an antagonistic Nature. Spectators felt no sympathy for the animals being victimized because they regarded them as enemies and believed that their suffering was the penalty they paid for threatening human life.

The development of these spectacles in urban areas can also be linked to the sport hunting enjoyed by wealthy men who owned property in the countryside. Although hunting was for them a recreational activity, they maintained that they were doing a public service because the eradication of wild animals benefited the entire community. Few town-dwellers, however, could afford to travel to rural areas to hunt. Therefore, ambitious men, who competed with one another for political offices and power and who were therefore eager to win the support of urban voters, brought the excitement of the rural hunt to the city masses by producing arena spectacles that featured the destruction of animals. These exhibitions permitted even the poorest of urban residents to participate, at least as spectators, in the same activity enjoyed by their wealthy political leaders. And the political leaders, in turn, gained a reputation for being generous toward those less fortunate than them. Thus the Latin word *venatio* (plural: *venationes*), which is translated as "hunt," came to refer to either the killing of wild animals in the countryside or an urban exhibition of killing animals.

The rural and urban "hunts" were similar in the sense that their focus was on the killing, not the tracking and pursuit of the animals. In rural areas, the "hunters" picnicked and partied while slaves beat the bushes and drove the animals into entangling nets, where the "hunters" could easily spear them. Some wealthy property-owners built game preserves on their land. Here animals were lured to a location where food was put out for them every day. When the property-owner and his friends wanted to "hunt," they simply traveled to the feeding station and speared the animals. Thus it is not surprising that the urban "hunts," where people watched the staged destruction of animals confined in an arena, were considered counterparts of rural "hunts." One Roman author, Varro, commented that the sight of wild animals gathering at the feeding station of a game preserve was as beautiful as the "hunts" staged in the city of Rome. For him and his readers, the similar and impressive element in each location was the fact that a large number of wild animals had been assembled and would be killed by clever humans. Both the rural and the urban hunts offered opportunities to prove that humans can outwit Nature.

The urban spectacles became a very popular entertainment, and upper-class politicians were therefore eager to sponsor them as a means of gaining public support. In 51 BCE, for example, one ambitious young politician, Marcus Caelius Rufus, wrote to his friend, Marcus Tullius Cicero, who was then governor of a province in southern Asia Minor, and asked him to capture some leopards and ship them to Rome. Caelius was anxious to outdo the spectacles of his political rivals by offering the audience a show of exotic beasts. Producing a dazzling spectacle could convince city residents that the sponsor was sympathetic to popular interests, had the military and political connections necessary to obtain animals from foreign lands, and was thus worthy of holding a position of great authority. Cicero replied, however, that there were few leopards left in that region. The exchange of letters between Cicero and Caelius informs us about the efforts that the political elite made to win popular favor. However, Cicero's remark about the scarcity of leopards also reveals the enormous impact that the collection of animals for urban arenas had on local populations of wild species. In 55 BCE, for example, Pompey produced spectacles at which he exhibited about 400 leopards, 600 lions, 20 elephants, and one rhinoceros, the first seen in Rome. And, in 46 BCE, Julius Caesar sponsored a display of 400 lions and, for the first time in Rome, a giraffe. The slaughter of large numbers of wild animals was justified not only as entertainment, but also as a method of turning dangerous and "useless" wilderness into areas that were productive and safe for farmers, ranchers, and their livestock. One author, Strabo, notes with approval that Numidia, a region of North Africa, could now be settled and cultivated by humans because so many lions had been killed.

The appearance in Rome of exotic species, such as lions, leopards, elephants, and the occasional giraffe, rhinoceros, hippopotamus, and crocodile, was a consequence of the relentless expansion of the Roman Empire. Roman administrators, soldiers, and businessmen who were posted in conquered territories recognized that the strange creatures they encountered in far-off lands would thrill Roman audiences. A thriving business developed of capturing, transporting, and selling animals for presentation at arena spectacles. The slaughter of these animals before cheering crowds, like the slaughter of native rabbits and foxes, provided proof of the ability of humans to dominate Nature. However the exhibitions of animals that had been collected in lands at the fringes of the Empire, whose residents had been forced to submit to Roman rule, also served as evidence of the superiority of the Romans over all other humans. The capture of strange, large, and fierce animals and their transport over great distances was a dangerous and very expensive operation. The Romans' ability to bring vast numbers of exotic animals to the city—solely for the purpose of providing pleasure to urban residents—was verification that their state was indeed powerful and prosperous and that it thus deserved to rule all others.

Spectators may have regarded the exotic animals as representatives of the regions from which they had been imported. The appearance of leopards, for example, in Roman arenas may have reminded the audience that the Roman state dominated the affairs of Asia Minor. And when spectators saw elephants, a species that had been used in battles against the Romans by the army of Carthage (in north Africa), they may have thought about Rome's crushing defeat of the Carthaginian people. Most human prisoners of war were not killed but rather sold as slaves; the state benefited from this system because the money from the sale went into the treasury. However, when animals from conquered areas were brought to Roman arenas to be killed, their torment and execution may have appeased the lust of the masses to exact vengeance on human opponents. The slaughter of the animals was a re-enactment of the process by which the people of conquered areas had been vanquished. The animals served as substitutes for human enemies, and their destruction thus provided spectators with the opportunity to witness and rejoice in the imposition of Roman justice on the rest of the world.

The men who killed the animals were skilled and well-trained performers who were called *venatores* ("hunters") and *bestiarii* ("beast-fighters"). Like other performers, such as actors, chariot drivers, and gladiators, they were often slaves or former slaves, but despite their very lowly position in society, they would win the respect of the audience if they performed bravely. Because the purpose of the spectacle was to demonstrate the superiority of humans, it is probable that most *venatores* and *bestiarii* emerged victorious from their hazardous battles against wild beasts. If the animals "won" many of the battles, it would be difficult to argue that humans were dominant. In any case, the odds were stacked against even the most ferocious of animals because they were, not infrequently, severely weakened by the injury, illness, and stress caused by their capture and transport. At the spectacles that celebrated the opening of the Flavian Amphitheater, at least one bull had to be prodded with fire to make it face an elephant, and a rhinoceros had to be goaded to attack a bear. At spectacles produced in 281 CE, one hundred lions were brought to the Circus Maximus, but they refused to leave their cages; they were therefore simply slaughtered in their cages. And in 393 CE, bears that were shipped to Rome for a spectacle arrived very weak from starvation and stress.

Exhibitions of humans killing animals and of animals killing other animals were two types of arena spectacles. A third type was the display of animals killing humans. These displays were, in fact, public executions of people who had been judged guilty of a capital crime. In contrast to the *venatores* and *bestiarii*, these people were condemned to be killed in the arena and they were therefore not trained and not given protective equipment or weapons. Sometimes they were even tied to a stake so that they could not defend themselves at all from the attack of the animals. This very painful and very public method of execution, called *condemnatio ad bestias* in Latin, was intended not only to satisfy the community's desire for revenge but also to serve as a warning to potential wrong-doers of the excruciating death that awaited criminals. People subjected to this form of execution had been convicted of flouting the laws that the rational community had established to restrain bestial behavior in humans. Like predatory animals, they had harmed community members, and they appeared, by their actions, to prefer the life of lawless, irrational animals to membership in human society. Consequently they forfeited the protection offered by society and made themselves liable to the same type of destruction that animal pests suffered. In the eyes of the community, the condemned person was indeed a beast, and he was therefore released to the company of his "fellow" beasts. In other spectacles, the audience watched animals—a bear and a rhinoceros, for example—attack one another in an exhibition of the chaos and violence of Nature. Now they watched as a human predator was attacked and killed by his fellow beasts. His death proved that the natural world, in which he had chosen, by his lawlessness, to live, was savage.

Condemnatio ad bestias was an expensive method of execution because of the costs associated with capturing and shipping the animals, but it was very gratifying for the audience. Spectators might believe that they were not responsible for the criminal's death; they had simply placed him in Nature and witnessed the consequences. In turn, they could congratulate themselves for recognizing that adherence to the code of civilized behavior protects humans from the brutality of Nature. In addition, the method of execution validated the correctness of the community's decision to punish the condemned person as an animal. The sounds that he made as the attack progressed were non-verbal, animal-like shrieks and whimpers. The inarticulate utterances confirmed that the condemned was, in fact, a beast. In the final process of the execution, the flesh of the condemned was eaten by the animals, and he was thus converted truly to bestial flesh. Spectators watched the torment with the same amusement and detachment that they watched their animal enemies being abused. Thus the condemned paid the penalty for his lawlessness not only by giving his life but also by providing entertainment for law-abiding citizens.

By the first century CE, such executions were often designed to appeal to spectators who demanded novelty. In one case, for example, scenery that resembled a forest was placed in the arena. Animals of several species were released there and also a condemned man. The man was costumed as the mythical figure Orpheus. Orpheus was such a skilful musician that he had charmed even wild animals to approach him without fear or aggression. This myth signifies the power of music, which is able to pacify even ferocious beasts. It also suggests that humans can live in harmony with other animals and can tame them by nonviolent methods. In the execution in the arena, however, the condemned man was mauled to death by the animals. The audience was thus doubly delighted; they saw the execution of someone who had threatened their society, and, in addition, they were entertained by a novel inversion of the traditional myth of Orpheus. Unlike the mythical Orpheus, who was able to live peaceably among the animals, the arena "Orpheus" was killed by them. The "snuff play" in the arena instructed spectators that wild animals are enemies, that we cannot live peaceably with them, and that they must be dealt with ruthlessly.

Spectacles of humans killing animals and of animals killing humans and other animals were popular throughout the Roman Empire. The wide-spread interest is attested to by the numerous representations on art objects and in literature. The arena was considered a place that brought into sharp focus the boundaries between order and chaos, culture and nature, and human and animal. Here the human community assembled to witness and celebrate its ability to eliminate threats to its security. The spectacles of torment and death offered Roman audiences the assurance that their state could defeat the chaos of Nature.

Further Resources

Coleman, K. M. (1990). Fatal charades: Roman executions staged as mythological enactments. *Journal of Roman Studies, 80*, 44–73.

Jennison, G. (2005). *Animals for show and pleasure in ancient Rome.* Philadelphia: University of Pennsylvania Press (reprint of 1937 edition, Manchester).

Shelton, J.-A. (2004). Dancing and dying: The display of elephants in ancient Roman arenas. In R. Egan & M. Joyal (Eds.), *Daimonopylai: Essays in classics and the classical tradition* (pp. 363–82). Winnipeg: University of Manitoba Centre for Hellenic Civilization.

Toynbee, J. M. C. (1973). *Animals in Roman life and art.* Ithaca, New York: Cornell University Press.

Jo-Ann Shelton

■ History
Colonial America and Domestic Animals

When we think about key figures in the history of colonial America, what typically comes to mind are explorers, Pilgrims, or Native Americans—not swine, cattle, or horses. Yet livestock played a vitally important role in colonization. They sustained English settlers busy establishing towns and plantations along the eastern seaboard with meat, milk, and muscle power. Indeed, English colonization would not have been so swift or so successful had the settlers not brought livestock with them. But these same domestic animals posed problems for Indians unfamiliar with such creatures. Including livestock in the story of early America complicates a familiar narrative by revealing the extent to which human-animal interactions were central to that history.

Until Europeans arrived, there were no livestock in North America. Christopher Columbus first transported horses, cattle, swine, sheep, and goats to Caribbean islands on his second voyage in 1493. Virtually every colonizing expedition thereafter included domestic animals as well as people. This essay will explore two key aspects of this larger story of colonial-era human-animal interactions as they pertained to relations between Indians, English settlers, and livestock in the seventeenth century. The first theme concerns the Indians' reception of the strange beasts that arrived along with shiploads of colonists. The second addresses the ways in which livestock husbandry shaped England's ideas about its imperial enterprise.

First, the introduction of European animals to America had a significant impact on Indian lives, requiring adjustments every bit as important as those Indians made to European peoples' presence. Native Americans had never seen pigs, cows, horses, or any of the other creatures the colonists brought with them, and thus they had to come to terms with the physical presence of strange creatures in a landscape otherwise thoroughly familiar to them. Equally important, Indians encountered new ideas about animals. To the colonists, a domestic animal was a piece of property under the control of an owner, to be used as the owner wished. In a broader sense, English settlers believed that ownership of domestic animals accorded with the biblical instruction, written in the book of Genesis, for humans to exercise dominion over all living things on the earth. To the English, these ideas were so ingrained as to be beyond question. To Indians, they were, at the very least, a puzzle.

Native Americans thought quite differently about the indigenous American creatures with which they were familiar. They understood animals to be distinct from people, but not necessarily subordinate to them. Certain animals, particularly game animals, were thought to have *manitou,* or spiritual power. In order to be successful in their pursuit of game, hunters were obliged to show respect for that power by performing certain rituals. Powhatan hunters in Virginia, for instance, made special offerings of blood, deer suet, and tobacco to the animals' guardian spirits to thank them for allowing game to be killed.

The concept of domestication as the English understood it was essentially unknown to Algonquian Indians living along the east coast of North America. Native peoples in other parts of the Americas did have a few domesticated animals (turkeys in Mexico, llamas and guinea pigs in Peru), and all native peoples had dogs. The dogs kept by the eastern Algonquians, however, seem mainly to have been scavengers roaming in native villages, the loosest form of association that can be called domestication. Unlike Indians living elsewhere in North America, eastern Algonquians do not appear to have hunted with dogs, used them to carry burdens, or eaten them (except, perhaps, in times of famine).

Algonquians were likewise unacquainted with the notion that living beasts could be private property—a legal status central to the colonists' definition of domestic animals. Village residents seem to have exercised a loose communal ownership of the dogs in their settlements. Otherwise, animals only became property after they were killed—the property of the people who killed them.

Such ideas—that living animals had spiritual power and were "wild" and unowned—differed dramatically from the English definition of domesticated creatures. At first, before the colonists were numerous enough to challenge Indians on the issue, and when many encounters between Indians and livestock occurred without colonists being present, Indians attempted to incorporate the new creatures into their world literally on their own terms. Native Americans initially called English animals by the names of the indigenous beasts the new animals most closely resembled. Thus Powhatan Indians in Virginia used their word for "small bird" for chicken, and Narragansetts in Rhode Island called pigs woodchucks because both creatures were about the same size and rooted in the ground. By emphasizing similarities rather than differences, Indians found ways to absorb new animals into their way of life.

Indians may also have detected spiritual power in English animals, although the evidence is fragmentary. This may have been what John Smith, an early leader of the Virginia colony, meant when he described the first encounter between Powhatan Indians and an English boar. The Indians, Smith reported, thought that the creature was "the God of the Swine."

Native American efforts to apply familiar names and ideas about animals to European livestock as a way of incorporating them into their mental and physical worlds were short-lived. A combination of factors—principally the massive immigration of colonists on the one hand and high mortality among Indians due to epidemics of Old World diseases on the other—tilted the balance of population and power in the colonists' favor. Increasingly numerous colonists expected Indians to regard livestock as they themselves did—as private property placed on the earth for human use. Even more, colonists wanted Indians to see that keeping livestock was an essential component of a civilized way of life.

This introduces a second theme: the role livestock played in English imperial ideology during the seventeenth century. English farmers had worked with livestock longer than anyone could remember and were excited to see meadows up and down the eastern coast of North America where imported animals could graze. English colonists fully expected that their accustomed dependence on livestock could thus persist in the New World. Even more important, they believed that, by bringing domestic animals to America, they legitimized England's claim to empire.

England's imperial ideology in the seventeenth century largely developed in opposition to Spain's colonial experience. From England's perspective, Spain offered a thoroughly negative example of colonization. Not only was Spain's empire Catholic, it was also, according to the English, founded upon the subjugation of native peoples through intimidation and unspeakable cruelty. England claimed that its empire would be quite different. It would be Protestant, devoted to the improvement of Indians, and agricultural rather than military in its purpose.

Central to England's imperial ideology was a concept derived from Roman legal theory (*res nullius*), which argued that "empty things," including land, remained common property until they were put to use. At that point, they became the private property of the user. Agriculture was, to the English, the best way to use land. Because Indians tilled only a small proportion of New World land, colonists regarded everything else (such as woodlands used for hunting and gathering) as "empty," or available for English farmers. As long as "we leave them sufficient for their use," declared Massachusetts Bay Colony's first governor, John Winthrop, "we may lawfully take the rest." And the colonists would improve their land in ways that Indians did not—that is, by grazing livestock on it.

Raising livestock as well as crops, in English minds, constituted civilized agriculture because these activities increased the extent of improvements made on the land and promoted sedentary settlement. Assuming that they would import English agricultural practices, and not just animals, colonists expected to demonstrate for Indians the superiority of the English agrarian regime. They anticipated building farms with enclosed pastures, using animals to work the land and manure to fertilize the soil, thus making improvements that legitimized English claims to ownership. Because the Indians "inclose noe Land, neither have any setled habytation, nor any tame Cattle to improve the Land by" (to quote John Winthrop again), native claims to territory were of lesser force. But once they saw how civilized farmers worked, Indians presumably would change their ways, acquiring and using livestock as well. As Roger Williams, the founder of Rhode Island, put it, the Indians would thus move "from Barbarism to Civility."

By basing their imperial ideology on this agricultural foundation, colonists not only legitimized—in their own minds—their claims to land but also offered the possibility of peaceful coexistence with Indians who learned how to farm in the proper way. But events did not turn out as planned. Colonists had not anticipated that they would have to adapt their livestock husbandry to suit New World conditions. But they did have to change, and the changes they made generated friction, not harmony, with Native Americans.

During the seventeenth century, colonists simply did not have the labor resources to raise livestock as English people did at home. There were not enough workers to enclose pastures, build fences, plant forage crops such as clover, and save manure. Colonists were far too busy cutting down trees, clearing brush, building houses, planting food crops, laying out roads, and performing all the other tasks involved in creating new settlements. So they adapted to new circumstances by conserving labor and making more use of the abundant land they regarded as being at their disposal. They built fences around planting fields, not pastures, and turned their animals loose to graze. Livestock grazed at large year-round in the southern regions. New England's colder climate required that they be brought back to farmsteads for the winter.

Natural meadows and woodlands in eastern America, however, produced less nutritious forage than the clover and other plants found in cultivated English pastures. As a result, the colonists' animals needed to spread out over a great deal of land to find enough to eat. One modern estimate suggests that a free-range cow needs as much as five to fifteen acres of woodland to sustain itself. Thus as colonial livestock herds proliferated, they required substantial amounts of land upon which to range. This explains one of the paradoxes of early American history—the frequent complaints from colonists that they were running out of room, even though their towns or plantations were barely a few years old and contained far more land than could be found in much more populous English villages. Many of these complaints specified that the "room" they needed was for livestock, not planting fields.

Colonial-style livestock husbandry thus demanded relentless geographical expansion that only intensified over time as new generations of colonists matured and wanted land for their own animals. That expansion, of course, occurred at the expense of Indians. Native Americans not only had to deal with the inevitable problems of animal trespass that free-range husbandry generated but also had to contend with colonists who wanted to buy land, who seized it in payment for debts, who acquired it through treaties, or who simply sent their animals to graze on Indian lands without first obtaining legal title to them. Because decades of contact with colonists had not in fact encouraged Indians to adopt the colonists' way of life as their own, English settlers remained convinced that they were the only New World inhabitants to use the land properly and, thus, that their claims superseded those of Indians. Blind to the adaptations they had made in their own husbandry practices, colonists continued to regard livestock as symbols of civilization—even as Indians had come to resent them as agents of empire.

The depth of Indian resentment was fully revealed in two cataclysmic conflicts that occurred in 1675–76: King Philip's War (or Metacom's War) in New England and Bacon's Rebellion in Virginia. Both were highly complicated events, yet they shared one common feature. The behavior of roaming livestock and the related factor of English expansion were primary causes of conflict. At the outbreak of war in New England, John Easton, Rhode Island's deputy governor, approached Metacom (known to the English as King Philip) and asked why the Indians fought. In response, Metacom supplied a list of grievances that included unceasing difficulties with livestock. Even when Indians sold land to the English, Metacom complained, the colonists' livestock continued to stray beyond the bounds of their towns and invade Indian lands. As Metacom put it, the Indians had expected that "when the English bought land of them that they would have kept their cattle upon their own land."

Bacon's Rebellion in Virginia began as an Indian war that widened into a revolt against the colony government. When the dust had settled, colonists admitted that their thirst for land and the uncontrolled foraging of their livestock had helped to spark hostilities. Conceding that "the Violent Intrusions of divers[e] English into [Indian] Lands" had forced Indians "by way of Revenge, to kill the Cattel and Hogs of the English," Virginia's magistrates subsequently required colonists from then on to settle at least three miles away from Indian villages. Needless to say, this policy was not successful in halting the expansion of English settlement or in keeping livestock within the bounds of English plantations.

In conclusion, livestock played several critical roles in the drama of seventeenth-century English colonization. The animals made it possible for colonists to create a semblance—if not a replica—of English agrarian society in America and served as symbols of English cultural superiority. Livestock enabled the colonists to extend their dominion over the New World with remarkable speed and thoroughness. To Indians, animals that at first posed conceptual puzzles ultimately became targets of wrath as they helped colonists to dispossess native peoples. Human-animal interactions, therefore, were central to the story of early American history in ways that we are only beginning to appreciate.

Further Resources

Anderson, V. D. (2004). *Creatures of empire: How domestic animals transformed early America.* New York: Oxford University Press.

Cronon, W. (1983). *Changes in the land: Indians, colonists, and the ecology of New England.* New York: Hill and Wang.

Crosby, A. (1972). *The Columbian exchange: Biological and cultural consequences of 1492.* Westport, CT: Greenwood.

———. (1986). *Ecological imperialism: The biological expansion of Europe, 900–1900.* New York: Cambridge University Press.

Virginia DeJohn Anderson

■ History
Harvesting from Honeybees

Humans have a deep-seated desire for sweet foods and drinks. For most of the history of our species, the only intense sweetener available year round was honey. Our ancestors thus had a powerful incentive to learn how to first exploit and then to manage the industrious but aggressive honey bees—insects that turn the nectar of flowers into a thick syrup that never spoils. At first this interaction was unpleasant for all concerned, but now the harvesting of honey has taken on many of the attributes of farming.

Cave paintings as much as 15,000 years old show that humans knew even then that smoke has a partial (and inexplicable) calming effect on honey bees (*Apis mellifera*). Equipped with a rope ladder to scale a cliff, one drawing depicts a woman covered with some kind of protective garment (animal skins, probably) as she "smokes" the entrance. In other sketches, the harvester has climbed a tree (the preferred habitat of bees) to break into the nest. In either case, the small entrance hole used by the foraging bees must be greatly enlarged in order to tear out the honey-filled comb. Not only is the colony robbed of much of its food reserves, but the gaping wound left behind opens the hive to other predators, and exposes the colony to the elements. Without an intact cavity to protect the hive, and insulate the bees from cold weather, the colony is unlikely to survive.

Domesticating honey bees was a clear priority once humans had switched from hunting to herding, and from gathering to cultivation. No longer nomadic, societies could not afford to make fatal raids on the nearby colonies: soon all the hives in the vicinity would be gone,

An illustration depicting Virgil writing on the life of bees, from a manuscript by Servius Grammaticus. ©Giraudon/Art Resource, NY.

and with them the supply of honey. Instead, our more settled ancestors needed to farm bees—an impossibility until the "technology" could be developed. What was required was hives that could be opened, robbed in a very restrained way, and then reclosed so the bees could set about replacing the lost comb and restocking it with honey. Moreover, these artificial cavities had to be structures colonies would *choose* to live in, since the group could always depart (swarm) and find another, more acceptable place to live if the human-supplied home was not to their liking.

The first historical records of successful artificial hives are from Egypt, where 10,000 years ago clay cylinders were constructed and stacked like logs. One end of each cylinder was closed with a wall containing a small entrance hole for the bees. The disc covering the other end was removable, allowing the beekeeper access to the comb from the rear. Grooves were scratched on the inside top of the cylinder parallel to the two end plates, which encouraged the comb-building bees to construct the honey comb in crosswise sheets. The beekeeper would smoke the hive, remove the back, and take the comb nearest the rear. Once resealed, the bees would rebuild the missing comb and begin filling it again. Of course, the beekeeper had to be careful not to remove too much honey: the bees use it to keep themselves warm during the winter, and if the colony runs out before the blooming season begins again, it will die.

In other places the cylinders were fashioned from logs, bark, or terra cotta, and in impressive numbers. Based on tax and other records, tens of millions of such hives must have been in use. In one sacrifice to the Nile god, for instance, Ramses III was able to offer 30,000 pounds of honey. Tradition holds that Egyptian beekeepers floated banks of hives up and down the Nile, following the changing seasons. Some "migratory" beekeepers in North America do the same, using flatbed trucks and following a route that takes them from the Deep South up into Canada. The occasional road accident involving these trucks is always big news.

In Northern Europe, however, beekeepers initially preferred to let the bees find their own homes. After locating a colony, the beekeeper would cut a door in the back of an existing cavity; he or she would then, year after year, climb up the back of the tree, smoke the hive, open the colony's chamber, and remove some of the comb. Enterprising beekeepers also refurbished existing but unoccupied cavities to make them more attractive to bees; this involved altering the internal volume, plugging leaks, and reducing the size of the entrance to one a colony could readily defend. Bee trees were individually owned, and sometimes equipped with elaborate traps to capture potential nest robbers, most especially bears. Awkward and inefficient as this technique might seem, parish records from a single church in Novgorod in 1156 show that the residents paid a tax of 350,000 kilograms of beeswax, the produce of perhaps 350,000 colonies.

But cavities high in trees, even with back doors, are inefficient to harvest, and when the trees in the forests without colonies were being cut, damage and exposure were risks to the bee trees. Beginning about 1100, Europeans finally began switching to freestanding hives. They developed the classic inverted-basket design known as a "skep." Straw is used to create a thick rope, and then the rope is wound first into a cylinder, and then finished at the top as a curving cone. The adjacent strands of rope are tied together to make the structure sturdy. Skeps are light and portable and provide their occupants with good insulation. Any number of older buildings have niches in south-facing walls which once held ranks of these inexpensive hives. But skeps are both difficult and wasteful to harvest. The usual techniques—gassing with burning sulfur or dunking in water and then scraping everything out—are inevitably fatal to the colony.

This problem was partly overcome with the invention of an amazing variety of two- and three-piece skeps, which allowed the separate harvesting of, say, the bottom third of the comb. In a two-part stacked skep, for instance, the bottom part of the cylinder (the "eke") was about a quarter of the combined height. The beekeeper could draw a fine wire across the junction between the two segments, severing the bottom section of the comb from the comb above.

The modern commercial hive was invented by Rev. L. L. Langstroth about 1850. After extensive study and experimentation, Langstroth realized that bees always leave two bee diameters between combs; this is so workers can move about on the adjacent surfaces without colliding with each other. He also discovered that gaps only one bee diameter are left open as passageways elsewhere—between the comb and the side of the tree cavity, for

instance. Any openings larger than two diameters between combs or one diameter between combs and another surface are generally filled in with wax. His hive had rectangular frames of just the thickness necessary for bees to build a comb with a two-bee gap between; and the frames were exactly one bee diameter smaller along all four edges than the box that held the frames. As a result, the top could be removed from the box, and the frames would slide out one by one.

But despite the advances in technology, one unusual fact stands out: unlike chickens, cattle, and corn, bees have not been domesticated in any real sense. Their morphology and behavior is the same as that experienced by the honey gatherer in the ancient cave paintings all those thousands of years ago. It is humans, in fact, that have done the accommodating; it is our behavior, and the morphology of the domiciles we create for them, that have changed.

Over the past 400 years, the commercial importance of honey bees has declined dramatically. The discovery by the Western world of sugar cane, sugar beets, and corn syrup has left honey as

Beekeepers robbing a hive for honey. Unlike many other species humans have lived with for hundreds of years, bees have never been domesticated in any real sense. Courtesy of Shutterstock.

an expensive niche sweetener. Beeswax was once the high-priced, sweet-smelling, clean-burning alternative to tallow for candles. But then humans discovered how to extract lanolin from wool, wax from the carnauba plant, and paraffin from oil. Electricity was the final blow. Beeswax is still prized for its resistance to sagging in heat, and is irreplaceable in lost-wax sculpture and batik painting—but these are tiny markets.

But although honey bees are no longer necessary for either sweetness or light, they remain the dominant pollinating insect for many commercially important flowering plants. Beekeepers are likely to earn more money for "servicing" groves of almond trees, acres of blueberries, fields of sunflowers, and plantations of tomatoes than from selling the honey their insects produce. But now honey bees are becoming rarer, and beekeeping is a less profitable business: another insect, the tiny verona mite, has spread through much of the world and lives inside the breathing tubes of honey bees. Foraging bees find it hard to respire and can no longer be as busy as once they were. Perhaps, after the current period of severe decline, new strains of mite-resistant honey bees will evolve. In the meantime, researchers are hard at work trying to "domesticate" other species of bees to help honey bees with the business of agricultural pollination.

See also

Living with Animals—*Honeybee and Human Collaboration*
Living with Animals—*Honeybee Society and Humans*
Living with Animals—*Pesticide Effects on Honeybees*

James L. Gould

Honeyguides: The Birds that Lead to Beehives

James L. Gould

Humans in Africa have depended for thousands of years on a colorful bird, the African honeyguide (suitably named by ornithologists *Indicator indicator*) to lead them to bee hives. Once the nest is found, the humans smoke the hives to calm the bees, break the colony open, and then gather the honey. But, most importantly, they leave behind substantial pieces of comb for the bird to eat. The honeyguide is one of the few species in the world that can digest wax, a near-certain sign that it depends on beeswax for its sustenance.

Although the behavior had been observed and written about long before Theodore Roosevelt published his memorable account, skeptics doubted that a mere bird could be capable of knowing where it was in a forest relative to a hive, or that it could understand how to lead people there. The usual explanation was that the ravenous bird made hunger calls, then fled in fear from humans when they approached; but when, during its meandering escape, it chanced upon a colony, it went after the bees. Humans simply took advantage of the situation. All the rest of the "cooperation" story was a romantic fairy tale—or so they said.

In fact, humans are exploiting a highly evolved partnership between honeyguides and so-called honey badgers (the ratel, *Mellivora capensis*). Ratels are powerfully built, intelligent predators with massive claws and very dense fur—characteristics that serve them well when they attempt to break into and rob a honey bee colony.

The honeyguides provide a number of signals understood by their human collaborators, and presumably by the ratel as well. The bird, happening upon a ratel, flies to a prominent perch and utters one of its repertoire of "follow-me" calls (including one that sounds like a tree being torn open). If the "badger" (or human) shows no interest, the calls become more insistent, and the bird will follow the potential helper with repeated entreaties until either getting its would-be aide's attention or giving up and going off to look for other, more cooperative partners.

Natives claim (and studies have verified) that the height from which the bird calls is a reliable if approximate indicator of the distance to the hive, and the direction the bird flies when humans begin following correlates well with direction. During these flights, the time between the honeyguide's disappearance and return is a good guide to the distance to the nest. Again, ratels probably understand these relationships too, though for them it may be more a matter of instinct than learning.

Careful studies show that individual birds know the location of several colonies in the vicinity, and guide humans (and, we must suppose, ratels) to the closest. Honey-hunting humans working with honeyguides find hives three times as fast as those working on their own, making this a valuable collaboration. That humans came to understand signals intended for ratels, and that honeyguides have the wit to use humans as hunting partners, says much for the flexibility of both species.

History
The Ravens of the Tower of London

The ravens in the Tower of London are a remarkable example of modern totemism, in which a people and its destiny are identified with a group of animals. A group of six or more ravens is kept at all times in the Tower of London and cared for by the Ravenmaster, one of the Yeoman Warders or "Beefeaters." The ravens are among the foremost tourist attractions in Britain and, for that matter, the world.

Visitors are told by tour guides and brochures that the ravens have been at the Tower for many centuries at least. According to the standard account, Royal Astronomer John Flamstead complained to King Charles II (reigned 1660–85) that ravens were interfering with his observations. At first the king wanted to get rid of the ravens, but then he remembered an ancient prophecy that Britain would fall if the ravens left the Tower. He ordered the wings of the ravens to be clipped in order to prevent them from flying very far. As to the astronomer, he and his observatory were moved to Greenwich.

Despite a lack of any evidence, this account was almost universally accepted through much of the latter twentieth century. At the beginning of the twenty-first century, Geoffrey Parnell, Keeper of Tower History at the Royal Armouries, and Boria Sax, a scholar of human-animal relations, researched the history of the Tower Ravens and simultaneously discovered that they dated back only to late Victorian Times. The first references to ravens in the Tower in print are from the 1890s. Wild ravens had vanished from London around the middle of the nineteenth century, but one suddenly appeared in Kensington Gardens, and the press speculated that it might have escaped from a collection of ravens in the Tower of London. In the early twentieth century brief references to the ravens began to appear in books about the Tower. The ravens were generally used to dramatize a plaque commemorating those executed at the Tower of London, which, though the exact location of the executions is unknown, is called "the scaffold." Yeoman Warders, serving as tour guides, would entertain visitors with accounts of how ravens would devour the bodies of those who had been beheaded.

The legend that Britain would fall if the ravens leave the Tower appears to have originated around the end of World War II. The common peril from aerial attacks helped bond the citizens of London to the ravens. The actual source of the legend was probably the Stag Brewery located in Pimlico, just a few kilometers from the Tower of London. Records in the British National Archives reveal that the brewery kept a raven as mascot, and the workers had come to believe that the bird protected them from bombs. The mascot died in 1944 and, in summer of that year, the brewery requested another raven from the Tower. After some lively discussion, the request was turned down, though the Tower agreed to help the Stag Brewery import a raven. The legend that ravens are protectors appears to have been transferred from the brewery to the nation itself.

Two of the few ravens at the Tower of London that survived the war were mysteriously killed shortly afterward, perhaps because of a superstition that the croaking of a raven foretells death. For a short time there were no ravens left, but a new contingent of young ravens was brought in for the reopening of the Tower to tourists in January 1946.

The idea of the ravens as a talisman to defend the country may have been deliberately spread by the Yeoman Warders, in order to win protection and support for the birds as national pets. In 1988, Ravenmaster John Wilmington established a breeding program for ravens in connection with the public zoo. The Tower continues to be marketed to tourists as a "house of horrors" today, with emphasis on tortures, beheadings, and ravens as scavengers of human flesh. The present ravens, however, are nevertheless beloved for their playful intelligence, and even antics such as occasionally distracting a visitor to

The ravens at the Tower of London with the Royal Guard in background. Legend dictates that if the ravens ever leave the tower, the tower and England will fall. Courtesy of Shutterstock.

steal his sandwich never seem to be resented. One raven named Thor likes to climb on top of a wall and startle visitors by saying something that sounds very much like "good morning." When new ravens, generally injured birds that would be unable to survive in the wild, are brought in, the news is announced on Web sites and in newspapers throughout the United Kingdom. Dolls, refrigerator magnets, coasters and other objects depicting ravens are sold at souvenir shops in the Tower.

The idea that ravens in the Tower of London guard the United Kingdom of Great Britain may not be ancient, but it is, despite commercial exploitation, an authentic bit of folklore. Like many national myths, it originated during a time of severe crisis, when danger moved people to look about intently for signs from providence. That the destiny of Britain depends on a group of birds may sound like a "primitive" superstition, which is why people have long assumed it was at least many hundreds of years old. The modern origin of the legend suggests that the psychology of people of today may not be so profoundly different from that of their ancestors.

Despite its recent origin, the legend of the Tower Ravens serves to connect the people of Britain with their remote past and, ultimately, with the natural world. The fable may have been influenced by the medieval legend of the Celtic god Bran, whose name means "raven." According to medieval Welsh manuscripts, the head of Bran is buried in London, perhaps at the site of the Tower, and protected Britain against invasion until it was dug up by King Arthur.

Like all legends, however, the tale of the Tower Ravens will need to be reinterpreted as the culture of Britain evolves and national identity takes new forms. As ravens, vanished in much of Britain but now protected, gradually expand their range, the ravens in the Tower of London may come to represent the ecological heritage of the nation, without which life will be impoverished.

Further Resources

Granz, J. (Ed. and Trans.) (1976). Branwyn, daughter of Lyr. In *The Mabinogion* (pp. 66–82). New York: Dorset Press.

Kennedy, M. (2004, November 15). Tower's raven mythology may be a Victorian fantasy. *The Guardian.* [Available online at http://www.guardian.co.uk/uk_news/story/0,3604,1351360,00.html]

Sax, B. (2005a, January). Black birds of doom. *History Today, 55*(1), 38–39. [Available online at http://www.historytoday.com/dt_main_allatonce.asp?gid=30768&g30768=x&g30758=x&g30026=x&g20991=x&g21010=x&g19965=x&g19963=x&amid=30213802]

———. (2005b). The Tower ravens, *ISAZ Newsletter, 28,* 5–8.

Boria Sax

History
The Turkey in History

Perhaps more than any other animal, the turkey symbolizes the profound ambivalence that human beings have toward nonhuman animals. Thus seen, the turkey provides an opportunity to explore the mixed and often contradictory emotions that people bring to many animals. The character that is often imposed upon specific animals, such as the turkey, may reflect genuine characteristics of that animal, while at the same time distorting and obscuring the animal's true nature and personality in a caricature having little resemblance to reality. In such cases, the anthropomorphism that derives from humankind's evolutionary kinship with other animals, enabling people to perceive animals accurately on the basis of shared traits, gives way to an anthropomorphism that imposes human fears and desires on animals whose actual life bears little or no relation to human fantasies. In Western society the turkey figures simultaneously as a sacrificial victim, a figure of fun, and a sacred player in America's mythic drama about itself as a nation. In *More Than a Meal: The Turkey in History, Myth, Ritual, and Reality* (2001), I explore the many conflicting and bizarre roles assigned to the real turkey who stands behind them.

An American Paradox

The turkey is not America's official national bird. Congress adopted the bald eagle of North America for this purpose in 1782. However, the turkey has become an American symbol, rivaling the bald eagle in actual, if not formal, significance. The turkey is ceremonially linked to Thanksgiving, the oldest holiday in the United States. Yet, unlike the bald eagle, the turkey is not a symbol of prestige or power. Nor, despite frequent claims, is there any evidence that Benjamin Franklin seriously promoted the turkey as the national bird—more "respectable" than the bald eagle—except as a passing jest in a letter to his married daughter, Sarah Bache, on January 26, 1784, two years after the bald eagle had already been adopted.

Although the turkey has a long history of involvement with Native American, Colonial American, and European cultures, in the twentieth century the wild turkey was invoked primarily in order to disparage the domestic, factory-farmed turkey. For example, the consumer newsletter *Moneysworth* wrote the following on November 26, 1973: "When Audubon painted it, it was a sleek, beautiful, though odd-headed bird, capable of flying 65 miles per hour. . . . Today, the turkey is an obese, immobile thing, hardly able to stand, much less fly."

Before Europeans Arrived

Turkeys were an integral part of the Native American cultures and continental landscape encountered by the Europeans in the fifteenth and sixteenth centuries. They occupied North, South, and Central America, the West Indies, and Mexico. They roamed through the hardwood forests of the northeastern United States from New England to southern Ontario. They ranged from Pennsylvania and Ohio into the Midwestern prairies and the Great Plains, and south from Maryland through coastal Virginia into Florida, where they lived along the wooded streams. Turkeys were numerous in the Ohio Valley and in the southwestern regions of what would become Oklahoma, Texas, New Mexico, Arizona, and southern California. Their ancestral homes included the Ponderosa pine forests of Arizona, the Yucatan Peninsula of southeastern Mexico, the Rio Grande of the south-central plains, and the Chesapeake Bay area of the Mid-Atlantic states.

Turkeys and Native Americans

Native Americans used turkeys in various ways. The eastern Indians hunted wild turkeys but did not domesticate them as did the Pueblo Indians of the Southwest, the Aztecs, the Mayans, and other pre-Columbian inhabitants of Mexico and Central America. In such places, turkeys had been domesticated—procured in the wild, propagated, raised, and restrained—for a thousand years before the arrival of the Europeans for food, feathers, and sacrificial purposes.

Use of Turkeys in Mexico and Central America

Wild turkeys living in the Sierra Mountains were caught and raised by the Mexican Tarahumaras, a people of southern Sonora and Chihuahua, who removed wild turkey eggs from the nest and placed them under their own brooding hens. At night, the turkeys slept on top of their houses and in nearby trees. The Tarahumaras were said to have a turkey dance and to use wild turkeys in sacrificial feasts. At one of their burial sites, a human body was recovered with a wad of cotton between the legs mixed with blue jay, woodpecker, and turkey feathers.

When the Spanish explorer Hernando Cortes and his men entered Mexico in 1519, they found domesticated turkeys throughout the Aztec Empire. Aztec ambassadors offered them turkeys, humans, cherries, and maize bread to eat when they arrived. The Mexican markets were full of both live and cooked turkeys and turkey eggs, and Aztec families raised turkeys for food in their gardens next to their houses. At certain festivals an entire turkey was served in a tamale (minced meat and red peppers rolled in corn meal) wrapped in palm leaves.

Poor people ate turkeys only on special occasions, and the average family ate its own birds or shopped at the market. By contrast, the Aztec emperor's royal household and the lords of the states and towns consumed huge numbers of turkeys exacted as tribute from the communities they ruled. For example, in 1430, the lord of Texcoco required one hundred turkeys daily, and everyone in the town of Misquiahuala had to give to the Aztec emperor Montezuma one turkey every twenty days. Montezuma's meals typically included some three hundred guests as well as a thousand guards and attendants. In contrast to the profusion of people within the royal household, George Vaillant describes how "outside the kitchen door squatted patiently a threadbare group of countrymen from whose carrying bags swayed the mottled heads of the trussed turkeys [skewered or bound by the wings for cooking] which they had brought as offerings for the royal larder."

In addition to human consumption, huge numbers of turkeys were demanded for Montezuma's captive raptors and mammalian carnivores. The raptors alone got 500 turkeys each day, including a turkey for each large eagle. A.W. Schorger speculates that between "Montezuma's menagerie and his large household, it seems safe to assume that his levy was one thousand turkeys per day, or 365,000 yearly."

Turkeys were ritually sacrificed in Mexico. For example, the Mayans of southeastern Mexico sacrificed and beheaded female turkeys as part of their feasts for the gods that controlled the earth's waters. The Zapotec people of central Mexico sprinkled turkey blood on their newly planted fields. In Mexico and Central America, as in the Southwest, turkeys were regarded as symbolic manifestations of earth, rain, and fertility and were ritually sacrificed to manipulate these elements.

Use of Turkeys in the Southwest

Archeology indicates that turkeys were domesticated in the Southwest between 700 and 1100 CE. Turkeys were taken from the Ponderosa pine forests and penned or kept in caves by the cliff-dwelling Pueblo people. Caves containing ancient turkey droppings, desiccated adult and young turkeys, turkey eggs, and loose and tied turkey feathers have been identified, and artifacts of turkey bones, beads, and bird callers have been found in these caves.

Spanish records of the sixteenth-century Southwest emphasize the use of turkeys' feathers over their use as food. Turkey feathers were used to make clothes, blankets, pouches, costume ornaments, and necklaces. The Pueblo Indians made prayer sticks, masks, and headdresses out of turkey feathers. As a result of live plucking for these purposes, the turkeys in the Hopi villages were said to have a "ragged aspect." Among the Hopis, feathers from the turkey tail were put on the backs of two prayer sticks tied together to represent male and female, and sticks adorned with turkey plumes were placed in the fields and in pools of water in an effort to bring rain.

Use of Turkeys by the Eastern Woodland Indians

The Eastern woodland Indians hunted turkeys with bows and arrows made of turkey feathers, bones, and spurs. The skill of these bow and arrow hunters consisted not in long-range shooting but in silently stalking and tracking wounded animals over long distances. Turkeys were lured by callers made of turkey wing bones, and the heads, skins, and tail feathers of turkeys were used as decoys to attract live turkeys in order to kill them. Hunters imitated the male bird's gobble and the call of a turkey poult (a young turkey) for its mother. In addition to bows and arrows, the Eastern woodland Indians used nets and snares, drove turkeys into pens and trees, caught them with baited hooks, and used blow-guns made of arrows plugged with thistledown at one end and puffed on at the other end.

Turkeys and the Europeans

Starting in the fifteenth century, the Europeans and their descendants conducted a full-scale assault on turkeys and other birds. John Bakeless describes how in presettlement days, turkeys "Up the Missouri" looked tamely "down from the treetops at canoes, passing down the stream." But in the seventeenth century, the colonial traveler John Josselyn wrote that turkeys, which had formerly flourished in flocks of "threescore" in Maine, were all but destroyed within twenty-five years there.

Across the continent turkeys were hunted unsparingly and sold for such things as a pound of lead shot, a bag of salt, or a bunch of pins and needles. As army troops, hunters, settlers, cowboys, and assorted travelers pushed through the country, hunting turkeys indiscriminately and tearing down the forests in which these birds had lived for thousands of years, turkeys vanished altogether or retreated to isolated and impenetrable areas of the southeastern United States, like the bottomland swamps of Alabama. Roosts that had once been "black" with a thousand to three thousand turkeys settled in the trees for the night, in an area a quarter of a mile wide and a mile long, were emptied as men, in the words of A.W. Schorger, took "all they wanted."

Schorger describes in detail how turkeys disappeared under the relentless pressure from market and sport hunters who killed and crippled vast numbers of turkeys and left countless numbers to rot as they went. Often hunters would eat some of the turkeys they killed, but soon, as a member of a party traveling through Oklahoma wrote in 1832, "they despised such small game & I have seen dead turkeys left behind on marching." In 1832, after turkeys had already been eliminated from many of their former ranges, it was noted that in some places turkeys were still so numerous that they were easily killed "beyond the wants of the people." By 1813, Connecticut had no wild turkeys; by 1842, Vermont followed suit. By 1920, wild turkeys had been eliminated from eighteen of the original thirty-nine states of their range, and from Ontario, Canada, according to *The Wild Turkey: Biology and Management,* edited by James G. Dickson (1992).

In their letters, journals, and memoirs, men wrote about the excitement they felt during a turkey hunt or upon seeing a single turkey or a flock of turkeys. Though the observation was often keen, the slaughter was sentimentalized in a melodramatic style that continues in turkey sports writing today along with the scientific management approach. Virtually exterminated by the 1920s, the turkey was not only restored, "but to a record high population that is growing with no end in sight," according to Seth Borenstein writing in the *Albuquerque Journal* in 1998.

A Civil War—era photograph published by the National Wild Turkey Federation in *The Wild Turkey: Biology and Management* tells much of the story of the wild turkey in America. It depicts rows of dead turkeys strung upside down on a clothesline at an army campsite. Men slaughtered cartloads full of turkeys, often shooting them at roost when they were resting and sleeping, a practice that led to the term "turkey shoot" signifying a "simple task or a helpless target . . . considered to be stupid and easy to catch."

What the naturalist John Muir wrote of the passenger pigeon in the nineteenth century was equally true of the turkey: "Every shotgun was aimed at them." In anecdote after anecdote, the idea conveyed by A. W. Schorger is that whatever the circumstances, when a man saw a turkey or a flock of turkeys, he got his gun. Even if he found the turkeys somehow engaging, he still killed them all if he could, or took a few potshots at the flock. Foreign visitors practiced shooting birds on shipboard, believing, as Albert Hazen Wright relates, that they should have "such excellent sport in America shooting wild turkies."

Whole American communities gunned down turkeys for eating the grain. In Ohio, people used clubs to drive turkeys from the wheat fields, and circular hunts were organized to exterminate the birds because they ate the corn. Schorger quotes an Illinois resident who wrote in 1937, "One of my earliest and most vivid recollections was of the day when everybody combined to slaughter the last immense flock of Wild Turkeys. They enticed so many tame Turkeys away and were so destructive to the crops, that their extermination was decreed by the grange, churches, and the general public."

The Turkey as a Holiday Meal

Following a mild summer after the first terrible winter during which half the Mayflower company died, the Pilgrims, who arrived at what is now Cape Cod in November of 1620, gathered their first small harvest and held a feast with the Indians. But while the table was full of birds that probably included turkeys, there is no specific record, as George Willison writes, "of the long-legged 'Turkies' whose speed of foot in the woods constantly amazed the Pilgrims."

In his history of Plymouth Plantation, Governor William Bradford does not mention the Thanksgiving celebration of 1621, the date of which is unknown, and he refers to the turkey only by noting that "besides waterfowl there was great store of wild turkeys, of which they [the Pilgrims] took many, besides venison, etc."

Although by the 1850s the turkey had become a traditional part of the Thanksgiving holiday in New England, the turkey did not become a Thanksgiving main dish outside New England until after 1800, any more than did Thanksgiving itself. Long before being established as the Thanksgiving Day bird in America, the turkey appeared on Christmas tables in England. First shipped to Europe from Mexico by the Spanish in the early sixteenth century, turkeys were bred in Renaissance England, raised on country estates, shot in royal hunting parks for sport, and served at various royal and ecclesiastical functions. The transplanted bird was then brought back to America to become the forerunner of modern domestic turkeys.

Turkeys quickly entered the slaughter markets and households of England. P. E. Jones writes that regulations issued in 1513 forbade the poulterers of Southwark to permit their "hens, ducks, turkeycocks, or any other kind of poultry to go into the streets to 'rayse upp the myre and mucke to the common anoyauance.'" The turkey appeared as an English household meat in Gervase Markham's 1615 handbook *The English Housewife,* together with chickens, pigeons, partridges, peahens, and "such like" as "lesser land fowl." Markham included turkey recipes and described the proper way to roast turkeys "with the pinions folded up, and the legs extended." According to Richard Ryder, the typical eighteenth-century household slowly bled turkeys to death "suspended upside down from the kitchen ceiling." William Howitt's eyewitness account of this practice in a nineteenth-century kitchen appears in my book *More Than a Meal*.

Walking Turkeys to Slaughter

Before World War II, turkeys not butchered on the farm or shot in the wild or on an estate were driven to the nearest terminal on foot. In Europe as early as 1691, Cardinal Perron described people driving turkeys from France to Spain in flocks, "like sheep." In eighteenth-century Europe, turkeys were typically walked one hundred miles or more. The northern counties of England drove thousands of turkeys on foot to the London markets each fall, and in America, prior to the use of trucks in the 1950s, turkeys were driven eight to ten miles a day on foot through terrain ranging from densely wooded mountain trails to treeless Texas plains on journeys of fifty to two hundred or more miles. Typically, a drive of 20,000 turkeys employed four to six drovers, in relays of forty drovers in all, carrying long whips with strips of flannel tied to the ends which they used to "flick" the birds in line. At night the turkeys roosted in trees and were regrouped in the morning, though some birds disappeared into the woods and fields and did not return.

As well as being walked to markets, turkeys were slaughtered on farmsteads, a practice that continues today during the holidays alongside industrialized mass production and slaughter of turkeys. Turkeys were slaughtered for home consumption and for buyers in nearby towns. Before they were killed, turkeys were, as they still are, starved for approximately twelve hours to reduce the spillage of stomach contents during killing. While many farmers simply chopped the bird's head off with an axe on a tree stump, as shown in the March 1930 *National Geographic Magazine,* other methods, like wringing the birds' necks, were used together with the customary and commercial practice of suspending farm fowl head down by their feet from a shackle and killing them in a procedure that consisted first of throat-cutting and bleeding out, followed by "sticking or braining" the bird through the roof of the mouth to induce paralysis so the feathers could be plucked more easily.

Birds slaughtered but not eaten on the farm were delivered directly to customers or transported by rail to "commission men," who sold them to city retailers. In many places, just before Thanksgiving, a "turkey day" was held when turkeys slaughtered the day before were taken to town the next morning to be bid on. By 1930, turkeys and other poultry were being shipped by rail in specially designed cars, and by 1956, these cars had been replaced by trucks.

The Turkey as a Figure of Fun

Reflecting stereotyped attitudes, the word "turkey" has become a slang term for failure and vanity. As a generalized term of derision, the earliest recorded use of the word comes from columnist Walter Winchell who, in the November 1927 issue of *Vanity Fair,* said that the word "turkey" means "a third rate production." Since then, "turkey" has become a byword for mockery of just about anything pretentious enough to be considered laughable, from government officials to overrated movies. The turkey's humorous image derives largely from the courting behavior and sexual characteristics of the adult male bird during the spring mating season, during which time his appendages swell and he struts, fights with other toms, gobbles loudly from the treetops, and blazes forth his sexual condition in the pulsing red and blue colors of his face.

Terms related to "turkey" have been put to similar use. "Gobbler," the Yankee term for a male turkey, comprises age-old echoes of jabber, chatter, babble, and gabble, and "gobble" is an all-purpose word for noisy nonsense talk and voracious swallowing, or greed. The term *gobbledygook* is attributed to a U.S. House Representative from Texas, Maury

Maverick, who denounced bureaucratic jargon as "gobbledygook language" based, he said, on "the old bearded turkey gobbler back in Texas who was always gobbledy-gobbling and strutting with ludicrous pomposity."

"Turkey-cock" similarly is an epithet for the absurdly arrogant, pompous, boasting type of man, combined with stock images of the barbarous Turk. During the sixteenth century, when turkeys were first shipped from America to Europe, and a Turkish invasion of Europe seemed possible, the word *Turk* became a stock term for a type of exotic male savagery considered both barbarously ferocious and uproariously funny. Indeed, three centuries before any turkeys appeared in Europe, the word "turkey" was being used to describe exotic birds from Asia, because the Turkish Empire was the main European trade route to the East through which exotic birds such as peafowl were transported to the European continent in trade. Even after it became generally known that the turkey was an American bird, the idea clung in eighteenth century Europe that the turkey came from Turkey.

The Turkey in Transition

In America, the word turkey was used by the Irish and others to signify an Irish immigrant in the United States. In James T. Farrell's 1932 novel *Young Lonigan,* the character Dooley is described as "one comical turkey, funnier than anything you'd find in real life." The idea of the comical turkey persists in the litany of sarcasm that accompanies the piety of Thanksgiving each year in the United States, when newspapers and other media poke fun at the "Thanksgiving Day bird" along with the human "turkeys" in power, and holiday rituals include, or have included, everything from throwing turkeys off scaffolds and out of airplanes to forcing them to participate in turkey "Olympics" and in White House "turkey pardoning" ceremonies, as described in *More Than a Meal.*

America celebrates its heritage paradoxically by feasting on a bird reflexively despised by mainstream culture as stupid, dirty, and silly, a misunderstanding reinforced by the turkey food industry, which alternates between caricaturing the turkey as a ludicrous "personality" versus representing the bird as an anonymous "production animal." Stock photos of thousands of debeaked turkeys crowded together awaiting slaughter in nondescript sheds reinforce the popular idea that turkey are worthless except as objects of sport and meat.

Even so, the derogatory turkey stereotype is starting to modify. In the last quarter of the twentieth century, the creation of farmed animal sanctuaries and turkey-adoption programs offered new opportunities for people to get to know turkeys differently from the demeaning stock versions of the bird. Partly in response to these encounters, a growth in vegetarianism is occurring in the United States and elsewhere. At the same time, the avian sciences are debunking the prejudice against birds in general, and ground-nesting birds such as turkeys and chickens in particular, as "primitive." Avian scientists are calling for a whole new birdbrain nomenclature based on the now overwhelming evidence that birds' brains are complex organs that process information in much the same way as the human cerebral cortex, findings summarized by The Avian Brain Nomenclature Consortium in *Nature Neuroscience Reviews* in 2005.

An irony of the low esteem in which domestic turkeys have been held is that, as wildlife biologist William Healy points out, much of what is known about the wild turkey's intelligence is based on work with domestic turkeys whom Healy defends from the charge of stupidity by observing that genetic selection for "such gross breast development that few adult males can even walk," fuels the fallacy that such creatures are "stupid."

A further irony is that the wary turkey that dominates modern hunters' discourse is not exactly the bird the early European explorers and colonists encountered. As John Madson writes in the *Smithsonian,* "Wild turkeys, as the first settlers found them, were as trusting and unwary as they were plentiful." From the seventeenth through the nineteenth centuries, wild turkeys were characterized repeatedly as showing the same kind of friendly curiosity towards people that modern visitors often discover with surprise and delight when they meet domesticated "food production" turkeys at farmed animal sanctuaries. "They often sat with their young on my fences so trustingly that I found it difficult to bring myself to shoot them," said one person typically of the wild turkey's amiableness towards the settlers.

It remains to be seen whether modern experiences and the advancing sciences of avian cognition and ethology will lead people to rethink, as did naturalist Joe Hutto in the course of raising young turkeys to adulthood, many of their attitudes and presumptions about "the complexity and profoundly subtle nature of the experience within other species." As the single most visible animal symbol in America, the de facto symbol of the nation and "icon of American food," the turkey highlights the growing conflict in Western culture between the age-old presumption that animals exist solely for humans to exploit and the view that nonhuman animals are kin to humans with value and autonomy in their own right.

Further Resources

Avian Brain Nomenclature Consortium. (2005). Avian brains and a new understanding of vertebrate brain evolution. *Nature Neuroscience Reviews, 6,* 151–59.

Bakeless, J. (1961). *America as seen by its first explorers: The eyes of discovery.* New York: Dover Publications.

Bigelow, J. (1904). *The works of Benjamin Franklin in twelve volumes.* New York: G.P. Putnam's Sons.

Borenstein, S. (1998, November 26). Turkey lovers rejoice as wild bird population rebounds. *Albuquerque Journal.*

Bradford. W. (1981). *Of Plymouth Plantation 1620–1647.* New York: Modern Library.

Davis, K. (2001). *More than a meal: The turkey in history, myth, ritual, and reality.* New York: Lantern Books.

Dickson, J. G. (Ed.). (1992). *The wild turkey: Biology & management.* Harrisburg, PA: Stackpole Books.

Healy. W. M. (1992). Behavior. In J. G. Dickson (Ed.), *The wild turkey: Biology and management.* Harrisburg, PA: Stackpole Books.

Hutto. J. (1995). *Illumination in the flatwoods: A season with the wild turkey.* New York: Lyons & Burford.

Jones. P. E. (1965). *The worshipful company of poulters of the City of London: A short history.* London: Oxford University Press.

Josselyn, J. (1988). A critical edition of *Two voyages to New England,* P. J. Lindholdt (Ed.). Hanover: NH: University Press of New England.

Madson, J. (1990, May). Once, he was almost a "goner" but now Old Tom's a "comer." *Smithsonian.* 54–62.

Markham, G. (1986). *The English housewife,* M. R. Best (Ed.). Montreal: McGill-Queen's University Press.

Ryder, R. D. (1989). *Animal revolution: Changing attitudes towards speciesism.* Cambridge, MA: Basil Blackwell.

Schorger, A. W. (1966). *The wild turkey: Its history and domestication.* Norman: University of Oklahoma Press.

Teale, E. W. (Ed.). (1954). *The wilderness world of John Muir.* Boston: Houghton Mifflin.

Vaillant, G. C. (1941). *Aztecs of Mexico: Origin, rise and fall of the Aztec nation.* New York: Doubleday, Doran & Co.

Willison, G. F. (1945). *Saints and strangers: Being the lives of the pilgrim fathers & their families.* New York: Reynal & Hitchcock.

Winchell, W. (1927, November). A primer of Broadway slang. *Vanity Fair.*

Wright, A. H. (1914). Early records of the wild turkey. *The Auk: A Quarterly Journal of Ornithology, 31*(3), 334–58.

Karen Davis

■ Human Anthropogenic Effects on Animals
Human Anthropogenic Effects on Animals

Human Effects on Animal Lives

Humans are a unique species and a very curious and inquisitive group of mammals. We are here, there, and everywhere, and our intrusions, intentional or not, have significant impacts on animals, plants, water, the atmosphere, and inanimate landscapes. We are the most dominant species the earth has ever known. When humans influence the behavior of animals, the effects are referred to as being "anthropogenic" in origin.

The relationship between humans and animate and inanimate nature is a complex, ambiguous, challenging, and frustrating affair. Although we do many positive things for animals, as many of the essays in this encyclopedia illustrate, we also make the lives of animals more difficult than they would be in our absence, and we make environmental messes that are difficult to fix, a point also highlighted in a number of essays in this encyclopedia. On the positive side, the Introduction to this encyclopedia mentioned how animals are blessed and welcomed into our homes and how music was used to make the life of Suma, a grieving forty-five-year-old elephant in the Zagreb Zoo in Croatia, better after her friend Patna died. In addition, in October 2006 the German parliament unanimously voted to ban seal products from the country because of the way in which seals are clubbed to death during mass slaughters. And recently, Whiteface Mountain located in the Adirondack mountains in upstate New York changed the configuration and design of ski trails so as not to have a negative impact on an elusive bird called Bicknell's thrush who nests there. Bicknell's thrushes are not an endangered or even a threatened species but rather a "species of special concern."

Scientists are also becoming increasingly concerned about how we affect deep-sea communities (see "Venting Concerns" in the Further Resources list) that frequently do not receive this sort of attention, and ecotourism also has many sides to it and is getting more detailed attention so that we come to better understand the positive and negative aspects of our intrusions into animals' lives and the ecosystems in which they live (see "Ecotourism" by Constance Russell and Anne Russon, this encyclopedia, and "Good Gone Wild," listed in Further Resources). Although ecotourism can negatively affect animals and the communities in which they live—because animals avoid humans or because we damage their habitats—many countries depend on the money that is generated by ecotourism, so we must be ethical ecotourists, learn about the effects that we have, and do something to avoid being intrusive and destructive. For example, many people are concerned that ecotourism in "Darwin's paradise," the Galapagos Islands, is having a grave impact on the fauna and flora living there.

We also influence the behavior of the "urban" animals with whom we share our homes, so that many try to leave what was their land, but their presence also enriches our lives (see "Wildlife in Urban Areas"). We must remember that our land is their land too. Numerous animals live around my house in the mountains outside of Boulder, Colorado.

My neighborhood friends include numerous birds, lizards, and insects and some wonderful mammals—squirrels, chipmunks, mice, deer, red foxes, coyotes, black bears, and mountain lions. One summer morning, I noticed that the local red foxes were drinking from small pools of water that collected on the top of my hot tub after a brief rain during an unusually hot and dry period (see photograph in the color insert). They did this almost every morning. Over time, the foxes became less and less concerned about my presence and that of the dogs in the neighborhood. I could also see that as the foxes lost their fear and became increasingly bold, they also were more vulnerable to cars and to the presence of predators such as mountain lions and coyotes. One day, I left a pair of shoes outside my front door, and almost instantaneously, a fox came up to the door, grabbed one of the shoes, and backed up a few feet. I opened the door expecting the fox to show some fear, but rather, he stood his ground and looked at me as if to say "What are you doing in my home?" and then tore my shoe into little pieces. Previously, when I opened my door to look at these beautiful animals, they would retreat slowly and usually run away. These very same foxes play around my house, and once, my wife Jan and I watched them as they ran here and there and frolicked right in front of us as if we were not there at all.

When wild animals become accustomed to the presence of humans, it is called habituation, and numerous animals have changed their daily routines because of our intrusions into their homes. Often predators and their prey become bolder and less concerned, and this causes problems for everyone, humans and animals alike. Mountain lions, for example, have become very habituated in many communities in the western United States, and this has caused people to launch campaigns to rid themselves of these magnificent animals. Yet when humans take precautions to protect themselves and others, lions are not all that problematic, and attacks, while slightly on the rise, are still very rare. I once almost stepped on a male mountain lion while backing up and telling my neighbor that there was a lion in the area, and on another occasion, thinking that a tan animal running toward my car was my neighbor's dog, I opened the car door only to see that it was a lion, not a dog, who was prancing my way. Once, while I was sitting on my couch and reading, I saw a big black animal move slowly across my deck, seemingly without a care in the world. Then I heard some noise at my sliding glass door. I got up and ran to the door only to see a male black bear trying to open the door. When he saw me, he stepped back, looked at me, and walked off my deck and visited my neighbor's house and went to sleep under her hammock. Years ago, my friend who lives up the mountain from me left her kitchen window open when she went to town, and a bear came in and had a feast in her kitchen, leaving behind an incredible mess. We all laughed, thinking that this was one happy bear who had gotten a delicious and easy meal. But the wayward bear would be better off living away from human homes.

I have written more about habituation in my essay titled "Human (Anthropogenic) Effects on Animal Behavior" in my *Encyclopedia of Animal Behavior*. Recognizing that we can have such large effects on the behavior of animals, we must take precautions so that they and we do not suffer from their neighborly proximity. I try not to walk outside at night, and when I do, I take a flashlight and often carry bells to announce my presence. I am now careful to keep my doors and windows closed when I am not home, I do not feed the wild animals around my home, and I make sure not to leave garbage for them to feed on. It is usually easy to change our ways to accommodate the presence of the animals whom we displace as we expand our horizons.

Because of the widely varying settings in which we interact with animals, sometimes we just do not know what to do when human interests compete with those of other beings, which happens almost every second of every day worldwide. Many people claim to love nature and to love other animals and then, with little forethought, concern, or regret, go on to abuse them in egregious ways, too numerous to count. Many of the animals whom we want to study, protect, and conserve experience deep emotions, and when we step into their worlds, we can harm them mentally as well as physically. They are sentient beings with rich emotional lives. Just because psychological harm is not always apparent, this does not mean we do not do harm when we interfere in animals' lives. It is important to keep in mind that when we intrude on animals, we are influencing not only what they do but also how they feel.

Humans Leave a Large Ecological Footprint

Suffice it to say, our influence on the lives of other animals is ubiquitous with few bounds. We leave what scientists call a huge ecological footprint. We really do have *that* much of an effect on the lives of animals. We tend to forget that this land is their land too. It is difficult to be neutral because of who we are as a species, rather unique, large-brained mammals, with the ability to travel rapidly on Earth and to venture deeply into surrounding waters and far into space. People who worry about what humans are doing to animals are not being hysterical. Indeed, leading scientists are writing essays about this in many of the most prestigious journals and in this encyclopedia because they are very worried about the irreversible effects of human activities. Attitudes toward animals as expressed in religious doctrines, cultural traditions, political discussions, textbooks, classrooms, popular books, literature, poetry, movies, and the popular press also influence how we view and treat animals.

"Oh, I Didn't Know That!"

Often our influence on the behavior of animals and the unbalancing of nature is very subtle and long-term and surprising. And sometimes, we conveniently ignore our effects because we do not want to face up to them—they are what former vice president Al Gore calls "inconvenient truths." Mr. Gore was referring to the negative and wide-ranging effects of global warming and climate change that a few people still deny. Sometimes we exclaim, "Oh, we didn't know that we had this or that effect," but it is too late to do anything about it. For example, the effects of global warming, increases in atmospheric carbon dioxide, and climate change in general are influencing the lives of innumerable animals, although from day to day we do not see these changes, or they are happening in remote parts of the world (see "Global Warming and Animals"). The Monteverde harlequin frog and its cousin the golden toad in Costa Rica's Tilarán Mountains, once numerous for millions of years, have not been seen in almost twenty years.

We are also losing numerous species of migratory birds because of global warming. In an essay in *Time* magazine in May 2006 aptly titled "Bye Bye Birdies," Michael Lemonick reported that radar studies of annual migrations suggest that the number of birds flying along America's flyways may be down by as much as 50 percent over the past thirty years. For example, there is a large decrease in flycatchers because climate change makes them "late for dinner." It turns out that the flowering plants in the Netherlands flower earlier, and caterpillars show up earlier, so when these birds arrive in the Netherlands from West Africa, they are too late for a much-needed meal. Climate change also influences wolves because warmer temperatures influence vegetation, which in turn affects the wolves' prey, such as vegetarian elk.

Large animals who seem resistant to human intrusions are also very vulnerable. In the Democratic Republic of the Congo, the hippopotamus population was within a few months of extinction in October 2006 because the Mai Mai militia were hunting the few remaining animals unrelentingly for food and ivory (see "DR Congo Hippos 'Face Extinction,'" listed in the Further Resources). Also, at about the same time, Iceland announced that commercial hunting of whales could resume, and shortly thereafter, an endangered fin whale was killed (see "Bad Day for Whales"). Unsustainable killing is threatening numerous different species globally.

Elephants also are greatly affected by human intrusions into their lives. These amazing animals, who live in large groups led by an old female called the matriarch, are highly sensitive beings with long memories. In a cover story in the magazine section of the *New York Times* titled "Are We Driving Elephants Crazy?" Charles Siebert summarized how humans can cause elephant societies to become highly dysfunctional and violent and how these changes escalate over time to the point where elephants holding grudges and possibly suffering from posttraumatic stress disorder (PTSD) become serious threats to humans. Left to live on their own, elephants usually leave humans alone (see the entries "Elephants and Humans" by Gay Bradshaw and Lawino Abo under Living with Animals and "Human Emotional Trauma and Animal Models" by Christine Caldwell under Health).

Researchers are also interested in how environmental contaminants influence other animals. Environmental toxicologists have discovered that polychlorinated biphenyls (PCBs—industrial chemicals including those used in adhesives, fire retardants, and waxes) are widespread environmental pollutants that can disrupt reproductive and endocrine (hormone) function in fish, birds, and mammals, including humans, especially during embryonic development. Pesticides can have devastating effects on populations of wild animals ranging from mammals to insects. Dicofol, a pesticide that is contaminated with DDT (dichloro-diphenyl-trichloroethane), can influence the reproductive organs and population sizes of alligators. PCBs can also cause sex reversal in alligators. It also has been discovered that perchlorate, a chemical found in rocket fuel, can disrupt sexual development in fish and turn females into males. The females, called "macho moms," display male courtship and produce sperm. And in an article by Kenneth R. Weiss in the *Los Angeles Times* titled "Sentinels under Attack," it was reported that marine mammals such as California sea lions are dying along the coast of northern California because toxic algae infects their brain and causes strandings and mass die-offs. The animals become emaciated and disoriented and suffer from seizures. According to this essay, more than 14,000 sea lions, seals, and dolphins have turned up sick along California's coast in the last decade.

Webs of Nature Unite Us All: We're All in this Together

Humans and animals are integrated into complex webs of nature, a point stressed by Rachel Carson in her landmark book *Silent Spring,* which many claim triggered the modern environmental movement. Any ecological view that does not consider the depths and interconnections among all components of nature—human, nonhuman, and inanimate landscapes—is only a "partial ecology" rather than an "integrated ecology" and misrepresents the magnificent webs of nature that abound all over the place. A wonderful book titled *The Unlikely Burden and Other Stories* highlights an integrated view of human-animal relationships in essays written by people from all over the African continent.

Suffice it to say, and to risk being trite, the deep reciprocal interconnections among members of the earth community are such that we are all in this together, and we all need others we can lean on. A view of nature that sanitizes, reduces, and simplifies the complex interrelationships that exist is bound to tarnish and diminish the appreciation that people have for what is out there and result in a feeling of alienation. This feeling of alienation can then in turn feed back and produce more alienation, and we and other animals suffer.

Coexistence Is Difficult

Often we become at odds with the very animals with whom we choose to live when they become nuisances, become dangerous to us or to our pets, or destroy our gardens and other landscapes. Thus, we have to make difficult decisions about whose interests and lives to favor, theirs or ours. A more aware public no longer believes that human interests *always* trump the interests of other animals, so we have to factor in all of the variables to make the best choices on a case-by-case basis. For example, in some areas of Boulder, Colorado, people choose to coexist with prairie dogs, whereas in other locales, some people want to kill these family-living rodents because they have become a nuisance to those who want to build shopping malls, parking lots, soccer fields, and more homes. Killing prairie dogs, however, does not really solve the problems, and many believe we need to figure out what are the most humane solutions so that people can pursue their interests and prairie dogs do not have to suffer because of our arrogance and our inability to limit growth.

Humans are generally motivated to care about other animals because we assume that individuals are able to experience pain and suffering. Fortunately, very few people want to be responsible for adding pain and suffering to the world, especially intentionally. However, in our interactions with other animals, we often cause much intentional pain, suffering, and death, usually for human ends. In addition, because humans interact with animals in an increasing number of settings as we expand our own horizons, it is becoming more common to debate whether or not to cull (or kill) members of a species because they may be involved in the transmission of disease to other animals or humans. For example, badgers in the United Kingdom play a role in the transmission of bovine tuberculosis that infects cattle. A move to cull badgers to control the spread of this disease was met with substantial public resistance—96 percent of about 47,000 people polled throughout England said "no" to the planned cull, many favoring better farming practices (see "Public Says 'No' to Badger Cull," listed in Further Resources). Years ago, this sort of response was not very usual, people either ignoring the problem or favoring the well-being of humans or domestic livestock. This example along with the treatment of prairie dogs shows that as time passes, more and more people are showing concern for how we interact with other animals.

Research Effects on Animals: Applied Animal Behavior Science

Sometimes when we conduct scientific studies to learn about the behavior and ecology of other species, we unknowingly affect their lives. Some behavior patterns influenced by various research methods and other forms of human intrusion include nesting and reproductive activities, dominance relationships, mate choice, use of space, vulnerability to predators, and feeding and caregiving behaviors. Intrusions include such activities as urbanization (urban development and sprawl, the development of bodies of water, changes in vegetation, the need for more electric power) and recreational activities, including the use of snowmobiles and other off-road vehicles, photography, travel, and ecotourism. I have also written about many of these in my essay titled "Human

(Anthropogenic) Effects on Animal Behavior" in the *Encyclopedia of Animal Behavior*. In November 2003, it was reported that only two years after a new finch-like bird, the Carrizal blue-black seedeater, was discovered in Venezuela, its habitat was destroyed so that a hydroelectric dam could be built.

Models that are generated from these studies can be misleading because of human intrusions that appear to be neutral. Often our intrusions mean that we cannot really collect the data we need to answer specific questions. The topic of human-animal interactions is relevant to studies of applied ethology or applied animal behavior studies. Ethological studies in which the behavior patterns of animals are studied in detail, preferably under natural conditions, are needed because we must take into account just how our research influences the behavior of other animals; otherwise, we risk drawing the wrong conclusions.

Reintroducing Species: Making Room for Wolves

We also face difficult questions about our relationships with animals in conservation efforts—for example, when we try to bring back a species to a particular locale to restore or recreate an ecosystem from which animals disappeared years ago, usually as a result of human activities. We need to balance the fate of individuals, some of whom surely will die, with the fate of their species.

Reintroductions require massive human effort and large amounts of money. For example, we are a major factor in what goes right and what goes wrong, and people who wake up to reintroduced wolves at their door often are resentful and hostile (see "Wolf and Human Conflicts: A Long Bad History," this encyclopedia under Conservation and Environment). This is easy to understand, especially when these people have been living their lives and making their livelihood absent these predators. Also, the myth of the savage wolf precedes the presence of these magnificent, important, and maligned creatures, and this also makes it difficult for some people to accept their presence.

An excellent example is the reintroduction of gray wolves to Yellowstone National Park in Montana, an area in which humans exterminated wolves in the 1920s, because of their predatory habits. The project is considered by many people to be successful in that numerous wolves now roam the Yellowstone ecosystem. However, in the process, some of the wolves who were moved from Canada and Alaska have died, and the newcomers have killed numerous coyotes in various parts of the park. Did we do harm when we removed wolves from one area to bring them to another locale? Are we robbing Peter to pay Paul? Should we favor ecosystems and species over individuals? These are some of the difficult questions with which conservation biologists are faced. Some people argue that individual well-being should come before the fate of a given species or the integrity of an ecosystem, whereas others believe that it is all right to have a few individuals die for the good of their species.

Other questions also need to be considered because not everyone favors bringing wolves back to Yellowstone. Ranchers and farmers believe that predation by wolves is responsible for significant losses of livestock, although available data do not support this claim. Nonetheless, there is a move to remove wolves from the endangered species list so as to make it legal to kill them. The U.S. Forest Service is planning to ease limits on killing predators in the western United States, and the Wisconsin Department of Natural Resources recently was given permission to kill "problem" wolves. Furthermore, there are examples of "Judas wolves"—one wolf is collared, and then that wolf's pack is followed and slaughtered. Or an injured wolf is left as bait, and when the wolf's pack returns for him or her, they are killed. The reason for killing wolves centers on their predatory nature, so it is important to point out that only about 3 to 5 percent of more than 7.5 million sheep raised in 2004 were killed by predators, and recent studies have shown that killing wolves and coyotes does not have any positive effect on the survival of sheep or cattle. Better ranching methods, including control of disease and fencing off areas, work better to save livestock. Nonetheless, in the United States, a government group called Wildlife Services killed more than 82,000 mammalian carnivores in 2004 as part of predator control, using methods such as trapping, poisoning, and shooting these animals and gunning them down from aircraft. It is important to stress that the overmanagement and overcontrol mentality that we see in North America does not necessarily hold in other parts of the world. However, for carnivores, even within protected areas, conflicts with humans are usually the most important cause of mortality in adults. But lately in some areas, ranchers and conservationists are working together to limit land development, so that there will be more room for wolves and for ranches.

Consider also the reintroduction of Mexican wolves in New Mexico and how federal gunners are free to wipe out the Nantac pack despite the fact that these wolves have not stabilized or reached suitable numbers to increase the likelihood that they will survive. The federal predator-control program has been responsible for reducing the population of wild Mexican wolves from fifty-five at the end of 2003 to forty-four at the end of 2004 and thirty-five at the end of 2005. During May 2006, federal gunners killed eleven wolves, including six pups from one pack.

To sum up, the big questions with which we must be concerned include, "Is it permissible to move individual wolves from areas where they and other wolves have thrived and place them in areas where they might not have the same quality of life, for the perceived good of their species?" and "Is it permissible to interfere in large ecosystems that have existed in the absence of the species to be reintroduced, and is it permissible to remove animals from an ecosystem in which they play an integral role?"

We must also ask, "What are we doing and why are we doing it?" Is it possible to remove ourselves from the cycle of persecuting, killing, exterminating, and trying to bring back species? And if we cannot get out of the loop, do we need to stop trying to right the wrongs of the past? These are difficult and frustrating questions, but they will not disappear if we ignore them. Is it possible to make room for wolves and not have them suffer once again because of human intrusions into their lives? I bet that some young readers of this encyclopedia will be pursuing answers to these and other questions in the future.

What Can We Do?

When behavior and activity patterns are used as the litmus test for what is called "normal species-typical behavior," researchers need to be sure that the behavior patterns being used truly are an indication of who the individual is in terms of such variables as age, gender, and social status. If the information used to make assessments of well-being is unreliable, then it is likely that the conclusions that are reached and the animal models that are generated are also unreliable and can mislead current and future research programs. And of course, human errors can have serious negative effects on the lives of the animals being studied. Many believe that animal behavior scientists' research ethic should require that we learn about the normal behavior and natural variation of various activities so that we learn just what we are doing to the animals we are trying to study.

In addition to learning about how our intrusions influence the lives of animals, it is important to share this knowledge so that we do not inadvertently change them. Sharing involves disseminating information about what is called the "human dimension" to administrators of zoos, wildlife theme parks, aquariums, and areas where animals roam freely so that visitors can be informed of how they may influence the behavior of animals they want to see. Tourism companies, nature clubs and societies, and schools can do the same.

Many animal behavior scientists believe that the major guiding principle is that the lives of the animals whom humans are privileged to study should be respected, and when we are unsure about how our activities will influence them, we should err on the side of the animals and not engage in these practices until we know (or have a very informed notion about) the consequences of our acts. This precautionary principle will serve the animals and us well. Indeed, this approach could well mean that exotic animals that are so attractive to such institutions as zoos and wildlife parks need to be studied for a long time before they are brought into captivity. For those who want to collect data on novel species that are to be compared to other (perhaps more common) animals, the reliability of the information may be called into question unless enough data are available that describe the normal behavior and species-typical variation in these activities.

We must continue to develop and improve general guidelines for research on free-living and captive animals. These guidelines must take into account all available information. Professional societies can play a large role in the generation and enforcement of guidelines, and many journals now require that contributors provide a statement acknowledging that the research conducted was performed in agreement with approved regulations. Guidelines should be forward-looking as well as regulatory. Much progress has already been made in the development of guidelines, and the challenge is to make those guidelines more binding, effective, and specific. Fortunately, many people worldwide are working to improve our relationships with other animals.

Deep Ethology

Deep ethology, a concept that I developed, comes into play because we need to recognize that we are all integral members of a single earth community and that as moral agents with enormous brains and the capacity to do whatever we want, we have unique responsibilities to nature. We bring a lot to the party along with something special, namely the moral capacity to restrain our exploitive practices regarding other animals and a plethora of landscapes. A deep ethological perspective, like a view of ecology that stresses deep interconnections, allows us to be sensitive to the lives and experiences of animals and implores us to try to take their point of view in our interactions with them. We need to observe animals carefully, listen to their stories attentively, inhale their odors with zeal and some caution, and never forget that they, like us, have likes and dislikes, preferences, goals, beliefs, and feelings. My views, and those expressed by some others, on animal use are indeed restrictive, but we realize that we must make decisions that take into account people and animals. We are not nature's keepers if this means that we "keep nature" by dominating other animals using a narrow, anthropocentric agenda in which other animals are objectified—to be numbers and not names, transformed into points on a graph—and their worldviews discounted. Surely, the sharing of feelings should make us deeply reflect on how we cause so much enduring pain and suffering to other beings, human and other-than-human. Allowing ourselves, if we dare, to occupy the paws, heads, and hearts of other animals should, I hope, reduce the gap between "them" and "us."

That many animals have subjective and intersubjective communal lives—they live in social groups, and other animals are in their thoughts and feelings—and a personal point of view on the world that they share with other individuals seems beyond question. In his development of an "anthro-harmonic" perspective on human-nonhuman relationships, Stephen Scharper (see "Conservation and Environment—From Anthropocentrism to Anthropoharmonism"), who studies the relationship between religion and environmental ethics, notes that "intersubjectivity is a fundamental reality of all human existence." Harmonic means "of an integrated nature" that "acknowledges the importance of the human and makes the human fundamental but not exclusively focal." Working towards an anthro-harmonic understanding of human-nonhuman relationships in the future is a good road to travel.

What Should We Do?

Inquiries about how we interact with other animals raise a host of "big" ethical questions such as why care about other animals; who we are (or who we think we are) in the grand scheme of things; how we should go about wielding our almost limitless power when we interact with other individuals, populations, species, and ecosystems; whether there are any "shoulds" (yes, there are; however, just because we can do something does not mean we should); whether we should be concerned with the well-being of individuals, populations, species, or ecosystems; whether we can reconcile a concern for individuals with a concern for higher and more complex levels of organization?

First and foremost in any deliberations about other animals must be deep concern and respect for their lives and the worlds within which they live—respect for who they are in their worlds, and not respect motivated by who we want them to be in our anthropocentric scheme of things. Can we really believe that we are the only species with feelings, beliefs, desires, goals, expectations, the ability to think, the ability to think about things, the ability to feel pain, or the capacity to suffer? Other animals have their own points of view, and it is important to appreciate, honor, and respect them when we interact with them. Ethics and scientific research are not incompatible.

Dingos are wild animals who also spend a lot of time around people. Here, an inquisitive young female solicits food on Fraser Island, Australia. The editor of this encyclopedia is on the far left. Courtesy of Jan Nystrom.

Four members of an elephant herd being studied by Iain Douglas-Hamilton and his colleagues in the Samburu Reserve in Northern Kenya. Elephants form social groups called matriarchies, and individuals of different ages form very close social bonds with one another. Elephants experience a wide range of emotions, ranging from joy when they play to grief when they lose a friend. They also empathize with other individuals. I had the pleasure of visiting Iain Douglas-Hamilton in Samburu in July 2005 and was amazed by my first-hand experience of the deep emotional lives of these magnificent animals. Courtesy of Jan Nystrom.

The Best and Worst of Times for Animals

In many ways, these are the best of times and the worst of times for many species of animals—the best in that more and more people around the world are truly concerned about how we affect the lives of the animals with whom we share space, and the worst in that the global population of humans is increasing steadily at unprecedented rates, and there is less and less space for us to live in without intruding into the lives of other animals.

Humans are a powerful force in nature, and obviously, we can change a wide variety of behavior patterns in many diverse species. Coexistence with other animals is essential. By stepping lightly into the lives of other animals, humans can enjoy the company of other animals without making them pay for our interest and curiosity. There is much to gain and little to lose if we move forward with grace, humility, respect, compassion, and love. Our curiosity about other animals need not harm them. The power we potentially wield to do anything we want to do to animals and nature as a whole is inextricably tied with responsibilities to be ethical human beings. We can be no less.

Further Resources

Bad day for whales. (2006, October 28). *New Scientist*, p. 6.

Bekoff, M. (2000). Field studies and animal models: The possibility of misleading inferences. In M. Balls, A.-M. van Zeller, & M. E. Halder (Eds.), *Progress in the reduction, refinement and replacement of animal experimentation* (pp. 1553–59). Amsterdam: Elsevier.

———. (2002). *Minding animals: Awareness, emotions, and heart*. New York: Oxford University Press.

———. (2006). *Animal passions and beastly virtues: Reflections on redecorating nature.* Philadelphia: Temple University Press.

———. (2007). *The emotional lives of animals: A leading scientist explores animal joy, sorrow, and empathy and why they matter.* Novato, CA: New World Library.

Bekoff, M., & Jamieson, D. (1996). Ethics and the study of carnivores: Doing science while respecting animals. In J. L. Gittleman (Ed.), *Carnivore behavior, ecology, and evolution* (Vol. 2, pp. 15–45). Ithaca, NY: Cornell University Press.

Bekoff, M., & Nystrom, J. (2004). The other side of silence: Rachel Carson's views of animals. *Human Ecology Review, 11,* 186–200.

Black, Richard. (n.d.). Public says "no" to badger cull. *BBC News.* Retrieved from http://news.bbc.co.uk/2/hi/science/nature/5172360.stm

Caro, T. M. (Ed.). (1998). *Behavioral ecology and conservation biology.* New York: Oxford University Press.

Cronin, W. (Ed.). (1996). *Uncommon ground: Rethinking the human place in nature.* New York: Norton.

DR Congo hippos "face extinction." (2006, October 19). Retrieved from http://news.bbc.co.uk/2/hi/science/nature/6065796.stm

Festa-Bianchet, M., & Apollonia, M. (Ed.). (2003). *Animal behavior and wildlife conservation.* Washington, DC: Island Press.

Goodall, J., & Bekoff, M. (2002). *The ten trusts: What we must do to care for the animals we love.* San Francisco: HarperCollins.

Good gone wild. (2006, September 30). *Science News.* Retrieved from http://www.sciencenews.org/articles/20060930/bob9.asp

Margolis, M. (2006, October 16). Why the frogs are dying. *Newsweek International.*

Nicholls, H. (2006, October 14). Trouble in Darwin's paradise. *New Scientist,* pp. 8–9.

Pabari, D., & Luce, L. (Ed.). (2006). *The unlikely burden and other stories.* Nairobi, Kenya: Sasa Sema.

Raloff, J. (2006, August 12). Macho moms: Perchlorate pollutant masculinizes fish. *Science News, 170*(7), 99. Retrieved from http://www.sciencenews.org/articles/20060812/fob2.asp

———. (2006, October 27). Venting concerns: Exploring and protecting deep-sea communities. *Science News, 170*(15), 232. Retrieved from http://www.sciencenews.org/articles/20061007/bob7.asp

Scharper, S. (1997). *Redeeming the time.* New York: Continuum.

Siebert, C. (2006, October 8). Are we driving elephants crazy? *New York Times Magazine.* Retrieved from http://www.nytimes.com/2006/10/08/magazine/08elephant.html?ex=1160884800&en=b2676c7a2fa539e1&ei=5070&emc=eta1

Weiss, Kenneth R. (2006, July 31). Sentinels under attack. *Los Angeles Times.* Retrieved from http://www.latimes.com/news/local/oceans/la-me-ocean31jul31,0,1410884.story

Whiteface mountain and Bicknell's thrushes (2006, August 24). http://select.nytimes.com/gst/abstract.html?res=F40B12FA385A0C778EDDA10894DE404482

Wilmers, C. C., & Post, E. (2006). Predicting the influence of wolf-provided carrion on scavenger community dynamics under climate change scenarios. *Global Change Biology, 12,* 403–09.

Marc Bekoff

■ Human Influences on Animal Reproduction
Canine Surgical Sterilization and the Human-Animal Bond

It is a sad but true fact that the less we know about animals, the easier it is to project human beliefs on them. As society becomes more remote from nature, an inverse relationship increasingly exists between people's lack of animal-related knowledge and their willingness to project human beliefs on animals. And because people living in complex contemporary societies may simultaneously feel alienated from or threatened by interactions with other people, a companion animal may make a perfect surrogate on whom to consciously or subconsciously project beliefs regarding human-human concerns. Unfortunately, unlike wild animals on whom humans have traditionally projected beliefs, companion animals bred for dependency may fall prey to human attempts to physically or behaviorally manipulate the animal to validate those beliefs. Even more unfortunately, sometimes people may be so unmindful of the origin of any animal-related belief that they automatically assume that the belief and its consequences benefit the animal. To demonstrate how projected beliefs actually may represent the complex interaction of deep cultural memories, science, and contemporary human emotional needs in ways that may or may not fulfill animal needs, this essay explores the link between the American belief in canine surgical sterilization, high-quality canine health and behavior, and a high-quality human-canine bond.

Early Surgical Sterilization History and Politics

Because what happens in the present invariably arises from what happened in the past, we first need to understand the roots of canine surgical sterilization. It did not arise from the same practice in farm animals. In those species, the procedure was and is almost exclusively limited to male castration for five reasons:

1. A male-to-female sex ratio of approximately 1:1 results in more males than necessary to ensure offspring.
2. Properly handling larger and more reactive intact (unsterilized) male farm animals requires more knowledge and skill.
3. Intact male hormones impart an odor and texture to meat that many humans find unacceptable.
4. It is more cost-effective and energy-efficient to control population by castrating males than by performing ovariohysterectomies (OHEs) on females because of the male testicles' location outside the body wall. This simplifies the surgery as well as reduces the probability of postoperative complications.
5. Reproduction triggers the female's production of milk, a product that is often as valuable to humans as animal offspring.

In companion canines, however, sterilization as a means of population control initially focused on females. Later, as questions regarding the biological soundness of the approach grew, male castration was added to veterinary recommendations for companion animals. Even so, although veterinarians traditionally recommended spaying females prior to sexual maturity to avoid unwanted pregnancies, until the late 1960s, the standard recommendation was to postpone castrating males until they reached sexual maturity, thereby dulling castration's population-control effect.

If domestic canine surgical sterilization practices did not derive from farm animal ones, from where did they come? The answer to that lies in the answer to another question: in what other domestic species does the incidence of surgical removal of female reproductive organs exceed that seen in males? The answer is the human species. In the post–World War II United States, when canine overpopulation was becoming an issue, the veterinary medical profession was migrating away from the practices and philosophy underlying farm animal medicine and toward those of human medicine. Simultaneously, the more urbanized public's attitudes were migrating toward dogs as family members rather than as members of a separate species with its own species-specific needs. Consequently, in order to understand the origins of contemporary views of companion canine sterilization, we need to understand the origins of those same procedures in humans, as well as animals.

Views regarding these fall into one of two opposing groups:

1. Humans learned to surgically sterilize each other from doing the same to farm animals.
2. Humans learned how to surgically sterilize farm animals by doing it to each other.

The former represents the more widely held view. Because the procedure made animals more submissive, the idea that early humans would use it to similarly modify the behavior of human slaves and prisoners of war seems reasonable. As in farm animals, male humans were the primary targets because the surgery was less complicated and females had more value as milk-producers (i.e., wet nurses) and sexual partners. Also, although the claimed behavioral benefits of male animal castration have long been reported, no such claims were made for female surgical sterilization, which further negated it as a viable option.

In addition to using sterilization as a means to subjugate others, humans also paradoxically removed their own reproductive organs in an attempt to *increase* their own status. According to Matthew Kuefler, legends and ancient writings speak of men called *galli* who castrated themselves to honor a goddess known as Cybele/Kubala, Mater Deum, Magna Mater, or Caelestis, all of whom had links to the more familiar goddesses Aphrodite, Rhea, Isis, and Hera. These men were the forerunners of others who castrated themselves to prove their loyalty to higher-ranking religious, political, or military figures or to certain ideologies. Conceivably this process never attracted many women because of the previously mentioned likelihood of postsurgical complications, including death.

It is also important to our understanding of the history of canine neutering to note that the earliest descriptions of human castration refer to the removal of both testicles and penis. Early humans most likely first associated reproduction with the erect penis rather than the testicles, with the removal of testicles being accidental in those first crude castration attempts. Logic suggests that had those early humans decided to develop their surgical sterilization skills on members of another species, the dog—that smallest, most domesticated of domestic animals—would have been the ideal species on which to do this.

Why did this not occur? Why is widespread canine castration a relatively recent phenomenon? Those answers come from anatomy. Unlike the human penis, most of the canine penis is firmly held within a protective sheath that is incorporated into the body wall. Additionally, and unlike these organs in men, stallions, bulls, rams, and boars, the canine penis contains a bony structure, the *os penis* or *baculum*. These anatomical differences could have yielded such poor results in early attempts to castrate dogs that those doing it concluded it was not worth the effort. Thus, although failed attempts to castrate dogs could have preceded or followed human attempts to castrate other humans, this could explain why the surgical sterilization of male farm animals preceded that of dogs.

Ironically, unlike human slaves whose death from postsurgical complications may have been of little concern, castration of early farm animals almost surely was done to control these animals because they were so valuable. Consequently, it seems reasonable to suggest that farm animal castration occurred after humans recognized the key role the more readily removed testicles played in reproduction. A single ligature around the spermatic cord controls bleeding and permits the removal of each testicle in an older animal. Younger ones may be castrated simply by tying a ligature around the scrotum (the sack that holds the testicles), thereby negating the need for any surgery at all. However, by the time this discovery was made, the dog's role as "man's best friend" may have been sufficiently established that it precluded the need for canine castration as a means of control.

Nonetheless, the fact that other- or self-removal of reproductive organs does not commonly occur in nonhuman animals would seem to negate the premise that early humanoids did this to each other or themselves. On the other hand, exposure of the genitalia to signal submission routinely does occur in social mammals. Although dogs normally will roll over on their backs, members of other species will bow with their exposed external genitalia facing the perceived more dominant individual.

Although higher-ranking members of the same species who trigger such a response could easily remove the submissive animal's exposed genitalia, they do not do this for three probable reasons. First, social animals depend on each other for their own survival, so any display that undermines the survival of another pack member could undermine one's own. Second, packs are dynamic; today's subordinates may be tomorrow's leaders, part of whose value to the group lies in their ability to reproduce. Third, the physiological changes accompanying submission may serve as such potent indicators of lesser rank that further action is unnecessary.

Chief among the physiological changes are the messages transmitted and reinforced by pheromones, powerful communicating hormones found in animal urine, feces, and other bodily secretions. The submissive posture not only provides visual evidence that the animal displaying it poses no threat; it also exposes urogenital and anal orifices with their wealth of pheromonal scent data supporting this same message. Simultaneously, the more dominant animal may directly and indirectly affect the subordinate's behavior by secreting pheromones that communicate higher rank as well as suppress the other's sexual behavior, trigger abortion in females impregnated by other males, or prevent subordinates from settling too close.

The loss of olfactory ability that accompanied the evolution of flatter human facial features conceivably undermined any benefits of scent-based communication sufficiently to force evolving humans to find other ways to limit the numbers and control the behavior of competitors. Castration might then have emerged as a viable option, particularly as those submissive postures made those organs so readily available. If this coincided with increasing numbers of humanoids or humans competing for the same scarce resources, the permanent removal of a subordinate from the gene pool would be of little immediate consequence. Social animals also routinely use submissive displays to ingratiate themselves to more dominant animals to gain the latter's protection. It is not such a great leap from these submissive animal displays to human self-castration to achieve that same purpose.

But although enslavement or devotion were regarded as acceptable reasons to remove male human reproductive organs, medical reasons emerged as the most common reason for removing female ones. Soranus of Ephesus performed the first recorded medical hysterectomy in the second century CE. By the 1950s, hysterectomies were one of the most commonly performed human surgeries, whereas human castration for medical reasons was, and remains, relatively rare.

When Human Sterilization Beliefs Meet Canine Physiology and Behavior

Next we must ask whether any corollaries exist between the reasons given for human sterilization and those used to justify this procedure in companion animals. To determine this, consider the following list of benefits people often associate with canine surgical sterilization:

1. Freedom from euthanizing or finding homes for unwanted puppies
2. No bleeding or other inconveniences associated with female estrus
3. Elimination of all diseases associated with the organs removed
4. Decreased incidence of prostate gland problems in males
5. Decreased incidence of breast cancer in females
6. Decreased aggression in male dogs
7. Freedom to let dogs run loose without fear of them reproducing
8. Positive response from veterinarians, trainers, and other animal-care professionals

Many contemporary Americans sterilize their dogs to keep them from reproducing and ensure their health and good behavior. Additionally, sometimes the pressure from those advocating sterilization is so intense that some dog owners may sterilize their animals to win the approval of those people, not unlike the ancient *galli*. Although it is tempting to dismiss such societal pressure as inconsequential, this is not necessarily the case. Pressure exerted by veterinarians, trainers, and other animal-care professionals can be significant. In addition to being labeled "irresponsible" by animal-care professionals, those with intact animals may be banned from dog parks, doggy daycare centers, and other social gatherings.

Notice that only three of the perceived benefits—numbers 3, 4, and 5—directly pertain to the health of a sterilized animal. Obviously, removing reproductive organs eliminates disease in those organs. The question is whether sterilization is warranted given the relatively low incidence of these diseases and/or their high response to treatment. The incidence of periodontal disease in pet dogs is higher; would we advocate the removal of healthy teeth in young animals to prevent periodontal disease from occurring at some time in the future?

Contrary to popular belief, castrated male dogs may develop prostate cancer; the widespread belief that this cannot happen may actually lead to delayed diagnosis of this condition. Castration can reduce the incidence of prostatic hyperplasia—the increased growth of normal tissue—but this condition also responds to a combination of medical and behavioral therapy.

The relationship between canine OHE and breast cancer strikes a responsive chord among the many women who take the family pet to the veterinary clinic more often than their male counterparts. Additionally, enough American women either have had hysterectomies or know enough others who have had the surgery to consider it almost normal. Simultaneously, many women have had similar experiences relative to breast cancer and naturally perceive it as a problem to be avoided at all costs. Because of this, the idea of removing a female dog's reproductive organs for preventive medical reasons makes sense to them. However, although early studies did link breast cancer to intact female canine reproductive status, later studies revealed that multiple factors contribute to the disease, including the animal's age, weight, and conformation. And although studies of some shelter dog populations reveal a correlation between intact status and aggression in male dogs, these results are not supported by studies of owned aggressive dogs seen by behavioral consultants.

Once we recognize that any links between intact status and certain health and behavioral benefits may be tenuous, then we must ask whether any problems associated with the loss of reproductive organs justifies those benefits. To that end, we may turn to the growing body of information devoted to the nonreproductive effects of sex hormones. For example, researchers have found estrogen receptors in the pituitary gland, kidneys, adrenal glands, osteoblastic (bone-forming) cells, prostate gland, lungs, urinary bladder, and areas of the brain associated with learning and memory, as well as in the reproductive organs. What those receptors tell us is that estrogen plays a role in the normal function of all those different organs. Similar studies reveal testosterone's effects throughout the body.

Evidence that canine problems may be associated with the loss of reproductive hormones includes the fact that 20 percent of spayed female dogs will develop urinary incontinence, with reports of similar problems in males increasing. A study of osteosarcoma (bone cancer) linked its increased incidence in rottweilers to pediatric neutering. Relative to negative behavioral changes, studies link castration to hyperactivity and a higher incidence of canine cognitive dysfunction ("canine Alzheimer's").

In the past, we could claim that surgical sterilization posed no problems for the animal because science lacked the knowledge and technology to provide evidence that proved otherwise. Now that this increasingly exists, however, it is no longer possible to make such statements.

Still, as tenuous as a direct relationship between sterilization and quality canine health and behavior may appear to be, it would be foolish to maintain that the procedure is of no value for the animal whatsoever. The biological raison d'être is reproduction, not the possession of gonads. Female physiology evolved to support mating and pregnancy anytime that nursing or age did not negate that possibility; male reproductive physiology and behavior is equally complex and use-dependent. Having an intact, but unused reproductive system may create as many, albeit different, problems as the lack of one. However, until we objectively explore the subject of sterilization, we will never know.

A Bond-Based Response to the Dilemma

At this time, no obvious solution to the canine sterilization dilemma that would meet all canine and human needs exists. However, conceivably, we could find one if as a society we were willing to do the following:

1. Eliminate any emotion-based, unsubstantiated beliefs we attach to canine reproduction. That means recognizing and accepting that some of the symbolism we attach to canine reproduction may have nothing to do with dogs at all; it might relate to our desire to impose our beliefs on others, our beliefs about our own sexuality, or other beliefs we project upon companion animals.
2. Engage in objective debate to find answers to extremely difficult questions such as

   • If all the responsible pet owners spay and neuter their dogs, who will determine the makeup of the canine gene pool? Do we care about that? Should we care about that?
   • Does the freedom from euthanizing unwanted puppies compensate for any pain or suffering experienced by owned sterilized dogs with problems related to reduced reproductive hormonal levels or shrinking canine breed or species gene pools?
   • Are we as a society willing to invest in the research necessary to develop and implement more dog-friendly alternatives to canine overpopulation problems?

Canine sterilization was chosen as an example of how human beliefs projected on animals arise and how they may affect the animal because it illustrates how complex this relationship may be. However, it is by no means unique. *Every* human-animal interaction possesses this potential, and this is especially true in the case of house pets, with their constant presence and responsiveness to humans. Under those circumstances, the temptation to reduce the animals to animate symbols of human beliefs may prove overwhelming, particularly for those lacking concrete knowledge of the animal's own needs that might temper this desire. To prevent this from happening, all beliefs attached to animals should be periodically reexamined in the light of new research that might challenge their wisdom or validity. By doing so, pet owners can ensure that a vibrant, mutually rewarding human-companion animal bond will exist.

Further Resources

Drickamer, L. C., Vessey, S. H., & Jakob, E. M. (2002). *Animal behavior: Mechanisms, ecology, evolution.* New York: McGraw-Hill Higher Education.

Guy, N. C., Luescher, U. A., Dohoo, S. E., Spangler, E., Miller, J. B., Dohoo, I. R., & Bate, L. A. (2001). Risk factors for dog bites to owners in a general veterinary caseload. *Applied Animal Behaviour Science, 74,* 29–42.

Hart, B. L. (2001). Effect of gonadectomy on subsequent development of age-related cognitive impairment in dogs. *Journal of the American Veterinary Medical Association, 219,* 51–56.

Kuefler, M. (2001). *The manly eunuch: Masculinity, gender ambiguity, and Christian ideology in late antiquity.* Chicago: Chicago Series on Sexuality, History, and Society.

Rawlings, C., Barsanti, J. A., & Mahaffey, M. B. (2001). Treating incontinence in spayed female dogs. *Journal of the American Veterinary Medical Association, 219,* 770–75.

Schneider, R. (1970). Comparison of age, sex and incidence rates in human and canine breast cancer. *Cancer, 26,* 419–26.

Sonnenschein, E. G., Glickman, L. T., Goldschmidt, M. H., & McKee, L. J. (1991). Body conformation, diet, and risk of breast cancer in pet dogs; a case-control study. *American Journal of Epidemiology, 133*(7), 694–703.

Myrna Milani

■ Human Influences on Animal Reproduction
Human Influences on Animal Reproduction

For animals who live their lives directly under the control of humans, one of the most important forms of influence that we exert is control of the animals' reproduction and family relationships. This influence is exerted in order to achieve the number and the type of animals that are needed for various human requirements, such as food, work, commerce, entertainment, research, or companionship. Ever since humans have consciously bred animals to achieve certain characteristics, we have put considerable technical and emotional resources into this enterprise and have shown a range of sometimes-contradictory attitudes. These attitudes and goals encompass the unashamedly instrumental or commercial, the aesthetic, the sentimental, the anthropomorphic, the exploitative, the recreational, and even the political.

The following describes and evaluates some examples where human control of animal reproduction has been carried out most deliberately and with most effect. One example is in farming, where humans have bred farmed animals such as cows, pigs, and chickens to meet both economic and aesthetic goals. The second example is companion animal breeding, in particular the creation of dog breeds for both work and aesthetics. What are the advantages and disadvantages, for the animals and for modern human society, of this influence? How far has this type of human influence affected the underlying behavior and emotions of these animals?

The physical and sometimes the psychological characteristics of animals kept directly under human control are selected not by the evolutionary pressure of the environment but by the needs and choices of humans. The reproductive choices that animals would normally make for themselves are made instead by the animals' owners. Human influence extends to when the animals breed, which animals breed and which do not breed, how many young are produced and in what physical and social environment, what social relationships between parent and offspring exist, and how the genotype and phenotype of the animals may be changed. The widespread use of reproductive technologies such as artificial insemination (and less frequently, embryo transfer) means that one highly valued bull, for example, can be the biological father of hundreds of thousands of calves in several continents, altering and reducing the gene pool of the entire breed. The widespread use of one selected pedigree dog for breeding can have a dramatic impact on the appearance, and possibly health, of the breed as a whole. Although it is surely true that humans also influence the breeding behavior of wild animals indirectly, the control is almost total in the case of animals that are kept to be used exclusively for human use.

Farmed Animals

Human control of farmed animal reproduction has led to very large increases in production of meat and milk, with productivity increasing most steeply over the last thirty-five years. Human influences have created animals that are more productive but also less able to survive and reproduce in a natural setting without human intervention. Human-created effects are not always lasting, however. The basic motivations of farmed animals, inherited from the wild ancestors, are still present, and in some cases, the animals can revert to a semi-natural state.

In herds of wild and feral cattle that scientists have studied, adult females would normally have one calf and one yearling with them. A calf would normally suckle for at least eight months or until the next calf is born, and the herd's calves often stay together in a "creche" guarded by the herd. Cows and calves show signs of distress when they are separated even up to six months after the birth and can maintain their relationship even after the calf has grown up.

Selective breeding by humans has specialized domestic cattle into those used for producing milk (dairy cows) and those used for producing meat (beef cattle). Dairy cows have been specialized to put most of their physiological effort into producing milk in their very large udders and tend to be thin animals. The amount of milk produced for human use by specialized dairy cows (such as the Holstein breed, which now dominates in developed countries and is increasingly being exported to developing countries) is about ten times what a calf would need. The highest-yielding dairy cows now produce 10,000 kilograms or more of milk a year. The average milk yield per cow is eight times higher in North America than in developing countries, where specialized breeds may still be a minority. Beef cattle, in contrast, have been bred to put most of their physiological effort into fast growth and heavy musculature, and they produce only enough milk to rear a calf.

In modern farming, the production of large quantities of milk for human use requires control of the reproduction of the cow and her relationship with her calf. A dairy cow in commercial farming is required to have one calf a year, to ensure that she lactates for the next ten months. If she fails to become pregnant, she is considered uneconomic and is likely to be sent for slaughter. But commercial dairying also requires that the calf be removed from its mother a few days after birth, breaking the emotional bond that has formed between them. The calves are then reared away from their mothers, and if they are reared for veal production in veal crates, they are reared in isolation from others of their kind. (The use of veal crates for calves is prohibited in the European Union from January 2007, on grounds of animal health and welfare.) Because dairy breeds are selected for high milk production, not for muscle, the male calves of dairy breeds are often considered useless for beef production in developed countries and are shot at birth.

Highly specialized dairy cows have such high physiological demands on their bodies that they are more likely to suffer from painful lameness, mastitis, and low fertility. Often, they are worn out and in poor health after having produced only two or three calves, in comparison with traditional breeds of cows that can last for fifteen lactations. In this sense, the breeding strategy adopted by humans, which produces in the short-term high milk yield from a cow, is also costly from the point of view of creating healthy and long-living cows. A similar paradox exists for beef breeds; the most heavily muscled beef cows, such as the Belgian Blue breed, generally require the costly procedure of cae-sarean section in order to give birth.

Equally dramatic changes have been made in the control of the reproduction of commercial pigs (hogs). Wild and feral pigs live in small groups of a few sows and their litters. When she is about to give birth, a sow walks away from the herd and builds a nest of grass, sticks, and leaves to cover herself during birth and suckling for the first couple of weeks. The mother and piglets then join the rest of their herd, and the piglets become integrated into the group gradually. Sows wean their piglets gradually up to sixteen to seventeen weeks of age.

The aim of commercial pig farming is to rear and sell the maximum number of piglets per sow per year, with a steady supply throughout the year. Human control of pig reproduction has achieved big increases in the productivity of pigs. Maximizing production means control of the sow during pregnancy, birth, lactation, and weaning. The most productive pig breeds are the commercial hybrids that are fast-growing, large, and lean. These sows have large litters of around twelve piglets, in comparison with around four to six piglets produced by her ancestor the wild boar. Nearly all sows in developed countries at least are artificially inseminated. In order to monitor and control the sows during pregnancy, many sows are kept in sow stalls (gestation crates), narrow stalls that prevent them from turning around or even lying down easily. (These gestation crates are prohibited in the European Union from 2013, on grounds of animal health and welfare.)

In the search for productivity, selective breeding may now have created sows that are too large for their many tiny piglets, making it more likely that some of them get crushed to death when she accidentally lies on them. To try to solve this problem, most sows, when they give birth and are suckling their piglets, are kept in farrowing crates, narrow stalls that prevent the sow from turning around and prevent the piglets from coming close to her, except to enable them to reach her teats. In order to reduce the time to the sow's next pregnancy, the piglets are weaned and removed from their mothers at a time when naturally they would still be suckling and are still very dependent on their mother socially. In Europe they are removed around three to four weeks of age, but in North America, this can be done as early as two weeks of age.

How do the sows and piglets react to these interventions? It seems that the pigs try to maintain their natural behavior, and to that extent, we can say that these types of human intervention in their reproduction stress the animals but do not change their motivations or their emotions. Sows have not lost their very strong motivation to build a nest, and they make the same movements to try to do so even in a bare farrowing crate. Piglets have not lost their need for their mothers. Abrupt early weaning and mixing with unfamiliar pigs stresses the piglets and results in a high incidence of diarrhea and other disease. Dan Weary and David Fraser at the University of British Columbia observed that in the first few days after weaning, the piglets call constantly for their mother.

Farmed chickens are descended from the wild jungle fowl. In natural conditions, a hen builds a hidden nest and lays a small clutch of eggs and then stops laying and incubates the clutch. She communicates with her chicks even before hatching, and after hatching, the mother spends her time protecting, warming, and teaching the chicks. In commercial production, hens lay 300 eggs continuously during a year. Chicks are reared in tens of thousands from eggs incubated in hatcheries, without ever seeing a parent bird.

The human selection of chickens, by specializing the birds into laying breeds and meat breeds, has immensely increased the availability and cheapness of both chicken meat and eggs. But it has also caused biological anomalies on perhaps the largest scale yet seen in the human uses of animals. Laying hen breeds have very little breast muscle development, the muscle needed for meat production. In commercial hatcheries, the just-hatched chicks of laying breeds are sexed, and the male chicks are killed at one day old (approximately 368 million per year in North America and 416 million a year in the EU25 alone).

The economics of large-scale meat chicken farming depends on the chickens' speed of growth, their quantity of breast muscle, and their efficiency at converting food into muscle. The application of breeding technology to developing commercial hybrid chickens during the period since the 1960s has resulted in chickens designed to grow at a speed that puts them just on the edge of biological viability to the age of typically five to seven weeks, when they are ready for slaughter. Their mortality even during this short lifetime is seven times that of laying hens of the same age. These birds are normally unable to reach adulthood in good health unless their food intake is severely restricted because their skeletal and heart development cannot keep up with their growth rate if they are allowed to eat as much as they want. Human control of chicken breeding in the service of human needs has thus created animals that can be seen either as maximally productive or, alternatively, as biologically unviable and even, in the case of male layer chicks, commercially worthless.

Companion Animals

Human intervention has created the large numbers of breeds and types of dogs, cats, horses, and other animals that have been kept for cooperation with people in work, in sport, and for companionship. Many enthusiasts devote their lives to the creation and refinement of these animal breeds.

Approximately 400 dog breeds have been created so far by humans over the course of hundreds of years, all of them believed to be descended from the gray wolf. Modern dog breeds include extremes of size and shape very far removed from the wolf ancestor. Dogs were bred to have short legs to chase animals underground, to have strong jaws for guarding or fighting, to be large and strong for hunting large animals. Even in modern urban society, where nearly all dogs are kept as companions rather than for work, people still appear to prefer dogs of defined breeds. In Europe, typically three-quarters of the dogs owned are pedigree dogs, sometimes called "purebred" dogs, rather than mongrels.

In modern society, the most important characteristics of the dogs are their appearance rather than their working ability. Dog breeds have been refined and defined into "breed standards" by the breed societies and Kennel Clubs of the world, which can include not only very detailed physical characteristics but also frankly anthropomorphic qualities such as "noble," "regal," "dignified," "aristocratic," "proud," "courageous," "wise," "gay," or "elegant." New breeds are still being designed for the requirements of modern urban life, such as tiny "teacup dogs" as accessories for celebrities and bald dogs bred for people suffering from allergies. The enthusiasm for dog breeding sees the largest dog show, Crufts in Britain, attract 22,000 dogs and 120,000 visitors over four days. But dog breeding has also created painful dilemmas about the health and welfare of the dogs.

The emphasis on breed standards and breed purity can give the impression that pedigree dogs are in some sense of higher quality than dogs of a thoroughly mixed breed, but that is far from being the case. As animal scientists from the University of Sydney, Australia, have commented, "breeders and scientists have long been aware that all is not well in the world of companion animal breeding." Veterinarians are aware that certain dog breeds have a much-increased risk of inherited or breed-related disease than the general dog population. While breeders strive to perfect an ideal dog type or develop a new breed, two serious problems can arise. These are inbreeding and the development of breed standards that call for unnatural and inappropriate body conformations. Inbreeding is almost inevitable in breeds that have only a relatively small number of dogs, and for numerically large and established breeds, it is common to use only a fraction of the dogs for breeding, in order to maintain a uniform appearance.

Inbreeding (sometimes called "linebreeding") decreases the genetic diversity of the breed and increases the effect of deleterious genes. Many breeds, including Labrador and Golden Retrievers, German Shepherd dogs, and Rottweilers, suffer from a high incidence of hip and elbow dysplasia (disorders of bone growth that lead to painful arthritis and lameness). Between a quarter and a third of the world's dog breeds have inherited eye diseases, including painful and blinding conditions such as glaucoma and degeneration of the retina. Several breeds that carry the piebald or merle genes for coat color have inherited deafness. These conditions can be disabling and lead to euthanasia.

It is generally recognized that some of the ideal standards for how dogs of particular breeds should look have resulted in features that cause health problems. Very flat faces and short noses (such as for the bulldog and Pekinese) make breathing difficult and increase the risk of heart problems; legs that are too short in proportion to backs (such as for the dachshund) increase the risk of painful spine problems; loose skin and skin folds on face and body (such as the shar-pei) lead to irritating and painful dermatitis between the folds; ears and hair that are too long may prevent dogs from keeping themselves clean without human help; very long hair covering the eyes may make a dog timid or defensive. In the case of the bulldog, the massive head size means that puppies often have to be born by caesarean section. Dogs bred for herding, guarding, or chasing behavior can be frustrated by the restrictions of modern urban living conditions, with resulting behavior problems.

These examples illustrate some of the ways in which human control of animal reproduction has had paradoxical results. Our influence on animal breeding has affected both animal lives and human lives profoundly. Human intervention has brought what some see as great benefits for food production, aesthetics, and leisure activities. It has also in several respects reduced the fitness or usefulness of the animals themselves and is viewed by some as an example of undue human domination of the lives of other animals. Whatever viewpoint is taken, the evidence must make us question to what extent our intervention operates to the benefit of the animals.

Further Resources

Advocates for Animals. (2006). *The price of a pedigree: Dog breed standards and breed-related illness.* Edinburgh: Advocates for Animals. Available online at http://www.advocatesforanimals.org.uk

Clutton-Brock, J. (1999). *A natural history of domesticated animals* (2nd ed.). Cambridge, UK: Cambridge University Press.

Gough, A., & Thomas, A. (2004). *Breed predispositions to disease in dogs & cats.* Oxford: Blackwell.

Keeling, L. J., & Gonyou, H. W. (Eds.). (2001). *Social behaviour of farm animals.* Wallingford: CABI.

McGreevy, P. D., & Nicholas, F. W. (1999). Some practical solutions to welfare problems in dog breeding. *Animal Welfare, 8,* 329–41.

Rauw, W. M., Kanis, E., Noordhuizen-Stassen, E. N., & Grommers, F. J. (1998). Undesirable side effects of selection for high production efficiency in farm animals: A review. *Livestock Production Science, 56,* 15–33.

Weary, D. M., & Fraser, D. (1997). Vocal response of piglets to weaning: Effect of piglet age. *Applied Animal Behaviour Science, 54,* 153–60.

Webster, J. (2005). *Animal welfare: Limping towards Eden.* Oxford: Blackwell.

Jacky Turner

■ Human Perceptions of Animals
Animal Minds and Human Perspectives

The Problem of Subjectivity

Despite many years of close observation of animals, the ethologist Niko Tinbergen concluded that regardless of what we think we know about the subjective mental experiences of other species, because we cannot directly observe these experiences, we can make no claims as to their existence. Yet the question of whether other animals can be considered to be mindful and to have subjective characteristics such as emotions and mental states is a crucial one in scientific work on animal awareness, animal welfare, and the study of human-animal relationships. Within the context of these relationships, we often feel the need to make reference to the inner lives of animals, and the popular literature is replete with references to canine loyalty, feline cunning, and elephant emotion.

Current models of animal behavior and cognition tend to employ a more mechanistic framework and language (a good example is Shettleworth, 2001). From this perspective, animals are often portrayed as objects upon which various genetic, social, and environmental forces act and whose behaviors are objective and observable but whose mental states and subjective experiences are not. We can trace this view directly back to Descartes, who assumed not only that should animals be considered "destitute of reason" (i.e., without a rational soul), but also that in fact it was impossible for us to ever really access their innermost thoughts and feelings. For Descartes, and for some writers today, the assumption that animal behavior is meaningful is problematic because even machines can perform clever behavior. How do we know whether such behavior is accompanied by any inner subjective understanding, or whether, like a laptop computer, complex actions are performed without any experience of an inner world?

Faced with this problem, society's common approach to bridging the gap between our knowledge of objective and subjective worlds has been to assume that the expression of similar behaviors in different species indicates similar mental experiences. This argument is termed the "argument by analogy" and is brought into play in both scientific and lay interpretations of behavior. The argument by analogy can sometimes seem appropriate, as when we interpret the cries of an animal caught in a trap as indicating pain, but can also be misleading, for instance in the assumption that the "bared-teeth" expression of a chimpanzee (which is somewhat like a human grin) indicates pleasure (when it is more likely to indicate fear). Critics of the argument from analogy claim that the main difficulty with it is that there is ultimately no fail-safe correlation between thought and behavior such that we could always use the latter to uncover the former. At root, it is claimed, subjective experience always has to be inferred from behavior, rather than being expressed through it.

Many people would consider this disengagement with the subjective nature of experience unsatisfying, not least because it seems much more difficult to discuss the meaning of behavior if it is stripped of any reference to subjectivity. The great psychologist Ivan Pavlov, anxious to rid psychology of mentalistic terminology, actually began imposing fines on the workers in his laboratory each time they used terms such as "intent" and "mind" (Krementsov & Todes, 1991). Our Cartesian intellectual heritage has ensured that attempts to discuss subjective experience are often seen as difficult or impossible or as necessitating contested anthropomorphic language or sentiments. Yet philosophical arguments about the scientific status of subjective interpretations of behavior often lie behind the metaphors and rhetoric used in the debate about animal minds. This debate is commonly portrayed as featuring "believers" and "sceptics," as involving "killjoy" interpretations waging war on "sentimental" anthropomorphic accounts or as being a battle between "naïve" observers or laypersons and "hard-nosed" natural scientists.

Expressions of Mind in Social Relationships

In an attempt to take a more objective, rational stance toward understanding other species, people often assume that more detached, mechanistic, and quantitative interpretations of behavior, devoid of any reference to intention, motivation, or emotion, are always more accurate (the position of Kennedy, 1992, is a good example). In many animal behavioral studies, the focus is often on general patterns of behavior across a species rather than on close and intimate observation of a few individuals. Sociobiological approaches, for example, provide little information about the inner experience of animals, focusing as they do on the adaptive function of behaviors. The individual as a unit of study is important in sociobiology only in so much as it is a vehicle for the action of a specific genotype.

But in addressing the deeper question of how other animals experience the world, we cannot avoid the subjective point of view of the animal, for animals are not just objects that experience the world in a physical way; they are also subjects that act and think and move in a social sphere. As the philosopher Wittgenstein pointed out, understanding other minds is actually a social process. The meaning of a particular behavior, and its subjective flavor, is not something that hovers in a vacuum; rather, behaviors acquire meaning through their positioning within the context of social relationships. Although it is certainly true that it is not an easy task to perceive the world from another animal's perspective, it is in our relationships with animals that we come closest to apprehending their subjective point of view and have the best chance of understanding their experience. The relatively simple act of walking a dog, for instance, employs a mutual understanding of the intention of each partner through the tension on the leash. As Eric Laurier and his colleagues have suggested, the intention of dog and human within such an interaction is not obscure, and it is not wholly subjective. Rather, intention (mind) is expressed and embodied within the context of the walk as a social interaction. Far from being a form of naïve anthropomorphism, attributing mental states, intention, and emotion to animals is actually a fundamental (if sometimes unconscious) way in which we construct our relationships with other species and through which we respond physically and emotionally to them.

The primatologist Barbara Smuts gives a nice account of the difficulties and the rewards of exploring the inner world of another species. By immersing herself in the daily life of a troop of wild baboons and learning the social meaning of behavioral signals and gestures, she discovered a richer awareness of the social and personal worlds of this species. Smuts charts the process by which her presence acquired different meanings for the baboons; from treating her initially as an object of threat, the troop members gradually began directing gestures and vocalizations at her: in other words, they treated her as a subject rather than an object—as a being whose actions had some meaning for them (Smuts, 2001).

Toward an Embodied Perspective on Animal Minds

In conceptualizing thought as something that arises out of social relationships, researchers such as Smuts are more in tune with early traditions in psychology that described how the development of thought depends on social relationships. This is very different from the current paradigm that characterizes the way in which many psychologists think about the mind today; current models of cognition tend to be based on computational metaphors, where thinking is seen as a form of information-processing, and concepts and categories are considered to be abstract and linguistically based. From this perspective, thought is seen as essentially mental rather than physical: it is something that happens in brains rather than in bodies. And the logical implication of this position is that although thought may cause behavior, it is private, subjective, and hidden from scrutiny.

In contrast to this position, however, more recent work in the field of cognitive science has begun to offer a more useful perspective that promises to help bridge the gap between what we know of objective and subjective worlds. Researchers such as George Lakoff and Mark Johnson and Francisco Varela and his colleagues have suggested that far from being abstract mental computations, the concepts that people have are quite obviously structured by their sensory-motor and perceptual experiences. For example, all humans and nonhumans share experiences of physical force because we are all physical bodies subject to environmental and climatic pressures. As such, awareness of such aspects of force as intensity, origin, and direction (such as being subject to a blow) are probably shared by many species. As humans, we also have a more abstract concept of force, where we use metaphors to help us express experiences that are similar to those we have when subject to physical forces. These metaphors are apparent when we use phrases such as "her eyes bored into me" or "he gave her a sharp stare." What we are doing here is mapping one experience (our experience of physical force) onto another experience (our experience of visual gaze). Here we are conceptualizing gaze as being like a type of physical pressure or force, and indeed it can seem this way; we all know what it is like to be on the receiving end of an aggressive stare from somebody. It is in this sense then that we conceptualize gaze as some kind of force, not an actual physical force, but a *social* force. In humans, such metaphors reach across many human cultural and linguistic boundaries. It is likely that other species share at least the rudiments of this same concept, given that for many species such as apes, monkeys, cats, and dogs, visual gaze is an important element in structuring social relationships.

This more embodied account of mind may provide a more useful tool for the understanding of animal awareness because it rejects the dualistic division between mental and physical, mind and body, objective behavior and subjective thought. From a more embodied perspective, such distinctions become meaningless because it is clear that mind is not separate from behavior, not locked up inside the head; rather, it is structured through physical and social interaction in the world, and it is expressed through such action.

Take the concept of "dominance" as an example. This concept is used to describe patterns of behaviors in many different species. It is clear that dominance is not just an abstract concept but is clearly embodied and expressed in behavior, evident to anyone who has watched a dog grasp the neck of a subordinate and force it to the ground or anyone who has seen a male chimpanzee, hair bristling, perform an aggressive display that culminates in him jumping over the crouching figure of a subordinate. The similarities in many different dominance displays are striking. For many animals, including humans, dominance is mapped out in a spatial way: dominant individuals endeavor to make themselves bigger, higher, or more central than their rivals. Although humans are able to think about concepts such as dominance in ever more abstract ways, such concepts at root express bodily experience in the world and are composed through the dynamism of social life. If this is so, then the inner awareness of other animals, the truth of how they see the world, is not a closed book; rather, thought, emotion, and intention are embodied in the way in which animals move, gesture, and handle social relationships.

If we hope to make real gains in the study of animal minds, we need to recognize that the philosophical distinction that science makes between objective and subjective knowledge is problematic, particularly in its implication that we can have no intimacy with the way other animals experience the world. Because rationality and even language seem to be grounded in basic sensory-motor experience, such attributes cannot be seen as purely human characteristics, qualitatively different from any aspects of animal experience. Essentially, the mental and emotional worlds of humans and other animals arise and develop from their physical and social interactions in the world. Social interactions are the matrix out of which mind emerges, and the study of human-animal relationships is therefore of central importance in attempting to fully understand animal minds.

See also

Anthropomorphism
Culture, Religion, and Belief Systems—*Religion's Origins and Animals*
Human Perceptions of Animals—*Sociology and Human-Animal Relationships*
Sentience and Cognition—*Animal Consciousness*
Sentience and Cognition—*Animal Pain*
Sentience and Cognition—*Descartes, Rene*

Further Resources

Bekoff, M., & Allen, C. (1997). Cognitive ethology: Slayers, skeptics and proponents. In R. W. Mitchell, N. S. Thompson, & H. L. Miles (Eds.), *Anthropomorphism, anecdotes and animals* (pp. 313–34). Albany: State University of New York Press.

Crist, E. (1999). *Images of animals: Anthropomorphism and animal mind.* Philadelphia: Temple University Press.

Dutton, D., & Williams, C. (2004). A view from the bridge: Subjectivity, embodiment and animal minds. *Anthrozoös, 17*(3), 210–24.

Kennedy, J. S. (1992). *The new anthropomorphism*. Cambridge: Cambridge University Press.

Krementsov, N. L., & Todes, D. P. (1991). On metaphors, animals, and us. *Journal of Social Issues, 47*(3), 67–81.

Lakoff, G., & Johnson, M. (1980). *Metaphors we live by*. Chicago: University of Chicago Press.

———. (1999). *Philosophy in the flesh*. New York: Basic Books.

Laurier, E., Maze, R., & Lundin, J. (2002). *Putting the dog back in the park: Animal and human mind-in-action*. Retrieved from the Department of Geography and Topographic Science, University of Glasgow, Web site: http://130.209.26.101/online_papers/dogspotting.htm

Nagel, T. (1974). What is it like to be a bat? *Philosophical Review, 83,* 435–50.

Sanders, C. R. (1993). Understanding dogs: Caretakers' attributions of mindedness in canine-human relationships. *Journal of Contemporary Ethnography, 22*(2), 205–26.

Shapiro, K. (1990). Understanding dogs through kinesthetic empathy, social construction, and history. *Anthrozoös, 3*(3), 184–95.

Shettleworth, S. J. (2001). Animal cognition and animal behaviour. *Animal Behaviour, 61,* 277–86.

Smuts, B. (2001). Encounters with animal minds. *Journal of Consciousness Studies, 8*(5–7), 293–309.

Varela, F. J., Thompson, E., & Rosch, E. (1991). *The embodied mind: Cognitive science and human experience*. Cambridge, MA: MIT Press.

Wemelsfelder, F. (1999). The problem of animal subjectivity and its consequences for the scientific measurement of animal welfare. In F. L. Dolins (Ed.), *Attitudes to animals: Views in animal welfare* (pp. 37–53). Cambridge: Cambridge University Press.

Diane Dutton

■ Human Perceptions of Animals
Cryptozoology (Bigfoot, the Loch Ness Monster, and Others)

Cryptozoology is the study of unknown and mysterious animals (often called "cryptids"). The best-known cryptids are probably Bigfoot, the Abominable Snowman, and the Loch Ness Monster, though there are hundreds, perhaps thousands, of such creatures currently on the margins of mainstream, accepted science. If such creatures exist, they undoubtedly have implications for the ways in which we humans categorize the world and think of our own nature (as well as our relationships with other animals). If they do not exist, cryptids still have much to teach us about how we conceptualize the world, how we see ourselves, and how science is related to culture and belief systems.

"Cryptozoology" is the English translation of the French term *la cryptozoologie*, which is itself based on Greek roots. From the Greek *kruptos* ("hidden/unknown"), *zoon* ("animal"), and *logos* ("study/discourse/reason"), "cryptozoology" is, literally, "the study of unknown animals." There have been cultures and individuals interested in cryptozoology for centuries, but the word itself was coined in the 1950s by Bernard Heuvelmans, a Belgian zoologist living in France who would go on to become one of the most important field workers and encyclopedists of cryptids in the twentieth century. Like many of his generation, scientist and nonscientist alike, Heuvelmans was inspired to begin thinking seriously about mysterious animals by an article written by Ivan T. Sanderson in January 1948 for the *Saturday Evening Post* titled "There Could Be Dinosaurs." Sanderson, himself a zoologist (and someone to whom Heuvelmans gives credit for coining the term "cryptozoology" at roughly the same time he did), took seriously the idea that some species of dinosaurs might have been able to survive the mass extinction of their time and could be hiding in the vast, dense, unexplored jungles of the world.

Cryptozoology is something of an interdisciplinary pursuit, marking a space where anthropology and sociology are often used toward the aim of uncovering mysterious animals. Two decades after Sanderson inspired the public to consider a modern dinosaur, popular culture also began to take up the stories of creatures that had previously only been thought to exist in indigenous cultural history and myth. The Loch Ness monster gained notoriety in the 1970s (though locals had stories of Nessie dating back centuries), as did the Sasquatch (Bigfoot's Native American Indian alias, a name marking a creature that had long been part of local stories and mythology in the Americas). Although still operating more or less outside mainstream science, the hunt for cryptids was on, and the romantic and exciting nature of the chase surely inspired many young people to consider studying zoology, history, anthropology, and other related fields in the sciences and humanities.

The question of cryptozoology's status as a science per se is controversial. Often, mainstream rejection is based on the dual beliefs that (1) the world has been fully explored and thus offers no new surprises (certainly no new important animals to classify), and (2) cryptozoologists do not proceed in a manner befitting accepted science in that they "want to believe" rather than adopt skepticism and thus are willing to let anecdote, myth, legend, folk knowledge, and imagination stand in for evidence. Interestingly, both assumptions are debatable.

A preserved skull and hand claimed to be that of a yeti, or abominable snowman, on display at Pangboche Monastery, near Mount Everest, in Nepal. ©Ernst Haas/Getty Images.

This shadowy something is what many claim to be a photo of the Loch Ness monster in Scotland. ©AP/Wide World Photos.

That the world has been fully explored is itself something of a myth. Vast stretches of land of all climates (jungle, glacier, desert, and so on) and water (both fresh and salt) fortunately remain virtually untouched by humans. There are, perhaps, more things in the heavens and on the Earth than are dreamt of in our philosophies. Furthermore, what were once thought to be mythical creatures have, indeed, from time to time been discovered and documented. The strange Pygmy hippopotamus (*Hexaprotodon liberiensis*) was discovered and shown to be real by Hans Schomburgk; the assumed mythical giant panda (*Ailuropoda melanoleuca*) was found by Ruth Harkness. Although no mile-long Kraken sea monster has yet been located at the bottom of the ocean, the giant squid (*Architeuthis*) has been shown to exist; and it is directly because of W. Douglas Burden's fascination with and desire to go in search of dragons that the Komodo dragon (*Varanus komodoensis*) was discovered in the 1920s. During the twentieth century, in fact, hundreds of species and other taxonomic groups were described or discovered, with cryptozoologists often having the last laugh. To choose just four examples from the last fifty years, we have seen the discovery of (1) a miniature shrimp (*Neoglyphea inopinata*) caught in the Philippines in the 1950s that turned out to be a member of a family thought extinct since the Eocene (50 million years ago); (2) the megamouth shark (*Megachasma pelgios*) found in 1976 near Hawaii, which turned out to have, indeed, a mega-mouth and a mega-body to match (4.5 meters); (3) the sao la (the oryx-horned antelope or *Pseudoryx nghetinhensis*), a new, large (larger than 100 kg) Vietnamese bovid species discovered in 1992; and (4) the *Lophocebus kipunji*, a species of African monkey surprisingly discovered in 2003 in southern Tanzania. Sometimes, at least, yesterday's myth can become tomorrow's new well-accepted member of the community.

What worries most mainstream scientists, perhaps, is that cryptozoology, for the most part, embraces a methodology that refuses to separate different sorts of cryptids into those that are more fanciful and those that seem to have a reasonable chance of being recognized as a new species at some point. Cryptozoology, that is, covers—or has covered in the past—everything from fairies and mermaids to satyrs and centaurs, from Bigfoot and modern dinosaurs to coelacanths and orangutans. There are, to be sure, more chances of uncovering either a modern individual from a species once thought extinct (e.g., a type of dinosaur) or even a creature that is a proposed off-shoot of some other known animal (e.g., a Sasquatch) than there is of ever encountering a griffin, goblin, or gnome. Yet the sense in which cryptozoology is open to any possibility is itself an important reminder of both the nature of science and the nature of our relationships with other animals in the world and in our imagination. Rigor, thorough testing of hypotheses, collaboration with and confirmation by peers, and a commitment to empirical evidence must remain the focus of any scientific investigation, yet science of all types does well to rethink long-accepted facts, to celebrate curiosity, and to afford practice of its craft on the most difficult sorts of questions rather than stifling such inquiry. And it is surely the case that our thinking about animals—and our relationships to them—is typically best when we, too, are open and curious, willing to consider the scientific, anthropological, psychological, and cultural implications of having our well-accepted categories of classification between *us* and *them*—however we define such terms—called into question.

Further Resources

Eberhart, G. M. (2002). *Mysterious creatures: A guide to cryptozoology*. Santa Barbara, CA: ABC-CLIO.

Grumley, M. (1974). *There are giants in the Earth*. New York: Doubleday.

Halpin, M. M., & Ames, M. M. (Eds.). (1980). *Manlike monsters on trial*. Vancouver: University of British Columbia Press.

Heuvelmans, B. (1995). *On the track of unknown animals*. London: Kegan Paul International.

Keel, J. A. (1975). *Strange creatures from time and space*. London: Spearman.

Krantz, G. S. (1992). *Big foot-prints*. Boulder, CO: Johnson Books.

Levy, J. (2000). *A natural history of the unnatural world: Selected files from the archives of the Cryptozoological Society of London*. London: Carroll & Brown.

Markotic, V., & Krantz, G. (Ed.). (1984). *The Sasquatch and other unknown hominoids*. Calgary: Western.

H. Peter Steeves

■ Human Perceptions of Animals
Freaks—"Human Animals"

In the sixteenth century, museums on the European continent began to display so-called "lusus naturae" (literally, "sportive actions of nature"), which included human beings with physical deformities. By Victorian times, interest in "living curiosities" had spread to the United Kingdom and the United States, giving rise to institutions like P. T. Barnum's American Museum in Manhattan. Soon after, "freak acts" were largely relocated to less respectable venues such as dime museums, traveling circuses, and sideshow carnivals (outlawed in the United Kingdom in 1886), the biggest of which was Samuel W. Gumpertz's Dreamland Circus Sideshow at Coney Island, where real-life "freaks" were exhibited alongside a number of elaborate hoaxes, most famously that of the "Feejee mermaid."

Many of these acts were given animal nicknames that served to highlight their category-defying features, thus arousing the fascination of audiences. Banners and promoters would invite passers-by to "step right up" and witness "curiosities" such as Jo-Jo the "Dog-Faced Boy," Betty "Koo Koo the Bird Girl" Green (aka "the Stork Woman"), Dickie the "Penguin Boy" ("looks and walks like a penguin," no doubt an inspiration for the Penguin character depicted by Danny DeVito in the Tim Burton–directed film *Batman Returns*), Stanislaus "Sealo the Seal Boy" Berent, Julia "the Ape-Woman" Pastrana, Alzoria "the Turtle Girl" (aka "Walrus Girl" and "Pig Woman"), Grace "the Mule-Faced Woman" McDaniels, Johanna "the Bear Girl" Dickens, Minnie "Ha-Ha" Woolsey (aka "Koo Koo—the Blind Girl from Mars"), Alice "the Bear Woman" Bounder, Ella "the Camel Girl" Harper, Lionel the Lion-faced Man, Crocko the Crocodile, Leona "the Leopard Girl," and various "frog princes and princesses," "caterpillar boys and girls," and "lobster men and women."

Often, promoters would invent some fantastic story purporting to explain how the "freak" in question came to become "half-animal," but unlike the animal men and women of mythology (such as centaurs and mermaids), these people were, needless to say, not hybrids of any kind, but humans with various kinds of birth abnormalities. For example, Grace McDaniels suffered from facial tumors, Leona "the Leopard Girl" had vitiligo (which results in patches in the skin), so-called "lobster man" Grady Franklin Stiles suffered from a hereditary condition known as ectrodactyly, and Stanislaus Berent had a deformity (whose cause remains unknown) that made his arms look like the flippers of a seal.

Of all these "attractions," the most famous was undoubtedly Joseph Carey Merrick (1862–90), named the "Elephant Man" because of his rough skin and facial deformity (which one promoter creatively "explained" by dreaming up a story about Merrick's mother being trampled by an elephant during her pregnancy). Merrick was a charming, sensitive, and intelligent Englishman who could not get much work because of his appearance and thus reluctantly resorted to becoming a "sideshow." Originally thought to have suffered from elephantiasis (a blockage in the lymphatic system that causes enormous swelling), he was later said to have had neurofibromatosis (a common genetic disorder of the nervous system that causes large tumors to grow on nerve tissue). However, more recent research suggests that what Merrick actually had was severe Proteus syndrome (a far rarer—though still existing—disease that causes bone, skin, and head overgrowth).

Many twentieth-century sideshows appeared as themselves in Tod Browning's 1932 film *Freaks* (banned in the United Kingdom for thirty years), in which the performers of a traveling show avenge a cruel trapeze star, who is shown to be the real monster in the film. This idea was later taken up by David Lynch, who in his moving 1980 film *The Elephant Man* portrays Merrick's manager Tom Norman as being far more monstrous than his well-mannered client. Relatedly, the term "freak" is nowadays typically used to refer to someone with psychological abnormalities rather than physical ones. In some cases, such as that of Dennis "Stalking Cat" Avner, who has spent thousands of dollars in body-modification surgery to transform himself into a tiger, the former appears to have led to a voluntary bringing-about of the latter. The term "animal freak" has recently also been used to refer to animal lovers who take their activism to violent extremes.

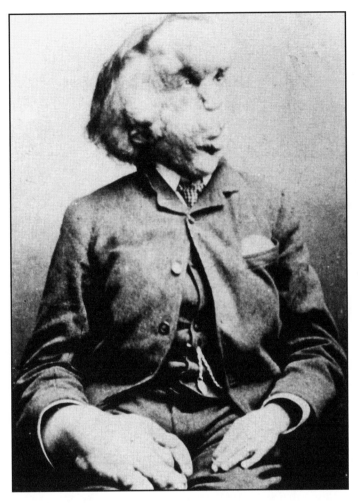

Joseph Cary Merrick, Victorian England's famous "Elephant Man," shown in a photo from the Radiological Society of North America. ©AP/Wide World Photos.

From the end of the nineteenth century onward, various disability activists have protested against the exploitation of the physically disabled by sideshow promoters. Although this was certainly true in many cases, it is equally true that some "freaks" were treated kindly by their managers. However, neither this nor the fact that many sideshow acts entered the profession voluntarily (often because they rightly thought that it would make them very rich), and in some cases even fought against those who tried to ban their profession, necessarily undermines the view that voluntary exhibitionism is, like the voyeurism it relies on for its existence, a vicious activity.

Although a few people with physical abnormalities, such as contemporary celebrity Tallon "Batboy" Crawford, proudly preserve the sideshow tradition, others—such as Cathy Berry, who in 2003 appeared as "Lobster Girl" in HBO's acclaimed television series *Carnivale* and as "Lobster Woman" in Tim Burton's film *Big Fish* (based on Daniel Wallace's book, which has a certain resonance with Burton's own book *The Melancholy Death of Oyster Boy and Other Stories*)—have found a different kind of vocation within the entertainment industry.

Further Resources

Anderson, E. (2006). *Phreeque.* http://phreeque.tripod.com/

Bogdan, R. (1988). *Freak show: Presenting human oddities for amusement and profit.* Chicago: University of Chicago Press.

Howell, M., & Ford, P. (2001). *The true history of the Elephant Man.* London: Allison & Busby.

Constantine Sandis

■ Human Perceptions of Animals
Gender Differences in Human-Animal Interactions

In some ways, men and women are alike in their interactions with other species; in other ways, they are different. For example, in the United States, roughly the same proportions of men and women choose to live with companion animals, and boys and girls treat their pets similarly. But women are much more likely than men to donate money to animal protection organizations, volunteer at animal shelters, and participate in animal rights demonstrations. In contrast, many more men than women participate in blood sports such as recreational hunting, fishing, and cockfighting. In many areas of human life, the magnitude of male-female differences varies with the type of interaction. For example, there are small gender differences in verbal ability, moderately sized differences in aggression, and large differences in height. The same is true for human-animal relationships.

Attitudes

Gender is one of the most important influences on how people feel about the treatment of animals. Nearly all investigations of attitudes toward the moral status of other species have reported that as a group, women are more sympathetic toward animal welfare issues than men and are more likely to oppose the use of animals in biomedical and behavioral research.

Psychologists use the statistical concept of effect size to estimate the magnitude of group differences in attitudes and behavior. The effect sizes of gender differences in attitudes toward the treatment of animals are typically in the moderate range. (A moderate effect size means that about 70 percent of women score higher on measures of attitudes toward the treatment of animals than the average man. This is about the same effect size as has been found for gender differences in assertiveness.)

Greater sympathy in women for the treatment of animals is consistent across cultures. Linda Pifer and her colleagues asked adults in fifteen countries about their views on acceptability of biomedical research using dogs and chimpanzees. There were gender differences in fourteen of the fifteen countries; in all of these cases, women were more opposed to animal experimentation than men.

Attachment

Both men and women show deep emotional attachments to their companion animals. Males and females keep pets in similar proportions, and their grief reactions to the loss of a companion animal are generally comparable. Some investigations of attachment to pets have found statistically significant gender differences, with women showing higher levels of attachment. The effect sizes in these studies, however, have tended to be in the small size range (about 60 percent of women score higher on measures of pet attachment than the average man).

Animal Protectionism

More women than men become actively involved in the causes related to animal protection. Sociologists have found that the ratio of women to men among grassroots animal rights activists in the United States is about three to one. Not surprisingly, women also donate more time and money to animal welfare organizations such as the Humane Society of the United States and the American Society for the Prevention of Cruelty to Animals, and more women become involved in pro-animal activities such as purebred dog rescue.

The preponderance of women in organizations devoted to the well-being of other species is not a new phenomenon; it has been characteristic of the animal protection movement for over 100 years. Seventy-five percent of the members of the Victoria Street Society, one of the first antivivisection organizations in England, were women. But although most rank-and-file members of animal-protection groups are women, historically, a disproportionate number of movement leaders have been men. Even though the Victoria Street Society was founded by a woman, Francis Power Cobb, the first president and all of the vice presidents were men. The *Encyclopedia of Animal Rights and Animal Welfare* (1998) included twenty-eight entries describing the achievements of writers, organizers, and philosophers who were particularly notable for their contributions to animal protectionism; 75 percent of these individuals were men. This pattern has changed in recent years, as women such as Carol Adams, Mary Midgley, Ingrid Newkirk, Marjorie Spiegel, and Linda Birke have emerged as intellectual leaders and political organizers of the animal protection movement.

Violence toward Animals

Nearly all studies on the topic of animal abuse have reported that men are more likely to inflict cruelty upon animals than women. Accurate statistics, however, on gender differences in rates of cruelty are hard to obtain. Among the most difficult groups to study in this regard are non-clinical (normal) populations. When anonymously surveyed, male college students are more likely than females to admit that they intentionally hurt animals when they were children. Arnold Arluke has suggested that tormenting of other species in childhood is a fairly common form of "dirty play." In this regard, some studies have found that in children, female cruelty may be more common than is generally acknowledged. Thus, although more boys than girls torture animals for fun, the degree of this difference is uncertain.

In adults, the infliction of cruelty on animals is much more likely to be perpetuated by men than women. For example, men make up 90 to 95 percent of individuals charged with activities such as beating, torture, and mutilation of animals and bestiality. Although not technically considered a form of cruelty, recreational hunting is also a male-dominated activity, with a ratio of about ten male hunters to every female.

Two areas of animal cruelty do involve substantial numbers of women; both are pathologies of attachment to companion animals. The first is animal neglect in which animals are not given adequate food, shelter, or veterinary care. The second is hoarding, a form of obsessive-compulsive behavior in which socially unacceptable and unhealthy numbers of animals are accumulated, typically in the hoarder's home. There are nearly as many women as men charged with animal neglect each year, and there are two or three times more female than male hoarders.

Explanations of Gender Differences and Animals

Numerous explanations have been offered to account for the differences in cognition and behaviors between men and women, including differences in human-animal interactions. Researchers who take a biological perspective argue that human gender differences are the product of differing evolutionary pressures on the sexes (e.g., selection favored traits related to hunting ability in males and maternal behavior in females) or of hormonal and genetic influences. Many developmental psychologists argue that gender differences ultimately arise from patterns of social reinforcement, conformity to social roles, and early cognitive differences. Sociological and feminist theorists have proposed structural and constructivist interpretations of gender differences.

In some cases, proponents of these various views feel strongly that there is a single correct answer to the question, Where do gender differences come from? This view is almost certainly incorrect. Rather, each of these perspectives offers a partially correct, yet necessarily incomplete, explanation of gender differences in how we think and behave toward other species. Genes, hormones, social reinforcement, peers and family, social pressures, and political and economic power structures interact in complex ways to influence our interactions with animals.

Gender differences in relationships with animals can and do change over time. Take veterinary medicine. Fifty years ago, this field was almost exclusively a male preserve. Presently, in the United States, the majority of veterinarians are women, as are about 80 percent of first-year veterinary students. Similar shifts in gender differences have occurred in other aspects of human-animal interactions.

There are a few generalizations that emerge from studies of gender and the treatment of animals. First, females tend to be more sympathetic to the treatment of animals than males. Second, the sizes of gender differences vary with the type of interaction; they tend to be small in attachment to pets, moderate in attitudes toward animal welfare, and large in involvement in animal protection and abuse. Third, the average man and the average woman are more alike than they are different when it comes to the treatment of animals. However, large gender differences emerge when it comes to extreme behaviors; many more women than men show up at animal rights demonstrations, and many more men than women are arrested for animal cruelty. Finally, gender differences in human-animal interactions are caused by multiple factors operating at many levels.

Further Resources

Adams, C. J., & Donovan, J. (1995). *Animals and women: Feminist theoretical explorations.* Durham, NC: Duke University Press.

Arluke, A. (2002). Animal abuse as dirty play. *Symbolic Interaction, 25,* 405–30.

Bekoff, M. (Ed.). (1998). *Encyclopedia of animal rights and animal welfare.* Westport, CT: Greenwood.

Birke, L. (1994). *Feminism, animals and science: The naming of the shrew.* Buckingham: Open University Press.

Elston, M. A. (1987). Women and anti-vivisection in Victorian England, 1870–1900. In N. A. Rupke (Ed.), *Vivisection in historical perspective* (pp. 259–94). London: Routledge.

Kellert, S. R. (1996). *The value of life: Biological diversity and human society.* Washington, DC: Island Press.

Lippa, R. (2002). *Gender, nature, and nurture*. Mahwah, NJ: Erlbaum.

Lockwood, R., & Ascione, F. R. (1998). *Cruelty to animals and interpersonal violence*. West Lafayette, IN: Purdue University Press.

Pifer, L., Shimizu, K., & Pifer, R. (1994). Public attitudes toward animal research: Some international comparisons. *Society and Animals, 2,* 95–114.

Harold Herzog

■ Human Perceptions of Animals
Humanness and What It Means to Be a Human Animal

Animals have always occupied a central role in imagining what it means to be human. Different cultures and individuals have defined humans as similar to and different from certain nonhuman animals, even from all other species, in various ways. Cultural, religious, and moral convictions, as well as scientific theories, all contribute to different interpretations of humanness. In turn, these definitions have influenced the ways that people view and treat other animals, as well as the ways people understand themselves.

Recent research in evolutionary biology suggests continuity between human and nonhuman species, insofar as no clean line separates the last "primate" ancestor and the first truly "human" ancestor. Of course, like all animals, humans have certain unique characteristics and capacities. Various human qualities have been held up as particularly relevant in defining what it means to be human, often with the aim of setting humans apart from the rest of the animal world. These might be termed "oppositional anthropologies." On the other hand, some theories grounded in the natural sciences, as well as some cultural systems, highlight continuity and relationship with nonhuman entities. These might be called "relational anthropologies." This entry uses these two broad categories to examine a number of different theories of humanness and their significance for human-animal relations.

Humans in Opposition to Other Animals

Many otherwise diverse views of human nature share a common conviction that some quality definitively sets humans apart from all other species. Most of these theories of human exceptionalism assume, further, that humans are superior to other animals. Beyond these shared assumptions, various wide-ranging qualities are targeted as the feature that distinguishes humans from the animals.

Some scholars point to the domestication of both plant and animal species as the root of human exceptionalism. Domestication allowed humans to lead more sedentary lives and arguably set the stage for the first towns and cities. Decreased dependence on the natural habits and patterns of animals may be correlated with the beginnings of civilization.

The concept of a soul, conceived in early Egyptian and Greek philosophy and more formally institutionalized and separated from the physical realm in post-Nicean Christian philosophy, has historically received much of the blame for the alienation of humans from other animals. Philo was one of the first philosophers to combine a dualistic metaphysics drawn from Greek thought with Jewish theology to create a human person whose salvation lay completely outside the material realm. The soul set humanity apart from the rest of the animal world, and the ultimate destiny of humans was no longer tied up with other animals. The notion of a soul is related to the biblically rooted idea that humans were uniquely created in God's image.

Centuries of Christian theology have developed this concept. Saint Paul argued that the real mark of a Christian was internal and that the flesh was an unwelcome chain on the spirit. Later, Thomas Aquinas posited a structure of the human person wherein the highest, reasonable portion was a unique possession of mankind (not women or animals) and the only portion of creation that could directly converse with the divine. Some of these concepts reach back to and beyond Augustine and some other church fathers, but it was between the thirteen and fourteenth centuries that the philosophy reached to the ends of the ancient world, ushering in modernity. "Man" was moving swiftly away from other animals and their environments, through increasing capacity for abstraction and perception of human uniqueness.

The separation between the soul and the material body was famously formalized by Rene Descartes in the seventeenth century. Other Enlightenment thinkers further developed the notion that the material realm was anathema. Isaac Newton, despite referring to his work on physics as "natural theology," aided in the creation of a world where reason ruled. Reason, it was argued, was a distinctly human characteristic, and animals, of course, were intended by a distant creator for human use and improvement.

More recently, a number of scholars have argued that "nature" itself is a cultural construction. These "social constructionists," as they are often termed, generate a different sort of oppositional anthropology, which assumes that human creativity, language, and other cultural capacities are the locus of human uniqueness. Although this stance relativizes some aspects of humanness, it tends to strengthen the idea that humans are radically different from other species.

Ultimately, oppositional models reduce humanness to a specific quality or set of qualities that define human uniqueness. This not only separates humans from other creatures but also devalues the range of human capacities, including those that are shared with other animals. These definitions, in other words, narrow both our understanding of what it means to be human and our understanding of other creatures.

Humans Related to Other Animals

Although many Western religious and philosophical approaches in particular define humans as radically different from (and generally superior to) other animals, some worldviews highlight continuities and shared capacities.

Animism describes the belief, popular among "primitive" indigenous peoples, that animals, plants, and occasionally even inanimate objects are "enspirited." In this sort of worldview, nonhuman animals are central to defining human social categories and behaviors. These world models emerge from cultures in which humans depended on other animals not only for material sustenance but also for drawing meaning from and categorizing their worlds. Human social groups practicing totemism identify with a particular animal species. In such cultural systems, animals help to literally classify and organize human experience; they are not mere passive observers of human culture.

The growth of ethology, including allusions by those who study primates and aquatic mammals to these animals' morality or spirituality (i.e., Goodall, 2005; MacIntyre, 1999), suggests another research program that emphasizes the continuities between humans and animals. Likewise, the field of sociobiology highlights the genetic and developmental history of humans and thus, their similarities to other animals. Edward Wilson's seminal work *Sociobiology: The New Synthesis* suggests that human development includes adaptive responses to the environment, such as innate affinities for certain landscapes and animals and phobias toward other creatures. Definitions of humanness that emerge from this work do not minimize the distinctive qualities of *Homo sapiens,* but they do set difference in the context of shared evolutionary heritage and adaptations.

Some non-Western cultures and religions also emphasize continuity. Many Native American groups, for example, historically believe that nonhuman animals share a single culture with humans even though their physical natures are distinct. Thus, some Plains Indian cultures offer tobacco to nonhuman animals in return for their continued provision of the population. Relationships between the human and nonhuman worlds are often negotiated with the help of spiritual interpreters, or shamans. Nonhuman animals often appear as helpers to, or even alternative identities of, shamans. Among some tribes in the Amazon, for example, shamans are really jaguars who have assumed human form. The idea that humans and nonhumans have in common an essential substance that may take different physical forms reflects a definition of humanness in which people are related and similar to nonhuman creatures in important ways, even though a tension is usually maintained that allows for humans' need to kill other animals in order to live themselves.

The value of these perspectives is their novel approach to an enduring philosophical problem: the relationship between nature and culture. Many Amerindian views suggest that nature and culture are neither completely discontinuous nor completely entwined, but rather carefully negotiated.

In the end, it is clear that humans do share some biological continuity with nonhuman animals, but it is also equally certain that cultural systems impact how humans conceive of and perceive nonhuman animals. The oppositional and relational categories used to guide this article are reflective of both forces, with the relative strength of certain aspects of the human-nonhuman relationship ultimately determining the shape of the conceptual anthropology. In both cases, though, it is clear that humanness is defined largely in relation to nonhuman animals. It is perhaps possible that relational anthropologies, often relegated to marginal populations and further compromised by the globalizing forces of development, can be to some extent preserved and studied to provide both a diversity of perspectives on how human and nonhuman worlds relate and ultimately more species resilience as we practice "being human" in a world characterized by dynamic change.

See also

Ethics and Animal Protection
Sentience and Cognition—*Animal Consciousness*
Sentience and Cognition—*Descartes, Rene*
Sentience and Cognition—*Morality in Humans and Animals*
Sentience and Cognition—*Selfhood in Animals*

Further Resources

Douglas, M. (1973). *Natural symbols.* New York: Vintage Books.

Lease, G. (2005). Hunting and the origins of religion. In B. Taylor (Ed.), *Encyclopedia of religion and nature* (pp. 805–09). London: Continuum.

MacIntyre, A. 1999. *Dependent rational animals: Why human beings need the virtues.* Chicago: Open Court.

Midgley, M. (1995). *Beast and man: The roots of human nature.* New York: Routledge.

Peterson, A. L. (2001). *Being human: Ethics, environment, and our place in the world.* Berkeley and Los Angeles: University of California Press.

Shepard, P. (1996). *The others: How animals made us human.* Washington, DC: Island Press.
Wilson, E. O. (1975). *Sociobiology: The new synthesis.* Cambridge, MA: Harvard University Press.
———. (1984). *Biophilia.* Cambridge, MA: Harvard University Press.

Anna Peterson and Lucas F. Johnston

■ Human Perceptions of Animals
Perceptions of Animals in Medieval Europe

When envisioning animals in the Middle Ages, the first image that springs to mind is probably that of some mythical creature, such as a dragon or a unicorn—followed, perhaps, by an image of St. Francis of Assisi preaching to birds. A realist, scorning both myth and religion, might picture a brutish, miserable peasant raining blows on an even more miserable ass, as he slaves in the fields of an aristocratic overlord. Superficial histories overlook systematic scientific inquiry as a feature of medieval thought, but it certainly existed; no account of man and animal in the Middle Ages would be complete without a thorough review of Albertus Magnus's monumental *De Animalibus.*

All of these aspects (mythological-literary, religious, practical, and scientific) of human-animal relationships are interrelated, even in our own day, though our priorities are different and many people would deny the validity of traditional religious constructs altogether, failing to recognize the degree to which religiously based philosophical concepts—such as man's supposedly God-given dominion over nature—still inform modern behavior.

Except for a few wealthy urban dwellers, medieval man lived close to the animals on which he depended for his livelihood—often sharing the same cramped quarters, with nothing but a wattle partition separating the family from the livestock. To the extent that they were able, people treated productive animals and beasts of burden tolerably well, not because of any philosophical position, but because deviations from good animal husbandry spelled economic disaster, and there was not much margin for error. No one doubted that the larger mammals, at least, were distinct individuals, sharing many of the same emotions humans experience. This narrowness of the perceived distinction between man and beast is usually formulated as "people thought of peasants as little better than animals" but might more accurately be restated as "people thought of animals as little inferior to common people."

Literary and learned opinion concerning animals, prior to the thirteenth century, revolved around the bestiary. This form of literature ultimately derived from the *Physiologus,* a collection of animal lore compiled in Alexandria around 300 CE. The unknown author(s) of the *Physiologus* followed contemporary interpreters of the Old Testament, viewing natural history as a symbolic foreshadowing of Christ's birth, death, and resurrection. They selected thirty-six stories illustrating theological concepts to produce a handbook for preachers, and this handbook, disseminated throughout Europe, came to be regarded as an authoritative reference about exotic wildlife.

Since these bestiaries dealt mostly with animals unfamiliar to Europeans, readers had no way of assessing the accuracy of the information. It is not surprising that they believed equally in the existence of giraffes and unicorns. They read that a lion's whelps were born dead, and brought to life by the father—the King of Beasts—and that this symbolized salvation through Jesus Christ. The pelican of bestiaries, often depicted in religious art, killed its young and then revived them on the third day by piercing her flank to spill her blood on them.

Familiar beasts of farm and woodland appear infrequently in the *Physiologus*, and, when they do, the behavior ascribed to them is not based on accurate observation. Hedgehogs, symbolizing the devil, climbed grapevines and stole grapes, impaling them on their spines. The beaver, on the other hand, demonstrated resistance to the devil's snares by castrating himself when pursued by hunters—beaver testicles being in great demand for their medicinal value. In the descriptive catalog of animals that concludes *De Animalibus*, Albertus Magnus repeated—but questioned—tales from the bestiaries that conflicted with his own numerous observations of natural history.

Another very early treatise on animals is the *Etymologica* of Isidore of Seville (560–636), which takes as its premise that the original names of animals, given to them by Adam, were an essential part of their nature. More scholarly than the *Physiologus*, Isidore's writings illustrate the difficulties under which a medieval compiler labored. Because of a similar etymology, he concluded that a rhinoceros (of which he had a fairly accurate description from a classical source) and a unicorn were the same animal.

Beast fables were a popular form of secular literature in the twelfth and thirteenth centuries, and the fabulist's choices reveal a good deal about attitudes toward animals, as well as towards human society. Each animal species was seen as fitting into a hierarchy similar to that existing among humans, with the predators at the top and the herbivores at the bottom. The fabulist envisioned the lion, King of Beasts, and the eagle, King of Birds, as actually ruling their kingdoms. Herbivores, especially domesticated ones, were viewed as base and unworthy of esteem—but not wicked. Often, a fable's action hinged on damage resulting from an animal's deserting its usual station in the hierarchy. If a pig or an ass succeeded in passing himself off as a member of the royal court, it indicated that the royalty had become swinish and asinine. Medieval men knew from experience that the noble could become base; they less readily accepted that the base could become noble.

Canids (wolves, foxes, and dogs) exemplified evil in its various guises. The wolf and fox were predators that resorted to subterfuge and trickery, while the dog divested itself of any nobility by becoming a servant. Of course, people were free to idealize lions, with which they had no direct experience, while wolves still roamed Europe, threatening livestock and occasionally attacking humans. One of the miracles ascribed to St. Francis of Assisi involved taming a wolf that was terrorizing the Italian city of Gubbio. Some scholars suggest the wolf was actually a human bandit. Whether or not the story is true, it illustrates St. Francis's belief that all living creatures possessed an awareness of their creator, and were essentially good when in harmony with their surroundings.

Rather than confining the animal mirror of Christ's redemptive power to a few semi-mythological species, as in the *Physiologus*, St. Francis saw that story in all of nature. He viewed the actions and vocalizations of animals as forms of prayer, joining in chorus with birds and even with cicadas. Sometimes, his preaching to animals and birds was ironic, intending to shame a human audience that had earlier refused to listen to him; but, in other cases, he clearly attributed some measure of understanding to beasts and fowl. Saints' tales hinging on the power of a holy man or woman to communicate with animals are not restricted to St. Francis; St. Anthony of the Desert (fourth-century Egypt) preached to fish, and several early Irish saints had animal companions.

A saint might elevate animal nature to a higher spiritual plane by his preaching, but people in secular life took pains to preserve the distinction between beast and man, and anything appearing to partake of both natures was regarded with extreme abhorrence. The central nature of communion in the Catholic liturgy translated into an obsession with the moral and spiritual aspects of eating in general. People and animals did not share food. Observing certain rituals could restore a loaf upon which a mouse had nibbled to purity, but nothing could erase the contaminating influence of a dog. Flesh of carnivores and scavengers was universally proscribed. Swine, a forbidden food in the Middle East (where pigs are predominantly scavengers), were in medieval Europe, mainly a foraging species living off acorns. Feeding them on human garbage—especially slaughterhouse offal—rendered them off-limits for human consumption. To expose a human cadaver to be devoured by wild beasts was the ultimate degradation, and to feed on an animal which had eaten human flesh was tantamount to cannibalism.

Members of some religious orders abstained from eating meat altogether, either as part of a general avoidance of luxury consumption, or, in the case of the Franciscans, out of respect for creatures sharing so many human emotions and appetites. Everyone abstained from meat during Lent. During the Dark Ages, low population density allowed all classes of society to enjoy a diet high in animal protein, but by the twelfth and thirteenth centuries, hunting large game had become the prerogative of the nobility, and pasture for nonworking animals was too scarce to permit much meat consumption by peasants and urban laborers. Had the Black Death in the fourteenth century not reduced the population far below carrying capacity, modern European dietary habits would probably more closely resemble those of eastern Asia.

Most medieval thinkers started from the premise that man's sense of and relationship to God was the most essential part of his humanity, and framed their construction of animal nature according to perceptions of how that relationship differed among various categories of living things. St. Francis saw animals as active participants in the spirituality of the universe. He differed from most of his contemporaries in ascribing to animals a conscious capacity to participate in worship of the divine. At the opposite extreme, some modern animal behaviorists see in the apparent absence of any religious sense in primates in controlled settings proof that such a sense in humans is a mere cultural artifact.

Prior to the beginning of the thirteenth century, Neoplatonism dominated Western academic philosophy. Neoplatonic philosophy saw the natural world as an imperfect copy of an ideal universe existing in the mind of God, and thus considered the study of nature to be an unproductive source of fundamental insights. Neoplatonic scholars made significant contributions to mathematics and the physical sciences, but it was not until translations of Aristotle and Arabic commentaries thereon began to appear in Europe that the systematic study of nature became respectable.

In the Aristotelian view, all of nature is the handiwork of its creator, and the systematic study of nature is a legitimate approach to the ultimate goal of better understanding God. The greatest medieval proponent of the view (and indeed the only European to attempt comprehensive integration of observation of the animal kingdom with theory between antiquity and the mid-seventeenth century) was Albertus Magnus of Cologne (1197–1280), Dominican friar, church administrator, and patron saint of scientists in the Roman Catholic canon. Over the course of his long career, he wrote commentaries on all of Aristotle's recently translated scientific books, attempting to make them intelligible to students at European universities. Of these treatises, which cover physics, astronomy, mineralogy, geography, optics, and botany, the long treatment of animals is considered to be the most original.

Written between 1257 and 1263, Albertus Magnus's *De Animalibus* consists of three sections: a commentary on Aristotle's *Historia Animalium* and *De Generatione Animalium;*

a largely original treatise on the anatomy, physiology, and classification of animals; and a compendium of described species, including information from many classical and contemporary sources, as well as much original observation. The third section is a valuable source for social historians, containing, as it does, much practical information on medieval agricultural practices. The long entry on falcons and falconry is well-known and has been repeatedly translated and reprinted. There is a lively description of whaling in the Baltic Sea. The practical or theological significance of each species to human existence forms a prominent part of each entry. Implicit in all medieval treatises on nature is the assumption that God created the natural world for the benefit of man, and therefore any species must have its own unique use to us. If it appears to be useless or harmful, it is either because we have not yet discovered its virtues, or because it is out of place in the hierarchy of nature.

Book 20 of *De Animalibus* describes the physical properties of animal bodies. It is evident from this and the subsequent chapter that Albertus Magnus personally dissected animal cadavers and was a perceptive observer of comparative anatomy. Although he follows earlier authors in basing the outlines of his classification on habitat and mode of locomotion, he recognizes the fundamental differences between whales, bony fishes, and octopi, and ranks them according to a scale of "perfection" corresponding closely to the modern concept of evolutionary complexity.

Every physical body is made up of the four elements: earth, air, water, and fire, commingled in different proportions to give the body its unique physical characteristics. Just as small differences in the concentration of elements in a human body affected the production of the four humors to which they corresponded and thus governed the temperament of a person, different proportions of elements determined the behavioral characteristics of an animal, as well as its physical appearance.

The soul is the force governing the sequential combination of elements during fetation (development) to produce a living being capable of differentiation, growth, and change. Albertus Magnus and his contemporaries envisioned a hierarchy of souls, the highest of which belongs to man alone. At the bottom is the vegetative soul, which governs growth and differentiation; this belongs to all living beings, including plants. Above that are the animal soul, conveying movement, and the sensory soul, governing reaction to stimuli. All animals, even earthworms and sponges, possess vegetative, animal, and sensory souls. The sensory soul of a mouse, which has eyes and ears, is more perfect than that of an earthworm. As one progresses further up the scale of "perfection," one encounters animals with memories and the ability to learn from experience. At the top of the scale of animal perfection is the monkey, with an anatomy very much like our own, a seemingly limitless ability to mimic our actions, and no apparent capacity to reason abstractly or profit by instruction apart from experience.

Albertus Magnus even included a "missing link" in his scheme of animal perfection—the Pygmy, who constructed houses and fashioned tools, but had, according to classical authors, the barest rudiments of articulate speech. The descriptions with which he was familiar were probably based both on pygmies and chimpanzees, with no little fanciful elaboration. To his credit, he placed his missing link far outside the boundaries of direct experience, and never proposed that any adult human, individually or as an ethnic group, possessed anything other than the uniquely human rational soul.

Animal souls, in his view, arose from the physical constituents that made up an animal body, whereas the uniquely human rational soul was a direct gift from God, instilled into the human body late in the physical development process after the vegetative and sensory souls were already in place. According to this view, subsequently elaborated by Thomas Aquinas, human fetuses, very young infants, and the profoundly retarded lack rational souls, which does not (as is sometimes claimed) make them worthless in the scheme of God's creation. It does, however, release a human who is fundamentally incapable of discerning right from wrong from the onus of eternal damnation.

Going beyond the Biblical concept of man as a being specially created by God to be qualitatively different from any animal, and attempting to find an objective, observable difference represents a great advance in scientific thought. Having seized upon rationality as the crucial difference, Albertus Magnus was not blind to evidence that animals might behave rationally. He observed that sheep became fearful when they saw a wolf, even if the wolf was not behaving in a threatening manner, and decided that they were capable of sensing intentionality—that is, they were reacting to a direct stimulus (the wolf's malice) rather than predicting a sequence of events. The debate over whether an animal, such as a sheep, is capable of predicting a sequence of events and acting accordingly is still going on.

Following Arabic commentators on Aristotle, Albertus Magnus believed deviations from the intrinsic developmental plan of an organism to be due to astrological influences. Like does not invariably beget like, in humans or among animals, and ascribing this to the motions of the sun, moon, and planets did not seem as farfetched to the medieval mind as it does to a modern scientist, especially as Christian philosophers from St. Augustine onward considered astrological phenomena to be visible signs of an unseen divine force, rather than the actual agents. With respect to the physical bodies of animals, this thirteenth-century scholastic concept of animal variation and its causes appears to allow for a certain amount of evolutionary change, mediated, of, course by a divine intelligence that is constantly at work in the world.

Between the modern view, that science, theology, and moral philosophy are entirely separate disciplines—and that to mix science with either of the other two hopelessly compromises it—and the medieval view, that the three are inextricably, obligately commingled, there is surely a middle ground. Medieval philosophy at least provides a framework within which animals can be respected as individuals with certain intrinsic rights, while also being seen as having been created for the service of mankind. It deserves more attention than it currently receives from people concerned with the ethical treatment of animals.

Further Resources

Armstrong, E. A. (1973). *Saint Francis: Nature mystic*. Berkeley, Los Angeles, London: University of California Press.

Baxter, R. (1998). *Bestiaries and their users in the middle ages*. Stroud, Gloucestershire, and London: Courtald Institute.

Creager, A. N. H., & Jordan, W. C. (Eds.). (2002). *The animal/human boundary: Historical perspectives*. Rochester, NY: University of Rochester Press.

Magnus, A. (1999). *On animals: A medieval summa zoologica*. Translated and annotated by K. F. Kitchell and I. M. Resnick. Baltimore and London: Johns Hopkins University Press.

Roberts, L. D. (Ed.). (1982). *Approaches to nature in the middle ages*. Binghamton, NY: Center for Medieval and Early Renaissance Studies.

Salisbury, J. (1996). Human animals in medieval fables. In N. A. Flores (Ed.), *Animals in the middle ages: A series of essays* (pp. 49–65). New York and London: Garland Publishing.

Weisheipl, J. A. (Ed.). (1980). *Albertus Magnus and the sciences: Commemorative essays*. Toronto: Pontifical Institute of Mediaeval Studies.

Martha Sherwood

■ Human Perceptions of Animals
Social Construction of Animals

The meanings of animals seem to be fixed and enduring. The tenacious persistence and widespread acceptance of these meanings suggest that they are cultural phenomena—part of the normative order of the society in which they occur. Much like other cultural phenomena—love of country, motherhood, or the success ethic—the meanings of animals are passed from generation to generation. They are social constructions.

Consider which animals are regarded as wild and which as tame. At an early age, we learn—by watching Disney movies, reading fairy tales, and listening to our parents—that a "wild animal" can be a tiger in the jungle, an elephant in a zoo, a squirrel living in the backyard of a suburban home, an ownerless dog that roams the neighborhood, or a mean-spirited, raunchy person looking to pick a fight in a bar. As a social designation, *wildness* comes to mean "distance" and "danger," with *tameness* its converse. Many learn what a "tame" animal is by owning one themselves. Parents often acquire pets for their children, who themselves, in turn, attribute humanlike qualities to these animals and protect them from the dangers that lurk outside in the world of nature. The result is that children come to view what constitutes a wild or tame animal as hard and fast fact whose meaning is given—external to human culture and social process. Yet we know, sociologically, that "facts" can vary, because in different places and times people will assign them different meanings.

In contemporary American homes, for example, it is taken for granted that dogs will be regarded in a certain way. A puppy is transformed into a make-believe or pretend family member that is named, fed, groomed, dressed, photographed, talked to, mourned, slept with, given birthday parties, and taken to therapists for behavior problems. According to Lucy Hickrod and Raymond Schmitt, this process begins when a dog is taken into a home (1982). Naming the new pet begins its transformation from a generic puppy into a specific member of the family. The name affords the dog an identity and makes it easier to talk about, and it also directs activities toward the dog as though it were part of the family. Acquiring a status in the family is contingent on family members' willingness to meet the pet's needs. Pets that do not obey house rules or that are considered too difficult may be given away or euthanized. If pets survive this probationary period, many family members develop intense feelings for them. During this *engrossment stage*, personal qualities—such as loyalty or humorousness—are often attributed to pets, who are seen as being more consistent in displaying these attributes than are most humans.

After becoming engrossed in their dogs, most owners come to the realization that they are treating their pets as genuine family members. As they become aware of their feelings for their animals, owners often are amazed at how intensely family members care about the pets, even though they are "only animals." Soon they begin to communicate their feelings for their pets to people outside the family, so they, too, can participate in this definition of the animal as family member. This may entail introducing pets to newcomers by mentioning their names and discussing their personal histories, as well as nonverbally communicating this status through fondling, special dressing, or the like. These *tie signs* minimize the social boundaries between pet and human, thereby demonstrating the pet's special position to strangers.

Even after pets die, their intimate connections to families may be remembered when stories are shared about the animals' exploits—or when later pets are given the same names. Since human emotional responses to the death of a pet can be as intense as those precipitated by the loss of any family member, it is not surprising that there is a modest but growing interest in burying pets in animal cemeteries, in order to maintain this connection. That dead animals are typically treated as kin can be seen in the intense public disapproval generated by media reports that Roy Rogers stuffed his faithful steed, Trigger, so that it could be displayed in Rogers's living room. This negative reaction, no doubt, arose because the public regarded the horse as a pet or member of the Rogers family. Presumably, for these people, the stuffing of a pet—rather than a hunting trophy—seemed to deny an emotional tie, by turning Trigger into an object.

However, a pet is still something less than a full-fledged family member, because of ever-present *frame breaks*. Bystanders, media presentations, and certain situations constantly call into question this definition of the pet as a family member and reinforce its definition as an animal or a toy. Signs reading No Pets Allowed or Beware of Guard Dog, as well as instances when pets nonchalantly vomit in living rooms, eat their own excrement, or mate in public, are reminders that they are, at best, make-believe members of families.

Yet, dogs in another setting might be anything but adjunct members of the family. In the context of a dog track, they are racing machines. This construction requires impersonal identities, and the dogs are assigned special names and numbers. Their official names appear in the programs but are almost never used, except when announced with their position numbers at the checkpoint. These official names, like the names of racehorses, exist outside everyday human usage, and their meanings are clear only to those deeply involved in the race world. These names are not even anthropomorphized, but usually refer to abstract images or emotional states—such as Peaceful Darkness, Fine Style, or Surprise Launch. When names are humanized, they are almost always in the possessive form, such as Tara's Dream or Bobby's Showtime, suggesting that the names apply to the owners and not the dogs. Transforming the dogs into machines reduces their identities to numbers that appear on racing blankets, starting gates, programs, handlers' armbands, and monitors displaying betting odds. The numbers are also used by track announcers—when dogs pass through the checkpoint and during the race—to indicate their positions, as well as by bettors who shout numbers, not names, as they cheer on their choices.

Standard handling practices also help to construct a numbered machine, suppressing the dogs' personalities. For example, when the dogs are presented to interested bettors in the paddock, steel bars keep onlookers about five feet from the thick glass behind which dogs, handlers, and judges do their work. The distance and glass muffle all sounds, although one can still hear the barking of unmuzzled dogs in cages turned away from the public's view. When in view, they are muzzled and tightly controlled. Muzzling partially covers their faces, restricts their barking, and gives them a badge of human dominance. Handlers rarely look at, talk to, or touch their dogs, and the exceptions only point to the more pervasive construction of these animals as machines. Occasionally, when handlers talk to each other, they might quickly pet or scratch a dog's head or neck. This touching looks more like a reflex or afterthought, because they neither look nor talk to the dog as they do it, nor do they try to solicit any response from the dog. For the most part, the dogs themselves have little response. No dog, for instance, responds by licking a handler's hand or jumping up on the handler. Indeed, the greyhounds rarely initiate any interaction with each other: they are trained not to do so, and the handlers stop any attempt. At the rare times when people try to interact with the greyhounds, the handlers immediately restrain the dogs and ignore the people, almost as though nothing had happened. Even when the dogs and the handlers are not busy, such interactions are prevented.

The meanings of social constructions can change over time. What a group regards as "wild" can, at another time, be regarded as "tame." For example, our conception of primates has developed dramatically in the twentieth century, so that their place in the modern order has changed from being exotic and wild to being tame and almost human. Anthropologist Susan Sperling claims that several factors account for this shifting view. As postwar America grew increasingly interested in the complex cognitive and social abilities of animals, images of primates, in particular, were remodeled to become more humanlike. Anthropological models of evolution started replacing "primitive" human groups, such as the Trobriand Islanders, with nonhuman primates—making the latter our "ancestors." These models, along with observational studies of primates, were dis-seminated to the public—in magazines such as *National Geographic,* nature shows on television, and movies such as *Gorillas in the Mist*—giving the impression of extreme similarity between the species. Baboon troops, for instance, were uncritically viewed as microcosms of human society, because they, too, had social characteristics, such as a "division of labor." Compounding the effect of this research were field studies, such as those of Jane Goodall (1998), in which chimpanzee subjects were given human names and their personalities described in human terms. Additional anthropomorphization came from researchers who studied the acquisition of language by apes and treated their animals like foster children who could talk and live in human settings. The conse-quence, contends Sperling, was the "obliteration" of the border between humans and nonhumans.

With such boundary blurring, it is not surprising that what were once wild animals may now be regarded as pets. A case in point is a project that allowed laypersons to assist with research and conservation efforts with wild orangutans in Borneo. Accord-ing to Constance Russell (1995), some people had expected to have a "cuddly" experi-ence with these animals and, not surprisingly, experienced the apes as "children" in their interaction with confiscated infants, which were to be sold as pets but were being rehabilitated in a clinic. Although they were initially forbidden to have physical contact with the infants, many of the tourists expressed an intense desire to hold them. Many oohed and aahed when seeing the animals and commonly described them as "cute"— or as "sweet little ones." When permitted physical contact, all the tourists felt very for-tunate to have the opportunity, saying that it "profoundly" affected them. Some tourists even competed for the affections of certain infants, as they sought to "babysit" them or were reluctant to break off contact, because they felt "needed" by the young animals. No longer seen as alien or strange creatures, as they might have been years ago, these primates were related to in the only way that made sense to these tourists; they defined the primates as pets or quasi family members of human society.

In fact, some wild animals are literally transformed, over time, into pets of a sort. Elizabeth Lawrence contends that this century has witnessed "a remarkable American social phenomenon" in the transformation of wild bears into tame and civilized stuffed teddy-bear dolls that hardly resemble their natural forebears. Although the teddy bear is obviously an inanimate object, it is now often seen and treated as though it were a pet or companion animal. Indeed, Lawrence goes so far as to say that "teddy bears are ani-mals that nearly, but not quite, become people." Child and adult owners attribute per-sonalities, thoughts, feelings, and behaviors to their teddy bears and, not surprisingly, report that their dolls make them feel comfortable. In return for the "spirit of caring and unconditional affection" provided by the dolls, owners cherish them. This conversion of nature into culture, according to Lawrence, has resulted in a "counterpart of opposi-tions" between the teddy bear and the living bear. The former is "tame, dependable, neutered, civilized, humanized and sanitized," while the latter is "wild, unpredictable, uncontrolled, aloof," and dangerous (1998). Lawrence's analysis highlights the power of social constructions to alter what we think of as normal or natural—they do nothing less than shape our consciousness.

Animals in zoos, for instance, are animals that most people will never see in their natural state, but will only read about or imagine. Taken from their natural context, these animals are in a human frame, while their natural habitat is transformed into our dream of a human-animal paradise. They become creatures of leisure, given food, rather than having to hunt and fight among themselves for live game. Even the "prey" is transformed, so that it cannot be identified as a specific animal—meat is thoroughly butchered—and zoo animals are not allowed to eat fellow captives. Zoo animals also live in harmony, never struggling over territorial matters with other animals. They live in an environment built by humans: a constructed world that shrinks entire continents into acres and often combines different species in the same exhibit, even though they may live far apart in their normal climatic zones.

The artificiality of this zoological paradise, as Boria Sax observes, renders the traditional dichotomy of wild versus domestic animals invalid. There, symbols of captivity, such as cages and cold, cement floors, are increasingly being eliminated, while animals in nature are being carefully observed and controlled through devices such as concealed cameras and radio collars. Because of modes of modern captivity, animals can be closely approached and admired in ways that are impossible in the wild. This proximity means that humans are not unobtrusive to zoo animals. Although it is not known exactly how their behavior changes, instead of ignoring or being habituated to visitors, zoo animals respond to and interact with them. Captive bears, for instance, are often encouraged by the public to beg. What zoo visitors see, then, is a culturally falsified version of how these animals actually behave in the wild.

The result is often a captive, wild animal that is regarded as a human in animal skin. Perhaps one of the best examples of this transformation is the giant panda. Many of the panda's physical features—the round head, large eyes, and vertical posture—facilitate its anthropomorphization. The panda is not a distant animal; it has become a cuddly friend, given a human personality and human needs and emotions. Once perceived in this manner, the distinction crumbles between what we regard as wild and tame. Thus, visitors often have a fondness for zoo animals, sometimes naming them, "adopting" them, giving them tea parties, or playing with and touching them in children's zoos and aquaria's petting tanks. For their part, animal handlers often form even closer relationships with individual zoo animals, treating them as typical companion animals.

If social constructions can tinker with people's consciousness of the "natural" order, they can just as easily toy with other basic distinctions that humans make—including those between life and death. An example of this is the taxidermist who tries to make dead animals appear animated or just as they were in life. Unlike those in real life or even in the zoo, the stuffed wild animal can be examined closely and handled without fear. Some of these transformations into inanimate life are intended to be instructional, as in the case of museums that display animals faithful to their "natural state," in order to teach the public or stimulate interest in zoology. Other transformations symbolize human domination over nature, as hunters do who display their trophy animals on walls or floors. In contrast to the hunter's trophy, some pet owners stuff and display a deceased animal as a memorial to its faithful companionship. This practice was common—though the source of some controversy—among middle-class Parisian dog owners during the nineteenth century.

Stuffed animals or animal skins may also be used as product trademarks, home decoration, or fashion statements to project certain images and capture attention. When used as fashion statements, they can, to some degree, transform their owners' personae. Fur coats for women, for instance, in addition to being used as status symbols suggesting affluence, can offer a reflection of animal spirit or passion by projecting an exotic or erotic image, while leather coats on men can convey a macho attitude. In this way, the distinction between wild and tame becomes sufficiently blurred, to allow some humans to entertain the idea that they are a little "wild."

In some instances, humans may transform the natural order by reshaping the very disposition and biology of animals. Jeffrey Nash notes, for example, that over centuries, the English bulldog was bred to be a one-hundred-pound muscular fighter with a deep chest, powerful jaw, and setback nose for the sport of bullbaiting. Breeders were unconcerned with beauty, while giving special attention to such traits as savagery and insensitivity to pain. Despite the eventual outlawing of bullbaiting and bulldogs, dog lovers resisted the edicts and used selective breeding methods to preserve all the bulldog's traits, except for its ferocity. The result, over a few generations, was the exaggerated appearance and pleasant personality of the modern English bulldog (1989). What the human consciousness takes for granted as innate biology is, at least in the case of the bulldog, the result of decades of social and generic construction.

Further Resources

Hickrod, L., & Schmitt, R. (1982). A naturalistic study of interaction and frame: The pet as family member. *Urban Life, 11*, 55–77.

Lawrence, E. (1990). The tamed wild: Symbolic bears in American culture. In R. Browne, M. Fishwick, & K. Browne (Eds.), *Dominant symbols in popular culture* (pp. 140–53). Bowling Green, OH: Bowling Green State University Popular Press.

Nash, J. (1989). What's in a face? The social character of the English bulldog. *Qualitative Sociology, 12*, 357–70.

Russell, C. (1995). The social construction of orangutans: An eco-tourist experience. *Society and Animals 3*, 151–70.

Sax, B. (1997). The zoo: Prison or paradise? *Terra Nova, 2*, 59–68.

Sperling, S. (1988). *Animal liberators: Research and morality*. Berkeley: University of California Press.

Arnold Arluke and Clinton Sanders

■ Human Perceptions of Animals
Sociology and Human-Animal Relationships

The discipline of sociology, which, very broadly defined, is the study of society, has obstructed from its purview the roles that nonhuman animals (hereafter referred to as *animals*) play in human societies and the associated importance of human-animal relations. This marked oversight has been acknowledged by several scholars (Agnew, 1998; Arluke & Sanders, 1996; Bryant, 1979; Beirne, 1995; Bryant, 1992). For instance, in the opening of their book, *Regarding Animals*, Arnold Arluke and Clinton Sanders (1996, p. 2) remark, "Although there is an enormous literature about animals by novelists, journalists, philosophers, biologists, psychologists, and animal behaviorists we have been disappointed that there is so little by our fellow sociologists." This oversight on the part of sociologists is largely the result of the anthropocentric nature of the discipline itself (Beirne, 1999). The term *anthropocentric* refers to the exclusion of nonhuman animals from consideration, and the term speciesism is used to refer to the resultant discrimination against animals, due to their membership in a nonhuman species.

In order to understand and challenge the pervasive anthropocentrism in sociology, we must understand how and why it developed this way. The exclusion of animals from the sociological purview is not accidental; rather, it has historical roots. The discipline developed in opposition to biological and social Darwinist theories, and sociologists consequently sought to distinguish their work and their discipline from the realms of biology and nature. As a result, they constructed a rigid culture-nature dichotomy and concerned themselves strictly with the former domain. This dichotomy rationalizes and perpetuates the belief that humans are unique and vastly different than animals: humans are believed to belong exclusively to the realm of culture, whereas other beings are considered simply as nature (Noske, 1989). It is important to note, however, that membership in the realm of the cultural has not been equally afforded to humans—privileged men have been associated with culture, whereas women have been largely identified with nature. This assumption that the social and the cultural are exclusively human is fostered by the belief that social relationships depend upon the verbal use of language (Arluke & Sanders, 1996). The result of these assumptions is the pervasive anthropocentrism and speciesism in sociology, still evident today, which has resulted in an unwillingness to attend to human-animal relationships. Examining the historical, theoretical grounding of these assumptions is useful in understanding and challenging the anthropocentric sociology of today.

John Bellamy Foster (1999) locates the root of resistance to nonhuman nature concerns within sociology in the "early barrier erected between society and nature, sociology and biology—dividing the classical sociologies of Marx, Weber, and Durkheim from the biological and naturalistic concerns that played a central role in the preclassical sociology of the social Darwinists" (p. 367). The roots of anthropocentrism within sociology extend across the micro-macro theoretical continuum. A detailed account of this anthropocentric foundation is clearly beyond the scope of this essay; however, in order to provide insight into the groundwork laid by the classical theorists, the works of two notable theorists will be discussed. On the structural side of the continuum, the work of Karl Marx is examined, paying attention to the ways in which he dichotomized "men" and animals. At the interactionist end of the spectrum, the work of George Herbert Mead is examined, and it is demonstrated that Mead constructed much of his theory of symbolic interactionism by juxtaposing humans and animals, reinforcing the culture-nature dichotomy. Using the works of these theorists, it is demonstrated herein that classical social theory propagated the anthropocentrism still prevalent in sociology today.

The Writings of Karl Marx: Macro Anthropocentric Theorizing

Of the distinctions between humans and other animals Marx delineated, the most significant are, arguably, his claims that only humans are capable of willful production, language, and consciousness. A brief examination of each is instructive. In *The German Ideology*, Marx (and Engels) spent significant time drawing distinctions between humans and animals. Marx asserted that production is the earliest distinguishing factor: "Man [sic] can be distinguished from the animal by consciousness, religion, or anything else you please. He begins to distinguish himself from the animal the moment he begins to produce his means of subsistence, a step required by his physical organization. By producing food, man [sic] indirectly produces his material life itself" (Marx & Engels,

1994, p. 107). He did concede, in *The Economic and Philosophic Manuscripts of 1844*, that animals are capable of production; however, he maintained that human production and animal production are inherently different: "Admittedly, animals also produce. They build themselves nests, dwellings, like bees, beavers, ants, etc. But an animal only produces what it immediately needs for itself or its young. It produces one-sidedly, while man [*sic*] produces universally. . . . An animal produces only itself, while man [*sic*] reproduces the whole of nature" (Marx & Engels, 1988, p. 77). The implication here is that not only is human production—which is said to be universal and free—to be distinguished from animal production, but also that the latter is inferior.

Marx furthered the dichotomy in his subsequent discussion of human consciousness. He asserted that in its early development, human consciousness is solely concerned with the immediate surroundings and that relations with others are severely restricted. He characterized this stage of consciousness as *animalistic*: "Men's relations with this consciousness are purely animal, and they are overawed like beasts. . . . This beginning is as animalistic as social life itself at this stage. It is the mere consciousness of being a member of a flock, and the only difference between sheep and man [*sic*] is that man [*sic*] possesses consciousness instead of instinct, or in other words his instinct is more conscious" (Marx & Engels, 1994, p. 117). Marx's denial of animal consciousness is the outcome of his contention that consciousness and language arise from relations with others and his assumptions that animals are incapable of social relations. As a result, he also asserted that animals are incapable of language: "Language, like consciousness, only arises from the need and necessity of relationship with other men (My relationship to my surroundings is my consciousness). Where a relationship exists, it exists for me. The animal has no 'relations' with anything, no relations at all. Its relations to others does not exist as a relation" (Marx & Engels, 1994, p. 117). Marx clearly considered humans significantly different from and superior to animals, reinforcing the culture-nature binary and a perceived impermeable barrier between the two.

Implicit reification of the culture-nature dichotomy can be found in Marx's discussion of his concepts of estranged labor and species-being in *The Economic and Philosophic Manuscripts of 1844*. Therein his critiques of capitalist production implicitly assume a need to reaffirm the culture-nature binary. Marx argued that, as a result of worker estrangement (or alienation), proletarians are dehumanized. He delineated four aspects of the estrangement of labor: the estrangement of the worker from the product of his labor; the relation of the worker to his activity as alien; the estrangement of the worker from his species-being; and the resultant estrangement of men from men. In his discussion of the estrangement of the worker from the product of his labor and from his species-being, Marx frequently discussed how such estrangement reduces workers to animals. He explained that, as a result of the alienation of the worker from the product of "his" labor, "man [*sic*] (the worker) no longer feels himself to be freely active in any but his animal functions—eating, drinking, procreating, or at most in his dwelling and in his dressing-up, etc., and in his human functions he no longer feels himself to be anything but an animal" (Marx & Engels, 1988, p. 74). Thus, Marx concluded that under the capitalistic division of labor, whereby human-labor activity becomes an end instead of a means and the laborer is no longer freely active, the laborer is dehumanized and is free only in his "animalistic" functions.

Marx subsequently explained that through labor under capitalism, "man" becomes estranged from his species-being. Marx considered "man" a species-being because he considers his entire species and treats himself as a universal and free being (Marx & Engels, 1988, p. 75), which is clearly related to his conceptualization of production. He argued that labor under capitalism estranges "man" from nature and from himself, and transforms the life of the human species into a means of individual life. That is, "man" is no longer concerned with his species; rather, his concern is limited to his individual self, and he is thus no longer a species-being. Under capitalism, "man" becomes nothing more than his labor, which, according to Marx, reduces him to the level of an animal:

> The whole character of a species—its species character—is contained in the character of its life activity; and free, conscious activity is man's [sic] species character. . . . The animal is immediately identical with its life-activity. It does not distinguish itself from it. It is its life activity. Man [sic] makes his life-activity itself the object of his will and of his consciousness. He has conscious life-activity . . . . Conscious life-activity directly distinguishes man [sic] from animal life-activity. It is just because of this that he is a species being. Or it is only because he is a species being that he is a Conscious Being, i.e., that his own life is an object for him. Only because of that is his activity free activity. Estranged labor reverses this relationship, so that it is just because man [sic] is a conscious being that he makes his life-activity, his essential being, a mere means to his existence. (1988, p. 76)

Later in this section, Marx sums up how the estrangement of labor serves to dehumanize proletarians, asserting that "in tearing away from man [sic] the object of his production . . . estranged labor tears from him his species life, his real species objectivity, and transforms his advantage over animals into the disadvantage that his inorganic body, nature, is taken from him" (Marx & Engels, 1988, p. 77). Therefore, according to Marx, under capitalism, as a result of estranged labor, the life activity of humans is no longer free; it becomes merely a means to existence, and people can no longer distinguish themselves from it. In these ways, humans are "reduced" to, or are even below, the level of animals.

Examining these examples from Marx's writings demonstrates that he distinguished between humans and animals in at least three ways: on the bases of production, consciousness, and language. Accordingly, Marx and Engels's proclamation, at the end of *The Communist Manifesto*—"The proletarians have nothing to lose but their chains. They have a world to win. Workingmen of all countries unite!" (1998, p. 268)—can be regarded not only as a call to liberate the proletarians from capitalism, but also to liberate them from the inferior realm of nature, which Marx implicitly asserts they were relegated to under industrial capitalism.

The Writings of George Herbert Mead: Micro Anthropocentric Theorizing

This tendency toward reifying the culture-nature dichotomy in classical sociological theory is also found at the micro end of the continuum. The work of George Herbert Mead is a more explicit example of dichotomizing humans and animals and reinforcing the culture-nature dichotomy. In fact, Flynn (2000, p. 100) asserts that the exclusion of animals from sociological theorizing and research is due to Mead's influence in particular. There are likely numerous aspects of Mead's work that have contributed to the anthropocentrism prevalent in contemporary sociology. He built up much of his theory of symbolic interactionism by juxtaposing humans and animals, which becomes most evident in his descriptions of some of his key concepts.

Gestures are extremely important in Mead's theory. A gesture is the first overt phase in a social act (Mead, 1910, p. 397), which serves as the stimulus for other social acts (Mead, 1956, p. 154). Mead conceded that animals, too, can have conversations of gestures (1910, p. 398). He maintained, however, that conversations of gestures are different among animals and humans; specifically, he dichotomized gestures as insignificant (those that are instinctive and consequently lack meaning) and significant, placing animal gestures in the former category (1956, p. 155). Humans, however, are believed to be uniquely capable of significant gestures, because they have language capabilities (1956) and possess consciousness. For Mead, consciousness is the ability to be self-aware, and his notion of consciousness is directly related to his conceptualization of gestures:

> Gestures may be either conscious (significant) or unconscious (non-significant). The conversation of gestures is not significant below the human level, because it is not conscious, that is, it is not self conscious (though it is conscious in the sense of involving feeling or sensations). An animal as opposed to a human form, in indicating something to, or bringing out a meaning for, another form, is not at the same time indicating or bringing out the same thing or meaning to or for himself; for the animal has no mind, no thought, and hence there is no meaning here in the significant or self-conscious sense. (1956, p. 168)

There are numerous conclusions that Mead drew based upon his assumptions that animals are incapable of significant gestures and consciousness, one of which is that animals lack minds and intelligence. In defining the mind, Mead explained that what he considered "characteristic of the mind is the reflective intelligence of the human animal which can be distinguished from the intelligence of lower forms" (1956, p. 180). For Mead, intelligence is synonymous with reflective intelligence. He defined reflective conduct as that which takes the future into account (1956, p. 181) and asserted that "reflective behavior arises only under the conditions of self-consciousness and makes possible the purposive control and organization by the individual organism of its conduct" (1956, p. 169). He also asserted that humans can select from alternate responses to a stimulus and can delay their reactions, whereas animal reactions are said to be instinctive and immediate (1956, pp. 173–77). Mead's claim that animals lack consciousness and intelligence led him to conclude, as Karl Marx did, that animals cannot control their environment and, consequently, cannot produce as humans can. "The human being as a social form actually has relatively complete control over his environment. The animal gets a certain slight kind of control over its environment; but the human form, in societies, can determine what vegetation shall grow, what animals shall exist besides itself" (1956, p. 30). Additionally, he considers this control necessary for the development of the mind (1956, p. 41).

In addition to the requirement that an organism be able to control its environment to develop a mind, Mead adds that being a minded being requires being able to take the role of "the other," which he maintained animals are incapable of, because they lack language (1956, p. 159). He explained, "I know of no other form of behavior than the linguistic in which the individual is an object to himself, and, so far as I can see, the individual is not a self in the reflective sense unless he is an object to himself. It is this fact that gives a critical importance to communication, since this is a type of behavior in which the individual does so respond to himself" (1956, p. 206).

Mead also saw language as a necessary part of the development of the self. He viewed the self as emerging out of social interaction through language and explained that

> The importance of what we term 'communication' lies in the fact that it provides a form of behavior in which the organism or the individual may become an object to himself. It

is that sort of communication that we have been discussing—not communication in the sense of the cluck of the hen to the chickens, or the bark of a wolf to the pack, or the lowing of a cow, but communication in the sense of significant symbols, communication which is directed not only at others but also at the individual himself. (1956, p. 203)

According to Mead's logic, it is impossible for animals to develop selves because they lack verbal language, minds, and consciousness and are therefore incapable of symbolic interaction. The implication is that sociality is possible only among humans and within the realm of culture. Thus, in building up his theory of symbolic interactionism, Mead constructs a seemingly impermeable boundary between culture and nature.

There are individuals who disagree with Mead's conclusion that animals are incapable of symbolic interaction. For instance, in the context of his research on animal abuse and domestic violence, Flynn (2000) argued that animals take on specific roles within the family, that they are perceived as minded and emotional beings by humans, and that they are therefore capable of symbolic interactions. Similarly, in 1931, Theodor Greiger (as cited in Teutsch, 1992, pp. 67–73) published a paper on the animal as a social subject, wherein he argued that whenever there is evidence of "thou-awareness" or intimacy in a human-animal association, a social relationship exists, even if it is not equally understood and equal in all respects. These positions clearly contradict not only Mead's assertion that animals are incapable of symbolic interaction, but also Marx's assertion that animals are incapable of social relations. These assertions by Marx and Mead are based upon their shared assumptions that animals lack consciousness, language, and various other capabilities. By measuring animals against humans, Mead and Marx conclude that the former are deficient. It is this sort of human exceptionalism that must be challenged within the discipline of sociology. It may not be that animals are deficient, but rather that they are different than humans. They may lack a language and consciousness identical to that of humans but, nonetheless, be capable of social relations with each other and with humans, as Flynn and Greiger assert. If this is the case, then there cannot be as rigid a divide between culture and nature as classical—and even contemporary—social theorists would lead one to believe.

Conclusion

It has been demonstrated that, in various ways, notable classical social theorists rigidly dichotomized culture and nature. Although their theoretical contributions to the discipline of sociology are quite different, Marx and Mead made similar contributions to the anthropocentric nature of the discipline: they juxtaposed humans and animals, and drew numerous alleged distinctions between them, such as denying animal consciousness, language, and sociality. The consequence of the acceptance and reinforcement of the culture-nature dichotomy in classical sociological theory is an anthropocentric tendency that continues within the discipline today. As a result, animals are perceived as residing outside of human society, and the speciesist sentiment that animals are unworthy of being included in sociological investigations persists. It is precisely within the discipline of sociology that such investigations should be undertaken; however, whether the discipline can transcend its anthropocentrism remains to be seen. Flynn (2000, p. 125) has remarked: "It will be interesting to see if sociology, which as a discipline has exposed and fought against social inequality based on gender, class, or race, will accept this challenge to end speciesism and include animals in its sphere of study." Arguably, the best place to start such an endeavor is in questioning the anthropocentric assumptions upon which the discipline itself has been built.

Further Resources

Agnew, R. (1998). The causes of animal abuse: A social-psychological analysis. *Theoretical Criminology, 2*(2), 177–209.

Arluke, A., & Sanders, C. R. (1996). *Regarding animals.* Philadelphia: Temple University Press.

Beirne, P. (1995). The use and abuse of animals in criminology: A brief history and current review. *Social Justice, 22*(1), 5–31.

———. (1999). For a non-speciesist criminology: Animal abuse as an object of study. *Criminology, 37*(1), 117–47.

Bryant, C. D. (1979). The zoological connection: Animal-related human behavior. *Social Forces 58*(2), 399–421.

———. (1992). On the trail of the centaur: Toward an amplified research agenda for the study of the human-animal interface. In E. K. Hicks (Ed.), *Science and the human-animal relationship.* Amsterdam: SISWO Publications.

Flynn, C. P. (2000). Battered women and their animal companions: Symbolic interaction between human and nonhuman animals. *Society & Animals, 8*(2), 99–127.

Foster, J. B. (1999). Marx's theory of metabolic rift: Classical foundations for environmental sociology. *The American Journal of Sociology, 105*(2), 366–405.

Marx, K., & Engels, F. (1988). *The economic and philosophic manuscripts of 1844.* (M. Milligan, Trans.). Amherst: Prometheus Books.

———. (1994). The German ideology. In L. H. Simon (Ed.), *Karl Marx: Selected writings.* Indianapolis: Hackett Publishing Company.

———. (1998). The communist manifesto. In L. Panitch & C. Leys (Eds.), *The communist manifesto now: Socialist register 1998.* New York: Monthly Review Press.

Mead, G. H. (1910). Social Consciousness and the Consciousness of Meaning. *Psychological Bulletin, 7,* 397–405.

———. (1956). George Herbert Mead on social psychology. In A. Strauss (Ed.), *George Herbert Mead on social psychology.* Chicago: University of Chicago Press.

Noske, B. (1989). *Humans and other animals: Beyond the boundaries of anthropology.* London: Pluto Press.

Teutsch, G. M. (1992). Traditional sociology and the human-animal relationship. In E. K. Hicks (Ed.), *Science and the human-animal relationship.* Amsterdam: SISWO Publications.

Amy J. Fitzgerald

■ Hunting, Fishing, and Trapping
Falcons, Hawks, and Nocturnal Birds of Prey

The reciprocal relationships between mankind and vertebrate "beasts" had variable "cultural temperatures," depending on the various human historical periods. In some, falconry played a peculiar role, being rather widespread, both in time and geographically.

The first historical evidence of humans using birds of prey for hunting comes from an Assyrian bas-relief dated between 722 and 705 BCE, which was found in the ruins at the ancient site of Khorsabad (Dur Sharrukin, Iraq) during the excavation of the palace of King Sargon II. In China, records describing falconry date back to 680 BCE (Ch'u Kingdom); in Japan, the first record dates back to around 720 CE. In feudal Japan (and in medieval Europe, see below), falconry—or *takagari*—was a popular sport and a status symbol among nobles. Within the samurai class, strict restrictions (based on rank) regulated hunting. The few defensible evidences indicate that this sport probably already existed in Persia and Arabia at a much earlier time than in Japan. Around 400 CE, with the increase of the commercial exchanges—especially between Arabia, Europe, and the Far East—falconry reached the Mediterranean countries.

The historical period between 500 and 1600 CE saw the acme of interest in falconry, with the sport becoming the most popular and revered among all the social classes in Europe. At the beginning of the thirteenth century, the Emperor Frederick II of Hohenstaufen (1194–1250) made popular the art of falconry within the European aristocracies (1983), and falconry became a status symbol in medieval society. A well-quoted text (*Boke of St Albans,* written in 1486 and attributed to Dame Juliana Barnes, or Berners, Prioress of Sopwell nunnery) describes that, in medieval English society, specific bird species were used for hunting in each social rank (e.g., emperor: golden eagle, vulture, and merlin; king: gyrfalcon; prince: female peregrine; duke: rock falcon; earl: peregrine; baron: male peregrine; knight: saker; squire: lanner falcon; lady: female merlin; yeoman: goshawk or hobby; priest: female sparrow hawk; holy-water clerk: male sparrow hawk; knave, servant, children: Old World kestrel).

Falconry is still rather common, as witnessed by the multitude of private, voluntary, and spontaneous associations still enthusiastically active. However, the most intriguing, among contemporary enterprises, remains the Saudi Arabia King (in Arabia, falconry dates back thousands of years, probably much earlier than 720 CE), German (around 600 CE), and British (about 875 CE) associations. For example, about 3,000 falcons are employed for falconry on the Arabian Peninsula each year, with saker falcons (*Falco cherrug*) representing at least 70 percent of the captive population in Arabia. In the United States, we can mention the North American Falconers' Association (NAFA), which was founded in 1961. Various British societies are active, the most prestigious being the Richard III Society, founded in England in 1924, and the British Falconers' Club, founded in 1927 by the surviving members of the Old Hawking Club. Unfortunately, some criminal activities, such as damaging the wild stocks of raptors (mainly due to the smuggling of eggs and illegal capture of nestlings and young-adult birds) represent the dark side of the falconry business. The good side is the high level of veterinary standards established in the past two or three decades (and still constantly rising).

From an ethological perspective, the "training" of a falcon, a hawk, or—more rarely—an owl is a matter of reflection. It in fact appears that, at a variance with training practice for mammals (especially pets), the exploitation of the nocturnal predatory repertoire of raptors (especially the most commonly used species, such as Peregrine, *Falco peregrinus;* Saker, *Falco cherrug;* Gyrfalcon, *Falco rusticolus;* Lanner, *Falco biarmicus;* Goshawk, *Accipiter gentiles;* and Sparrowhawk, *Accipiter nisus*) is under strict control, according to a few main variables. These variables include hunger of the trained subject; familiarity with the trainer; familiarity with the hunting environment; coping with unexpected "extraordinary" events, particularly those evoking medium or high levels of fear; and familiarity with the prey. For this latter, rather relevant variable, early predatory experiences—particularly for birds caught as young adults or adults—are a major determinant. Especially in the past, experienced trainers were able to force some individuals (e.g., sparrowhawks) to prey on very large-sized prey.

The most common species utilized in the falconry practice can be classified as "long-winged" or "short-winged" species, depending on adaptation to their natural hunting environments and prey repertoire. Long-winged species (genus *Falco,* such as peregrine, saker, lanner, and gyrfalcon) mainly hunt other birds in flight, pursuing their prey over considerable distances. The falconer uses long-winged species while hunting in open spaces (such as desert or grassland), so that he can keep the falcon in sight. Short-winged hawks (genus *Accipiter,* such as goshawk and sparrowhawk), hunting their prey from trees, mainly search for ground quarry, such as rabbits, hares, and bird species close to the ground. The falconer uses these birds of prey while hunting in woodlands.

In the last thirty years, a great number of innovations have been introduced in falconry practice. One of these is the creation of hybrid falcons. In order to get the best performance out of the birds exploited for falconry, crossbreeding techniques are developed to produce hybrids that have the best qualities (e.g., survival, size, speed, endurance, and agility) of each purebred captive species. Hybrids are, for the most part, produced by artificial insemination. Today, gyr-saker and gyr-peregrine hybrid falcons are the most common hybrids utilized in falconry. However, hybridization is not limited to falcons: in 2001, Scottish biologists and falconers successfully crossed (by artificial insemination) a male golden eagle (*Aquila chrysaetos*) with a female steppe eagle (*Aquila nipalensis*). Another innovation in falconry practice has been the increasing use of bird-of-prey species with limited natural range, such as Harris's hawks (*Parabuteo unicinctus*), which have excellent agility and a natural tendency to hunt cooperatively with two or more other birds (in contrast, most other raptors cannot work together, as they tend to attack each other). Since about 1980, Harris's hawks have been widely bred in captivity (especially in the United States and United Kingdom), as they have a submissive temperament and are easily trained.

At a variance with both mammals and "lower" vertebrates, the relationship of the raptor with his (or her, more rarely) trainer is a very delicate equilibrium. Raptors are highly fearful, and, especially for young-adult-caught birds, the first impact with the trainer is crucial. For this reason, in the past a trainer would disguise himself as an ox, for example, wearing a dressed hide, with the head or other special mask included. Also crucial is the immediately following period, when the raptor has to progressively adapt to the trainer's presence. Often, the raptor is maintained in semidarkness for appropriate time periods. A common medieval practice was *seeling*—namely, sewing up the eyelids so that the raptor was temporarily blindfolded while it recovered from the initial stress due to interaction with its trainer; raptors' eyelids were progressively unstitched as training advanced.

Hunger is a relevant motivation: a raptor having feed or prey may not respond to the lure, which is a fake prey that it has been trained to feed on and which remains the main link between trainer and raptor. By making the lure more mobile (e.g., by rotating it)—hence, more attractive for raptor—the trainer may regulate raptor flight distance.

It is surprisingly aesthetic to see the trio (raptor, trainer, and lure) in action. It seems that the two minds are maintained in empathic contact by the movements of the lure. But who is really the trainer? Is it the experienced falconer or the natural attention of the hungry bird—never losing sight of the lure, its main and most usual source of food?

In some central Asian areas, such as Kazakhstan, Kyrgyzstan, and Mongolia, golden eagles are traditionally trained and used for hunting wolves, foxes, and gazelles.

In the European tradition, Little owls (*Athene noctua*) were also trained, mainly by boys, as little falcons. The "mobbing" reaction naturally and powerfully elicited by nocturnal birds of prey (their shapes exert a pull on dozens of birds, in direct proportion to their sizes) has been regularly exploited to attract edible or "dangerous" (i.e., competing with humans for eggs, nestlings, or adult quarries) bird species. Hunting for migrating skylarks (*Alauda arvensis*) by trained Little owls was a typical practice in villages and towns rising along their migratory routes.

Therefore, in the European tradition, various owl species, from the large-sized eagle owl (*Bubo bubo*) to the Little owl, were "trained" to move (e.g., around the apex of a baton skillfully handled by the hunter; remote control of the baton movements was also obtained using a pulley system) to be particularly attractive. In fact, a "dancing" live decoy, rotating its head up to 180 degrees or flapping its wings at intervals, elicits a very high level of attraction of diurnal birds. When they approach, the hunter may easily hit them by slings, arches, guns—or catch them using well-positioned nets.

See also

Zoos and Aquariums—*Amphibians and Zookeepers*

Further Resources

Beebe, F. (1984). *Falconry manual.* Blaine, WA: Hancock House.

Beebe, F. L., & Webster, H. M. (2000). *North American falconry & hunting hawks.* Western Sporting: http://www.westernsporting.com/Merchant/

British Falconers' Club: http://www.britishfalconersclub.co.uk/

Dutch Flying Hawk Team: http://www.roofvogels.nl/nederlands/index.htm

Glasier, P. (1998). *Falconry & hawking.* New York: Overlook Press.

Il Portale Italiano di Falconeria: http://www.falconeria.org/

The International Association For Falconry: http://www.i-a-f.org/index.html

Wood, C. A., & Fyfe, F. M. (Trans.). (1983). *Art of falconry; being "De arte venandi cum avibus" of Frederick II of Hohenstaufen.* Stanford: Stanford University Press.

Enrico Alleva and Nadia Francia

■ Hunting, Fishing, and Trapping
Fishing and Human Attitudes

Since the last decades of the last century, human attitudes toward fishing have become more than just a matter of academic interest. They have become critically important from the point of view of the survival of marine, natural environments, and also of human economies that are dependent on fishing. Until quite recently, people had considered fish to be an inexhaustible resource for human use and fishing was viewed as one of the most natural and traditional of human activities, whether for food, trade, or recreation. But increasing evidence about overfishing is changing perceptions. The use of factory ships, trawl nets that are hundreds of meters wide, and large-scale coastal fish farming have made commercial fisheries one of the most intense forms of human exploitation of a living natural resource, with far-reaching environmental and economic consequences. Debate rages between scientists, policy makers, governments, and the fishing industry about how far our attitudes and practices need to change. At the same time, scientists are discovering more about the nervous systems and behavior of fish and providing evidence that, like other vertebrate animals, fish are capable of experiencing pain and fear and are capable of learning and of remembering. These two strands of concern—about the economic and environmental effects of large-scale fisheries and about the welfare of individual fish—are now coming together to start some people to reappraise their views of fish and fishing.

The collapse of the populations of commercial species of fish as a result of overfishing during the last fifty years, the damage to the seafloor by bottom trawling (in effect, plowing up the seabed to dislodge bottom-dwelling fish), and the large-scale wounding and killing of other fish, invertebrates, birds, and mammals that are accidentally caught in trawl nets (so-called by-batch and discards), are becoming better known as important environmental issues. Charles Clover, in his book *The End of the Line,* estimates that around one third of what is caught worldwide is thrown back into the sea, dead or dying, and that the number of animals such as whales, porpoises, turtles, and birds that are damaged but not caught, plus the number destroyed on the seabed, are equivalent to half the tonnage of animals that are hauled from the sea in the nets. This type of evidence that current fishery practices are unsustainable has led many well-informed consumers and restaurants to question, for the first time, where and how their fish has been caught. But it is still probably true to say that most public concern about the ethics of fishing centers on the damage to the marine ecology and the collateral destruction of birds and mammals, and that most people feel less concern about the possible suffering caused to fish by fishing practices, traditional and new.

Our past attitude toward fishing probably owes a considerable amount to our past attitude to fish. Traditionally, fish were seen as food resources that, when alive, were barely, if at all, sentient, and the welfare of individual fish was not a subject for concern. While scientific studies are widening our knowledge, many people are still much less concerned about the potential of human activities to cause suffering to fish than to most other classes of vertebrate animals. Most people are fascinated by the variety and beauty of fish in their natural environment or in aquaria, but we inevitably have less interaction with them than we do with nonaquatic animals. This gives people less opportunity to judge whether they behave as though fish are conscious of their environment and of what happens to them. The commercially important fish that we use for food, for fertilizer in farming, and for feeding to farmed fish in aquaculture are usually invisible to us.

Have we underestimated the potential for fish to suffer from our treatment of them? Recent scientific studies would suggest that we have, but many sport anglers and professional fishermen—and even a minority of scientists—disagree. Despite the scientific evidence, it is still fairly common to hear anglers say that fish feel no pain. According to this view, a fish hooked in the mouth and struggling on the end of a line, or suffocating from lack of oxygen on a boat deck, is behaving in a purely reflexive manner and feels no pain or distress. People who hold this view maintain that fish do not have a brain structure that could produce consciousness, and therefore any of their behavior that may look like attempts to escape, distress, fear, or pain is, in fact, happening without any conscious sensation. On the other hand, it is an important point, in considering public attitudes to fish, that the legal systems of most countries that have animal protection laws do not take the view that fish are nonconscious animals. Fish, as vertebrates, are normally included in animal-welfare laws throughout the world, and are also included in the U.S. federal regulations on the use of animals in research. This inclusion implies that the consensus of scientific and public opinion in these countries is that fish are, indeed, capable of suffering. Some jurisdictions also include certain invertebrates that are caught in fisheries—such as octopuses, squid, crabs, and lobsters—in the scope of animal protection laws, again implying that they can feel pain.

Although the consensus in most developed countries is that fish can feel pain, it has to be admitted that much of the way we treat them in commercial and home life suggests that this attitude is not yet firmly anchored in our society. Most people think very little about the way in which fish for eating are caught and killed in commercial fisheries.

Commercial Catching and Killing

A review written for the guidance of the fisheries department of the New Zealand Ministry of Agriculture and Forestry suggests that many commercial fishing practices could cause considerable distress and, possibly, pain to fish. Bottom trawling, for example, involves dragging a very large bag-shaped net across the seabed. The ship overruns and tires the fish that are swimming ahead of the net to escape it, by outlasting the fish's stamina and glycogen supplies in their muscles. The exhausted fish are forced into the back (cod-end) of the net, where it tapers to a funnel. Here the fish become constricted and start to panic, beating their tails and colliding with each other and with the net, causing skin and tail damage. A large proportion die in the cod-end when their gills become compressed and they asphyxiate; a two-to-four-hour trawl can result in a 29 percent to 61 percent mortality. As the nets are hoisted out of water, the pressure on the fish can be so high that it squeezes out their gut contents. Although the net mesh size is often designed to allow small fish to escape, once they are in the crowded net, they are also likely to be crushed or to suffocate, and a high proportion of the fish that escape are damaged and die later. Gill nets rely on fish swimming into a stationary net and becoming trapped by their gills or gill-cover (the operculum). As they struggle, the net material can cut through their skin. Longlines (often up to 60 km long) are stationary lines carrying many baited hooks. The fish may therefore be held hooked for some time before the line is hauled up. In trolling (where the baited hooks are dragged by a ship) the fish can become entangled in the line or get hooked through an eye or a gill. Some fish species are kept alive in seawater pens on ships after they have been hauled in, where a significant proportion die from their injuries, while other species are left to suffocate in air or are bled to death on the ship. Harpooning is still used to catch some large fish (such as swordfish). It is quite common for fish to be cut open or have their throats cut while they are still alive—and presumably conscious.

Understanding of Sensation and Intelligence in Fish

Some of the ways that we catch, store, farm, and kill fish are hard to reconcile with the theoretical understanding that fish have a nervous system and physiology that makes it very likely that they are capable of feeling pain. Scientific research shows that fish have pain receptors (nociceptors), for example, in their mouths and faces, and that their peripheral nerves that potentially convey pain signals to the brain are similar to those of humans. Although it appears that the mechanism of conscious pain perception in the fish brain is still not understood, fish behave as though they feel pain when they are subjected to a painful stimulus that would cause pain to a human. As with many animals in laboratory experiments, fish can be trained by using electric shocks, and their pain reaction to the shock can be removed by giving them morphine, a substance that blocks pain signals to the brain. Studies by Lynne Sneddon and her colleagues in Scotland have shown that fish perform specifically pain-related behavior; when rainbow trout were injected with either acid or venom in their lips, their gill-beat almost doubled, and they lay on the gravel and rocked their bodies in a similar way to the rocking behavior seen in distressed mammals. The fish injected with acid also rubbed their lips against the gravel or against the sides of the tank. When they were given morphine, their gill-beat and the rocking and rubbing was greatly reduced, suggesting that it relieved their pain.

Pain perception is, of course, only one aspect of consciousness, and it seems likely that we have underestimated the complexity of fish's behavior in other ways, as well. One of the most important is their capacity to learn. As in other vertebrates, learning plays an important role in the development of a fish's behavior, and scientists are becoming increasingly interested in studying it; an issue of the journal *Fish and Fisheries* devoted to how fish learn cited 500 research articles on the subject. From the point of view of these scientists, the fact that fish have small brains should not prejudice our thinking and lead us to assume that they are unintelligent. According to Keven Laland and his colleagues at the Centre for Social Learning and Cognitive Evolution at the University of St. Andrews, Scotland, "Gone (or at least obsolete) is the image of fish as drudging and dim-witted pea brains, driven largely by 'instinct' . . . Now, fish are regarded as steeped in social intelligence, pursuing Machiavellian strategies of manipulation, punishment, and reconciliation, exhibiting stable cultural traditions and cooperating to inspect predators and catch food" (Laland, Brown, & Krause, 2003).

According to these scientists, it is not surprising that fish are capable of learning and memory; these abilities are adaptive in that they allow fish to improve their success in foraging for food and avoiding predators. Research also shows that fish recognize other familiar fish in their shoals and are aware of the social status of others. Similarly, many fish respond to the alarm cues emitted by fish of different species when a predator appears, and this is part of the way that they learn to recognize predators. The neural basis of their learning has been found to be very similar to the cognitive and neural processing in land vertebrates.

The rapid development of scientific knowledge about the cognition and behavior of fish is challenging the old view that they are cold-blooded semi-robots without conscious awareness. In particular, the evidence that fish can feel pain—and try to avoid it—now seems to be clear. How far have these findings changed human attitudes to fishing? The answer seems to be: rather little as yet. They may at least lead to questions of some of the common practices in fisheries, which have not yet changed in recognition of new understanding of the sentience of fish and other marine animals.

See also

Conservation—*Finland and Fishery Conservation Issues*
Hunting, Fishing, and Trapping—*Sport Fishing*

Further Resources

Advocates for Animals. (2005). *Cephalopods and decapod crustaceans: Their capacity to experience pain and suffering.* Edinburgh: Advocates for Animals. (download at http://www.advocates foranimals.org.uk)

Clover, C. (2004). *The end of the line: How overfishing is changing the world and what we eat.* London: Ebury Press.

Kurlansky, M. (1997). *Cod: A biography of the fish that changed the world.* New York: Walker Publishing Company.

Laland, K. N., Brown, C., & Krause, J. (2003). Learning in fishes: From three-second memory to culture. *Fish and Fisheries, 4,* 199–202.

Lymbery, L. (2002). *In too deep: The welfare of intensively farmed fish.* Petersfield: Compassion in World Farming Trust. (download at www.ciwf.org.uk/publications)

Sneddon, L. U. (2003). The evidence for pain in fish: the use of morphine as an analgesic. *Applied Animal Behaviour Science, 83,* 153–62.

Jacky Turner

■ Hunting, Fishing, and Trapping
Humans as the Hunted

Our primary relationship with animals during the vast 7-million-year period that hominids (the taxonomic family which includes modern humans and their fossil relatives) have walked this planet has been one of predator-prey interaction. A wealth of data confirming *Homo sapiens* and our extinct ancestors as the *prey* in these human-animal relationships is only now coming to light.

Ancestral humans as prey directly contrasts with a widely held belief in Western culture that our species has always been the dominant form of life, since standing up on two legs and bipedally striding away from our primate relatives. "Humans as Hunters" is the pervasive and popular scenario for human evolution, accepted both by the general public and many in the scientific community. The idea that human nature originated from hunting, violence, and cannibalism was proposed by Raymond Dart, a South African anatomist who found the first fossil of an African "ape-man" in the 1920s; his theory gained momentum in the 1960s, through best-selling books such as *African Genesis* by Robert Ardrey. Since then, everything from our upright bipedal stance to monogamy, territoriality, tool use, technology, and male aggression has been linked to this hunting paradigm. Yet, when the scientific evidence is objectively analyzed, there is no support for "Man the Hunter" as a valid evolutionary theory. In fact, the first unequivocal evidence for large-scale, systematic hunting, according to archaeologist Lewis Binford in *The Cambridge Encyclopedia of Human Evolution,* is available from paleoarchaeological sites possibly only 60,000–80,000 years old—in excess of 6 million years after the first hominids appeared.

Two kinds of evidence are appropriate for attempts to reconstruct the ecology and behavior of our earliest ancestors. The most important and legitimate body of evidence consists of the fossil record—the actual skeletal remains, plus tools and footprints left by early hominids, as well as fossil materials that give us clues concerning the environment (other animals, plants, water sources) in which these creatures lived. Secondary evidence that is less reliable but, nonetheless, offers insights for the reconstruction of early hominid lifestyles centers on the behavior of nonhuman primates living under similar ecological conditions to those of our earliest ancestors.

Our research has substantiated that nonhuman primates, including our closest genetic relatives, the chimpanzees, undergo substantial predation, particularly from large cats such as lions and leopards. Other primate species that inhabit edge and savannah zones—the environments of early humans—may sustain predation at nearly the same rates as more typical prey such as antelopes and gazelles. We compiled an auxiliary corroboration of this mode of investigation by looking at current predation levels on modern humans living in areas where predators have not been eradicated. For example, in the Sundarbans delta between India and Bangladesh, 612 people were killed by tigers between 1975 and 1985. Crocodiles are estimated to seize approximately 3,000 people per year in sub-Saharan Africa. In Uttar Pradesh, a state in northern India, leopards killed 95 residents and wounded 117 during the decade 1988–98. Chinese biologists contend that 1,500 farmers are killed annually by brown bears on the Tibetan Plateau. According to Hans Kruuk, the carnivore zoologist who wrote *Hunter and Hunted: Relationships between Carnivores and People,* wolf predation is still a fact of life in rural parts of eastern Europe, and local people wonder that no one in the West is aware of the human toll taken by wolves. Black bears, lions, cougars, spotted hyenas, Komodo dragons, and pythons are other species implicated in human predation.

Fossil evidence gives vigorous support to the theory of early humans as prey. Alan Turner, in *The Big Cats and Their Fossil Relatives*, estimates there were many more large carnivores during the period of human evolution than exist today, and we know from their fossilized remains that these carnivores were larger in size than the living members of the cat, dog, and hyena families. The incontrovertible fossil evidence supporting humans as prey to large carnivores spans sites in western Europe, eastern Europe, Asia, and Africa with a temporal record that stretches from 6 million years ago up to 250,000 years ago. The oldest fossil that exhibits marks of predation is a 6-million-year-old hominid discovered in 2001 by Brigitte Senut (and fellow paleontologists from the Museum National d'Histoire Naturelle, Paris) in the Tugen Hills of Kenya. Senut announced in an interview on the Essential Science Indicators Web site that several bones of this fossil hominid exhibit carnivore tooth marks and suggested that the individual may have been eaten by a leopard-like animal. Also unearthed in Kenya was a 900,000-year-old hominid skull found at Olorgesailie by Richard Potts and a Smithsonian Institution team. The Olorgesailie skull shows bite marks along the individual's browridge; in a report on MSNBC.com, Potts hypothesized that the individual was mauled and killed by a lion or other carnivore.

Innovative research by C. K. Brain, a South African expert in taphonomy (the process of how fossils come to be), connected a group of hominids who lived 1.8–1.0 million years ago, known as robust australopithecines, to predation by extinct relatives of the leopard. In his book, *The Hunters or the Hunted?*, Brain described how he linked two round holes in the back region of a robust australopithecine cranium, discovered in a South African cave, to the huge fangs of an extinct leopard. He knew that leopards, both millions of years ago and now, drag their prey into trees to protect the carcass from being stolen by other carnivores. In dry regions, trees usually grow near water sources such as sinkholes; slowly, over time, the sinkholes developed into caves that hid the remains of many bones falling from trees as leopards ate their prey.

A large number of saber-toothed cat species lived from approximately 8.5 million years ago to as recently as 10,000 years ago and were undoubtedly fierce predators of hominids. David Lordkipanidze and other scientists from the Republic of Georgia (featured in a 2002 *National Geographic* report) unearthed a 1.75-million-year-old site under the medieval city of Dmanisi. Besides finding fossilized remains of two saber-toothed cat

Drawing of several species of extinct cats as expected to appear, based on uncovered fossils. ©Christina Rudloff.

species, along with wolves, bears, and leopards, the Georgian scientists retrieved one hominid skull which exhibits holes corresponding to saber-toothed cat fangs and one hominid lower jaw revealing gnaw marks from a large carnivore.

While saber-toothed and other cat species flourished when early hominids walked the planet, as a group they never reached the extent or diversity of the hyena family. Over one hundred species of hyenas existed during the time of hominid evolution! These hyenas were mega-carnivores; some were gigantic in size, some were fast running, pack-hunting animals, and some had such immensely powerful jaws that they are called "bone-crunchers." The extinct hyena species that may have had a long-term deadly relationship with many of our ancestors was *Pachycrocuta brevirostris*—a short-faced, huge-jawed, 440-pound predator that hunted in packs. In 2001, Noel Boaz and Russell Ciochon wrote in *Natural History* about their reinterpretation of hominid remains from the Zhoukoudian cave in China, in light of research on this extinct predator. In the 1920s, Zhoukoudian cave was found to contain the skulls of forty-five *Homo erectus* individuals that had been grotesquely disfigured. These fossils of our direct human ancestors, dated at about 450,000 years, all had facial areas removed and the bases of their skulls enlarged. Chinese and German paleontologists attributed the manipulations to cannibalistic activity on the part of *H. erectus* seeking to remove the brains of fellow hominids. The theory of cannibalism as a lifestyle of our ancient ancestors was accepted as fact until Boaz and Ciochon found evidence that the Zhoukoudian cave cache was really the lair of *Pachycrocuta*. The cave's hominid crania exhibit all the signs of biting, chewing, and skeletal manipulation of a modern-day hyena attacking the face of primate prey to get at the tongue and the lipid-rich brain. Two other fossil sites illustrate widespread hyena predation on early humans. The oldest hominid fragment from western Europe consists of a small—but controversial—piece of skull found when a *Pachycrocuta* den was excavated at Orce, a 1.6-million-year-old site near Granada, by a Spanish archaeologist named José Gibert. Testing for the presence of human albumin in the fragment confirmed its hominid status, an important consideration since the Orce fragment indicates that western Europe may have been colonized by early humans much earlier than previously estimated. Far away from Orce in time and space is a skull found at Florisbad, South Africa. Hilary and Janette Deacon described how the 250,000-year-old cranium exhibits the clear impression of a hyena canine tooth on its forehead in their book, *Human Beginnings in South Africa*.

Raymond Dart, mentioned earlier, did not realize that his most famous fossil, the "Taung" child (named after the quarry in which it was unearthed), was a victim of predation. Eighty years after Dart's discovery, researchers minutely examined the 2-million-year-old skull of this juvenile australopithecine and judged it likely to have been killed by a huge eagle. Birds of prey, just like carnivores, might also have been larger and more rapacious in the past than the living specimens of today. Only the harpy eagle of South America and the African crowned hawk-eagle give us some indication of the type of raptors that might have preyed on young hominids. The crowned hawk-eagle, in particular, is known to take prey nearly five times its own weight. It is also the only bird of prey that has been witnessed attacking a young human in what appeared to be a predation attempt. Lee Berger and Ron Clarke of the University of Witwatersrand in South Africa first proposed, in 1995, that the Taung child was an ancient victim of raptor predation. Recent research on eagle predation carried out in the Tai Forest, Ivory Coast, by W. Scott McGraw, Catherine Cooke, and Susanne Schultz support this contention. Skulls of forest monkeys that fall victim to crowned hawk-eagles are usually left intact, but intense manipulation is exhibited by nicks, lacerations, and perforations inflicted during the eagles' efforts to get at the brain

A reconstruction of a leopard biting a Homo erectus. ©Russell Ciochon.

through thin facial bones. The great pressure of the talons leaves signature marks; the sharp talons can move through thin bones the way a can opener slices through steel, creating a small bone flap with each turn of the talon. When the findings of the Tai Forest study were applied to the Taung child, the same markings were visible on the ancient skull as those found on recent monkey kills.

Based on the remarkably complete skeleton of the fossil australopithecine named "Lucy" who lived between three and four million years ago, early hominid ancestors were medium-sized primates, with the females weighing approximately sixty pounds and males weighing about thirty to forty pounds more. They did not possess inherent weapons, such as teeth or claws, to fight off the many predators that shared their habitat, and because they lived in edge environments that incorporate open areas and wooded forest near rivers, they—like other primates alive today—were very vulnerable to predation. According to C. K. Brain, rates of predation were likely just as high in our early ancestors as they are in modern species of primates. Because of these predation rates, it is probable that hominids utilized certain strategies to protect themselves, strategies that are similar in many ways to modern primate species living in the same types of environments. Based on living primate models, these strategies were probably focused on group living (safety in numbers because more eyes and ears are alert to the presence of predators), social and interindividual cooperation, complex communication of alarm vocalizations that eventually evolved into language, versatile locomotion (use of trees and the ground), selection of sleeping refuges where group members came together at night in tall trees or cliffs, multiple protective males acting as expendable sentinels, and increased brain-to-body ratio resulting in cognitive skills that endowed our ancestors with the ability to monitor their environment, communicate with other group members, and sometimes stay one step ahead of predators by implementing defenses before the predator had launched a full-fledged attack.

Many circumstances have been proposed as a catalyst for the evolution of humans—competition for resources, intellectual capacity, male-male conflict, and hunting. The evidence, however, at this point seems to suggest that predation pressure—our interactions with other, larger, more ferocious species—may have played a major role in molding the human species.

Further Resources

Berger, L., & Clarke, R. (1995). Eagle involvement in accumulation of the Taung child fauna. *Journal of Human Evolution, 29,* 275–99.

Binford, L. (1994). Subsistence—a key to the past. In S. Jones, R. Martin, & D. Pilbeam (Eds.), *The Cambridge Encyclopedia of Human Evolution* (pp. 365–68). Cambridge: University of Cambridge Press.

Boaz, N., & Ciochon, R. (2001). The scavenging of "Peking Man." *Natural History, 110,* 46–51.

Brain, C. (1981). *The hunters or the hunted?* Chicago: University of Chicago Press.

Conroy, G. (2005). *Reconstructing human origins: A modern synthesis.* New York: W.W. Norton.

Deacon, H., & Deacon, J. (1999). *Human beginnings in South Africa: Uncovering the secrets of the Stone Age.* Walnut Creek, CA: AltaMira Press.

Fox, M. (2004 July). *Was pre-human a failed 'experiment'?* Retrieved from http://www.msnbc.msn.com/id/5343787/

Gibert, J., Campillo, D., Arques, J., Garcia-Olivares, E., Borja, C., & Lowenstein, J. (1998). Hominid status of the Orce cranial fragment reasserted. *Journal of Human Evolution, 34,* 203–17.

Gore, R. (2002). New find: The first pioneer? *National Geographic* 8/02 [addendum, no page numbers].

Hart, D., & Sussman, R. (2005). *Man the hunted: Primates, predators and human evolution.* New York: Westview Press.

Klein, R., & Edgar, B. (2002). *The dawn of human culture.* New York: John Wiley & Sons.

Kruuk, H. (2002). *Hunter and hunted: Relationships between carnivores and people.* Cambridge: Cambridge University Press.

McGraw, W., Cooke, C., & Shultz, S. (2006). Primate remains from African crowned eagle (*Stephanoaetus coronatus*) nests in Ivory Coast's Tai Forest: Implications for primate predation and early hominid taphonomy in South Africa. *American Journal of Physical Anthropology, 131,* 151–65.

Senut, B. (2001 December). *Fast breaking comments by Dr. Brigitte Senut.* Retrieved from http://www.esi-topics.com/fbp/comments/december-01-Brigitte-Senut.html

Turner, A. (1997). *The big cats and their fossil relatives.* New York: Columbia University Press.

Donna Hart and Robert W. Sussman

■ Hunting, Fishing, and Trapping
Hunting Beliefs

It is typically said that hunting is a basic or inherent part of human nature, in that it is natural or instinctual, that humans are predatory animals and have hunted for millions of years, and that hunting is necessary to reduce the number of animals and to prevent habitat destruction, as well as to save animals from starvation. Hunters deny that they are killers and insist that hunting is something more than simply killing. Some believe that these themes, not always clearly delineated or explained, often are more like a cluster of ideas where one idea is used to corroborate another.

Hunting is an Inherent Part of Human Nature

Some writers say that hunting is an inherent part of human nature. A particular version of this assertion holds that hunting is a basic drive or an instinct, although the exact meaning of *instinct* varies and is often neither consistent nor clear. Such authors argue that because drives or instincts are not a matter of choice, moral judgments are inapplicable.

If hunting *was* an instinct, it would be repeated without variation and would be common to all members of the species. Neither of these characteristics of an instinct apply to sport hunting. And even if hunting fulfilled basic biological or psychological needs, it would still be possible to reject it, for humans are capable of repressing drives for the sake of achieving more important goals.

The fact that hunting methods and practices vary—that hunting is thoroughly ritualized—shows that it is governed more by culture than by basic drives or instincts. What is hunted, who hunts, how the hunt takes place, and when and where the hunt occurs depends upon the culture. Because we can choose whether or not to take advantage of this cultural opportunity, moral judgments are applicable.

Humans have Hunted for Millions of Years

Many authors claim that humans have hunted for millions of years and, indeed, that hunting was the spur to our evolutionary development as modern *Homo sapiens*. They claim this despite the fact that researchers involved with human origins have fossil evidence to reject this idea and label it a myth.

Even if these scientists are mistaken and protohumans survived only by hunting, it would not be sufficient to show that *recreational* hunting was an inherent part of our nature. Furthermore, slavery, torture, war, and rape may have existed for millions of years, but few would assert that this long history is a sufficient reason to claim that such practices are inevitable. Similarly, there is some evidence that early humans were cannibals, but those who want to justify hunting on the basis of prehistory conveniently ignore this—as they should, for if there is one human trait that is outstanding, it is that human behavior is extremely flexible.

Humans are Natural Predators

The argument that humans are predators might be considered part of the more general argument that hunting is natural, but to liken human hunting to predation requires that we ignore a number of facts. Biological or physiological predators—mammals, raptors, fish, and insects—all have lethal weapons, such as cutting teeth, sharp beaks, claws or talons, and paralyzing or poisonous stings, with which to kill their prey. Lacking these weapons, humans do not share a predator's basic physical or biological characteristic. Thus equating hunters to natural predators is a myth. One could claim, however, that humans are cultural predators: we have used our brains to produce lethal weapons. The production of such weapons is not, however, the result of an inevitable biological heritage, but rather a conscious choice—and thus is open to praise or blame.

Somewhat similarly, it is sometimes argued that hunters must duplicate the role played by the "biological predators" they have killed. Predators, even opportunistic predators, primarily kill the sick, the old, and the young—the very animals that hunters seek to avoid. Furthermore, since prey animals tend to hide illness, it is thought that humans lack the discrimination necessary to spot them. Researchers have concluded that hunters probably could not replicate the actions of natural predators, even if they wanted to.

Hunting is Necessary

Those who argue in favor of hunting often assert that hunting is necessary because (1) humans have to eat, (2) an overpopulation of animals will destroy the environment, and (3) an overpopulation will cause the animals themselves to suffer and starve. First, humans have to eat, but we do not have to eat meat—as is proved by the often-long and healthy lives of many vegetarians. Second, the claim that in the absence of predators, prey populations will skyrocket and destroy their habitat if humans do not "manage" them rests on the false belief that predators control prey numbers. The reverse is true: the number of prey animals controls the number of predators. This makes sense, for prey are food for predators and food is one of the main factors that influence the number of animals that will be born or that can survive. Research has also shown that in the absence of predators, prey populations stabilize their numbers by what may be called *intrinsic factors* or *self-regulatory mechanisms*. Third, sometimes the myth of the necessity of hunting is expressed in terms of animal welfare: the hunter must kill the animal so that it does not suffer or starve. This argument, resting on the belief that it is kinder to kill the animal with a bullet or an arrow than to let it starve to death, involves the belief that hunted animals are not wounded, but die instantly and thus humanely. It is also dubious that hunting will prevent animals from starving. Animals may starve for a number of reasons. For example, deer born late in the year may starve during the winter, regardless of hunting pressures.

Recreational Hunting Reduces Animal Numbers

Nothing would seem more commonsensical than to believe that recreational hunting reduces animal numbers. Killing animals reduces the total number. Such a seemingly obvious assertion would seem difficult to argue with, but knowledge of population dynamics explains why recreational hunting as it is practiced in the United States does not reduce numbers.

Many mammals possess innate mechanisms that push reproduction to a maximum when a large number of the population is removed, but the food supply remains constant: multiple births may increase, and the age at which females breed for the first time may be lowered. We can generalize these intrinsic mechanisms: with a sufficient food supply or low density, reproduction will increase and mortality will decrease; with a high density and increased competition for food, reproduction decreases and mortality increases. These mechanisms explain, for example, how a deer herd can stabilize itself in the absence of predators or how a herd can increase despite hunting pressures.

Commercial hunting, as opposed to recreational hunting, can, of course, reduce numbers or even cause extinction. There are numerous examples: the passenger pigeon no longer exists due to commercial over-hunting, while bison in the United States were almost driven to extinction.

Hunting is More Than Simply Killing

Although many hunters acknowledge the thrill of hunting, the rush of adrenaline—sometimes called *buck fever*—they deny that hunting is motivated by bloodlust or that they are mere killers. Many hunters talk about being one with nature, and this may well reflect what the hunter thinks or imagines, but it seems unlikely that the modern hunter, with his high-tech bow or gun, razor-tipped arrows, camouflage clothes, portable tree stand, and so forth, can consider himself a rival to the hunted animal and can at least temporarily return to a simpler time. In this justification, modern hunters have to believe that their lives are threatened, that they are hunting a fierce or wily animal, that the odds in their favor are severely limited, and that the animal has some sort of "fair chance." Unless protecting her young, in most cases the hunted animal does not threaten the hunter's well-being and seeks only to flee. If discovered, often the only chance the animal has is the incompetence of the hunter.

None of these justifications for hunting seem to hold up under rational scrutiny. Yet they are repeated again and again. Moral judgments apply to hunting, just as they do to many other human activities that are a matter of choice. The only rational basis for hunting would rest on the idea that might is right—an idea that few people find acceptable.

Priscilla N. Cohn

■ Hunting, Fishing, and Trapping
Hunting Recruitment Initiatives

Recently there has been a perceivable recruitment effort to increase participation in sport hunting (not to be confused with hunting undertaken strictly for sustenance). The major reason for this effort is, undoubtedly, recognition that the sport-hunter population and the influence thereof are declining, at least partially due to the efforts of the animal rights movement. According to the Michigan Hunting and Fishing Heritage Task Force (2007), the number of hunters in that state peaked in the late 1980s and has shown declines and failure to grow since, reflecting trends in the rest of the country. Numerous interests are negatively affected by the decline in the sport-hunting population, such as weapons suppliers and manufacturers, hunting-gear suppliers and manufacturers, sportsmen's clubs and groups, the hospitality industry in popular hunting getaway areas, farmers selling crops commonly used to bait animals (where legal), and, finally, state governments, which receive revenue from hunting licenses. These primary groups have a vested interest in ensuring that the number of hunters increases (or at least ceases to decline), so that the revenue generated by the activity continues to flow in, and favorable hunting policies prevail.

There have been two noticeable efforts to actualize the goal of increased participation. The first effort consists of getting populations traditionally not involved in hunting interested in it. The largest population excluded from hunting has historically been women, and there has recently been a noticeable drive to recruit them; however, to a lesser degree, there has also been a move to include other traditionally excluded populations, such as disabled individuals and members of minority groups. The second effort entails recruiting prospective hunters when they are young, resulting in efforts directed at youths. A detailed examination of these initiatives everywhere they are being implemented is not possible here. Consequently, this essay uses the state of Michigan (which is third in the nation in the number of licensed hunters and has clearly articulated its recruitment plans) as a case to provide an overview of these recruitment initiatives. As will be demonstrated, this is a case of state-sponsored intervention in human-animal relationships.

Prior to examining the specific initiatives, it is necessary to provide background information on why they have been deemed necessary in the first place. One of the best articulations of the need to recruit new hunters is found in the *Report of the Hunting and Fishing Heritage Task Force* (commissioned by Governor Engler (R) in 1995). The first lines of the report read as follows: "Studies have proven that there is a noticeable decline in the percentage of the population taking part in hunting and fishing in this state and that similar trends are mirrored in other parts of the country. A collective effort has arisen here and elsewhere to make the citizenry aware of the important historical perspective of our outdoor heritage." To this end, the Hunting and Fishing Heritage Task Force "will establish a plan to promote fishing and hunting, promoting the historical importance of the outdoor sportsperson as conservationist" (Hunting and Fishing Heritage Task Force, 2007).

Included in this report were several recommendations made by the Task Force, the first of which was the creation of an Information and Education Division of the Michigan Department of Natural Resources, which was subsequently established in 1997. The objective of this new division was for the Department of Natural Resources (DNR) to "develop and implement an *aggressive marketing strategy*. . . . and thereby encourage more citizens to become involved in these outdoor activities" (Hunting and Fishing Heritage Task Force, 2007, emphasis mine). The Task Force also recommended securing adequate funding for the new DNR Information and Education Division. This initiative was, therefore, considered important enough to create an entirely new division of the DNR and specify that funding is dedicated to it. A further recommendation of the Task Force was the establishment of the Hunting and Fishing Heritage Defense Fund. The stated purpose of this fund is to defend and protect the future of hunting, fishing, and trapping.

The funding of the Hunting and Fishing Heritage Task Force, the new Information and Education Division of the Michigan DNR, and the Hunting and Fishing Heritage Defense Fund can be considered an investment in maintaining the hunting population and securing continued revenue from them. According to Mertig and Matthews (1999), "Natural resource agencies are demonstrating increased concern with attracting women into natural-resource-based recreation. This concern has been fueled in part by a slow down in the recruitment of new hunters and anglers . . . *a significant source of agency funding*" (494, emphasis mine). In fact, the Michigan DNR has a very large financial incentive to ensure that the hunting population ceases to drop. For instance, their budget for the 2001–02 fiscal year was $250.1 million, $43.6 million of which came directly from hunting and fishing licenses. Therefore, the Michigan DNR received just under 20 percent of their budget for the 2001–02 year directly from these licenses. The revenue generated by hunting is even greater when one takes into account miscellaneous revenue generated by hunters, such as that in the form of state park camping and entrance fees.

The existence of the task force, and the actions taken as a result of their recommendations, demonstrates quite clearly that tax dollars are being used to bolster the hunting population. The following statement from the report also demonstrates that this new mandate is at least partially in response to the influence of the animal rights movement, which advocates alternative human-animal relationships to that of predator-prey: "The

Hunting and Fishing Heritage Task Force acknowledges the threat of the organized anti-hunting and the animal rights movement to our hunting and fishing heritage" (2007). Aware of the continued threat of the animal rights movement, the declining number of hunters, and the associated decline in their revenue, the Michigan DNR has instituted policies aimed at what it refers to as "protecting hunting heritage." Recruiting more people into the activity appears to be their chosen means to that end. The recruitment of populations currently underrepresented in sport hunting has been an especially strong focus of the interested parties.

The largest group historically excluded from participation in sport hunting is women. The Michigan DNR has taken action to recruit from this large pool by establishing a local chapter of Becoming an Outdoors-Woman (BOW). The program, designed specifically to recruit women into hunting, was established in 1991, and within five years of its inception, approximately 20,000 women had participated in BOW workshops (Stange, 1997). In recognition of the potential gains, the Hunting and Fishing Heritage Task Force recommended developing partnerships between DNR-sponsored programs, such as BOW and similar groups. The Michigan DNR and the Task Force appear to consider the BOW program a particularly valuable recruitment tool, useful in targeting specific underrepresented groups, and a brief examination of the organization demonstrates why this is the case.

BOW undertakes recruitment in a variety of ways. They hold more than eighty general workshops per year in the United States and Canada, wherein women learn to hunt, among other activities. BOW is also targeting specific types of women who have been particularly underrepresented in the sport, such as women of color, low-income women, and women with disabilities. To this end, they have developed a series of conferences, referred to as "Breaking Down Barriers" conferences. In 1999, the conference specifically addressed the lack of participation by minority and low-income women. The title of the conference was "Introducing Women of Color and Low-Income Women to Natural Resource Based Recreation: Barriers and Strategies." As a result, they organized pilot workshops in three different states, targeting minority and low-income participants. Furthermore, in at least one state, BOW is planning instructor training for minorities. They have also employed the strategy of diversifying their publicity images of women to include racial/ethnic minority women (which is demonstrated on its Web site) and have made a point of inviting racial/ethnic minority women in person to join them in order to make them feel welcome. The Breaking Down Barriers conference in 2002 targeted women with disabilities, and, afterward, BOW decided to form an Accessibility Committee to implement pilot workshops to test the strategies and plans that were identified by the conference participants. All of these programs serve at least two functions: they not only serve the purpose of increasing the participation of women in hunting; they also serve to improve the public image of the sport, which is viewed as reaching out to racial/ethnic minorities and low-income and disabled women.

One noticeable characteristic of these workshops and other tactics of recruiting women is that they generally appeal to perceived differences between men and women, especially male and female hunters. This is evidenced by the descriptions of programs designed to introduce women to hunting, such as: "The instructors and staff are patient and supportive. The participants share in the success of each group member. This is a noncompetitive situation where each individual can learn at her own pace. . . . The emphasis is on the enjoyment that goes with the social side of outdoor activities" (www.michigan.gov/dnr). Furthermore, in the description of an upcoming BOW pheasant hunt, participants are told to bring their favorite pheasant recipes and advised that much time will be spent on cooking and tasting. This emphasis on women's "unique" concerns is also illustrated by the following workshop alumni quotes that BOW chose to put on their homepage: "There wasn't one person who felt intimidated by their lack of knowledge" and "I can't believe what this weekend did for my self-esteem." It is difficult to envision such descriptions of programs aimed at men.

Not surprisingly, private interests have also become involved in recruiting female hunters and capitalizing on their increased participation. According to Stange, "Industry spokespeople and wildlife agency personnel have simultaneously taken to characterizing women as the 'future of hunting,' given the rise in female-headed single-parent households and the shift in American demographics from a rural/suburban to an urban focus" (Stange, 1997, 178). Merchandisers have been actively targeting this new sector of the market, increasingly offering hunting and angling clinics for women and selling hunting gear and clothing for women. Additionally, outfitters are offering all-women trips and safaris to developing countries (Stange, 1997, 178).

A brief examination of the efforts of one company further demonstrates how they are marketing to women and emphasizing perceived differences between women and men. Cabela's is a company that markets and sells hunting gear. Although it may not have the political interests that the Michigan DNR and BOW do, the company nonetheless has similar economic interests, since it would benefit from an increase in the hunter population and being able to sell products to women (and youths). Cabela's not only has hunting clothing available in women's sizes—it has an entirely separate catalog devoted to women. However, an examination of Cabela's catalog reveals that the selection in women's sizes is much more restricted and the depictions of women and their clothing are very different than those of men. For instance, the description of men's sizes of Cabela's Boar Hide Field Pants and Chaps says they are "plenty tough to take on thorns, briars or brush" (2002a, 116), whereas the women's size apparently "protects your pants and legs from brush and other hazards" (2002b, 116). So, whereas the men's are "plenty tough," the women's are described as protective. Further, new items have been designed specifically for women, such as Cabela's Military Weight Polartec Power Stretch, which "are equipped with Cabela's exclusive QRS (Quick Release System) so completely disrobing is unnecessary when nature calls. Seams are flat-lock stitched to prevent chafing" (2002b, 127). Further, in Cabela's women's catalog, items not present in the general catalog are featured and marketed towards women, such as jewelry; "The Wild Side of the Kitchen" product line; and other accessories, such as hair dryers and curling irons that plug into car lighters, pads to go under bra straps to protect women from gun recoil, and leather cheek pads to protect cheeks from recoil.

There has clearly been a confluence of interests in increasing the participation of women in sport hunting. Private companies, such as Cabela's, are anxious to exploit this increasing market. And, as demonstrated, the state of Michigan also has financial interests in this regard, as well. They have consequently established a local chapter of BOW. The Hunting and Fishing Heritage Task Force additionally recommended in its report "expand[ing] upon the theme of 'Becoming an Outdoors-Woman' to include single-parent households, young women and girls, and other targeted groups" (Hunting and Fishing Heritage Task Force). The purpose of recruiting women into sport hunting is not only to increase their own participation, but it also has a recognized trickle-down effect: "Women are often the vectors for family participation in outdoor recreation—a key ingredient in the successful recruitment of youths into outdoor activities" (Mertig and Matthews, 1999, 494). More specific ways of recruiting youths have also been developed.

In addition to recruiting youths indirectly via the recruitment of their mothers, the Michigan DNR has created direct incentives for sport hunters to include youths—as young as twelve—in their hunting activities. For instance, the state has implemented daily priority drawings for hunting parties that have at least one youth, meaning that if a group of adults has at least one youth with them, they may be eligible to hunt for special game and at times generally not permitted.

In addition to giving special hunting opportunities when youths accompany hunting parties of adults, the state has also created special youth hunting seasons that take place prior to the beginning of the regular hunting season. For instance, there are specially designated youth hunting weekends in September for waterfowl and deer. These preseason hunting weekends give youths an opportunity to hunt before other hunters are in the woods, giving them greater access to game animals, and, the hope is, increased interest and future participation in the sport.

In addition to these special hunting opportunities implemented by the DNR, the Hunting and Fishing Heritage Task Force recommended creating outdoor, hunting, fishing, and trapping education programs that could be utilized by schools, clubs, associations, and other children's groups. As a result, in 1999, the Youth Education Outreach Curriculum was implemented. The LAP program (Learn from the past, Appreciate the present, and Preserve our outdoor heritage) was also developed, and, according to the Michigan DNR's Web site, Michigan fourth-grade teachers use this program with their students. Thus, even those children who do not come from hunting families are being introduced to sport hunting as a way to "preserve their heritage."

The emphasis on preserving hunting heritage and the recruitment initiatives undertaken by the state are the result of the growing realization that the hunting population's spending and political power are steadily declining, at least partially due to the influence of the animal rights movement. According to Lyle Munro, "in liberal democracies, a strengthened public good will toward animals has *compelled opponents of Animal Rights to adopt novel tactics* in their campaigns to defend the use of animals" (1999, 3, emphasis mine). This essay has demonstrated that, in at least one state, the government has become involved in developing such "novel tactics" to defend and preserve the heritage of—and revenues from—hunting animals for sport. Their preferred tactics have been clearly articulated in the recommendations made by the Hunting and Fishing Heritage Task Force. Although over time the specific tactics might change, it is safe to assume that the effort to recruit underrepresented populations and youths into sport hunting will continue, if not intensify. For the foreseeable future, the struggle for survival in sport hunting will continue for those at both ends of the gun.

Further Resources

Becoming an Outdoors-Woman: http://www.uwsp.edu/cnr/bow

Cabela's, Inc. (2002a). *Cabela's master catalogue fall*. Edition II. Sidney, NE: Cabela's Inc.

———. (2002b). *Women's outdoor fall*. Sidney, NE: Cabela's Inc.

Hunting and Fishing Heritage Task Force. (2007). http://www.michigan.gov/dnr/0,1607,% 207-153-10371_14724-35372—,00.html

Mertig, A., & Matthews, B. E. (1999). *Women at play: Examining the resource-based recreational activities of Michigan women*. Paper presented at the American Wildlife and Natural Resources Conference, Session 6: Women at Play.

Michigan Department of Natural Resources. (2007). http://www.michigan.gov/dnr

Munro, L. (1997). Framing cruelty: The construction of duck shooting as a social problem. *Society and Animals, 5,* 137–54.

———. (1999). Contesting moral campaigns against animal liberation. *Society and Animals, 7,* 35–53.

Stange, M. Z. (1997). *Woman the hunter.* Boston: Beacon Press.

Amy J. Fitzgerald

■ Hunting, Fishing, and Trapping
Man-Eating Tigers

Despite their size and hunting prowess, tigers seldom attack people. Hunters such as Jim Corbett, hired to exterminate the rare tigers that turned into man-eaters, almost invariably discovered that such tigers suffered some sort of disability—usually a previous gunshot wound that prevented the big cat from hunting natural prey such as wild boar, deer, or monkeys. Very rarely do healthy tigers attack people. In these instances, the victim is usually a child or a woman squatting while washing. Experts suggest this may be a simple mistake: the tiger may assume a shorter-looking-than-usual human is a monkey.

Only in one place on earth do healthy tigers regularly hunt and eat humans. In Sundarbans, a 10,000 square kilometer mangrove swamp along the Bay of Bengal in Bangladesh and India's West Bengal, some 500 tigers hunt and eat as many as 300 people a year. (The exact number is unknown because most of these deaths go unreported; Indian and Bangladeshi forestry department officials provided this estimate.)

Though they are Royal Bengals, the same subspecies of tiger found throughout the Indian subcontinent, these tigers' behavior is markedly different. Sometimes swimming out into ocean waves, a tiger will follow a boat like a dog chases a car. Or the predator will lie in wait for hours, concealing itself in riverside vegetation, carefully planning an attack in broad daylight. It might leap on board a boat, take a man in its mouth, swim to shore, and disappear into the forest.

Why the tigers do this is a mystery. But there are several theories.

Perhaps, some have suggested, Sundarbans' tigers are so aggressive because of their unusual habitat. This is the only mangrove forest on earth inhabited by tigers. Virtually no fresh water is available save dug rainwater ponds. German biologist Hubert Hendricks, who studied tigers on the Bangladeshi side of Sundarbans, suggested that drinking salty water causes liver and kidney damage, making tigers irritable. Others have theorized that tigers acquired a taste for human flesh from scavenging corpses. For many centuries (before the recent construction of the Ferraka dam) the river Ganges brought the tigers incompletely cremated human remains from the Calcutta burning ghats. Yet another theory was put forth by the late secretary of the Smithsonian, S. Dillon Ripley. Possibly, he suggested, the tigers may have learned to raid fishermen's nets—and then cut to the chase, skipping the fish and eating the fishermen instead.

The local people, however, claim that hunger has little to do with it. Two important facts suggest they may be right. First, the attacks seldom occur in the villages, where people would provide a steady source of tiger food. Second, if the tigers depended on humans for meat, they would eat many more of them than they do. One Indian expert calculated that if people comprised a major portion of the tigers' diet, then Sundarbans tigers would need to kill 24,090 people per year.

The local people, generally low-caste people of both Moslem and Hindu faiths have a different explanation for the man-eating—one that is preserved on both sides of Sundarbans in poetry, song, and myth, and in a yearly celebration or puja honoring the powers of Daksin Ray, the Tiger God.

Daksin Ray, the story goes, owned the natural riches of Sundarbans—its fish, timber, and wild bee honey—long before people came to the delta. When the people arrived, they asked the Tiger God to share his wealth with them. He agreed to do so, provided that the people showed him proper respect. If they did not, the god would enter the body of a tiger and eat them in retribution.

For many years, scientists dismissed the people's stories as silly superstition. But at least some of the legend seems confirmed by a statistical correlation. Some 90 percent of attacks analyzed by West Bengal Forest Department managers occurred while the victims were illegally sneaking inside the perimeters of the Core Area of the Tiger Reserve—an area set aside for tigers, which poachers disrespectfully invade. Here, contrary to the people's own laws and traditions, the victims were illegally harvesting wood, fish, or honey. And it is for this reason that the man-eating tigers are not hunted in Sundarbans, not even in retribution: According to the people's legends, the tigers are entitled and in fact expected to defend these resources.

The legend has served the people of Sundarbans well. Despite being shared by two of the poorest and most overpopulated countries on earth, Sundarbans' forest is the largest mangrove forest remaining in the world today. The delta provides its people with rich fishing. Its trees shelter the land from violent cyclonic storms. The reason the forest is still standing, and its fishing grounds still healthy, some say, is that the area is protected by 500 man-eating tigers—effective, albeit ruthless, eco-police, defending resources upon which both animals and people depend.

Possibly, some experts suggest, the man-eaters of Sundarbans kill in defense of their territory. Normally tigers mark their territories by spraying urine or piling feces at boundary lines. But because Sundarbans is a mangrove swamp at the edge of the sea, the rise and fall of the tides constantly sweep away these territorial markers. Possibly this makes the tigers here hyper-territorial, exceptionally eager to attack and defend their riches—much as the ancient legend suggests.

Further Resources

Chakrabarti, K. (1991). *Man, plant and animal interaction*. Calcutta: Darbari Prokashan.

———. (1992). *Man-eating tigers*. Calcutta: Darbari Prokashan.

Chaudhury, A. B., & Chakrabarti, K. (1989). *Sundarbans mangrove ecology and wildlife*. Dehra Dun: Jugal Kishore.

Corbett, J. (1946). *Man-eaters of Kumaon*. New York: Oxford University Press.

De, R. (1990). *The Sundarbans*. Calcutta: S.K. Mookerjee, Oxford University Press/India.

Jackson, P. (1990). *Endangered species: Tigers*. London: Apple Press.

Montgomery, S. (1995). *Spell of the tiger: The man-eaters of Sundarbans*. Boston: Houghton Mifflin.

Tilson, R. L., & Seal, U. S. (Eds.). (1987). *Tigers of the world*. Park Ridge, NJ: Noyes Publications.

Wray, A. (Producer). (1996). *Spell of the tiger* [TV documentary.] Washington, DC: National Geographic Explorer.

Sy Montgomery

■ Hunting, Fishing, and Trapping
Sport Fishing

Many humans interact with fish on a regular basis, although for most people, this is not an intimate relationship, what with fish being cold blooded, slimy, and inhabiting an alien, water world in which humans travel with difficulty. Despite that, fish have been a mainstay of human diets for time immemorial. They have driven the symbolism and life rhythms of entire cultures, such as those for many of North America's Pacific Coast First Nations whose year revolves around the Pacific salmon. In the Western world, dieticians and health gurus are telling us that if we want to lead long, happy lives, we need to eat more fish rich in heart-friendly omega-3 and omega-6 fatty acids. But for the most part, we modern humans undergo our interactions with fish at the seafood counter at the local supermarket, where our piscine "friends" generally arrive filleted and skinned from industrial commercial fisheries or aquaculture operations.

It wasn't always this way. For millennia the primary interaction people had with fish was entering the fish's world and coming up with ways to catch them. The modern, technology-driven fishing fleets of today are a far cry from the one-on-one struggle that for most of human history dominated the capturing of fish. However, at some point during human history, people consciously or unconsciously came to the realization that the process of fishing was pleasant, even spiritual. Out of this was born the pastime of sport fishing, the quest of an individual angler armed with a fishing rod to capture a fish.

The earliest sport fishing record we have, at least in the English language, is Dame Juliana Berners *The Treatyse of Fysshynge with an Angle*. Dame Juliana was reportedly a nun and prioress from an abbey in Hertfordshire, England, but there is dispute over whether or not she actually existed. Some believe that the name is a pseudonym for the true author who wished to remain anonymous. The book definitely exists, is written in the English language style of the fifteenth century, and appeared in 1496. The *Treatyse's* primary purpose was to inspire people to go sport fishing, but it also was the start and inspiration for the voluminous English-language angling literature that continues to pour forth to this day.

Consistent with the prioress theory, Dame Juliana starts her treatise by quoting the parables of Solomon, noting in particular that a healthy, happy, righteous life flowed from a beauty of spirit ("a good spyrite maketh a flouring age that is a fayre age and a longe"). She believed that to achieve that beauty a person needed to pursue activities that nurtured the spirit ("a mery occupacion which may rejoice his harte, and in which his spirites may haue a mery delyte"). Not for her the contemporary popular pastimes among the noble-born of hunting, hawking, or fowling, which were "laborious and greuous (grievous)" occupations and did not get people out of bed early enough to be "holy, helthy & happy." Angling was the ticket, and in her how-to book she takes prospective anglers with simplicity and great accuracy through the equipment and techniques needed, on a species-by-species basis, for catching fish with a fishing pole. She even includes a description of the first reported artificial flies and the materials needed to tie them.

Sport anglers today are more or less divided into two major groups: those with "hardware" and those devoted to "fly fishing." Hardware fishermen use a variety of artificial metal lures and/or baits to try and entice a fish to get caught. The equipment is primarily designed to securely hook and retain a fish, and the intent is to take it home and eat it.

The fly fisherman approaches the sport differently. Fly fishing is full of social hierarchies, elaborate rituals, and techniques that have to be perfected in order to become a "respectable" fly fisherman. You must master fly tying, which requires artistic capacities, manual dexterity, and a house full of esoteric materials such as jungle cock feathers and polar bear hair that can be woven into the "dress" of an effective artificial fly. You must equip yourself from head to toe, including waders, a fishing vest stuffed with tools, and a stylish hat. You need to obtain a fly rod and reel and through patience and hard work develop the motions that cast a nearly weightless fly accurately to the places in the water where the fish are lying. Being a fly fisherman can have curious impacts on people's psyche, as Fen Montaigne noted for Atlantic salmon fly fishermen: "In the angling world, there is no snob like an Atlantic salmon snob. And while being mindful not to tar all Atlantic-salmon fishermen with the same brush, the truth is this: many devotees of the "sport of kings" are insufferable, elitist, tweedy, name-dropping bores" (p. 41).

Fly fishing goes on in unlikely places, under unlikely circumstances, and with unlikely species. Atlantic salmon anglers were among the first wave of westerners to enter Russia when the Soviet Union dissolved. They were seeking the undisturbed rivers of the Kola Peninsula, and in those turbulent times some of them found themselves being escorted back out of the country at gun point. Fly fishing sport camps have been established in the Amazon River basin for Peacock Bass, and at least one of them has been overrun by guerillas, with the anglers escaping to the jungle. Salt water fly fishermen prize bonefish, and some are now even pioneering techniques for catching sharks!

Sport fishing is big business. In North America people spend millions of days and billions of dollars each year on fishing trips. These expenditures create valuable employment in rural areas for guides and small businesses, such as hotels and restaurants, and play to the traditional nature-oriented skills of people in these regions such as boat handling and river navigation. Since people take care of the things that they value, the economic benefits of sport fishing provide a powerful incentive to conserve fish populations and maintain clean water.

Recent surveys of recreational anglers consistently show that the thing they value the most is not catching a fish. Rather, it is the joy of being in the natural world and the gentle pace of life on the water. They are seeking to massage their spirits, which is what Dame Juliana recommended over 500 years ago.

Some anglers so prize the fishing experience and the conservation of fish populations that they can no longer bring themselves to kill a fish that they have caught. This has given rise to the practice of live release (also know as catch-and-release). Simply put, live release means that you treat a fish gently as you reel it up next to your boat or into a net, that you remove the hook as quickly as possible preferably without taking the fish out of the water to minimize stress, and that you then let it swim back into the wild. Many studies have shown that many species of fish treated this way will survive, reproduce, and even be caught again by anglers a second or more times. However, although live release has proved to be a successful and valuable conservation tool, it has not been without controversy.

Humans have to eat, and most societies accept the capture of fish for consumption as an ethical and necessary human behavior. However, recently some people have questioned the ethics of live-release fishing, irrespective of the conservation and water quality benefits that the presence of a sport fishery can bring. If you are not going to consume the fish, then is it cruelty to capture them by impaling them on a metal hook, forcibly coerce them up to wherever the angler happens to be positioned, and then release them to the wild to try and do the same again? A key component of the cruelty argument revolves around fish "awareness" and whether or not they feel pain. The available scientific evidence is conflicting and contradictory. Some hold that the neural system and brain of fish are not sufficiently developed to experience pain and awareness (Rose,

2002). However, recent experiments generated results that were consistent with fish detecting and nonreflexively attempting to avoid noxious stimuli and pain (Sneddon, 2003; Sneddon et al., 2003). Scientific work is ongoing in this important field, and there is a great deal at stake.

See also

Hunting, Fishing, and Trapping—*Fishing and Human Attitudes*

Further Resources

Berners, D. J. (1496). *The Treatyse of Fysshynge with an Angle.* Published online by Risa Stephanie Bear: http://darkwing.uoregon.edu/~rbear/burners/burners.html

Economic and Policy Analysis Directorate. (2003). 2000 Survey of recreational fishing in Canada. *Economic and commercial analysis report No. 165.* Canadian Department of Fisheries and Oceans.

Montaigne, F. (1999). *Hooked: Fly fishing in Russia.* London: Phoenix.

Muoneke, M. I. (1994). Hooking mortality: A review for recreational fisheries. *Reviews in Fisheries Science, 2,* 123–56.

Rose, J. D. (2002). The neurobehavioral nature of fishes and the question of awareness and pain. *Reviews in Fisheries Science, 10,* 1–38.

Sneddon, L. U. (2003). The evidence for pain in fish: The use of morphine as an analgesic. *Applied Animal Behaviour Science, 83,* 153–62.

Sneddon, L. U., Braithwaie, V. A., & Gentle, M. J. (2003). Do fishes have nociceptors? Evidence for the evolution of a vertebrate sensory system. *Proceedings of the Royal Society, B, 270,* 1115–21.

Whoriskey, F. G., Prusov, S., & Crabbe, S. (2000). Evaluation of the effects of catch-and-release angling on the Atlantic salmon (*Salmo salar*) of the Ponoi River, Kola Peninsula, Russian Federation. *Ecology of Freshwater Fish, 9,* 118–25.

Fred Whoriskey

■ Hunting, Fishing, and Trapping
Trapping Animals*

American history is filled with images of adventurous trappers braving the wilds of colonial North America and paving the way for settlement of the continent. These images persist, invoking notions of the pioneer spirit. The legacy of the fur trade, however, tells quite a different story.

The Early Fur Trade

Commercial trapping for wildlife in North America began during the initial occupation by European explorers and colonists, although it wasn't until 1581 that the first ship arrived on the continent with the purpose of delivering animal furs to Europe. Profits

*This entry has been adapted from a chapter entitled "Trapping in North America: A Historical Overview" from the book *Cull of the Wild: A Contemporary Analysis of Wildlife Trapping in the United States* published by the Animal Protection Institute (2004).

from that voyage were staggering, and fur traders recognized the potential wealth to be made from trapping wildlife and selling furs. The ethics of exploiting wildlife for economic gain were not considered, much less debated: Profit was the motivating factor.

By 1620, nearly 100 fur traders operated around Chesapeake Bay. Fur trading had become one of the most lucrative industries of the New World, and North American furbearers were being trapped in unprecedented numbers to satisfy the whims of European fashion.

The quest for fur led to the exploration of the western United States in the seventeenth and eighteenth centuries and was the impetus behind the Lewis and Clark expedition of 1803. Trappers, called "mountain men," replaced the Native American trappers with whom earlier explorers had bartered. Men such as Jim Bridger, Kit Carson, and Jedediah Smith blazed into fame, along with fur companies known as the American, Missouri, and Northwest. These trappers and traders traveled to western Canada and southern California in search of fur. Generations of settlers followed the trappers' land and water routes and colonized the West. During this period, millions of buffalo, antelope, bear, otter, beaver, fox, and wolf were slaughtered for their fur, hides, other body parts, or for no reason at all. Rotting carcasses remained, littering the prairies and plains.

In the eighteenth and nineteenth centuries, the continent's teeming populations of beaver, otter, fox, and other furbearing animals seemed inexhaustible, and trapping "seasons" and "bag limits" did not exist. Never in U.S. history had animals been slaughtered in such astonishing numbers. In some areas, beaver, wolverine, pine marten, fisher, kit fox, and otter were trapped to the verge of extinction. Wolves and grizzly bears were virtually exterminated south of Canada, and the North Pacific sea otter population inhabiting the waters between Baja California and Japan was almost wiped out by the end of the nineteenth century—all to feed the growing fur trade. The invention of the steel-jaw leghold trap in 1823 by Sewell Newhouse gave trappers a potent weapon that helped to increase the killing. By 1830, when silk top hats replaced beaver pelt hats as the reigning fashion, the beaver population in the United States had already been decimated. It would be almost a century before beavers began to recover.

By the beginning of the twentieth century, the fur trade had ebbed, many wildlife populations were depleted, and a new consciousness emerged regarding the necessity for wildlife conservation and the ethical treatment of animals. Concerned citizens began pushing for legislative controls of consumptive wildlife uses. Protective laws regulating hunting and trapping of certain species, albeit minimal, were passed and state wildlife agencies were established with the mandate of "managing" state wildlife populations. The birth of the conservation movement and the establishment of laws and regulations limiting hunting and trapping were controlled largely by trapping/hunting interest groups. They filled the positions of power on commissions established to adopt and enforce wildlife laws. Although the numbers of hunters and trappers have declined precipitously and are now far outnumbered by nonconsumptive wildlife enthusiasts, they still dominate state fish and wildlife agencies.

Trapping in the 1990s

Although many people think fur trapping went the way of the buffalo hunter, the worldwide fur trade persists. The United States and Canada remain two of the largest trapped-fur–producing countries in the world along with Russia. In 1997, more than five million animals were trapped in the United States for their fur, according to state wildlife agency estimates. This figure considers only target animals, however: At least as many unreported nontarget animals may fall victim to body-gripping traps every year. The

United States lags far behind the rest of the world with regard to trapping reforms. More than eighty countries have banned the leghold trap, a device condemned as inhumane by the American Veterinary Medical Association, the World Veterinary Association, the National Animal Control Association, and the American Animal Hospital Association.

In 1995, member countries of the European Union banned leghold traps and sought to ban the import of furs from countries still using these traps. However, the United States continues to defend commercial fur trapping and the use of the leghold trap and even threatened the EU with a trade war over the issue. Despite increased public opposition to the use of cruel traps and decades of redundant research, leghold traps and other primitive trapping devices remain legal in most U.S. states and public land systems.

In the United States, commercial trapping steadily decreased during the 1990s due to reduced domestic demand for fur, plummeting pelt prices, and increased public awareness. Accordingly, sales of fur-trapping licenses have declined in many states. Fewer than 150,000 Americans commercially trap wild animals for fur. Millions of animals, however, continue to be trapped for the growing overseas luxury fur trade, and trapping for "nuisance" and "damage control" has increased dramatically.

Animal advocates have had some success banning or limiting certain traps and/or trapping practices at the local and state levels through the administrative and public ballot-initiative processes. From 1994 through 2000, voters in five states (Arizona, California, Colorado, Massachusetts, and Washington) passed ballot initiatives restricting the use of body-gripping traps for commercial and recreational trapping. These successes reflect a growing public perception that trapping is cruel, unnecessary, and unjustifiable. With such heightened controversy and increased public awareness, efforts to restrict trapping will inevitably continue.

The Status of Fur in the Twenty-first Century

It was hard not to notice the return of fur trim, collars, and novelty items in fashion magazines and New York runways at the turn of the twenty-first century. Conspicuous consumption was "in," fashion magazines told us, and political correctness was "out."

Boosting U.S. fur consumption has been the huge increase in fur production in China and other Asian countries where fur has only recently become a symbol of status and affluence. With a huge supply of cheap labor, China has become central to all aspects of the fur trade, including fur factory farming and the production of finished fur garments. According to industry sources, China is the second leading producer of caged animals for the international fur trade; in 2005 the country bred, raised, and killed 8 million mink and 3.5 million foxes on fur farms that entered into the international fur trade. China is also the largest manufacturer of finished fur products and garments in the world, producing 70 percent of the world's fur goods, and the top exporter for fur to the United States; half of all fur apparel entering the United States now comes from China.

This is an especially disturbing trend given that China has almost no enforced animal welfare laws or regulations, and the animal protection movement has been largely absent. Recent undercover investigations of Chinese fur farms has brought international outcry against the country. Scenes of foxes and raccoon dogs being skinned and dismembered alive have led some fur buyers and users, including British fashion house Burberry, to pledge not to buy or use fur from China.

Increased worldwide interest in fur-trimmed and fur-lined items also threatens to increase trapping pressure in North America and fur-farming globally. An estimated 90 percent of the foxes killed for their pelts are used as fur trim on designer clothing and accessories. Consumers appear less concerned about the social stigma associated with wearing fur if it is discreetly used as trim or lining. Animal advocates have historically been less inclined to target wearers of fur-trimmed garments than those wearing conspicuous full-length fur coats (see following table for how many animals are needed for these).

Number of Animal Skins Needed for a 40-Inch Fur Coat

Mink	60	Muskrat	50	Red fox	42	Raccoon	40
Badger	20	Lynx	18	Coyote	16	Beaver	15

Public Attitudes toward Trapping and Fur

Most Americans are unfamiliar with traps and trapping practices, and this lack of knowledge and the misinformation disseminated by trapping proponents can lead to an inconsistent public opinion on trapping. In 1977, pre-campaign polling showed 66 percent of Ohio voters supported a proposed statewide trapping ban. Before the vote, opponents of the ballot measure conducted an intensive media campaign delivering the message that trapping is essential to wildlife management and the protection of public health and safety. Six weeks after the first poll, 63 percent of voters cast ballots against the ban.

Despite limited public awareness of trapping, opposition to the use of leghold traps has remained constant over the past twenty years. In a 1978 national survey commissioned by the U.S. Fish and Wildlife Service and conducted by Yale University professor Stephen Kellert, 78 percent of respondents opposed the use of steel-jaw leghold traps. Eighteen years later, a national poll commissioned by the Animal Welfare Institute showed that 74 percent of Americans opposed the use of leghold traps. Several statewide polls conducted in the 1990s during anti-trapping initiative campaigns supported these findings.

In a 1986 survey of veterinarians conducted by the Animal Welfare Institute, 79.3 percent of all surveyed veterinarians opposed the use of the steel jaw leghold trap. More people are opposed to leghold traps in particular than to trapping in general or to killing animals for fur. Although 78 percent of respondents to Kellert's survey opposed the use of leghold traps, only 57 percent disapproved of killing furbearers for clothing. A 1995 Associated Press poll reported that 60 percent of respondents agreed it was "always wrong to kill an animal for its fur," and 64 percent approved of "most of the protests being made by animal rights groups against using animals to make fur coats." Attitudes toward trapping, as with other animal-related subjects, depend on the species of animal involved, whether pain and suffering are present, and the stated purpose of the activity. In a 1997 statewide survey of California voters, 81 percent opposed "allowing animals to be trapped and killed for the commercial sale of their fur." There was less opposition, however, to trapping for private property damage control (60 percent), for flood damage control (58 percent), and for protection of public safety (44 percent). Wearing or selling fur products is somewhat less objectionable to the public than the killing of animals for their fur. Although 50 percent of respondents to a 1993 *Los Angeles Times* poll indicated they "generally oppose" the wearing of clothes made of animal furs, only 32 percent of those surveyed by ABC News in 1989 said seeing someone wearing a fur bothered them because animals were killed to make it. And only 20 percent of those sampled for a 1990 *USA Today* poll thought fur sales should be banned.

Trapping for Damage Control

In addition to the animals killed for profit and recreation, millions more animals are trapped for "damage control" purposes each year by state and federal agencies, private wildlife control operators (WCOs), and individual landowners. The U.S. Department of Agriculture's Wildlife Services agency traps tens of thousands of predators, including coyotes, bobcats, bears, mountain lions, and foxes annually in the name of "livestock protection." This program, funded by U.S. tax dollars to benefit a small number of ranchers, relies heavily on leghold traps, strangulation neck snares, and other indiscriminate devices. With increased urbanization and human encroachment into wildlife habitat, conflicts between humans and animals have grown dramatically over the past thirty years, creating a growing industry focused on lethal control of suburban and urban wildlife. Countless raccoons, opossums, squirrels, skunks, and gophers are trapped and killed by WCOs with almost no state or federal oversight. Because most states do not require that animals trapped for "damage" or "nuisance" control be reported, the total number of animals killed for these purposes is unknown.

Trapping and other indiscriminate control methods have failed to solve human/wildlife conflicts because they generally ignore the underlying systemic problems and provide at most only a temporary remedy to the perceived problem. Public education, as well as the implementation of effective and humane wildlife management methods, is necessary to resolve human/wildlife conflicts over the long term.

Trapping for Wildlife Management

"Wildlife management," as currently practiced by state and federal agencies, revolves largely around the utility of wildlife to humans. Consumptive uses of wildlife in the form of trapping or hunting are often favored, even to the detriment of wildlife species. Economics strongly dictate when, where, and how animals are trapped, even when necessary biological data are lacking. When pelt prices rise, pressure on furbearers increases and in some situations, the size of a furbearer population in a given area can fluctuate depending on its economic worth. Allowing economics and the interests of consumptive wildlife users to dictate wildlife management has depleted populations of some species and created unnatural increases in others.

Over the last century, there has been a paradigm shift in the public's perception of wildlife. The majority of Americans no longer view wildlife as a resource to be "stocked," "harvested," "culled," and killed for profit. Reflecting a societal shift from a utilitarian perspective in our relationship with other animals to one that is more protectionist oriented, many people now believe that humans have a moral responsibility to incorporate ethical protocols in how we interact and coexist with other animals. However, state and federal agencies are slow to reflect this societal shift as evidenced by the fact that consumptive wildlife users—despite being a minority in the United States—continue to dominate agency staff, boards, and commissions.

Conclusion

Globally, more than 50 million animals continue to be killed for their fur. Although the number of wild animals trapped in the United States has decreased from nearly 14 million in 1987 to less than 4 million in 2005, increasing overseas fur markets and the growing popularity of fur trim could reverse this trend. Moreover, many former fur trappers, unable to profit from their trade, have switched to "nuisance" or "damage control" trapping, a fast-growing, highly unregulated industry capitalizing on increased urban/suburban con-flicts with wildlife and employing the same body-gripping traps used in fur trapping.

Ultimately, the public will determine the future of trapping and wildlife management in North America. Enhanced public education and strategic policy efforts can bring an end to commercial fur trapping and ensure that humane treatment and co-existence, not lethal control, become the guiding principles of wildlife conservation.

Further Resources

Animal Protection Institute: http://www.BanCruelTraps.com/

Fox, C. H., & Papouchis, C. M. (Eds.). (2004). *Cull of the wild: A contemporary analysis of wildlife trapping in the United States*. Sacramento, CA: Animal Protection Institute.

Nilsson, G. (1980). *Facts about furs*. Washington, DC: Animal Welfare Institute.

Camilla H. Fox

■ Language
See Communication and Language; Literature

■ Law
Animal Law

Animal law is one of the most prominent of the disciplines in the emerging field of human-animals studies. This article both describes this new field and then outlines some of the most basic features of existing law as it impacts nonhuman animals.

This newly emerged field covers much, ranging from the many ways in which legal systems have *in the past* dealt with nonhuman lives to the fascinating possibilities of how legal systems can *now* and *in the future* deal with animals. Since there are multiple options before each society, which one is chosen by any one society will say much about both its citizens' values and our human possibilities.

Animal law as an academic topic has exploded onto the legal education scene in the last decade. A key development was the decision of Harvard Law School to offer an animal class in 2000 after years of student-led efforts made it clear that more than a hundred students supported such a course being offered. Legal education in industrialized societies has for more than a century focused on the kinds of law that benefited businesses and the richer elements of society. Social justice concerns became more and more prominent in the second half of the twentieth century, however, thereby opening the door to a broader understanding of how law functions in a society.

Animal law requires one to tap resources outside the normal realm of traditional legal education, much as does advocacy for the poor, women, environmental issues, or other marginalized causes. In particular, studying how legal systems have treated nonhuman animals requires one to assess historical patterns, sociological realities, and cultural differences (how law functions in small-scale societies, for example, as it affects human-animal relationships is quite different from how law functions in large-scale industrialized societies).

The influential philosopher Immanuel Kant once observed that our concepts about the world will be empty if not informed at least in part by experiences of the actual world around us. Similarly, any study of animal law that goes forward without good information about the actual realities and lives of nonhuman animals will be empty—we need such information to assess what the character of our law has been and might in the future be. Unless based on such information, a law that impacts nonhuman animals can be harmful and unjust.

Without a sense of the particular past that we have inherited, we see less well both what is happening with our present laws and what might happen in the future. Further, without a sense of what other cultures have done regarding human relationships with other animals we fail to grasp our own possibilities and, as importantly, the nature of our own culture's views—are the views and laws we have inherited reasonable, or are they dysfunctional and unrelated to the realities of the animals we impact so heavily?

As an educational subject, animal law includes the different options for human-animal relationships used around the world. Such a course of study is a superior vehicle for asking whether the scheme now being used in our own society actually meets our present needs and desires or can be replaced with something better and more compassionate.

Animal law as a subject is enhanced when it is informed by sciences such as animal behavior. Familiarity with the realities of nonhuman animals is needed for any number of reasons—for example, it helps one see how some animals suffer from contemporary practices. Many sciences have helped us recognize that nonhuman animals are diverse and sometimes exceedingly complicated in their social ways, intelligence, and daily realities. In fact, we now recognize that some nonhuman animals are, in some ways, surprisingly like humans. But, based on our sciences, we also recognize that many nonhuman animals are dissimilar from humans. A fundamental question in animal law is what place these similarities and dissimilarities might have in our actions and attitudes toward Earth's other living beings.

Further, animal law is best studied when the teaching institution promotes awareness of the pervasive educational biases that lead most academic work to focus heavily on humans alone. Other animals are, generally, very unsympathetically studied in many academic institutions, with the result being that they are poorly known. The upshot of this bias is that education rarely prepares contemporary students to notice and take seriously the rest of life on Earth—animal law can naturally and fully engage the ways in which law and other parts of educational system help or hinder us in our understanding of nonhuman lives.

Legal systems reflect underlying political, social, economic, cultural, and religious dimensions of a society. Studying these underlying dimensions is crucially important to understanding why so many people value nonhuman animals even as others and the general legal system dismiss them in countless ways.

Animal law thus focuses on not only what we are and have been doing in our relationships with other animals but also what we (and, indeed, any human culture) can do in light of what we know through sciences and the humanities about the actual realities of other animals. For all of these reasons, animal law must be, if it is to accomplish its work well, deeply interdisciplinary. In fact, without constant engagement with other disciplines that study nonhuman animals carefully and have accumulated bodies of knowledge about the world and its animals (human and nonhuman), animal law can become an empty recitation or catalogue of our society's dominance over other animals. To be a full exercise and relevant to the real world around us, animal law needs to be wide ranging in its assessment of how humans in their societies interact with nonhumans, as well as frank about law's past and present elitisms.

Said simply, framing of laws that affect how we treat each other and the world around us is one of the ways humans shape their relationship with animals. The study of animal law is a perfect tool for calling out the bias of past laws regarding animals and, more generally, the shortsightedness, greed, and lack of moral vision of past lawmakers. Human relationships with others (whether human or nonhuman) have been fraught with dishonesties and shoddy thinking—oppression in the form of racism, sexism, wars, and economic domination has been common. Oppression and domination of nonhuman animals rivals and perhaps even exceeds the moral bankruptcy of human-on-human oppression.

Noticing and taking seriously the nonhuman lives in and near our human communities is one of the ways to help us see how humans through their "law" have impacted, often on the basis of arrogance and ignorance, the diverse nonhuman lives with which we share Earth. The questions that animal law asks will be, then, somewhat provocative, that is, it will call out (the original meaning of the word "provocative") features of what we have been doing.

Once we see well what we have been doing, we can ask what we *might* do in the future. If the world is, as the remarkable thinker Thomas Berry has said, "a communion of subjects, not a collection of objects," we can ask how legal systems have impacted the way we see this remarkable reality. Has law helped us see the world around us better? Or has law been the purveyor of greed, shortsightedness, and fundamentalisms such as "humans alone truly matter in this universe"?

The study of animal law thus has the capacity to open minds that have, for a very long time, been closed and empty of insights about the world in which we live. The upshot is that, when given its natural breadth and depth, animal law turns out to be more than just another legal field. It is a form of legal education that can proceed, unlike so many other legal discussions, with questions that challenge the human-centered biases that now prevail in most legal circles. When it does this, animal law can be the most pertinent of subjects, applicable to our real lives and exhilarating in the extreme.

Animal law can, of course, also be taught in typical law-school fashion, calling out only the human-centered norms that dominate our modern industrialized society. When it does the latter, animal law, like so many other law school courses that purport to talk about justice and dignity, fails—such teaching is both blind and empty.

The most basic part of existing law as it impacts nonhuman animals includes the following features. Most legal systems in today's industrialized societies treat nonhuman animals as mere property, holding nonhumans to be mere "legal things" that can be owned, just as a chair, a book, or a computer. In a theoretical sense, relegation to this "legal thing" category is quite important because humans are put in the separate category "legal person." Legal persons are the holders of rights, whereas it is commonly asserted that legal things, because they are mere property, cannot hold rights.

The property status of nonhuman animals in the legal systems of industrialized nations reflects what the English philosopher Mary Midgley has called the "absolute dismissal" of nonhuman animals. It is true that there are some protections for some nonhuman animals in some legal systems, such as anti-cruelty provisions, but there is an important debate over what such protections truly mean and for whose benefit they were enacted. Whatever the answer to such questions, these protections are often *not* enforced.

The fact that there are some existing protections makes it clear that legal systems can, if humans so choose, offer important protections for nonhuman animals. Legal systems can in fact offer several different kinds of protection. Offering specific "rights" to specific individuals is the best-known possibility, but there are other tools in the legal toolbox that can be used to protect nonhuman animals. An example of a nonrights tool is a prohibition on ownership—under such a law, even though the protected individuals do not themselves necessarily hold rights, fundamental protections are possible for nonhuman animals *if* this legal tool is enforced.

Various kinds of animals receive different kinds of protections depending on which general grouping of nonhuman animals they fall into. Wildlife is sometimes given, for example, protection from certain harms when the overall species is threatened with extinction. Farmed animals are given far fewer protections, though there often appear "on the books" various purported legal limits as to how the production animals can be transported and then slaughtered—but these protections are notoriously unenforced in many countries.

Research animals are the subjects of many laws, and in some ways this category can be understood as the most regulated of the categories mentioned here. But in the United States, for example, the vast majority of laboratory animals are excluded from legal protection by virtue of the exclusion of rats, mice, and birds from the scope of the federal government's Animal Welfare Act. The legal issues affecting companion animals are dealt with in a separate article—that this category of animals is the subject of much debate presently raises the question of whether it is these animals' connections to humans or, instead, their inherent qualities that are the basis for the emerging legal protections of companion animals. If the former, one can wonder if laws purporting to protect "pets" are really just another manifestation of the overall legal system's preoccupation with *human* interests.

What we do to other animals is, simply said, virtually always a matter of choice. Careful study of animal law lays out what we as individuals and as human communities have done in the past to all of the Earth's other living beings, what we do to them now, and what we might possibly do to, with, and *for* them in the future.

See also

Law—*Public Policy and Animals*

Further Resources

Curnutt, J. (2001). *Animals and the law*. Santa Barbara, CA: ABC-CLIO.

Waisman, S. S., Frasch, P. D. & Wagman, B. A. (Eds.). (2006). *Animal law: Cases and materials*, (3rd ed.). Durham, NC: Carolina Academic Press.

Wise, S. M. (2000). *Rattling the cage: Toward legal rights for animals*. Cambridge, MA: Merloyd Lawrence/Perseus.

Paul Waldau

■ Law
The Legal Status of Companion Animals

Society is subject to constant change. Because the law reflects the norms of the society adopting it, the law evolves as society changes. These observations apply to the evolution of the law concerning companion animals in the United States, and development in the law concerning animals over the last two centuries reflects changes in society's attitudes toward companion animals.

Western society has viewed animals traditionally as personal property, and it still does to a great extent. The first laws concerning treatment of animals were explicit in protecting animals as a type of personal property, the way the law protected other classes of personal property from destruction or damage by third parties. In essence, these laws protected owners from damages caused to their personal property, and they reflected a society not so much concerned about kindness toward animals as the sanctity of private property.

Over time, more and more people came to see companion animals as having qualities that distinguished them from other personal property. Many Americans now view their companion animals as part of the family unit. Additionally, philosophers and thinkers argue that a society that prohibits gratuitous cruelty toward its animals is more likely to treat its human members with care and respect.

The law reflects these changes in attitudes toward companion animals, but its evolution also illustrates that national consensus about the status of companion animals is not perfect. Given the diverging views of various individuals and interest groups about the status of companion animals, the law concerning them is subject to national debate. Pressure for additional change to the law exists from some quarters, and various interest groups have emerged with opinions about the more controversial areas of animal law.

These discussions reflect uncertainty in American society, and in turn in American law, about the precise status of companion animals.

Criminal Laws Prohibiting Cruelty toward Animals

American colonial law was often silent on the subject of animals, and legal issues concerning animals were resolved under general personal property laws. In 1641, however, colonists in Massachusetts decided that cruelty toward animals was sufficiently significant to merit its own consideration.

Therefore, the Massachusetts Bay Colony enacted the first law prohibiting cruel treatment of animals. The law was part of the colony's Body of Liberties. This law prohibited tyranny and cruelty against "bruite Creatures which are usuallie kept for man's use." Although the law was addressed specifically toward animals' treatment, it protected them only in relation to their usefulness to humans, not because society viewed them as having any intrinsic value. Notably, only animals kept for use (rather than those kept strictly as companions) were deemed worthy of mention.

Meaningful development of anti-cruelty laws in the United States occurred during the nineteenth century (particularly in its second half) and into the early twentieth century. Based on changing views about animals and arguments that a society's treatment of animals was a gauge of its ethics and morals, states adopted laws against cruelty that established the framework of protections for companion animals today.

Early in the nineteenth century, legislatures tended to concern themselves only with commercially valuable animals, such as horses, cattle, sheep, and swine. They made willful mistreatment of these animals (such as poisoning) punishable by fines and imprisonment, but these states' laws did not address treatment of animals with minimal or nominal market value, such as typical dogs or cats. Most jurisdictions' statutory schemes prohibited an individual from mistreating an animal only if the animal was owned by another person. The law did not address the way in which a person treated his or her own animal, reflecting a view that the animal was purely the owner's property to do with what he or she chose.

Slowly, the law developed to reflect social attitudes that needless cruelty to an animal was wrong, regardless of whether the animal was owned or commercially valuable. Societal consensus on this concept was reflected in adoption of more meaningful animal protection laws in the United States in the second half of the nineteenth century.

By 1860, several states had established laws that specifically forbade cruelty to animals even if committed by the owner. This change was significant because it reflected social consensus that owners did not have complete freedom to treat animals however they desired. In an era emphasizing the sanctity of private property, such laws represented public recognition that animals were somehow different than other personal property, such as wagons or furniture. Some behavior was so egregious as to justify government interference with the owner's private property interest in the animal. At approximately the same time, states began to prohibit specified acts toward all animals (including dogs and cats with no commercial value), not just against commercially valuable animals. Laws continued to protect the property interests of owners by prohibiting conduct that would impair an animal's monetary value (such as maiming another person's cow); the expansion of the law occurred when some states explicitly prohibited gross mistreatment of an animal even by the animal's owner (thus forbidding a person from maliciously beating his or her own cows).

Throughout much of the nineteenth century, animal cruelty was a misdemeanor. (A guilty party was jailed for only up to a year and/or subject to a fine. This penalty must be read in the context, however, of a culture where abandonment of a child was also treated as a misdemeanor by some of the same jurisdictions.) Strays and other domestic animals without owners had no protection by most states, but these laws changed piece-meal among the states to protect animals from wanton cruelty regardless of ownership.

As the law developed and reflected increasing social and cultural concern for animal welfare, some states began to regulate behavior more stringently in the later part of the nineteenth century and into the twentieth century. Their legislatures passed laws making not only overt acts of cruelty illegal but also criminalizing such behaviors as abandonment of animals (under specified circumstances) and failure to provide adequate food or shelter. States that had not done so previously increasingly prohibited mistreatment of all domestic animals (including dogs and cats) by not only strangers but also by the animals' owners.

Other changes in the law occurred in the later parts of the nineteenth century and into the early twentieth century. States passed laws prohibiting specific conduct that the majority of people agreed was cruel. Most jurisdictions made use of dogs in human-staged fighting illegal (and all states have done so today, and federal law also makes many types of animal-fighting ventures illegal). States passed laws imposing requirements for humane transport (an area also regulated increasingly by federal law), explicitly protecting animals from the more extreme forms of neglect and establishing enforcement mechanisms allowing officials to enter upon private property to enforce the anti-cruelty laws and seize abused or neglected animals. All of these laws represented government intervention into the ways people could treat animals and reflected an implicit social consensus that animals were worthy of special protections.

The definition of "cruelty" changed over time, both by changes in statutory language by legislatures and through interpretation by courts. At their most basic, laws against cruelty prohibited malicious beating or maiming without cause. Later, as David Favre, Professor of Law at Michigan State University, has recounted, many states adopted versions of a standard definition of cruelty including the following elements: "(1) human conduct, by act or omission; (2) which inflicts pain and suffering on a nonhuman animal; and (3) which occurs without legally acceptable justifiable conduct (legislative language or socially acceptable custom)."

Thus the law had progressed from meager protection of animals simply because of their status as commercially viable property to increased protection based on status apart from monetary value. Some philosophers, social activists, and jurists observed that this progress demonstrated American society's advancement and benevolence. They observed that the protection afforded to animals was socially beneficial in that it protected the animals from undue pain and suffering—itself a recognition that animals were capable of experiencing these sensations, a view that was not widely accepted in earlier centuries and that remained controversial in some quarters, yet did not interfere unduly with use or enjoyment of animals by humans.

The last seventy-five years have not seen the sweeping changes in animal protection laws that were seen in the middle nineteenth through early twentieth centuries. Each state in the union has established laws against animal cruelty, however, and there are variations among them based on regional and local norms that make reference to a specific state's laws pivotal to an understanding of animal protection in that jurisdiction.

However, certain general observations pertain to the evolution of animal protection laws over the last several decades. Laws protecting animals have become more specific and honed to reflect certain societal values. For example, most states now treat certain forms of animal cruelty as felonies, punishable by a year or more in prison and accompanied by more sizable fines. Fewer laws exempt "first-time offenders." Increasingly, state laws establish extreme neglect or failure to obtain veterinary care in certain situations as resulting in sufficient pain to be criminally actionable. Legislatures have worked to design effective prevention of and reaction to problems fairly recently recognized as potentially abusive, such as animal hoarding, sometimes seeking to pass innovative programs involving treatment as an alternative to or in addition to traditional punishment through jailing or imprisonment and fines.

Interest groups (often aligning themselves with the "animal rights" movement in the United States) have sought but often failed to achieve outlaw of some traditional practices, such as ear cropping and tail docking. They have also sought with some success to require counseling at an offender's own expense as part of the penalty for animal cruelty or abuse and have called for criminal dispositions to include prohibitions against possessing or residing with animals.

In some communities, local interests have sought to outlaw certain practices, such as declawing of cats. Such local ordinances are subject to challenge if they conflict with state laws.

Changes in the law concerning animal cruelty reflect changes in what society views as mistreatment of animals and its members' perceptions of what protections animals require. Further, today's advocates against animal cruelty are concerned not only with protection of animals but also cite evidence indicating that an individual who perpetrates animal abuse may well abuse other vulnerable beings, such as children. Current scientific literature supports the view that children who abuse animals are particularly at risk of perpetrating interpersonal violence as they age.

The Owner versus Guardian Debate

Reflecting property concepts, those who possess animals and have the right to decision making (within legal constraints) regarding those animals have been traditionally and are still predominately called "owners." Two groups of people suggest that this term is outdated. The first group notes that animals are more akin to family members and have interests as such; therefore, their caregivers are more comparable to "guardians" and should be known by that name. The second group of people advocating use of the term "guardian" instead of the term "owner" see the change in language more politically and as part of a campaign to establish rights for companion animals.

Interestingly, in everyday parlance, some individuals use the term "owner" and "guardian" interchangeably, and those individuals appear to have little or no social or political agenda in so doing. Those active in the debate concerning which word should be used argue, however, that the words are important indices of social and cultural attitudes and should be chosen carefully to reflect the status our society wishes to give dogs and cats. Several jurisdictions have considered changing the language of their laws, but to date only several localities in their ordinances and one state (Rhode Island) in its constitution have adopted language changing the word "owner" to "guardian."

There is no social consensus concerning which of these words is more appropriate. Further, a secondary debate over the significance of the language remains unresolved. Is this merely a change of usage or does it presage social and legal change in animals' status?

Some who advocate use of the word "guardian" argue that animals' caregivers should be called "guardians" to reflect the enhanced status of "companion animals." They suggest that this language emphasizes the animals' status as true family members or at least dependent wards deserving care that takes into account the animals' best interests.

Advocates for change from traditional language argue that use of the word "guardian" reminds those who keep animals to treat them with respect and recognize them as more than mere chattel. For example, Edwin J. Sayres, president of the American Society for the Prevention of Cruelty to Animals, suggests that "[t]he term 'guardian' accurately describes the relationship of perpetual care that is needed to teach children respect, compassion and kindness for domestic pets." Animal rights advocates often explicitly note that use of the term "guardian" elevates dogs and cats above status as mere personal property. Typically, these advocates also oppose commercial buying and selling of animals, therefore, and support compensation for damages to animals at beyond animals' fair market or replacement value.

Entities such as In Defense of Animals (IDA), sponsor of "the guardian campaign," argue that the benefits of "guardian" language include recognition of animals as individuals as opposed to "mere property, objects and things." Some state boldly that their goal is to alter the status of pets from property to personhood. IDA proposes that "[b]y thinking, speaking, and acting as guardians, we recognize that animals are individuals with feelings, needs, and interests of their own."

Conversely, the American Veterinary Medical Association and the American Veterinary Medical Law Association argue that "guardian" language will do nothing to enhance the well-being of animals and may indeed have harmful consequences. For example, such language could alter the ability of "guardians" to choose health care options for their animals in that under guardianship law (formulated with regard to humans as both guardians and wards) the guardian must fulfill fiduciary duties toward the ward. Proponents of traditional language fear that owners who could not meet these markedly enhanced duties toward their animals might well abandon or fail to seek veterinary treatment for them.

The Council of State Governments believes that additional legal complications could result and resolved that "guardianship statues would undermine the protective care that owners can provide for their animals and the freedom of choice owners [now] exercise, and could permit third parties to petition courts for custody of a pet . . . [if] they do not approve of the husbandry practices." Further, those who oppose implications that companion animals have rights closer to those of persons, as suggested by "guardianship" language, note that society may be ill-equipped to deal with the repercussions of defining pets as other than property in matters such as animal control, animal inspection and quarantine, and alleged animal abuse.

The National Association for Biomedical Research (NABR) puts the position against changes in language in its most succinct form: "While this campaign [to change usage from "owner" to "guardian"] is marketed as a feel-good exercise, this 'simple' change in language elevates animals above their current status as property—with potentially enormous legal implications." Such a change could impact ownership rights, the availability of animals for use in research, and the ability of government to address animal-related and zoonotic disease, among other matters, caution those who argue against such a change in status.

This unresolved debate illustrates that law and its evolution do not happen in a vacuum. Several factors influence the law, such as ethical concerns, economic realities, and political advocacy. Various interests, groups, and subgroups may clash markedly, as has occurred regarding the debate concerning use of "guardian versus owner," yet all may view themselves as wanting what is best for animals. And the law may provide an imperfect mirror of those social conflicts in situations such as debate over language, where the significance of the discussion is not always easy to ascertain, much less its implications for legal change.

Legal Status as a Reflection of Social Value

Because companion animals have assumed a more important role in individuals' lives, these animals' traditional categorization as chattel has been challenged directly in some quarters. The vast majority of United States courts continue to treat animals purely as personal property, particularly at the appellate levels, where legal precedents are established. Nonetheless, some lower courts have made awards several times beyond traditional valuations in cases where companion animals were killed or injured due to the fault of defendants. These anomalous cases indicate that the traditional valuation of animals under law is potentially subject to change but that any change is likely to be incremental and carefully considered in terms of cost to society.

Because the law treats animals, including pets or companion animals, as personal property, actions to address civil wrongs (called torts) lie against those who illegally take, destroy, or injure that property through negligent, reckless, or intentional conduct. Such torts include claims that a defendant has wrongfully converted an animal to his or her own use, claims alleging negligence on the part of another resulting in loss of an animal or diminution of the animal's value, and claims for veterinary malpractice.

A defendant found civilly liable must compensate the prevailing owner for the owner's loss. That loss has been traditionally calculated by property measures, and this remains the strongly dominant approach in the United States, where property measures of loss are applied in the vast majority of courts today.

Therefore, a defendant must compensate the animal owner for that owner's economic loss due to death, injury, or diminution of the value of the animal for which the defendant was at fault. Typically, this means that the liable defendant must compensate the animal owner by paying the fair market value of the animal to that owner. If fair market value cannot be ascertained, the measure of damages is repair or replacement value. Special pecuniary value to the owner may be compensated in a state where a specific statute supports such damages or in cases where neither fair market nor repair/replacement value can be ascertained. Because the animal is treated as property under tort law, the owner receives no compensation for sentimental value of the animal, emotional distress of the owner or animal, or pain and suffering of the animal or owner.

Not all people who treat their pets as family members believe that they should be classified as other than property for legal purposes. They cite arguments that the potential for greater damages would create more and increasingly complicated litigation, unpredictable awards and unjustified windfalls to plaintiffs, and rising veterinary costs (based on increased veterinary malpractice premiums passed on to consumers).

Some advocates argue, however, that companion animals have greater than simple economic value given their enhanced status in modern society and their intrinsic worth. These advocates seek passage of legislation to provide noneconomic and, in some cases, punitive damages to owners who prevail against defendants liable for animals' deaths or injuries. Given that legislatures have been slow to adopt such changes, these advocates argue that courts should extend the law in cases involving beloved companions' deaths or injuries so that their guardians are adequately compensated. Such advocates look to the courts to recognize what they see as animals' true worth even if the society is inadequately progressive in doing so through legislation.

Those who support such increased legal recognition for animals advocate that an expanded range of tort claims should be allowed concerning them. They contend claims should exist upon the same bases recognized for humans—causes of action such as loss of companionship (the animal equivalent to loss of consortium), intentional infliction of emotional distress, and negligent infliction of emotional distress. Further, upon finding of fault on either traditional or nontraditional grounds, some people contend that courts should allow awards for mental anguish and for pain and suffering of the owner and/or the animal. They argue for compensation based on the companion animal's sentimental or personal value (without any pecuniary basis) and compensation for the animal's intrinsic value. These advocates support legislation allowing for recovery of noneconomic damages for death or injury to companion animals and for punitive damages meant to punish wrongdoers, and such legislation has been passed in a very few United States jurisdictions.

Although some lower courts have allowed expanded claims and noneconomic damages, these decisions have not typically survived appeal. Nonetheless, trial court and appellate arguments for expansion of allowable claims and damages concerning animals underscore the significance of the legal valuation of companion animals to many members of our society. Eventually, a middle ground may emerge, under which the legal system calculates civil damages in animal cases by a third measure beyond mere fair market value (the current system) but using measures that do not provide the full range of emotional and related damages available in cases involving death or injury of a human.

Changes in the status of companion animals have been reflected in the evolution of our legal system's response to companion animal issues. The nineteenth century saw institution of a system of animal protection, which was honed and enforced in the twentieth century. Today's debates focus directly on how companion animals compare to humans in terms of social status and value.

See also

Ethics and Animal Protection—*Great Apes Project*
Law

Sylvia Glover and François Martin

■ Law
Legislation: History (International) of Animal Welfare Policy

The type of political system we are most familiar with today, modern liberal democracies, began to develop in the early eighteenth century. It took hundreds of years for modern democracies to slowly mature from the early English model into the type of representative democratic state common today. Throughout that process the political landscape changed in many significant ways. For example, in early modern democratic political systems only the very wealthy landed aristocracy was permitted to influence issues of governance, whereas today universal suffrage is generally considered an important democratic principle, and political representatives are paid a wage, meaning the poor are not institutionally excluded from direct political participation. These changes, and the many others that have taken place over the last three hundred years, have meant that the issues considered appropriate for governments to regulate and influence have also changed significantly. On the one hand, institutions that were once an integral part of early democratic societies, such as slavery and child labor, have now been largely prohibited by statute legislation—that is, laws created by governments. On the other hand, there has been a vast increase in the number, and types, of things governments seek to regulate. Whether we are aware of it or not, many facets of our lives, including how we drive our cars, what we are taught at school, and how we build our homes, are all subject to government regulation. This was not always the case.

Such political changes have had an impact on how humans may lawfully interact with and use nonhuman animals. Whereas once there were very few laws dealing with animals, animal welfare legislation is now widespread throughout the developed world and is becoming more common in emerging economies. Animal welfare legislation developed alongside other welfare reforms, such as the establishment of orphanages and women's shelters and the introduction of unemployment assistance. Animal welfare legislation was born of the belief that the state should intervene to protect the interests of those who aren't easily able to protect themselves.

In general terms, animal welfare legislation outlines the ways in which people must care for animals under their control. It often provides a legal definition of cruelty and also identifies which economic animal uses are permissible and which are not. For example, animal welfare legislation commonly allows calves to be slaughtered for meat, but in some jurisdictions it is not permissible to use a calf for roping at a rodeo. In cases where animal welfare legislation is structured in this way, it indicates that the government views the slaughter of calves for meat as a legitimate or socially acceptable use of that animal, but the use of calves for roping as inappropriate or unacceptably cruel. An argument could be made for viewing both the slaughter of calves for meat and the use of calves in rodeos as cruel. However, one could equally argue that both uses are perfectly legitimate. Variations of this nature, where one activity is allowed and another is prohibited, are common in animal welfare legislation. In general, such variations may be understood as a broad reflection of the values underpinning the society for whom, and by whom, the legislation was created.

However, to return to the example of calves being legally allowed to be slaughtered for meat, although such use is routinely permitted under animal welfare legislation, it is also common to have laws in place that prescribe how animals are to be transported to the abattoirs and how the slaughter is to be carried out. Such regulations imply that although the slaughter of an animal for meat may be acceptable, the manner in which the slaughter is carried out will affect the animal's welfare, and therefore the government has a legitimate role to play in intervening to minimize any negative impact. The example of government regulations prescribing how slaughter should take place reveals another important feature of animal welfare legislation. That is, animal welfare legislation is precisely that—welfare legislation. It does not seek to prescribe rights and it does not seek to challenge the notion that animals are property items. Rather, animal welfare legislation seeks to limit the property rights humans exercise over animals by prohibiting certain "cruelties" and prescribing minimum standards of care.

The Origins of Modern Animal Welfare Legislation

In 1641 the Massachusetts Bay Colony passed a legal code called The Body of Liberties. The code outlined a long series of regulations and protections that were in many ways ahead of their time. Indeed, as well as clauses intended to protect vulnerable humans, such as children and refugees, the code contained two animal welfare provisions. The first protected against "Tirranny or Crueltie towards any bruite Creature" and the second required that cattle in transit receive appropriate rest stops.

However, the first piece of comprehensive modern legislation dedicated to animal protection was an act of British Parliament. Although animal welfare legislation is now common, early legislators fought hard to have animal welfare deemed an issue worthy of government intervention. For many years those who advocated introducing laws to protect animals from certain activities, such as overuse or fighting, were ridiculed and accused of seeking to bring the British Parliament into disrepute. In 1821 *The Times* newspaper reported the following:

> [W]hen Alderman C. Smith suggested protection should be given to asses, there were such howls of laughter that *The Times* reporter could hear little of what was said. When the Chairman repeated this proposal, the laughter was intensified. Another member said Martin would be legislating for dogs next, which caused a further roar or mirth, and a cry 'And cats!' sent the house into convulsions. (cited in Turner, 1964, p. 127)

Furthermore, it was often argued at the time that the introduction of animal welfare legislation would result in an unacceptable intrusion by the Government into the private lives of its citizens.

Richard Martin (1754–1834), the Member of Parliament referred to in *The Times* article, was the person responsible for the first piece of modern animal welfare legislation that successfully passed through the British Parliament. Attempts to create animal welfare legislation were made as early as 1800. However, it was not until 1822, after five unsuccessful attempts, that the first animal welfare law was enacted. The Act's formal title was an Act to Prevent the Cruel and Improper Treatment of Cattle, but it was commonly referred to as Martin's Act. The act stated that a penalty would result "if any person or persons shall wantonly and cruelly abuse or ill treat any Horse, Mare, Gelding, Mule, Ass, Ox, Cow, Heifer, Steer, Sheep or other Cattle" (3 Geo IV, c71). However, the Act was in fact rather limited in scope, in practice applying only to cattle employed as beasts of burden in large urban centers. Furthermore, the court subsequently ruled that Martin's Act did not apply to bulls (Radford, 2001, p. 44). The Act also excluded dogs, even though they were also commonly used to pull carts and transport goods during that period. However, Martin's Act became the basis for modern animal welfare legislation and was reformed and extended many times until 1911, when the Protection of Animals Act was created in order to modernize and streamline all the various pieces of animal welfare legislation that had been brought into effect in the intervening years.

Richard Martin, who was nicknamed "Humanity Dick," was a passionate reformer who worked to protect the interests of vulnerable humans, including slaves. However, he is most remembered for his work to defend animals. Two years after Martin's Act came into effect the Society for the Prevention of Cruelty to Animals was formed. Richard Martin, who prior to 1824 had already brought a number of private animal cruelty prosecutions himself, was one of the new Society's active supporters. The Society received Queen Victoria's patronage in 1840 and therefore became the Royal Society for the Prevention of Cruelty to Animals or RSPCA. The Society was established primarily to ensure that the United Kingdom's new animal welfare law was put into effect. The RSPCA did undertake education campaigns, but a key function of the new society was monitoring animal cruelty and enforcing Martin's Act. To that end the Society successfully prosecuted 149 cases of animal cruelty in its inaugural year (Radford, 2001, p. 42). Richard Martin introduced a number of other animal welfare bills into Parliament after 1822. Most importantly he fought to ban the practice of bull baiting. However, he was unsuccessful in that endeavor and in 1826 he left Westminster.

British animal welfare legislation slowly expanded to include other animals, such as dogs, bears, and birds. Animal protection laws were also extended to include animals in a whole range of situations. For example, laws were enacted that required slaughterhouse managers to provide feed and water to animals waiting to be slaughtered and that animals sold for slaughter could not be re-sold as laborers. Laws were also enacted to protect animals from harmful sports such as baiting and fighting. In 1840 the magazine *Rural Sports* described bull baiting in the following way:

> The animal is fastened to a stake driven into the ground for the purpose, and about seven or eight yards of rope left loose, so as to allow him sufficient liberty for the fight. In this situation a bulldog is slipped at him, and endeavours to seize him by the nose; if the bull be well practised at the business, he will receive the dog on the horns, throw him off, and sometimes kill him; but, on the contrary, if the bull is not very dexterous, the dog will not only seize him by the nose, but will cling to his hold till the bull stands still; and this is termed *pinning the bull*. What are called good game bulls are very difficult to be pinned, being constantly on their guard, and placing their noses closer to the ground, they receive their antagonist on their horne; and it is astonishing to what distance they will sometimes throw him. (cited in Fairholm and Pain, 1924, pp. 75–76)

It is interesting to note that the animal sports that were traditionally enjoyed by the working class, such as baiting and fighting, were banned in the first half of the nineteenth century. By contrast, fox hunting, which is traditionally an animal sport practiced by the upper classes, remained legal for a further century and a half. This apparent inequity was noted by legislators at the time. Indeed one of the early arguments used in opposition to a prohibition on bull baiting was that it was wrong to ban a working class pursuit while not applying similar restrictions to fox hunting. In 1874 the United Kingdom's first legislation was enacted that provided protection for wildlife. However, significantly, that act did not protect wildlife against hunting, chasing, or shooting.

The practice of creating laws that protect animals from cruelty gradually gained popularity in other regions. Three of the states that would later become part of the German federation, Saxony, Prussia, and Bavaria, passed animal welfare acts in 1838, 1851, and 1861, respectively. As with Martin's Act such new acts were often limited in scope and arguably more concerned with human sensibilities than with the suffering of animals. Yet such early legislation was an important stepping stone on the road to the development of modern animal welfare statutes. Britain and the other European colonial powers of the nineteenth century exported their animal welfare laws around the world as part of the colonization process.

Current Trends

Acts of law intended to protect animals from cruelty are now common throughout the world. However, it is not possible to speak of "animal welfare legislation" as a single entity because the detail of animal protection laws varies between countries. Indeed, in countries with a federal political system there are often many different pieces of animal welfare legislation operating simultaneously. This is because animal law is normally the responsibility of states or provinces. In a federal political system, it is common for the states or provinces to create animal welfare legislation that deals with everyday animal use and for the federal government to take responsibility for regulating the transport of animals into and out of the country. Even so, as with many other areas of government regulation, federal governments often influence the structure of state-based animal welfare legislation by mechanisms such as the provision of federal funding, which is conditional on certain animal welfare standards. Yet, despite some instances of federal government intervention, animal welfare legislation takes many different forms. Actions considered acceptable by one jurisdiction may be considered animal cruelty by another. For example, in some parts of Spain it is legal to injure and kill a bull as part of a bullfight. Yet, in many other countries, bull fighting has no cultural or historical significance and is therefore prohibited because it is considered to cause unacceptable suffering to the bull.

However, cultural differences are not the only reason why the detail of animal welfare legislation varies. Even within one state, ideas and beliefs change over time, and as they do so does the popular view of what constitutes animal cruelty. The practice of bull baiting was common throughout the United Kingdom until the mid-nineteenth century. When legislatures first sought to ban the practice, they met with considerable opposition. However, it is unlikely that legislation legalizing the practice of bull baiting would gain popular support in Britain today because cultural norms have changed a great deal. Nevertheless, it is not only cultural changes that result in amendments to animal welfare statutes. As with all government policy, behind-the-scenes political representatives are the target of rigorous lobbying activity. Both those who seek to extract profit from animals, such as farmers, circuses, and the racing industry, and those who seek to protect animals from such uses lobby members of Parliament. Political representatives are ostensibly elected to represent the people's will. Yet there seems little doubt that both animal welfare/animal rights groups and animal user lobby groups believe they have the capacity to influence both public perceptions and government policy. The considerable energy and resources each side of the debate puts into the task of lobbying government on animal welfare laws is evidence of this. Given the extent of lobbying activity, it is reasonable to conclude that the detail contained in animal welfare legislation is, at least in part, a reflection of the relative political and economic might of the various lobby groups seeking to influence government.

Even though the detail of animal welfare legislation varies from place to place, and within jurisdictions over time, it is possible to identify trends in the structure of animal welfare legislation and therefore discuss animal law in broad terms. The most striking feature of modern animal welfare legislation is its reliance on industry-based categorization. That is, animal welfare legislation does not tend to view animals as individuals, breeds, or species. Rather, the tendency is for animal welfare legislation to categorize animals based on their economic function. Different writers from around the world use different terminology to express this trend. However, we could say that animal welfare legislation tends to categorize animals as one of the following:

- Agricultural—raised for meat, dairy, eggs, or other products, such as feathers and fur
- Research—used for scientific research or education purposes
- Exhibited—used for entertainment such as in zoos and circuses
- Sports/gaming—used in competitions such as horse and greyhound racing
- Companion—maintained in the home as pets
- Wildlife—free-living native animals
- Feral animals—free-living introduced species

To demonstrate this idea, consider the example of a single rabbit whom we will call Bugs. Bugs is bought at a pet store and taken home as a present for a young boy. The boy is careless and leaves Bugs's hutch open. Bugs escapes and lives freely at the local golf course for a year. The manager then undertakes a trapping program to rid the golf course of rabbit pests. Bugs is trapped and sold, along with all the other rabbits, to a fur farm. The fur farming company then decides to move their business offshore and sells its remaining animals, including Bugs, to the local college where animals are used as part of an education program. Two years later the college decides to stop using animals in teaching and they donate all their animals, including Bugs, to a small local zoo. If all that was to ever happen to a single rabbit it would be an extraordinary journey! However, the key point to note is that at every stage of Bugs's life, the law would have protected him in different ways. When he was an agricultural animal the local animal welfare legislation may have ensured that he received food and water. When he was a research animal he may not have had any legal right to food and water, but the law may have said that he had to be provided with straw for nesting. As an exhibited animal he may have been legally entitled to live in a large cage, but when he was a feral rabbit he may have had no legal protection whatsoever. What this suggests is that legislatures do not begin with premises such as "what do rabbits need?" or "what would cause a rabbit to suffer?" Rather, animal welfare legislation is best viewed as seeking to protect animals from "unnecessary," or "unreasonable" suffering, while allowing animal industries to engage in their trade.

Robert Garner is a political scientist who specializes in animal issues. Also using the example of a rabbit, he describes the tendency of animal welfare legislation to categorize animals according to industrial use:

> [T]he level of protection afforded to an individual animal depends, not just—if at all— upon its needs and interests, but upon the institutional and legislative structure governing the particular use to which it is being put. To take one example, a rabbit raised for food would be subject to a totally different set of legislative criteria than would one utilized in a laboratory or one existing in the wild or one owned as a pet. (Garner, 1998, p. 21)

Although animal welfare legislation is now widespread throughout the developed world and becoming more common in the developing world, international law has been slow to take up the challenge of protecting animals on a global scale. There are a number of treaties and conventions in place that either deal with animal issues directly or include them in their terms of reference. Two of the best know are the International Convention for the Regulation of Whaling and the Convention on International Trade in Endangered Species of Wild Fauna and Flora (CITES). CITES came into effect on July 1, 1975, and seeks to protect species against the threat caused by international trade in wild animals and plants. Furthermore, the OIE or World Organization for Animal Health has developed international standards for the transport and slaughter of live animals, and the European Union has enacted a range of conventions intended to protect animals.

However, as with all international law, the regulations that exist are difficult to enforce. Furthermore, they tend to focus overwhelmingly on wildlife protection. Although the protection of wildlife can have a welfare aspect to it, it is more appropriate to view animal welfare legislation as something that applies predominantly to animals maintained in a captive state. The current lack of international animal welfare standards is problematic because, as with all capital in a global economy, animal industries move around the globe. For example, the majority of the world's fur is now produced in China. This was not always the case. Previously, European countries led the market in fur production. The ability of fur producers to move from Europe to China means that the animals farmed for fur are unlikely to receive the same legal protection they once did. This is because European Union countries have considerably stricter animal welfare laws than do Asian countries. The practice of animal industries moving around the globe in pursuit of weak animal welfare laws is in part a reflection of the fact that legally enforced animal welfare practices can be expensive for industries to implement and maintain.

What the Critics Say

Even though animal welfare legislation provides animals with protections that would have been unthinkable two hundred years ago, animal statutes have been criticized by many animal welfare/rights groups and some academics. One common criticism made of modern animal welfare legislation is associated with the tendency of legislatures to include a range of exemptions in the statutes. Employing strong language, U.S. legal theorists David J. Wolfson and Mariann Sullivan argue that the following is true in the United States:

> In the case of farmed animals federal law is simply irrelevant. *The Animal Welfare Act*, which is the primary piece of federal legislation relating to animal protection and which sets certain basic standards for their care, simply exempts farm animals, thereby making something of a mockery of its title. (2004, p. 207)

Another common criticism made against animal protection laws has to do with the common use of terms such as "necessary," "unnecessary," or "justifiable." The first British animal welfare statute did not employ such terminology, although it did refer to "improper treatment" (3 Geo IV, c71). Although legislators did not avail themselves of overly subjective terminology, some social commentators of the time did. In the June 1809 edition of the influential British publication the *Gentleman's Magazine*, it was argued that "few subjects in the whole compass of moral discussion can be greater than the *unnecessary* cruelty of man to animals" (cited in Radford, 2001, p. 38, emphasis added). Words such as "justifiable" and "necessary" are now very commonly written into animal welfare legislation, and some commentators have argued that such terminology can have a significantly negative impact on the effectiveness of animal welfare laws. One of the strongest critics of the use of "necessary" in animal statutes is U.S.-based legal professor Gary L. Francione. Prof. Francione argues that

> Some laws prohibit the "unnecessary" infliction of suffering, but such laws are useless, if, as is the case, no one is under a duty not to do any particular act; and indeed, virtually all acts involving animals are considered "necessary" as long as there is some identifiable human benefit. (Francione, 1996, p. 193)

Concern over the legislative use of "necessary" and "justifiable" has resulted in some animal advocates calling for an end to the use of animal protection legislation based on welfare principles and a move toward laws that create rights for animals. Such a shift would be consistent with the history and development of human protection legislation and critics argue it would afford animals stronger legal protection. However, animal protection legislation that creates legal rights for animals would arguably be as radical a development as the enactment of the first piece of animal welfare legislation was in the early nineteenth century.

Further Resources

Fairholm, E. G., & Pain, W. (1924). *A century of working for the animals: The history of the RSPCA 1824–1924*. London: John Murray.

Finsen, L., & Finsen, S. (1994). *The animal rights movement in America: From compassion to respect.* New York: Twayne Publishers.

Francione, G. L. (1996). *Rain without thunder: The ideology of the animal rights movement.* Philadelphia: Temple University Press

Garner, R. (1998). *Political animals: Animal protection politics in Britain and the United States.* Basingstoke: Macmillan.

Massachusetts Colony. *The Massachusetts body of liberties (1641).* Hanover, IN: Hanover College. http://history.hanover.edu/texts/masslib.htm

Meyer, C. (1996). *Animal welfare legislation in Canada and Germany: A comparison.* New York: P. Lang.

NSW Government. (1979). *Prevention of Cruelty to Animals Act.* NSW Government Printing Service.

Radford, M. (2001). *Animal welfare law in Britain: Regulation and responsibility.* Oxford: Oxford University Press.

Ryder, R. D. (1989). *Animal revolution: Changing attitudes towards speciesism.* Oxford: Basil Blackwell.

Singer, P. (1995). *Animal liberation* (2nd ed.). London: Pimlico.

Sunstein, C. R., & Nussbaum, M. C. (Eds.). (2004). *Animal rights: Current debates and new directions.* New York: Oxford University Press.

Turner, E. S. (1964). *All heaven in a rage.* London: Michael Joseph.

Wolfson, D. J., & Sullivan, M. (2004). Foxes in the Hen house—animals, agribusiness, and the law: A modern American fable. In C. R. Sunstein & M. C. Nussbaum (Eds.), *Animal rights: Current debates and new directions.* New York: Oxford University Press.

Siobhan O'Sullivan

Law
Public Policy and Animals

Five aspects may be used to introduce the field of animals and public policy. Part One opens the door to the field. The second and third parts ask straightforward questions about what we mean by the deceptively simple terms "animals" and "public policy." Part Four outlines a series of basic problems that one encounters often in this field. The conclusion explains why the study of "animals and public policy" is an important, groundbreaking field.

Part One—Questions about Public Policy and Animals

Studying "animals and public policy" not only encourages but also actually *requires* one to address a surprising range of themes. One ends up asking, for example, what *is now* happening in our relationships with other living beings. To explain what is now happening, one has to inquire about what has happened *in the past*, as well as what can happen *in the future*. These questions may seem simple at first, but as one engages them more and more, a remarkable range of new questions constantly emerges. This field is thus characterized by an interdisciplinary approach, that is, an approach that calls upon many other fields in order to evaluate the place humans have given to Earth's other animals in our formulations of "policies."

Another set of challenging questions arises when one tries to figure out *national* policy in industrialized countries—what role, for example, might experts such as veterinarians play in setting national policy on animals? Veterinarians are knowledgeable about many kinds of animals—are they and other science-based experts consulted when our society sets policy regarding particular animals?

When we go beyond national boundaries to worldwide policy, we stumble onto new levels of complexity. Complicated notions such as international law, trade, and cross-border politics and justice arise.

Most important, the field raises the profound issue of precisely *who* makes our policies toward other living beings. Everyone knows that local, national, and international governmental bodies make policy, but only a little exploration of the subject makes it clear that many others do so as well—consumers, researchers, corporations, unions, and the professions are only a few of the nongovernmental actors who develop policies regarding the living beings in and around our human communities.

With so many people contributing to policy formation, questions of consistency arise, as do questions of enforcement. The latter in particular are important in telling us what a society's *real* policy is regarding animals—sometimes, for example, laws that have been enacted are not enforced. The American political pundit Will Rogers once said, "People who love sausage and respect the law should never watch either one being made." In some ways, this tongue-in-cheek comment, which plays upon the messy and disorderly processes of making sausage and the law, applies just as fully to the chaotic world of policy formulation.

In this field, then, one studies our human communities' interactions with the world's other living beings. We begin with simple questions about this interaction, and as we try to answer these questions we encounter anything but simple answers. Answers must rely on many sciences, many academic fields, and, hopefully, commitments to both the truth and common sense.

Whatever answers we give to our basic questions about who makes policy and why, we will face important additional questions. How well do we now see other animals? Do those who set policy about animals know them well? How often have we hurt them because of our ignorance and arrogance? What kind of information about them should we have as we decide whether, as a matter of ethics and compassion, we should notice and take them seriously? Lastly, is it possible for us to live happy lives when they don't? Might it fulfill humans to care about other animals, or can we be fulfilled if we care only about ourselves?

The goal of the field of animals and public policy is to answer these and other similar questions.

Part Two—"Animals"

Perhaps the most fundamental question for this field is this—what and who are "animals"? This single, seemingly simple word describes, upon careful consideration, an astonishingly complicated group of living beings.

A convenient place to begin is the way that the term "animals" is used in scientific circles. Such uses *clearly* include humans—virtually all of us now agree that humans are mammals and primates, both of which are clearly animals.

In other circles, though, our scientific convictions seem to waver. We use, for example, the phrase "humans and animals" as if somehow all of the *other* living beings, but not humans, should be included in the term "animals." So at the very beginning of engaging "animals and public policy" we need to recognize that the most common uses of the term "animals" are not only *non*scientific, but actively and decisively *anti*-scientific.

There is, of course, a risk in using decidedly unscientific notions—will we be misled if we end up believing that such uses of the word "animals" are truly accurate? Whether or not one decides to honor the scientific insight that humans are animals, how we speak about the other living beings in this universe remains a politically charged issue because some people have *very* strong views about the appropriate terminology in this matter.

As one explores what people think about our animal cousins, it soon becomes obvious that most people bring to discussions about "animals" a surprisingly strong set of opinions. This is probably because, before any of us ever thinks about the other living beings on earth, we have *already* long been hearing about the world from our parents or others who raise us. In fact, by the time we begin speaking and thinking about animals, we already have in place an inherited set of ideas, words, and, sometimes, crazy notions about living beings from those who raised us.

Some of us break away from this inheritance, insisting that our obligation is not to advance what we have been taught but, instead, to notice and take seriously animals' actual realities. But, sadly, many simply take their inheritance in this area as an actual description of nature. This is tragic because in many of our inherited stories and ideas about animals are remarkable ignorances and biases regarding our fellow creatures.

Pointing out such ignorances and biases may be consistent with the spirit of science, but it is not, it turns out, always a popular thing to do. Lots of people don't want to be associated with, let alone called, "animals." So making the scientific argument "we are animals" is sometimes seen as radical—it can also exemplify that, as the English author George Orwell once said, telling the truth during times of universal deceit will be viewed by some as a revolutionary act.

Another way to grasp why the word "animals" is so complicated is to listen carefully to those around you—just how much do they *really* know about the animals they talk about? If they deny "animals" have feelings, is that assertion based on careful exploration of other animals' realities? Or are they merely passing along an inherited ignorance or bias? And when someone speaks of the birds that fly overhead, the deer and coyotes around your community, the whales and dolphins off the coasts of their country, or the dogs in our midst, do they really know these animals?

In scientific circles, much remains unknown about "animals" even if in policy circles people are not at all in doubt about what should be done. Consider the living beings that we are unable to see with our naked eyes—of them, the famous Harvard scientist E. O Wilson (1992, p. 5) says, "Five thousand kinds of bacteria might be found in a pinch of soil, and about them we know absolutely nothing." Wilson adds a description of how diverse unseen life can be—consider the "aeolian plankton":

> A rain of planktonic bacteria, fungus spores, small seeds, insects, spiders, and other small creatures falls continuously on most parts of the earth's land surface. (1992, p. 20)

Beyond these invisible micro-creatures, what of the "macro" animals, that is, those we can see easily? How much more do we know about them? Consider another comment by Wilson (1992, p. 4):

> Animals are masters of the chemical channel, where we are idiots. But we are geniuses of the audiovisual channel, equaled in this modality only by a few odd groups (whales, monkeys, birds).

About those who are "masters of the chemical channel, where we are idiots," perhaps we (and our policies) should be humble. About the small group who, like us, are "geniuses of the audiovisual channel," we might be able to say much more. They are, obviously, more like us than are, say, bats that use echolocation.

One thing we can assert with confidence about the other living beings in and near our homes is this—there are *lots* of animal "communities" out there in the world. Of these many communities, scientific literature speaks readily about some as very complex—for example, there are now commonly scientific discussions about cultures among chimpanzees, complex communication and learning in dolphin societies, social and emotional realities in wolf pack interactions, and on and on.

When we factor in how many species are unknown—literally tens of millions— there are lots of reasons to be humble about what "animals" are and experience. Another reason to be humble about how little *we* know of "animals" is this interesting feature about human interaction with "them"—some human cultures have been much more attentive to nonhuman animals than have the citizens of modern, scientific, industrialized cultures. This becomes clear when one asks certain groups (for example, many indigenous peoples) a simple question like "which living beings matter to you?" The list of animals that matter in "public policy" circles today usually includes those that we hold to be valuable *resources* for our own benefit, such as food animals or charismatic megafauna that we think our children would like to enjoy in the future (whales, pandas, bald eagles, and so on). In many other societies, people have a much longer list of valued animals than that that held by, for example, modern economists who often are integral players in the realm of public policy formation.

Some parents, of course, have trained their children to see other animals carefully— such training reveals that, as a philosopher once said, your ethics will be determined by the entities you are prepared to notice and take seriously (Clark, 1977, p. 7). Imagine what public policy would be like if those who framed it noticed other animals carefully and took them seriously. Imagine what public policy is like when such an approach is *not* taken.

With only a little investigation, then, those who study human-animal relationships notice that there are many different *human* visions of how we can live and interact with nonhumans. Further, notice how often people who have a broad, bold vision about living with lots of other animals suggest that we not only *can* do this but that we also *should* do so. In the area of "animals," this kind of ethical discussion among ordinary people has become increasingly common even though this kind of talk is not yet that common in traditional policy making circles.

Given that we can live in many different ways with other living beings, just what should *our* "public policy" be? Might we, for the sake of our children, strive in the future to create policies that bequeath to our heirs a fuller world than that in which we live? Might we do this primarily because we value other animals in and of themselves?

Whatever one's answer to the question about what we *should* do, consider how we go about getting enough information to have a robust, fair conversation about this issue. Does it make any sense to study only one culture's view of other animals, for example, that of Mexicans or Brazilians or Nigerians? This would narrow our understanding considerably.

What happens if we decide to study a number of different societies' policies, but we confine our study to only *industrialized* countries? If we engage only a few versions of our relationships with other-than-human animals, we end up with only a partial list of our options. The study of animals and public policy helps us see that our options today remain remarkably diverse—some are kind to other animals, and some are obviously much more human-centered.

If we study many different human visions of what our relationships can be with other forms of life, we must take an approach that some call "interdisciplinary." This is because our collective wisdom about "animals" is held in many different disciplines, among which are anthropology, cultural studies, ethology, religious studies, philosophy, and ethics.

Consider that the most ancient view of animals is that nonhumans are the bringers of blessings, even divinities. A modern version of a similar insight was recently stated by the theologian Thomas Berry (2006, p. 5):

> Indeed we cannot be truly ourselves in any adequate manner without all our companion beings throughout the earth. The larger community constitutes our greater self.

But consider a different story, namely, that which prevails in modern "policy" discussions about the appropriate relationship of humans and our cousin animals—this particular version is from the economist Leon Walras's 1883 classic *Elements of Pure Economics or The Theory of Social Wealth* (p. 54): "Man alone is a person; minerals, plants and animals are things."

This attitude is very prevalent today, but it can cause severe problems for any number of reasons. One is that life is interconnected—as Darwin suggested, life is in fact a dance of partners. Each form of life is involved in countless, inevitable actions and reactions to other forms of life. The tendency to see humans as separate from the rest of life fails to put us in our real context. Nonetheless, much policy is run as if humans can impact nonhuman lives in countless ways without risk to themselves or the general ecosystems in which all life must live.

Eliminating our "dance partners," as it were, is something humans are now risking—currently, the rate of nonhuman species being extinguished because of human action is in the range of 1,000 times the normal or "background" rate of naturally caused extinctions. It was reported in 2004 that 1 in 4 bird species is now "functionally extinct" (Proceedings of the National Academy of Sciences, December 2004). Nearly one-fourth of the world's mammals, one-third of the amphibians, and close to half of all turtles and tortoises are threatened with extinction (November 2004 Global Species Assessment by the World Conservation Union (IUCN)).

Even worse off are primate species, 1 in 2 is threatened with extinction, and large, oceanic fish, which in June 2003 were reported to have suffered a decline of 90 percent in the last 50 years. Worst of all may be our closest evolutionary cousins, the nonhuman great apes, which have been reported to have lost 93 percent of their numbers during the twentieth century.

At the same time that we diminish these fellow citizens of the world and our human communities increase the number of animals used for food production and for experiments, in other ways we exhibit the important ability to care about some other animals. With certain protected wildlife and, above all, with our companion animals, we've shown remarkable capacity to care. It is this peculiar dynamic of harm and protection that is explored by "animals and public policy" as it engages the many human relationships and possibilities with nonhumans.

Part Three—"Public Policy"

"Public policy" is taught in many places, and is for obvious reasons of great concern to many humans. But the notion of "public policy" is generally used in ways that are extremely inhospitable to concerns for any kind of nonhuman animals. This is because policy is taught at institutions that are relentlessly human-centered. Many individual academic subjects are also taught in this relentlessly human-centered way—economics, law, bioethics, religion, anthropology, and, of course, politics. Ironically, sometimes even those in the field of environmental studies assume that the real concern is what will happen to humans' environment, as if we alone occupied ecological systems.

The vast majority of theorists and political scientists who describe what "public policy" is or ought to be exhibit in their work unabashedly human-centered values. Interestingly, though, such theories often ring hollow even for some humans because, as most know, not *all* humans are beneficiaries of public policy decisions. Often, elite groups manipulate policy decisions to their own advantage.

Studying "animals and public policy" can help one see how biased past studies of public policy have been. Consider how narrow some simple descriptions of "public policy" are—sometimes people assert that "public policy is government action." This is far too simple, for policy includes much more than government actions. The "public" in "public policy" often includes, as already noted, a diverse, ever changing combination of consumers, corporations, unions, religious institutions, private landowners, and many other nongovernmental groups.

Whether the net effect of "policies" put into place by the actions and decisions of such a large and diverse collection of humans is good or bad for either humans or nonhumans is a complex subject. But generally one can say that the public policy, whatever its composition, needs to be informed about the Earth's other forms of life if we are to have a chance of making reasonable choices about our shared world. Contemporary public policy programs are, on the whole, virtually autistic about this commonsense, ecological insight. It is this shortcoming that can be addressed well by a careful study of "animals and public policy."

For all of these reasons, it should be clear that public policy will not be well taught if it is taught in ways that are focused exclusively on human facts and interests.

An example of an extreme imbalance that harms both humans and nonhumans is current official national policy in the United States promoting industrial agriculture and its "animal cities" in ways that promote only "the industrial values of specialization, economies of scale, and mechanization" and thereby forego "ecological values of diversity, complexity, and symbiosis" (Pollan, 2006, p. 161). The results of such a policy favoring factory farming are, according to Pollan, harmful to both humans and nonhumans.

"Animals and public policy" also studies the ways in which poverty or affluence affects views of our possibilities with nonhumans. Poverty can, of course, overwhelm our sense of ethics in protecting humans, nonhumans, the natural world, and its ecosystems. But, equally, affluence and arrogance can blunt our ethical sensibilities when consumerism becomes the dominant approach to the more-than-human world.

One who studies "animals and public policy," then, must know well the *human* situation and predicament. A second reason one must study humans is the cultural diversity of views about animals—some human cultures have exhibited an extraordinary ability to live with nonhuman animals. This is one reason that loss of indigenous wisdom regarding possible lifestyles that are in balance with the earth and its surroundings is tragic in the extreme. All of us—humans and nonhumans alike—are disadvantaged when human cultures are forced out of existence by our insensitivities, arrogance, and ignorance.

In general, then, the study of animals and public policy can make it clear why framing public policy only in human terms is *starkly inadequate* to the task of describing the world. As Frances Moore Lappe said, "Every choice we make can be a celebration of the world we want." This simple insight, which invokes our ethical abilities and imaginations, is crucial to recognition of all our possibilities in the realm of "public policies" toward Earth's other living beings. If this broad approach is taken to "public policy," and in particular if we notice and take other animals seriously as is done in the study of "animals and public policy," then we can experience "the power of grasping that the world could be other than it is" (Orwell, 1949, p. 211).

This field, then, begs a simple description of humans as ethical beings capable of answering our fundamental questions, "Who are the others?" and "What am I do to do with my finite abilities in regard to caring for such others?" The others can and do naturally include nonhuman animals and the larger ecosystems of which all of us are a vital part.

Part Four—Basic Dilemmas

As suggested previously, there are many interesting facts and features of the world around us that make the study of "animals and public policy" much more than a simple, minor subject. Two complications in particular are described here as one faces the challenges of studying "animals and public policy."

First, there is "the complications dilemma"—studying any kind of animal is complicated, and sometimes studying *nonhuman* animals is *far* more complicated than studying humans. This is because although we obviously can understand some features of some nonhumans' lives, it is equally obvious (and a matter of experience) that we do not, maybe even *cannot*, understand them in every respect and often not well at all.

Consider what we might understand of the nonhuman animals that are not near us in any sense—for example, sperm whales. These are deep-ocean creatures who dive easily to depths that are completely inaccessible to us even when we are aided by our most sophisticated technology. How do we learn about these creatures, who happen to have the largest brains ever? These animals have fundamentally different sensory apparatus than we do—they can echolocate and make sounds that we cannot hear. They thus live with each other on the basis of entirely different skills than we use when we live with each other. How do we, eminently social primates dominated by vision, learn about animals that are dominated by different senses such as sound or smell?

Animals with such different bodies and lives, though, are *in some ways* clearly *not* that different from us—some experience pain that is, physiologically, indistinguishable from our physical pains; many have families and live intensely social lives; and many have some of the forms of intelligence that we have. They bleed, they scream, and they are curious and sometimes friendly. Thus, with regard to some living beings, we seem to understand some things about them. We even adopt some of them into our family, even though we might still hold them inscrutable (think of cats). Through all the differences and similarities, our shared creaturehood is an invitation to try to understand *some* features of their lives.

Second, there is the entirely different dilemma of trying to ask unpopular questions. One cannot engage the world around us honestly, or participate in policy discussions about animals of any kind (whether human or nonhuman), without asking about the animals themselves. Science, ethics, and every one of humans' higher instincts (though not our more base instincts, to be sure) require us to inquire openly about the world around us. There are, literally, pieces of the "animals and public policy" puzzle *everywhere around us*—in the ever-present biological beings themselves, in our language, in our ecological realities, and, most relevant here, in our political risks.

But even if such "animal pieces" are everywhere around us, it is still a dilemma to talk openly about such things—this is because, frankly, discussions calling us to take other animals seriously are, in many circles, discouraged. In academic settings, insistence that harm to other living beings is an important *ethical* issue can be very harmful to one's prospects for advancement.

What makes some discussion possible is that "animals" (or to use the scientific phrasing, nonhuman animals) are now an increasingly popular topic. Thus, even if fundamentalisms denying the possible importance of nonhumans abound in some circles— whether secular, religious, business, government, or academic—a societywide "ferment" now exists on "the animal issue." More and more, people are studying animals, and more and more rigorously developed scientific findings are being announced regarding their actual lives and realities.

Part Five—Going Forward: Locating the Field of "Animals and Public Policy"

The study of "animals and public policy" is, then, for any number of reasons an important, groundbreaking field. The larger category of "human-animal studies" is growing, such that we have prospects of better understanding our history and possibilities with other animals. This surely requires many different approaches and disciplines, for our relationship with the rest of life on this planet is nothing short of extremely complicated and complex. But with emerging "think tanks" and academic programs studying "animals" more and more, and with the "ferment" on this subject increasing even as we continue with our environmentally destructive practices, the topic of "animals and public policy" will sit at the center of our attempt to understand ourselves. If, as Thomas Berry has said, "we cannot be truly ourselves in any adequate manner without all our companion beings throughout the earth," then the study of animals and public policy is an essential part of human self-discovery.

See also

Ethics and Animal Protection—*The Great Ape Project*
Ethics and Animal Protection—*Political Action Committees (PACs) for Animal Issues*
Ethics and Animal Protection—*Political Rights of Animals*
Ethics and Animal Protection—*Practical Ethics and Human-Animal Relationships*

Further Resources

Berry, T. (2006). Loneliness and Presence. In P. Waldau & K. C. Patton. (Eds.), *A communion of subjects: Animals in religion, science, and ethics* (pp. 5–10). New York: Columbia University Press.
Clark, S. R. L. (1977). *The moral status of animals.* Oxford: Clarendon.
Orwell, G. (1949). *Nineteen eighty-four, a novel.* London: Secker & Warburg.
Pollan, M. (2006). *The omnivore's dilemma: A natural history of four meals.* New York: Penguin Press.
Walras, L. (1954). *Elements of pure economics or the theory of social wealth* (W. Jaffe Trans.) from the 1883 edition of *Elements d'economie politique pure.* London: George Allen and Unwin.
Wilson, E. O. (1992). *The diversity of life. Questions of science.* Cambridge, MA: Belknap Press of Harvard University Press.

Paul Waldau

■ Law
Taiwan: Animal Welfare Law

As in many Asian countries, cruelty to animals can still be witnessed in Taiwan. Chickens are slaughtered in public at meat markets, dog meat is still sold, traditional Chinese medicine shops at times sell parts of endangered animals, and hunters still illegally trap monkeys from forests. Despite the widespread daily occurrences of animal cruelty, Taiwan has some of the world's most comprehensive laws governing the treatment of animals. The majority of the public in Taiwan tend to agree that animals should be treated fairly, like humans, based on moral or religious grounds.

Taiwan became the fifty-fourth country to introduce the law concerning the welfare of animals when it implemented the Animal Protection Law in November 1998. The law covers the treatment of all animals, including domestic and wild, and addresses maltreatment and abandonment as well as the use of animals in gambling. Fines against those who violate the law range from USD 60 to USD 75,000 and in some cases cruelty can lead to imprisonment.

According to the law, pets must be registered and given IDs by supervising agencies. Registration includes a complete record of the animal's birth, acquisition, transference, loss, and death. In addition, a program for sterilization and microchip implantation will be implemented for more effective management of pets. All citizens who keep animals, including private pet owners, veterinarians, pet stores, pounds, shelters, and so on, must supply the animals in their care with sufficient quantities of food and water. Safe, sanitary, and adequate living space with proper ventilation, lighting, and temperature must also be given. Owners may not abuse, harass, torture, harm, or otherwise maltreat their animals. Animal owners must provide proper medical care to their animals in the event of injury or illness. Any treatment or operation performed must be carried out by a qualified, licensed veterinarian. Pet owners who no longer desire the responsibility of caring for their pets are required to send them to a licensed shelter, and they cannot simply abandon them. Whenever animals are being transported, proper attention must be given to their food, water, excretion, environment, and safety. They must also be kept free from shock, pain, and harm during their journey. The euthanasia of animals must be carried out in a humane fashion by a trained veterinarian. Care should be taken to ensure that any pain felt by the animal is minimized.

Animal welfare activists continue to criticize that despite Taiwan's comprehensive laws, many of them only exist on paper due to lack of enforcement. Taiwan's Council of Agriculture, which is the government agency responsible for enforcing the law, has only six employees to cover the nation's capital city, Taipei City, and even fewer to cover other areas. Thus it's not astounding that the law enforcement is very weak. In addition, there are many legal loopholes in the law, so that individuals who harm animals are rarely prosecuted for the correct reasons. An example is the large number of stray dogs and cats wandering on the streets of Taiwan. This clearly shows that pet owners are rarely punished for abandoning their animals.

The stray dog issue has been publicized the most among all other animal-welfare issues in Taiwan. According to the animal welfare law, any dog found without a collar or identification chip must be held in a shelter for 10 days. If the dog remains unclaimed after this period it can be legally put to death by lethal injection. Some shelters ignore this regulation for financial reasons and euthanize the dogs only after four or five days. The animal rights group Life Conservationist Association is against the killing of any animal, and it argues that the way in which dogs are euthanized is contrary to the legal guidelines.

It further adds that dog shelters usually contract out the task to euthanize dogs, and it is sometimes done inhumanely without the guidance of a trained veterinarian.

Each year about 300 million chickens, 31 million ducks, 10 million pigs, 6 million geese, 364,000 turkeys, 260,000 goats, and 40,000 cows are slaughtered in order to satisfy Taiwan's growing meat demand. Although the majority of the animals are butchered in the 80 government-run slaughterhouses, some are killed at private places that still perform inhumane methods of slaughtering. According to the law, slaughterhouses must first shock the animal with an electric prod and then, when it's unconscious, fire a single shot from a bolt gun into the animal's head. If the slaughterhouse staff practices inhumane treatment, it faces a fine of USD 3 to USD 15,000 and the government can suspend the license to operate. Because animal rights organizations have put tremendous pressure via news and television media on the slaughterhouses to practice humane treatment to animals, they have been denied access at times to visit slaughterhouses.

Besides the domestic animals, their wild counterparts also suffer cruelty in captive facilities and in the wild. For example, an orangutan was seen chained to the floor for the sake of photography for visitors at a recreational park. Chained pig-tail macaques were also seen performing circus acts in a zoo, and one macaque was seen tied on a steel wire with a lock on a tree. Wild Formosan macaques at Mt. Longevity continue to face illegal trappers and hunters. Thus the government of Taiwan must realize the fact that the current law enforcement to reduce animal cruelty and relieve animal suffering is weak and not acceptable. I hope that the Council of Agriculture officials reorient their approach and collaborate with animal rights organizations, animal welfare professionals, and ordinary citizens to develop workable law enforcement strategies, and only then can cruelty to wild and domestic animals in Taiwan be minimized in the future.

Further Resources

Agoramoorthy, G., & Hsu, M. J. (2005). Use of nonhuman primates in entertainment in Southeast Asia. *Journal of Applied Animal Welfare Science, 8,* 141–49.

Minna J. Hsu

■ Literature
Animal Stories across Cultures

Animal stories abound across cultures today, but historically they also have played key roles in the formation of distinct human traditions. Informing ancient traditions as well as current practices worldwide, narratives of human-animal relationships serve as a primary means of accounting for correspondences within and across species. As varied as the people who share them and the animals they purport to describe, these kinds of stories also have influenced key changes in religion, literature, politics, and science.

In many societies, animal stories are used to explain social bonds. Legends preserved through oral traditions cast animals in a range of culturally defined roles, indicating how crafting stories about members of other species has long helped people to make sense of their own lives as well. For instance, across the Americas, various tribal origin stories depict dogs as direct kin, even crossing species lines as dog-wives (Arawak, Dogrib, Chipewyan, Ojibwa) and dog-husbands (Quinault, Tlingit, Haida, Nootka) to human progenitors. Although these stories clearly enable members of one culture to account for their own origins, other North American myths use the same kind of animal to mitigate relations with those outside the group.

Particularly when these ancient stories complement the daily and ritual practices of neighboring cultures, they show how stories of animals work to define cultures both internally and in contradistinction to each other. In other origin stories mythical dogs protect the ancestors from wolves (Cherokee, Penobscot), people (Cheyenne, Potawatomi), and even witches (Micmac). What is more, the ways in which mythical animal narratives change within cultures informs shifts in religious and other cultural practices, showing how belief systems not only document the development of these societies within mixed species communities but also reflect profound social and environmental changes.

When animal stories become part of religious doctrine, they illustrate why some species are considered sacred, whereas others are abhorrent, and, perhaps more importantly, define the ambiguities of particular human-animal relations. The sacred texts of monotheistic religions project a profound ambivalence toward animals through a combination of stories that mix reverence for loyal (often sacrificial) individuals, such as the guard dog Kitmir in the Koran, with abhorrence of packs of (especially unclean) animals, such as the devourers of the Old Testament's Jezebel. If the first meaning reflects the influence of pagan attitudes predating these religions, then the second introduces knowledge that is more specific to the burgeoning urban contexts of these religious communities, such as the increased threat of cross-species transmissible diseases such as rabies. More surprisingly, conflicting stories of animals at the heart of these religions lay the foundation for reform and revolution; at least, this is one way of understanding the medieval and renaissance development of the kinds of secular humanist philosophy that subsequently undermined absolute ecclesiastical Christian authority by providing (among other things) a compassionate rationale for petkeeping practices that apparently had persisted since prehistoric times.

Such accounts (and the practices based on them) may be meant to benefit humans, but as they mutate across the millennia their impact has been less predictable. The co-evolutions of still other religious stories prove far less ambivalent for their animal subjects. An Indian tradition dating from pre-Vedic times, the practices of Naga Panchami (or the Festival of the Snakes) have come under harsh criticism in recent years by wildlife conservationists. With the integration of this festival into the stories of Hinduism (especially the legends of Krishna that involve the god's feeding milk to a snake), its celebration in some regions has come to involve capture and mutilation, including deadly forced-feedings of milk, to wild snakes in large numbers. In this case, the fusion of disparate animal stories transforms a ritual based in reverence for animal lives into an instrument of their endangerment and possible extinction.

Structural elements of these narratives inform more recent developments in secular storytelling traditions. The exaggeration or isolation of a single characteristic in ancient depictions of animals as anthropomorphic characters models the animal heroes and villains of medieval and Renaissance literary traditions, including fables, bestiaries, and romances. In place of cross-species kinship ties to humans and more fabulous mythic powers, the main purpose of animals from Aesop through LaFontaine is to serve as metaphors or substitutes for humans, forming a pattern in which story animals more plainly appear to be creatures of human imaginations, who gain cultural value as tools for expressing and teaching humanist values. Often ancient stories such as Apuleius's *Golden Ass* retain a mythical and fabulous premise, here of the human protagonist's transformation to an animal, and even medieval fables such as the Persian Kalia wa Dimnah stories relay other animals' humanlike lives from the utterly imaginary perspective of animal narrators (in this case a pair of jackals). Although less magical in this sense, the Enlightenment development of children's literature continues in these traditions of redefining the animal as a perspective from which to gauge what is properly human. Such fictions today are often assumed to instill pro-animal sentiments. When animal characters are strictly and simplistically used to teach these ideals, however, they can do so to the detriment of understanding human-animal relations.

In literary history, one effect of these confusing and confused associations has been the devaluing of animal stories through the modern period, a response that also seems to reflect anxiety about the shifting status of human-animal relations in these texts. Although the morals of animal stories ostensibly concern human life, especially for the benefit of naïve or younger readers, their concern with dissimulation means that the lines between what are human as opposed to strictly animal properties become blurred. These aspects may be unsettling to readers who seek confirmation of human superiority in them, and as they become more pronounced in modern fiction they set in motion particular social and scientific developments of animal stories as well.

The history of literary criticism from the early modern period onward illuminates this problem. Metaphorical approaches to narrating animal perspectives flourish in satire and other forms of social critique, but these uses of animal masks come to anchor expressive and mimetic approaches to literary animal stories in, respectively, romanticist and formalist aesthetics as well. Although more literal animal stories are crafted to react immediately (and often strongly) against moralizing animal stories, in these literary traditions the animal remains first and foremost a reference point, a symbol/signal of the work of the poet/critic as gatekeeper of truth. They thus operate on the assumption that animal metaphors contain a higher understanding only about human life, a faith that begins to give way with the mid-twentieth century development of structuralist aesthetics and the consequent focus on the proliferations of meanings (not to mention purposes and forms) of literary animal narratives. In lieu of struggling over a single or true human meaning, these more recent approaches to animal stories focus on how they reflect the deeper ambiguities of human-animal relationships.

And it is hard to say whether these literary critical developments influence or reflect rising popular interests in a different kind of story, one that depicts animal lives through more plausible accounts of behavior and psychology as well as (and perhaps most significantly) posits animal independence from human control. The mid-eighteenth- or nineteenth-century origins of this kind of animal story remain the subject of some debate, as are other conditions factored into its rise. Whether stemming from a concern with realist aesthetics or less directly proceeding from misprisions, nonallegorical (or not so straightforwardly allegorical) animal stories increasingly concern literary critics from the mid-twentieth century onward. In recent years this kind of story has been instrumental to the critique of knowledge itself and its relation to the axiological and historical categories that come to define Western humanism, enhancing the work of deconstructive and more broadly poststructuralist animal studies in foregrounding the conditions under which animals in literary narratives for so long seemed only ever to speak of and for the human. Particularly through analyses of how film and other media open up new concerns with animal stories, these approaches have begun to question the ends of this human-centered understanding. Applying these methods, some scholars even have begun to argue for the significance of what animal stories do not (or refuse) to say.

Concern with anthropomorphism and especially anthropocentrism in animal stories certainly informs their proliferation by the early twentieth century, and these lines of inquiry more broadly trace the rising political stakes of animal representation. Styled as firsthand accounts of the everyday horrors of maltreated working animals and pets, respectively, Anna Sewell's *Black Beauty* (1877) and Marshall Saunders's *Beautiful Joe* (1894) are examples of bestselling novels solicited and promoted by animal advocacy organizations that proved influential in public policy changes concerning animal rights and welfare. The didactic purposes and younger audiences of these fictional domesticated animal biographies may hearken back to earlier humanist traditions, but key changes in the form set them apart from their predecessors. The focus on the animal (here as narrators) allows for the detailed development of their species-specific concerns of living in and around human societies, fostering a post-Darwinian awareness of human-animal interrelations and their limits that quickly becomes a central feature of individual wild animal histories.

Although these qualities are far from mutually exclusive, some have found it politically expedient to cast the development of animal stories in this way. Voicing a growing anxiety about the irreducibly literary and scientific credibility at stake in animal narratives, Charles G. D. Roberts in 1902 proclaims that the "best" animal stories eschew sentimentality and melodrama, and instead develop "a psychological romance constructed on a framework of natural science" (p. 24). Too often cited since to be read as a statement of personal taste, this judgment reflects broader social desires for animal stories not simply to convey the truth of humans or of animals but more profoundly to foster nonutilitarian values of human-animal relationships central to what were then innovative developments of wildlife conservation policies.

The sense for some that this potential was compromised in most animal stories, however, fueled the "nature fakers" controversy erupting at the turn of the century, pivotally involving U.S. President Theodore Roosevelt and illustrating how politically volatile readers' understandings of these elements in animal stories could become. The benevolent view of animal life anchored by strictly anthropomorphic storytelling techniques dominated the literary marketplace at the turn of the century, with not just the unprecedented success of the domesticated animal activist fictions of Sewall and Saunders but more generally through the popularity of naturalist story collections such as Ernest Thompson Seton's *Wild Animals I Have Known* (1898) and William J. Long's *School of the Woods* (1902). As suggested earlier, the popular success of this kind of animal story came at an immediate critical cost.

Premier American naturalist John Burroughs thought the latter so dangerous to the education about and subsequent legislation of the land that he publicly denounced it in 1903 for promoting distorted views of animal happiness (and natural benevolence) at the expense of the harsh realities of survival (and Darwinian fitness). Consequently, a war of words erupted among virtually all authors at the time with any stake in animal stories, with the deciding vote symbolically cast by Roosevelt (who coined the term "nature fakers"). In retrospect, however, the consequences for animal stories are less clear than the corollaries of these positions in human politics, which reflect opposing liberal Democrat and social-Darwinist Republican platforms of their day. And their influence continues to be felt in narratives of animal life at the center of conflicts of preservationist and sustainable land-use approaches to conservation today.

With the development of the discipline of ethology, animal stories arguably have had an even greater impact on understanding the relations of culture and science through the past century. Through bestselling nonfiction books such as *King Solomon's Ring* (1952), Konrad Lorenz exemplifies how paradigm shifts in the biological sciences pivotally involve telling convincing and compelling animal stories. Innovative biological theories (notably of Jakob von Uexküll) model the contents of ethological study, but Lorenz turns to modern fiction (specifically Jack London's dog stories) as a model for relating observations and graphic descriptions of complex human-animal group relations. Ethological studies increasingly have become the standard by which readers judge the authenticity of fictional animal stories, yet the influence of these scientific accounts circularly derives from their successful appeal to conventions developed through animal fictions.

Bound with leather and animal glue, printed on vellum, penned with quills, sometimes even inked in animal blood, and now featuring images and imprints of animals in virtually all environments (including new media), human documents at these most basic levels influence as much as they record evolving narratives of mixed species life, though some evidently are more self-reflexive this way than others. When the contents of these documents explicitly concern stories of human-animal relations, these texts reveal how central such relationships and their representations are to the continuation of our mixed communities. Animal stories may document and reflect our changing, shared conditions over time, but the fusions of artistic and scientific knowledge emerging in animal stories suggest how they will continue to serve as critical sites for ongoing negotiations of species life.

See also

Culture, Religion, and Belief Systems
Literature—*Nature Fakers*

Further Resources

Link, V. (1956). On the development of the modern animal story. *Dalhousie Review, 56*, 519–25.

Lippit, A. (2000). *Electric animal: Toward a rhetoric of wildlife.* Minneapolis: University of Minnesota Press.

Magee, W. H. (1964). The animal story: A challenge in technique. *Dalhousie Review, 44*, 156–64.

Nelson, B. (2000). *The wild and the domestic: Animal representation, ecocriticism, and western American literature.* Reno: University of Nevada Press.

Norris, M. (1985). *Beasts of the modern imagination: Darwin, Kafka, Nietzsche, Ernst, and Lawrence.* Baltimore, MD: Johns Hopkins University Press.

Roberts, C. G. D. (1905). *The animal story. The kindred of the wild.* New York: Sitt Publishing.

Scholtmeijer, M. (1993). *Animal victims in modern fiction: From sanctity to sacrifice.* Toronto: University of Toronto Press.

Wolfe, C. (2003). *Animal rites: American culture, the discourse of species, and posthumanist theory.* Chicago: University of Chicago Press.

Susan McHugh

■ Literature
Animals in Literature

Oxen that rattle the yoke and chain or halt in the leafy shade,
what is that you express in your eyes?
It seems to me more than all the print I have read in my life.

(Whitman, 1980, p. 58)

Animals in literature act as symbol and mirror, as economic product and subject of natural history. They are named and grouped but are rarely considered as individuals. They are subjects of desire by humans attempting to become a beast; they are soul mates and creatures across a great divide. They are worshipped and brutalized and simply described. They represent everything possible and rarely themselves. They find themselves inserted into poetry, prose, and drama by writers who want to understand animals and to understand their relationship to the human. Primarily, however, they are inserted into writing for the human to understand the human.

Rarely, writers engage the animal; more often it is used as a mirror. This may be because humans (especially in Western culture) tend to see animals as the "other," grouping them as types, naming these groups and caging them in generalities. It may be because the very act of writing essentially separates the human that writes from all other animal species. To write about the animal effectively requires awareness of the irony of attempting to create the animal from what has been used to cage it, what Pam Ore calls "keys hanging cold/on the gatehooks of language" (2005, p. 4). Added to these issues is the almost ubiquitous notion of a hierarchy of life, from the Great Chain of Being to the concept of universal oneness to the greatly entrenched, ethically simplistic, speciesism. Very little writing reflects the more egalitarian concept of supernature found in some Jainist, Mesoamerican, and Turkic groups. This hierarchy has resulted in a literature that is heavily weighted to a few taxa, primarily mammals and birds. There are a few invertebrate species found across literature, those typically associated with humans (e.g., bees, fleas, ticks), and a handful of reptile, fish, and amphibian species. However, all writing about animals, whether mirroring humans or actually elucidating the animal itself, provides a picture of how humans view and have viewed animals. In some cases, these texts provide an empathetic space where the gap in being is bridged, as in Marianne Moore's "The Paper Nautilus," which describes the animal creating her shell, a "perishable/souvenir of hope" (Moore, 1994, p. 122).

Ancient texts, oral folklore, and religious-philosophical musings tend to drive the perception, and therefore the portrayal, of animals in literature. Texts, such as the 5,000-year-old Sumerian cuneiform tablets and the 3,000-year-old Indus Valley text the *Rig Veda,* name animal groups, such as dogs, and give them roles in spiritual as well as everyday life. To the Sumerian dog, "a dream is a joy" (Sumerian Proverbs: collection 5: c.6.1.05 72) while the Indus Valley deity Yama has "two guardian dogs . . . who watch over men" (Rig Veda, 10.14.11). Ancient texts have often generated and reinforced the hierarchy of animals with the notion that humans "have dominion over the fish of the sea, and over the fowl of the air, and over every living thing that moveth upon the earth" (Gen. 1:28, King James) and the idea that humans must give care to the animals (such as in the Koran) because of their dependence on human agency. The tendency to group, name, and thereby simplify other species can be seen across ancient literature, "and whatsoever Adam called every living creature, that was the name thereof" (Gen. 2:19). Modern texts are influenced by these ancient texts in their integration of the character traits of particular animals as symbols, omens, and metaphors, such as the starling in William Shakespeare's *Henry IV,* the raven in Poe's "The Raven," and the whale in Herman Melville's *Moby Dick.* At this level, the animals illustrate something about their own nature as well as about the nature of the environment or society in which they find themselves.

Though all animal texts have the potential to engage the reader empathetically and emotionally with the focal animal or species, they often gloss over individual differences among animals within groups. This is a result of a desire to capture the essence of the other rather than its actuality and a lack of interest or awareness in variation in other species. The very unity of the species in which a single lion represents all lions results in a sense of immortality for the animal, thus linking it to the mythological. In contrast, a few literary works do engage variability within species, especially more modern texts. These texts, if they are memoirs or popular science (such as Jane Goodall's *Through a Window*, Mark Bittner's *The Parrots of Telegraph Hill*, or Frans de Waal's *Chimpanzee Politics*), document the individuality of the species from a descriptive angle. These books generally maintain constant awareness of the space between humans and animal subjects of study without overwhelming the text with a desire to become the beast (except in the case of books such as *Among the Grizzlies* by Timothy Treadwell).

Other modern texts, especially those considered fantasy novels, attempt to engage the animal from inside its consciousness—to become the beast in a fictional context. In the case of books such as *Watership Down* by Richard Adams, several individuals of the same species are portrayed as characters, imbued with the variation not typical of animals in other literary texts. The origins of these texts lie in folk and fairy tales as well as in the nineteenth- and twentieth-century surge in popular children's fiction involving talking animals (e.g., the Peter Rabbit books by Beatrix Potter, the Wizard of Oz books by L. Frank Baum and subsequent authors, *Charlotte's Web* by E. B. White, and *The Wind in the Willows* by Kenneth Grahame).

One such text, Paul Gallico's *The Abandoned*, navigates the risky waters of empathetic writing by making the protagonist both human and feline, in and out of dream worlds, thus engaging the common human spiritual practice of engaging a totemic creature through a different reality. This fundamental yearning to become the beast is found in many cultures. In cultures such as those of some Native American groups, early Turkic tribes, and Jainist groups, as well as in monist philosophies such as Hindu and Buddhist religions, humans move between their form and nonhuman animal form either within a single lifetime or between lifetimes. Similarly, much classical mythology engages transformation between human and nonhuman animal, as in the stories in Ovid's *Metamorphoses* and Homer's *The Odyssey*. In some cases of transformation, an intermediary is required to effect the change; in others, it is not. When Padraic Colum is an otter, he is "Lord/Of the River—the deep, dark, full and flowing River" (Colum, 1989, p. 46).

Authors that engage the idea of becoming the beast attempt to bridge the gap between human and nonhuman consciousness. To some, this gap is only a step wider than the gap between the consciousness of a human author and a human subject ("only connect"—E. M. Forster, 1921, p. 3). To others, the gap is insurmountable, and the attempt to bridge it results in anthropomorphism, or the assumption that the animal's cognitive and emotional space can be interpreted through the author's own experience. To some critics, anthropomorphic texts raise the specter that animals share similar consciousness and capabilities as humans. To others, anthropomorphic texts make animals a mirror of humanity and erase and downplay the animals' own undeniable and valuable differences from humans. Regardless of the criticisms, however, when done well, these works engage the reader's empathy for the animal they portray.

A different approach to the animal in literature occurs in allegories and satire such as *Animal Farm* by George Orwell. The origins of such forms of literature run further back into human history than do the empathetic fantasy novels. Although modern variants such as *Animal Farm* may borrow the empathetic approach, they generally use stereotype in order to create a symbol (e.g., pigs as gluttons). Allegory and satire are facilitated in cultures for which folklore and oral literature have included an extensive role for nonhuman animals and therefore contain a great store of implicit relationships between the animal and particular characteristics. In medieval Chinese poetry, for example, the egret often served implicitly to represent stately officials or intermediaries with the spiritual domain. Aesop's Fables, a conglomeration of fables gathered from ancient Greece onward but attributed to Aesop, represents a collection of tales that borrowed from and contributed to the cultural concepts of animals and their allegorical and satirical use.

In some cases, the oral history and folklore associated with a particular animal, though culturally specific, represents a more global tendency toward the use of an animal to characterize particular traits. The trickster character in literature, especially, exploits the rich cultural associations between an animal and particular characteristics such as humor, quick thinking, storytelling, rule bending, and rule breaking. Tricksters are represented by the fox in Europe, the monkey in India and Africa, the coyote or raven in North America, and the spider in West Africa. These characters developed through folklore and found themselves included in literary texts, dragging their associations along behind them. For example, the trickster Reynard the fox, who may have in part originated in Aesop's Fables, is referred to in texts varying from the twelfth-century *Ysengrimus* by Master Nivardeus of Ghent to the fourteenth-century "Nun's Priest's Tale" by Geoffrey Chaucer to the eighteenth-century *Reynard the Fox* by Johann Wolfgang von Goethe. Similarly, coyotes, rabbits, and ravens find themselves in works by modern writers such as Leslie Marmon Silko, Simon Ortiz, and Joy Harjo.

> Like Coyote, like Rabbit, we could not contain our terror and clowned our way through a season of false midnights. (Harjo, 1991, p. 128)

Trickster tales often use the actual animal to symbolize the trickster qualities, although some simply use the name of the animal (e.g., Maxine Hong Kingston's *Tripmaster Monkey*). In general, the tales tend to be far removed from the original animal itself because they are essentially dependent on the animal as cultural construct rather than as itself.

The Romantic poet Samuel Taylor Coleridge used cultural associations with the ass in his partially allegorical poem "To a Young Ass." This text, though capitalizing on the stereotyping typical of most allegory, attempts also to exploit the gain in sympathy that occurred for nonhuman animals post-Enlightenment. At this time, increasing urbanization and industrialization in the West resulted in a shift away from lifestyles that integrated human and nonhuman animals. The early German Romantic Novalis (Friedrich von Hardenberg) includes in works such as *Heinrich von Ofterdingen* the positive valuing of nature and nonhuman animals found in later Romantic writers. These writers—such as Coleridge, Percy Bysshe Shelley, and John Clare—tended to elevate the nonhuman animal (select species at any rate) to a "higher level" than humans. Although a general regard for nature has been considered the mainstay of Romantic literature, poets such as William Wordsworth created few if any pieces in which they focused on the nonhuman animal as anything other than part of the background. Although Novalis suggested a philosophical view relatively devoid of hierarchy, the work of other Romantic writers implicitly assumes a hierarchy of species. Intriguingly, the existential novel *The Metamorphosis* by Franz Kafka uses this implicit hierarchy to uproot ideas about what is human. The issue of hierarchy allows the text to reinforce the essential isolation that each human experiences in his or her society by intimately portraying a cockroach, typically considered "vermin," as the beast into which the protagonist, Gregor Samas, turns.

The shift in the perception of animals and the concurrent explosion in natural history studies that occurred during and after the Enlightenment resulted in a flood of popularly available writing by natural historians. This literature, such as works by Charles

Darwin and Jean-Henri Fabre, generally helped reinforce the changing notions regarding nonhuman animals. Whereas Darwin and colleagues emphasized the evolutionary relationships among, and the variation within, human and nonhuman animals, writers such as Fabre explored in intimate detail the workings of species not often considered by other writers. In his *The Life of the Fly*, Fabre describes and considers a parasitoid and victim: "What manner of life is this, which may be compared with the life of a night light whose extinction is not accomplished until the last drop of oil has burnt away?" (Fabre, 1919, 2002, e-text).

A class of literature focuses on relationships between humans and animals. Many such works focus on very specific animals and their human companions. These include joyful works such as T. S. Eliot's *Book of Practical Cats* and Christopher Smart's praise of his cat Jeoffry, who "counteracts the powers of darkness by his electrical skin and glaring eyes" (Smart, 2001, p. 40). Many poets, such as Lord Byron, Robertson Jeffers, Thomas Gray, and Eloise Klein Healy, have written laments and odes for animals that have passed on. Gerald W. Barrax's mother's dog encompasses all these laments because like "all pets, with no sense of justice" (Barrax, 2001, p. 692), she haunts him even after forty years.

Works of fiction by authors such as Arundhati Roy and Milan Kundera, and non-fiction by authors such as May Sarton, Paul Gallico, and Daniel Pinkwater, explore the relationship between narrator and companion animals. Many examples of classic juvenile literature (e.g., *Old Yeller* by Fred Gipson and *Where the Red Fern Grows* by Wilson Rawls) explore this relationship. In many of these pieces, the relationship is intense, positive, and often devastating. The bond that the human feels for the nonhuman animal is rendered immediate, and periodically, the author successfully describes aspects of the animal itself developing in the way the human characters are developed. There is always a space in these texts, however, where the ultimate mystery of the animal is maintained. In Milan Kundera's novel *The Unbearable Lightness of Being*, Karenin the dog, an emotional center for the book, is clearly a developed character but only through the eyes of the humans with whom he interacts. In works such as *All Creatures Great and Small* by James Herriott and *My Family and Other Animals* by Gerald Durrell, the human-animal relationship is more of an overview of an entire society, with the animals brushed into the construct—although often at a depth that allows the animals to exist as their own fully formed characters.

Some authors describe intimate relationships with specific animals that are based not on mutual caring and caregiving, but on more violent relationships such as bullfighting and hunting. Ernest Hemingway's texts show hunting as a thrill akin to that involved in the act of becoming the beast: for example, it "was wonderful to be eating the lion and having him in such close and final company and tasting so good" (Hemingway, 1999, p. 200). However, though both hunting and bullfighting are portrayed overwhelmingly positively in his texts (e.g., *Green Hills of Africa* and *Death in the Afternoon,* respectively), and the focus is entirely on the human and not the animal, there is a shift in his later works to an increasingly ambivalent sense of the actual moment of the kill.

Other literature is more consistent and explicitly negative about the human-animal space of interaction. In these texts, the danger for animals in interactions with humans may be directly addressed, as in the poem "Butterfly" by Chinua Achebe: "The gentle butterfly offers/Itself in bright yellow sacrifice/Upon my hard silicon shield" (Achebe, 1972, p. 4). Many authors, such as Mary Oliver, Barry Lopez, Pattiann Rogers, Diane Ackerman, Peter Matthiesson, Brigit Pegeen Kelly, and Gary Snyder, pursue these issues, sometimes succeeding in obtaining an image of the animal, but more often in portraying the emotional context of the narrator. In these texts, there is an apparent loss or distancing between the human and nonhuman animal as humans settle into full industrialization and technological innovation. These works detail less about the animals and more about their absence and the poverty of their absence—for example, "maybe the birds realized that Mexico City is dying, and have flown away before the final ruin" (Pacheco, 1993, p. 37).

The most explicit discussions of human-animal relationships involve those that openly address the philosophical framework under which humans make decisions regarding their relation to their animal cousins. These texts do not typically attempt to portray the actual animal, but rather, the impact of the human cultural construct on the perception and ultimate treatment of the animal. At times, however, the animal breaks through the text: "The life-cycle of the frog may sound allegorical, but to the frogs themselves it is no allegory, it is the thing itself, the only thing" (Coetzee, 2003, p. 217). These texts may concern themselves with language, as in Madeleine L'Engle's short story "She Unnames Them" or Pam Ore's book of poetry *Grammar of the Cage*. Alternatively, they may address the issue of animal rights directly, as do the works of J. M. Coetzee (e.g., *The Lives of Animals, Waiting for the Barbarians, Disgrace, Elizabeth Costello*). In the books centering on language, there is recognition that the act of naming and the act of language serve as constraints and cage and enable humans to eat, wear, conduct research on, or buy animals. In texts where animal rights are explicitly addressed, such as *Elizabeth Costello*, the open horror of what is actually done to animals is considered in the context of animal cognition and emotion. J. M. Coetzee's work is particularly intriguing because it allows for no redemption. As his protagonists move closer to grace in empathy and ethical awareness, they become marginalized and unable to effect moral conversions in others or even to save individual creatures, and thus, Coetzee refuses, in essence, to let the reader off the hook through any sort of emotional catharsis.

Ultimately, the contradiction—the worshipping but brutalizing of animals, the hierarchy and inherent speciesism, the confusion about what space there actually is between the human and the nonhuman animal—is made manifest in the nineteenth-century text *Moby Dick*, where Herman Melville's treatment of the whale ranges from symbol to allegory to subject of natural history to a source of ecstatic musings to a focus of all hatred to a source of unlimited wealth. Moby Dick is a unique whale, white and scarred and aggressive. He is something worth empathy and worth brutality, full of economic value. He is positive and negative and neutral. This text has baffled many a reader with its wild swings from a chapter on the color white to a detailed presentation of how a whale is rendered to the documentation of a whale's breeding behavior. In *Moby Dick*, a whale is a whale and then some. The white whale, though representative of its species, of nature, and of the other, is also itself a specific whale unlike all other whales—just as the unnamed Ishmael is unlike all other humans. *Moby Dick* is neither an animal rights nor a conservation text, but it is a true document of its time, the present time, and all times in the way in which the space between human and nonhuman grows and shrinks and changes shape to sometimes even allow brief moments of contact.

Further Resources

Achebe, C. (1972). *Beware soul brother*. Portsmouth, NH: Heinemann.

Barrax, G. (2001). All my live ones. *Callaloo, 24*(3), 692–93.

Coetzee, J. M. (2003). *Elizabeth Costello*. New York: Penguin.

Column, P. (1989). *Selected poems* (S. Sternlicht, Ed.). Syracuse, NY: Syracuse University Press.

Fabre, J.-H. (1919/2002). *The life of the fly; with which are interspersed some chapters of autobiography* (A. T. de Mattos, Trans.). Retrieved from Project Gutenberg, http://www.gutenberg.org/dirs/etext02/tlfly10.txt

Forster, E. M. (1921). *Howards End*. New York: Knopf.

Harjo, J. (2005). Grace. In J. Myers & R. Weingarten (Eds.), *New American poets*. Boston: Godine.

Hemingway, E. (1999). *True at first light*. New York: Scribner.

The Holy Bible, King James Version. (1999). New York: American Bible Society.

Moore, M. (1994). *Complete poems of Marianne Moore*. New York: Penguin.

Ore, P. (2005). *Grammar of the cage*. Los Angeles: Les Figues Press.

Pacheco, J. E. (1993). *An ark for the next millennium* (M. S. Peden, Trans.). Austin: University of Texas Press.

The Rig Veda (W. Doniger, Trans.). (1981). New York: Penguin.

Smart, C. (2001). To Jeoffry his cat. In L. Pockell (Ed.), *One hundred best poems of all time*. New York: Warner Books.

Sumerian Proverbs. (1998). *The electronic text corpus of Sumerian literature* (J. A. Black, G. Cunningham, J. Ebeling, E. Flückiger-Hawker, E. Robson, J. Taylor, & G. Zólyomi, Eds. & Trans.). Retrieved from http://etcsl.orinst.ox.ac.uk/

Whitman, W. (1980). *Leaves of grass*. New York: Penguin.

Jennifer Calkins

George Eliot and a Pig

The following excerpt by nineteenth-century English writer George Eliot (author of *Middlemarch, Silas Marner,* and others) is from her "Reflections of Ilfracombe" (1856) (published in *The Journals of George Eliot,* 1998, edited by Margaret Harris and Judith Johnston, New York: Cambridge University Press). Eliot, whose real name was Mary Ann Evans, and her lover, George Henry Lewes, were out for a stroll in Devonshire, near Ilfracombe, when they saw a pig. Lewes and Eliot were opposed in their reactions to the animal.

As we approached this cottage, we were descried by a black pig, probably of an amiable and sociable disposition. But as unfortunately our initiation in porcine physiognomy was not deep enough to allow of any decisive inferences, we felt it an equivocal pleasure to perceive that piggie had made up his mind to join us in our walk without the formality of an introduction. So G. put himself in my rear and made intimations to piggie that his society was not desired, and though very slow to take a hint, he at last turned back and we entered the path by the stream among the brushwood, not without some anxiety on my part lest our self elected companion should return. Presently a grunt assured us that he was on our traces; G. resorted in vain to hishes, and, at last, instigated as he says, by me, threw a stone and hit piggie on the chop. This was final. He trotted away, squealing, as fast as his legs would carry him; but my imagination had become so fully possessed with fierce pigs and the malignity of their bite, that I had no more peace of mind until we were fairly outside the gate that took us out of piggie's haunts. G.'s piece of mind was disturbed for another reason: he was remorseful that he had bruised the cheek of a probably affectionate beast, and the sense of his crime hung about him for several days. I satisfied my conscience by thinking of the addition to the pig's savoir vivre that might be expected from the blow; he would in future wait to be introduced.

■ Literature
Children's Literature: Beatrix Potter

Beatrix Potter (1866–1943), a beloved English author and artist of classic children's books, respected animals and went on to become a significant force for farm conservation and the preservation of English Herdwick sheep. As a writer, she created many of her fantasy Little Book characters, popular all over the world with young children, from observations of her real animal pets. The famous *Tale of Peter Rabbit,* the title character of which had a near escape from certain capture in Mr. McGregor's garden, was modeled after a Belgian rabbit Potter bought in London and christened Peter Piper. As a solitary young woman interested in natural history, Beatrix had tamed rabbits of all sorts, the first being Benjamin Bunny, also later made famous in an eponymous story. Peter was taught to do tricks for the amusement of visiting children to the Potter household, and he traveled everywhere with Beatrix. She drew him from every conceivable angle, understood animal anatomy and studied rabbit behavior, and for nine years was entirely devoted to him.

Beatrix wrote about Peter and his disobedient adventures first as a letter to a sick child in 1893. In 1901 she decided to use this letter as the basis for an illustrated book for children, but it was turned down by every publisher she tried. She expanded the picture letter, added more black and white drawings, and published it herself to give to friends for Christmas. She was insistent that the book be inexpensive and small enough for little hands to hold, with one picture to a page and the story text on the other. *The Tale of Peter Rabbit* proved so popular that Beatrix soon had to have more copies printed. Finally, she found a publisher for her book after she agreed to turn the illustrations into color. It was published in 1902, and the entire first printing sold out before publication.

Potter's genius in *The Tale of Peter Rabbit,* and in her subsequent stories about Squirrel Nutkin, Jeremy Fisher the frog, and Mrs. Tiggy Winkle the hedgehog washer woman was in combining the habits and environment of real animals with fantasy adventure. But however fancy their clothes or proper their speech, Potter's animals never lose their essential animal nature. In the end, Peter is just another little rabbit enticed by the lettuce in the garden, and Mrs. Tiggy is a hedgehog with prickles who sheds her apron and runs back to the safety of the hedge rows.

The popularity of Beatrix Potter's tales for young children for over a century has underscored the grounding of all her fantasy characters as true to nature first and as enlivening the imagination second. Potter was a serious student of natural history. The mice, guinea pigs, rabbits, hedgehogs, and even the fox and badger that she made famous in *The Tale of Mr. Tod* instructed children in the natural lives of those animals, all the while providing art and a matching story to entice and engage.

Beatrix Potter and Herdwick Sheep

In 1915 Beatrix Potter married a country solicitor, William Heelis, and neatly disappeared from the literary scene in Great Britain, writing few stories for children after that. In 1905 she had purchased a small fell farm, Hill Top, in Sawrey, in the north of England in what was called the Lake District. After her marriage, she lived and farmed there and became renowned as a breeder of Herdwick sheep. In the process, she developed a passion for preserving not only the unique landscape of the fells, but also the culture of fell farming and especially the survival of the unique and increasingly threatened Herdwick sheep.

Beatrix Potter pictured with her dog. ©Pictorial Press Ltd/Alamy.

Herdwick are not indigenous to the Lake District, but proved their merit in late medieval times as a breed uniquely suited to survive on the short grass of the rocky fells and hardy enough to endure the long, harsh winters. The Herdwick are a small, nimble-footed sheep. They are distinguished by being completely black as lambs, but turning gray in the body, their faces remaining black, as adults. As lambs they are taken by their mothers up to a specific pasture on the fells, known as the "heaf." Subsequent generations return instinctively to that same area on a fell where they were first taken. There they spend the summer and early autumn grazing. Herdwick sheep do not wander far from their heaf until being herded down for dipping, shearing, and breeding. Consequently, only a few shepherds are required to bring down a large number of sheep from the high fells, and fences are not necessary. Herdwick wool is very coarse and unsuited for the luxury wool markets, though it has been successfully used for rugs and for military insulation, and Herdwick meat is not considered of high quality. But they are one of only a few breeds of sheep that can live in the natural environment of the high fells, where they keep the vegetation in check and provide meat and milk for the farmers.

Beatrix Potter became enamored of these unique sheep. She understood that their survival, and that of the culture of fell farming, was threatened by the development of tourism in the early twentieth century. With the help of her farm manager, Potter bred and exhibited Herdwick sheep at the local agricultural fairs, soon raising a prize-winning line of Herdwick ewes. She bought her first large sheep farm in 1924, where she had thousands of Herdwicks, and a decade later, she purchased another large group of farms near Coniston Lake, half of which she designated for the immediate purchase of the National Trust, bequeathing the other half to the trust at her death. On each farm, she mandated the perpetual keeping of purebred Herdwick sheep.

As a sheep breeder and farmer, Potter was an innovator in animal husbandry. She sought out the latest remedies for historic skin infections, liver parasites, and foot rot. She took heroic measures to save every animal on her farms, whether dogs, sheep, cattle, geese, chickens, or cats. But if the creature was suffering and without cure, she quickly and kindly ended its life.

At her death in 1943, Beatrix Potter Heelis gave to the National Trust the largest gift to that time of land, farms, forests, and Herdwick sheep, nearly 4,300 acres in all. Her gift enabled cultural preservation and animal preservation to be considered equally important in framing conservation policy in Great Britain in the twentieth century.

Further Resources

Denyer, S. (1993). *Herdwick sheep farming*. London: The National Trust.

Lear, L. (2007). *Beatrix Potter: A life in nature*. New York: St. Martin's Press.

Potter, B. (1902/2007). *The tale of Peter Rabbit*. London: Warne.

Taylor, J. (1997). *Beatrix Potter. Artist, storyteller, country woman* (rev. ed.). London: Warne.

Linda Lear

■ Literature
Children's Literature: Cats

Many say that the world is made up of two kinds of people: dog people and cat people. Although the bond between cats and adult readers is often reflected by the number of real "watch cats" that actually live on the premises of small bookstores, the truth is that in children's literature, cat people have fewer books to choose from than do dog people. Nevertheless, children's cat books are found in a variety of literary genres

Similar to other animal-human relationships, cat-child relationships are expressed at different levels. The expression "as difficult as trying to herd cats" reflects the fact that cats are often independent and aloof. Nevertheless, for a child, a cat can be a nonthreatening pet. Of course, some cats are very aggressive and unfriendly, but those cats appear to be in the minority, especially in children's literature. The responsibilities of caring for a cat can provide wonderful opportunities for building character in a child. In addition, the experience of sitting quietly, stroking a cat, and triggering the cat's purr mechanism can be both soothing and spiritually uplifting.

The cat in children's literature can be either natural or anthropomorphized, and children seem to enjoy both types. For example, children's literature has a plethora of anthropomorphized cats, starting with the French traditional tale "Puss in Boots" (Lang, 1889). In the usual traditional tale format, the miller's youngest son is given a cat that proves to be a more valuable inheritance than his older brothers' inheritances. Puss in Boots is intelligent, crafty, and sneaky and, furthermore, speaks perfect human language. The youngest son ends up marrying the princess, thanks to his cat. In addition, nursery-age children have "The Three Little Kittens," "As I Was Going to St. Ives" (the man with seven wives, each with seven sacks, each with seven cats and each with seven kits!), and "Hey, Diddle Diddle" (the infamous cat and his fiddle), as well as many more cat rhymes to be found in various editions of Mother Goose books.

Often called "the first modern American picture book," Wanda Gag's classic black and white illustrated *Millions of Cats* received a John Newbery Honor Award in 1929. Through this story, children learn (albeit subliminally) that humility is more important and perhaps even more powerful than vanity.

Probably the most significant book in the effort to teach children to read and to enjoy reading at the same time is Dr. Seuss's *The Cat in the Hat* (1957). The Cat is, basically, an independent, self-centered, autocratic force that the children are unable to tame or even resist, until Mother comes down the front walk, and the Cat sets all of the chaos

An illustration from the classic Dr. Seuss book, The Cat in the Hat. *From left, Thing 2, Mr. Krinklebein the Fish, Thing 1, and The Cat in the Hat. Courtesy of Photofest.*

to rights again. This was the first book in the wildly successful Random House Beginner Book series.

Although a purist might argue that a comic strip is not literature, a realist would counterargue that it is. Nevertheless, even though he was not created especially for children, Jim Davis's Garfield has entertained children and adults since June 19, 1978, in newspaper comic strips, cartoon collection books, toys, and movies. The humor is over the heads of very young children, but that is usually not a problem because very young children do not need the text to tell them that the pictures are funny.

Kevin Henkes's Randolph Caldecott Medal winner, *Kitten's First Full Moon* (2004), provides the young child with insights into yearning and illustrates how persistence brings its own reward, even if the ultimate reward is not what one is yearning for or expecting.

Many children suffer their first experiences with death through their beloved pets. Children's literature can provide significant bibliotherapy to allow children to understand and come to grips with death. In Judith Viorst's *The Tenth Good Thing about Barney* (1987), a young boy works through his grief by thinking of ten good things about Barney, his deceased cat. Cynthia Ryland's *Cat Heaven* (1997) reassures children through her vibrant illustrations and lyric text that their deceased cats are going to be okay. The cats already know how to get to heaven, and God just loves cats.

Responsible cat ownership cannot start too early. Beverly Cleary's *Socks* (1973) begins with two siblings trying to sell a litter of "fresh kittens" for twenty-five cents each. When asked what they plan to do with the money they earn, the little sister answers:

> "Daddy says we should save up to have the mother cat shoveled, so she won't have kittens all the time," answered Debbie.
>
> "Spayed," corrected George. "She means he said we should have the mother spayed." (p. 14)

Socks, the kitten, is eventually purchased and becomes a cherished pet—until the birth of his new owners' baby. Socks and his human family are faced with a great deal of adjustment in the new dynamics, an adjustment much like the ones that many children themselves have to make after the birth of a sibling.

The transition from childhood to adulthood is not always easy. Pets often provide an outlet for children to pour out their very souls to a nonjudgmental other. The emotional bonding of a fourteen-year-old boy and his cat, appropriately named "Cat," in Emily

Neville's Newbery Award Medal winner *It's Like This, Cat* (1963) is clearly a healthy relationship that enables Dave to deal with the "old bull, young bull" battles between his demanding father and himself.

That children need to learn responsibility for their actions is made very clear in Paula Fox's Newbery Award Honor book *One-Eyed Cat* (1984). Eleven-year-old Ned is given an air rifle for his birthday, but his father places it in the attic until Ned is older and more responsible. Of course, the forbidden apple is always sweetest, and one night, Ned sneaks the rifle out of the house. He shoots once—at a dark shadow. To his horror, a few days later, he discovers a stray cat in his neighborhood. A stray cat with a freshly destroyed eye. Wracked with guilt, Ned tries to make up, to the cat and to himself, for his action and ends up a wiser young man. Concern for victims of wanton violence is a strong theme of this book.

The link between cats and the supernatural should be quite obvious. Through the mists of time, cats have been depicted in many supernatural guises. They have been identified as "familiars" for witches and warlocks, for example. In addition, the widespread belief in the mythical "nine lives" that cats supposedly possess is almost universal. Children's literature has many well-written fantasies centered on supernatural cats. Although Mary Downing Hahn's *Witch Catcher* (2006) takes place in West Virginia, the young hero, Jen, and her cat, Tink, must save her bewitched father, a newly released fairy, and the entire world of fairies from the evil-stepmother personified, Moura.

A promising start for a new series, The Edolon Chronicles, is Jane Johnson's *The Secret Country* (2006). Twelve-year-old Ben purchases a talking cat from Mr. Dodd's Pet Emporium and, in the company of his new cat, soon finds himself enmeshed in one adventure after another in an effort to save mythical creatures such as unicorns, selkies, and dragons from Ben's evil uncle and his henchman, Mr. Dodd. Children discover that all creatures need protection, even mythical ones. An already established cat fantasy series enjoyed by both adults and young people is Erin Hunter's *Warriors: The New Prophecy* (2004, with six books in the series at this printing). Although the cats communicate in human language (at least the reader can understand what the different cats say and think) and function a great deal as humans might in similar situations, they are not anthropomorphized to the extent that they look like "people in fur." The feral cats look just like regular house cats. They do not wear human clothing or wield human weapons; instead, their unique powers come from other worldliness. Children who have observed cats will recognize the feline movements and can easily suspend disbelief and accept that this must be how cats would think if they thought the same way that humans do.

Children's books about animals provide material that will enhance a child's understanding of animals, of nature, of the world, and of the child's place in that world. The average cat is neither large enough nor strong enough to pull a child out of the proverbial burning building, but heroes come in all sizes and forms. The cat in children's literature represents more than mere physical strength. Many a child has satisfied the human need for companionship and for something or someone to love through interaction with a significant literary cat. Cats in children's literature can also provide the child with a series of silly, funny, clever incidences that enhance the enjoyment of the interaction. Furthermore, because cats are seen as mysterious and aloof, a child can often gain self-confidence by observing how composed the cat appears to be.

Further Resources

Cleary, B. (1973). *Socks*. New York: Morrow.
Fox, P. (1984). *One-eyed cat*. New York: Dell.

Gag, W. (1928/1996). *Millions of cats*. New York: Putnam.

Geisel, T. (1957). *The cat in the hat*. Boston: Houghton-Mifflin.

Grover, E. O. (Ed.). (1915/1978). *Volland edition: Mother Goose*. New York: Rand McNally.

Hahn, M. D. (2006). *Witch catcher*. Boston: Clarion Books.

Henkes, K. (2004). *Kitten's first full moon*. New York: Greenwillow Books.

Hunter, E. (2004). *Warriors: The new prophecy #1: Midnight*. New York: Harper Trophy.

Johnson, J. (2006). *The Edolon chronicles: Vol. 1. The secret country*. New York: Simon & Schuster.

Kiefer, B., with Hepler, S., & Hickman, J. (2007). *Charlotte Huck's children's literature* (9th ed.). Boston: McGraw-Hill.

Lang, A. (c. 1889). *The blue fairy book* (pp. 141–47). London: Longmans, Green. [Lang's source: Perrault, C. (1967). *Histoires ou contes du temps passé, avec des moralités: Contes de ma mère l'Oye*. Paris.].

Lukens, R. J. (2007). *A critical handbook of children's literature* (8th ed.). Boston: Allyn & Bacon.

Savage, J. F. (2000). *For the love of literature: Children & books in the elementary years*. Boston: McGraw-Hill.

Temple, C., et al. (2002). *Children's books in children's hands: An introduction to their literature* (2nd ed.). Boston: Allyn & Bacon.

Viorst, J. (1987). *The tenth good thing about Barney* (E. Blegvad, Illus.) London: Aladdin.

Frances Gates Rhodes

■ Literature
Children's Literature: Dogs

It is only fitting that one of the first great works of literature in the Western canon, the *Odyssey*, should include the small dog story of Argos, faithfully awaiting the return of Ulysses. The story of Ulysses's faithful dog contains most of the elements recognizable in the best of the dog stories written over the years for children and adults. Argos, as with most fictional dogs, is a dog of extraordinary ability and total loyalty to the main protagonist. This capacity of the dog in literature, as in real life, to provide complete loyalty regardless of the desserts of the recipient is a major factor of what attracts children both to the literature of dogs and to the animals themselves.

Although the lives of children have changed enormously in the past century, the stories written for them about dogs have remained remarkably constant. The working or hunting dog remains a staple in these stories, from *Bob, Son of Battle*, written by A. Ollivant in 1898, to *Away to Moss*, written by Betty Levin in 1994—both stories of children dealing with disruption in their lives in the company of a working sheepdog. Even more remarkable, given how uncommon hunting has become, is the continued popularity of stories of hunting dogs. The Irish Setter stories of Jim Kjelgaard have given way to *Weep No More My Lady* (J. Street), *Sounder* (W. Armstrong), *Shiloh* (P. Naylor), and *Where the Red Fern Grows* (W. Rawls)—all stories in which a hunting dog holds the starring role.

A frequent motif in books for older readers finds the dog substituting for a missing parent. In these stories, the dog aids the main character, usually the oldest or only child, as the youngster deals with the problems the absent parent has left unresolved. These are all very much coming-of-age stories with the dog often substituting for the missing adult through his example of courage, steadfast loyalty, and confidence in the young hero. The dogs in these stories uniformly demonstrate a remarkable loyalty, often sacrificing their lives before the saga is complete, but this does not lessen the authenticity of their portrayal. Old Yeller (of *Old Yeller* by Fred Gipson) was a food thief and a crybaby, but when the family needed protection, he had the fortitude and courage to face any threat. Left behind by his father to protect young Jake, Jim Ugly (of *Jim Ugly* by Sid Fleischman) provides protection and guidance, acting as a sort of supernatural guide and demonstrating tracking skills far beyond the ability of any lesser dog. Winn-Dixie is a kindly, affectionate dog down on his luck who is rescued from the local Winn-Dixie grocery store by India Opal Buloni, a lonesome ten-year-old, in Kate DiCamillo's award-winning *Because of Winn-Dixie*. Through the magic of the dog's good-natured obtrusiveness, India manages to meet and make friends with several people and learn that if you just listen to what people have to tell you, there are opportunities for friendship all about you.

Loyalty lies at the heart of the best dog stories. The unconditional loyalty of the dog to the child gives that child the strength, support, and confidence to deal with the problems in his or her life and to grow and become a better person. *Lassie Come Home* by Eric Knight is the classic story of dog loyalty. Separated from her young master by her sale to a wealthy nobleman at a time of financial crisis in her family, Lassie never forgets her love for Ken. Repeatedly running away to return to her master, she is finally sent to Scotland, 400 miles from her home. Undaunted, she again runs away and returns to Ken. This time, recognizing that if he is to have the great collie, then he must have the man as well, the Duke hires Ken's father to manage his kennel. Hence, Lassie's loyalty not only ensures her living with young Ken, but also redeems her entire family.

A scene from the film version of Lassie Come Home *(1943), a classic story of dog loyalty. Directed by Fred M. Wilcox. Shown from left: Lassie, Roddy McDowall. Courtesy of Photofest.*

As contemporary life has become less rural, dog stories have become more suburban, and there has been less scope for the dogs to demonstrate their greatness. Increasingly, they have been portrayed as companions of unusual acuity. Newer dog books deal with lighter issues, oftentimes how the protagonist is to get the dog, as with *Shiloh* by P. Naylor, and then, once the ownership is achieved, how life changes, as in *Henry and Ribsy* (B. Cleary). The dog's loyal steadfastness often provides a link with the community and family for the young protagonist and increases his understanding of the world around him, as with young Marty in *Shiloh* and India in *Because of Winn-Dixie*. Through the difficult years of growing up, every child is subject to feelings of loneliness and of being misunderstood. The dogs in these stories act as spirit helpers and companions, providing guidance, solace, and protection.

The dog with human or superhuman intelligence and understanding, a fairly common theme in children's dog stories, reflects the expectation that any animal that can show as much compassion and understanding as a dog must be capable of fully understanding the issues and problems of the protagonist. Books for very young children that portray the dogs actually talking, such as *Martha Speaks* by Susan Meddaugh and Madeleine L'Engle's *The Other Dog,* show this understanding. Steven Kellogg's great Pinkerton books show that although Pinkerton's understanding greatly exceeds that of even the brightest Great Dane, he is still a dog with less-than-human acuity. Much of the humor of these books is based on his misunderstanding, as in his deciding in *A Penguin for Pinkerton* that he can hatch a football into a baby penguin. In other stories, the dog's understanding can exceed that of the protagonist, as in *Officer Buckle and Gloria* by Peggy Rathmann, where Gloria's understanding of how to teach the lessons Officer Buckle wishes to convey exceeds that of the human protagonist. Perhaps the Carl books by Alexandra Day best convey this superhuman understanding on the dog's part as Carl safeguards his young charge through a day of troubles and dangers.

The most memorable dog stories find the story within the dog's understanding of the world, such as *Harry the Dirty Dog* by Gene Zion, which portray the true-to-life, if a bit over the top, adventures of a terrier out on the town, or the wonderful Henry and Mudge books by Cynthia Rylant. Mudge is simply Mudge—a great mastiff whose very size, drippy jowls, and loyalty to Henry are sufficient for an endless number of stories for early readers that require no super-canine efforts on Mudge's part.

There are children's stories of dogs in which the dogs show all of the intelligence, understanding, and even language of people, with the dogs sometimes narrating their own stories, as in Avi's *The Good Dog* and Margaret Saunders's *Beautiful Joe*. These dogs, though fully cognizant and capable of telling their own stories, remain dogs. It is rare to find dogs dressed in clothes and portraying canine people, as with Ashley Wolff's Mrs. Bindergarten books. It is equally unusual to find the dog playing the villain. Those few dogs who deviate from this type, such as Rex in *Babe the Gallant Pig* by Dick King-Smith or Red Wull in *Bob, Son of Battle*, are usually found only to be misunderstood, damaged, or otherwise corrupted by their prolonged contact with evil people, and these dogs usually demonstrate at least one great strength before the book ends. The children's dog story is all about the dog as loyal companion supporting and aiding his master as the child grows into adulthood.

Nonfiction books for children of all ages offer a wide variety of stories and information. Some of the best of these involve stories of working dogs. Gary Paulson brings the same humor and wise storytelling skills to his nonfiction stories of dogsled racing that characterize his very popular fiction for middle school and older children. *Togo* by Robert J. Blake is a great picture book and nonfiction story of the Nome, Alaska, serum run of 1925. Dorothy Hinshaw Patent's story (*The Right Dog for the Job*) of the raising and training of a service dog informs the young reader with an interesting text and great photographs.

All good dog books deal with a fundamentally true aspect of the nature of dogs and the relationship that humans build with them. Even the most seemingly fantastical of dogs in literature for young children, such as Martha the talking dog or Pinkerton with his supercanine understanding, share aspects of real live dogs in their actions, understanding, and relationships with people. Just as people look to dogs for understanding and nonjudgmental acceptance, so too in children's literature do these most compassionate of animals provide understanding and love.

Further Resources

Day, A. (1996). *Good dog Carl*. New York: Little Simon.

DiCamillo, K. (2000). *Because of Winn-Dixie*. Cambridge: Candlewick Press.

Kellogg, S. (2001). *Penguin for Pinkerton*. New York: Dial Books.

Kjelgaard, J. (1949). *Big Red*. New York: Comet Books.

Knight, E. (1940). *Lassie come home*. Philadelphia: Winston.

Naylor, P. (1992). *Shiloh*. New York: Bantam.

Ollivant, A. (1898). *Bob, son of battle*. New York: A. L. Burt.

Ryland, C. (1996). *Henry and Mudge: The first book of their adventures*. New York: Simon & Schuster.

Beverly Lambert

■ Literature
Children's Literature: Horses

Human records illustrate the bond between horses and *Homo sapiens*. Children appear to instinctively intuit that bond, and books about horses remain popular. The widespread fascination with horses in children's literature extends well into the postindustrial technological context of modern American life. This is an age when horses are anomalies except for sports and pleasure. In the twenty-first century, the vast majority of American children have seen horses only on film or from a distance, perhaps in a parade; only a few have had the chance to actually stroke a live horse's warm hide, and even fewer have actually ridden a horse. Yet stories about horses remain a popular theme among young readers.

Where might children's fascination with horses come from, and why does it persist? Some think that horses awaken cultural memories of our ancestors from times when survival depended on the assistance of a horse. Others suggest that such large, powerful animals extend the child's own imagined size and strength. Bernard Poll believes that horse stories remain popular with children because "the horse is a larger mammal with whom the child can identify, one with whom he [/she] can rest with utter security" (Poll, p. 473). Donna Norton says, "horses and their owners, or would-be owners, usually have devoted relationships. . . . Sadness in many of these stories results when both horse and owner must overcome severe obstacles and even mistreatment" (Norton, 2007, p. 395).

The classic children's book *Black Beauty* (1877), by Anna Sewell, set in Victorian England, recounts the story of a beautiful black foal and the struggles he is subjected to after a careless young aristocrat remorselessly causes the knee injury and resulting blemish that leads Black Beauty's owner to sell him. Told in first person by Black Beauty, the story allows readers to suspend disbelief and understand what life must be like for the hapless horse. In fact, the book (never out-of-print since 1877) was instrumental in the passing of many anticruelty laws and actions in the British Isles and elsewhere.

Another significant horse book is the 1927 John Newbery Award Medal winner, Will James's *Smokey the Cow Horse*. The story begins with Smokey's birth on a Western range and his early years learning to survive and prosper in the wild. In the preface, Will James says, "Smoky is just a horse, but all horse, and that I think is enough said." He proceeds to write from Smokey's point of view. Will James wrote sixty-three "cowboy and horse"-themed books, many intended for children.

Dealing more with a horse than a child-horse relationship is Marguerite Henry's 1949 Newbery Award Medal winner, *King of the Wind*. Albeit romanticized, this is the story of the great Godolphin Arabian whose blood runs, even two hundred years later, through the veins of many of the racing world's superior thoroughbreds (e.g., Man o' War and Seabiscuit). When Sham ("Sun" in Arabic) is born, Agba, the mute stable boy, becomes his constant companion. The Sultan of Morocco sends six of his finest horses as a gift to Louis XV of France. Each of these horses is accompanied by one of the Sultan's best horse boys, who attends to the horse in his charge until the horse's death, at which point he returns to Morocco (p. 55).

Most horse-loving children have read Walter Farley's *The Black Stallion* (1941). All of the elements for adventure abound: a boy; a wild, high-spirited horse; a shipwreck; a desert island; a timely rescue; keeping the horse; winning an important horse race. What more could an adventure-loving child want? Apparently, Walter Farley discovered more

The Black Stallion is among many children's books about horses that have been made into movies. *Courtesy of Photofest.*

because he wrote an additional twenty books in the series, including one with his son, Stephen Farley, who wrote an additional two after his father's death.

Deciding whether some books are for children or adults can be difficult. In Mary O'Hara's *My Friend Flicka* (1940), for example, nine-year-old Ken longs for and begs to be given his own colt from his father's Wyoming horse ranch, but Ken is too flighty. When he finally proves himself, his choice is a female foal sired by the renegade stallion, Albino. Against his father's better judgment, Ken is allowed to keep the little foal, "Flicka" (meaning "maiden" or "little girl" in Swedish). Ken and Flicka grow up, suffer (both almost dying at one point), learn, and celebrate together in this magnificent book.

The sequel, *Thunderhead* (1943), finds Ken older (and wiser). Flicka gives birth to an angry, ugly white male colt—an obvious throwback to his renegade grandfather, Albino. Thunderhead turns out to be an unusually strong, speedy, intelligent horse, and he becomes a spectacular racehorse. Ken develops and matures along with Thunderhead.

The third book, *Green Grass of Wyoming* (1946), is the least well known. Escaping from his remote mountain valley ranch home, Thunderhead takes after his renegade grandfather in a spectacular way, stealing mares from various Wyoming ranchers. One of the stolen mares, "Crown Jewel," is a valuable racing filly. Ken, now a young man, goes after Thunderhead in a journey covering three states. One thing leads to another, and Ken's steadfast loyalty to his own horse is challenged by the feelings he develops for the lovely young owner of Crown Jewel.

Horse-crazy girls still read Enid Bagnold's *National Velvet* (1935). Fourteen-year-old Velvet Brown dreams of owning a horse. After winning a horse, Pie, in a raffle, she works hard to school this unruly creature into a racehorse she can ride to win the world's greatest steeplechase, the Grand Nationals. Sociohistorically, this novel foreshadows the actual entrance of women into racing as licensed jockeys, although at the time it was many a girl's dream.

The prolific Marguerite Henry wrote several excellent books based on actual events or horses (e.g., *King of the Wind*). Her Newbery Award Honor book, *Misty of Chincoteague* (1947), and the sequels, *Sea Star: Orphan of Chincoteague* (1949) and *Stormy, Misty's Foal* (1963), are based on the Chincoteague Volunteer Fire Department's annual Pony Swim and Pony Sale, which has taken place annually since 1925. *Justin Morgan Had a Horse* (1954) tells of the ancestor of the American horse breed, the Morgan. *Mustang, Wild Spirit of the West* (1966) recounts the story of Annie Bronne's saving wild mustangs from extinction when ruthless men were hunting them in order to turn them into dog food. *Black Gold* (1957) is the story of the Thoroughbred champion, "Black Gold"; *Born to Trot* (1950) is based on the true story of Rosalind, a world champion trotting mare. Henry even tells the story of an actual burro in *Brighty of the Grand Canyon* (1953). Henry may have been familiar with the delightful series of six or so "horse books" from England, which begins with the volume *Billy and Blaze* (1926), because Grandfather in *Sea Star: Orphan of Chincoteague* (1949) has a personal riding horse he calls "Billy Blaze."

More recent horse books for older children seem to have a different "feel" to them than did the earlier stories. For example, in Constance C. Greene's *Beat the Turtle Drum* (1976), Joss's longing for a horse leads to her tragic death. The family, especially the older sister, is left coping with the aftermath of the tragic accident. In Bill Wallace's *Beauty* (1988), Luke, the child of divorced parents, is staying with his mother on her father's Oklahoma ranch until she can "get on her feet" again. He befriends an old mare, but in his inexperience and thoughtlessness, he does not understand his grandfather's warning to be careful of her because of her age. When she suffers an unfortunate accident, Luke is to blame.

Each book in the *Treasured Horse Collection* (1997–99) has an accompanying horse model of the featured breed, which can be purchased to adorn the reader's environment (usually girls' bedrooms). The books all deal with girls who, in one way or another, are headstrong, frequently egotistical, or irresponsible. Each girl experiences a life-changing moment as a result of her interactions with a significant horse. Providing insights into the characteristics of different horse breeds, the storylines themselves, as well as additional breed facts after the conclusion of each story, teach as well as entertain. The subtitles are quite self-explanatory, succinctly describing the interactions leading to the life-changing moments: Deborah Felder's *Changing Times: The Story of a Tennessee Walking Horse and the Girl Who Proved That Grown-Ups Don't Always Know Best* (1999); Colleen Hubbard's *Christmas in Silver Lake: The Story of a Dependable Clydesdale and the Immigrant Girl Who Turns to Her for Comfort* (1997); Larry Bograd's *Colorado Summer: The Story of a Paint Named George and the Girl Who Strives to Make a Champion of Them Both* (1999); Colleen Hubbard's *The Flying Angels: The Story of a Vaulter Who Must Overcome Her Fear to Once Again Perform on Her Amazing Andalusian* (1999); Susan Saunder's *Kate's Secret Plan: The Story of a Young Quarter Horse and the Persistent Girl Who Will Not Let Obstacles Stand in Their Way* (1999); Deborah Felder's *Pretty Lady of Saratoga: The Story of a Spirited Thoroughbred, a Determined Girl, and the Race of a Lifetime* (1997); Carin Greenberg Baker's *Pride of the Green Mountains: The Story of a Trusty Morgan Horse and the Girl Who Turns to Him for Help* (1999); Deborah Felder's *Ride of Courage: The Story of a Spirited Arabian Horse and the Daring Girl Who Rides Him* (1999); Susan Saunders's *Riding School Rivals: The Story of a Majestic Lipizzan Horse and the Girls Who Fight for the Right to Ride Him* (1998); Jahnna N. Malcolm's *Spirit of the West: The Story of an Appaloosa Mare, Her Precious Foal, and the Girl Whose Pride Endangers Them All* (1999); and Jahnna N. Malcolm's *The Stallion of Box Canyon: The Story of a Wild Mustang and the Girl Who Wins His Trust* (1999).

Young readers are often "hooked" on series books, frequently formulaic, dealing with the same set of characters. Horse-lovers have several series to choose from, such as *Thoroughbred* (1991–2005), *The Short Stirrup Club* (1996–97); *The Saddle Club* (1996); *Winnie the Horse Gentler* (2002–04); and *Starlight* (2003). Again, the child-horse bond provides opportunities for the characters to experience emotional and social growth through interaction with specific horses.

In the almost 130 years since Anna Sewell wrote *Black Beauty*, children's social relation to both animals and society have changed. In the classic horse stories that still inspire children, families are intact, and children mature without facing the emotional difficulties of divorce or exposure to delinquent peers and without struggling with schooling. In more recent times, however, as children pass through the emotional path to maturity, animals and nature are often at a distance, and for many, only media and perhaps small pets provide a sense of the natural world, animal welfare, and a glimpse of animal rights.

Unlike the bear, the dog, the cat, and even the little red hen, the horse has largely escaped fanciful anthropomorphism in children's literature. Some horses do speak directly to humans (e.g., Jean Ekman Adams's *Clarence Goes Out West and Meets a Purple Horse*, 2000; Erica Silverman's *Cowgirl Kate and Cocoa*, 1995; and C. S. Lewis's *The Horse and His Boy*, 1954—of course, a talking horse in Narnia is truly a horse of another color!). At least one horse in children's literature plays baseball (Tim Egan's *Roasted Peanuts*, 2006), but a horse in a green suit (shades of Babar, the Elephant), sitting in a chair smoking a pipe and reading a newspaper, is just not to be found. Some observers think that the lack of "fingers" makes the horse an unlikely candidate for "us in fur," whereas others believe that the horse is basically too noble to be treated so.

The role of the horse in children's literature continues to be therapeutic and often joyful. Through children's literature, our cultural bonding continues with the horse as the forbearer of civilization. From prehistoric times, the horse has conveyed not only human bodies, but also human values, insights, and civilizing tools. The horse in children's literature provides children with that critical link with our cultural past.

However, because of issues related to the treatment of animals as objects or commodities, we cannot conclude that the horse story, across the board, has achieved a perspective on animal rights. The storylines, however, do reflect a sense of the animal self (Warner, 2004), which bodes well for the future.

Further Resources

Kiefer, B., with Hepler, S., & Hickman, J. (2007). *Charlotte Huck's children's literature* (9th ed.). Boston: McGraw-Hill.

Lukens, R. J. (2007). *A critical handbook of children's literature* (8th ed.). Boston: Allyn & Bacon.

Poll, B. (1961). Why children like horse stories. *Elementary English, 38,* 473–74.

Savage, J. F. (2000). *For the love of literature: Children & books in the elementary years.* Boston: McGraw-Hill.

Temple, C., et al. (2002). *Children's books in children's hands: An introduction to their literature* (2nd ed.). Boston: Allyn & Bacon.

Some Representative Books about Children and Horses

Adams, Jean Ekman. (2000). *Clarence goes out west and meets a purple horse.* Flagstaff, AZ: Rising Moon Books,

Andersen, C. W. (1936). *Billy and Blaze: A boy and his horse.* New York: Macmillan.

Bagnold, Enid. (1935). *National Velvet.* New York: William Morrow & Company.

Baker, Carin Greenberg. (1996). *Pride of the Green Mountains.* New York: Scholastic.

Bograd, Larry. (1999). *Colorado summer.* Milwaukee: Gareth Stevans.

Campbell, Joanna. (1991–2005). Thoroughbred series. New York: Harper Collins.

Egan, Tim. (2006). *Roasted peanuts.* Boston, Mass.: Houghton Mifflin.

Farley, Walter. (1941). *The black stallion.* New York: Random House.

Felder, Deborah. (1997). *Pretty lady of Saratoga.* New York: Scholastic.

———. (1999). *Changing times.* New York: Scholastic.

———. (1999). *Ride of courage.* New York: Scholastic.

Greene, Constance C. (1976). *Beat the Turtle Drum.* New York: Viking Penguin.

Henry, Marguerite. (1947). *Misty of Chincoteague.* New York: Simon and Schuster.

———. (1949). *King of the wind.* New York: Simon and Schuster.

———. (1949). *Sea Star: Orphan of Chincoteague.* New York: Simon and Schuster.

———. (1950). *Born to trot.* New York: Simon and Schuster.

———. (1953). *Brighty of the Grand Canyon.* New York: Simon and Schuster.

———. (1954). *Justin Morgan had a horse.* New York: Simon and Schuster.

———. (1957). *Black Gold.* New York: Simon and Schuster.

———. (1963). *Stormy, Misty's foal.* New York: Simon and Schuster.

———. (1966). *Mustang, wild spirit of the West.* New York: Simon and Schuster.

Hill, Janet Muirhead. (2002–03). Starlight Series. Norris, MT: Raven.

Hubbard, Colleen. (1997). *Christmas in Silver Lake.* Milwaukee: Gareth Stevans, 1997.

———. (1999). *The Flying Angels.* Milwaukee: Gareth Stevans.

Hudson, Jan. (1990). *Dawn rider.* Toronto: Harper Collins Canada.

James, Will. (1954/1926). *Smokey the cow horse.* New York: Charles Scribner's Sons.

Lewis, C. S. (1982/1954). *The horse and his boy.* New York: Harper Collins.

Mackall, Dandi Daley. (2002). *Winnie the horse gentler*. Wheaton, IL: Tyndale House.

Malcolm, Jahna N. (1996). *Spirit of the West*. New York: Scholastic.

———. (1999). *The stallion of Box Canyon*. New York: Scholastic.

Norton, Donna. (2007). *Through the eyes of a child* (7th ed.). Upper Saddle River, NJ: Pearson/ Merrill Prentice Hall.

O'Hara, Mary. (1940). *My friend Flicka*. Philadelphia: J. B. Lippincott.

———. (1943). *Thunderhead*. Philadelphia: J. B. Lippincott.

———. (1946). *Green grass of Wyoming*. Chicago: People's Book Club.

Saunders, Susan. (1996). *Riding school rivals*. New York: Scholastic.

———. (1999). *Kate's secret plan*. New York: Scholastic, 1999

Sewell, Anna. (1997/1877). *Black Beauty*. New York: Morrow.

Silverman, Erica. (1995). *Cowgirl Kate and Cocoa*. San Diego: Harcourt Paperbacks.

Wallace, Bill. *Beauty*. New York: Holiday House.

Frances Gates Rhodes

■ Literature
Children's Literature: Rabbits

Walk through the children's section of a library or bookstore and you'll find a bounty of books featuring bouncing bunnies: happy bunnies, sad bunnies, naughty bunnies, good bunnies, Mommy bunnies, Daddy bunnies, and herds of baby bunnies all looking terrifically cute and terrifically childlike. Indeed, rabbits are among the most popular animals used in children's stories, often serving as appealing stand-ins for human children. But more careful examination shows that the seemingly innocent representations of storybook lagomorphs (the order to which rabbits and hares belong) have a tremendous amount to tell us about the ways that both ancient and contemporary cultures have viewed the mysterious rabbit.

Rabbits as Children

In the very simplest depictions, storybook rabbits look and act like human children: they dress like children, talk like children, and, in some cases, even live in human-like habitations. Take Rosemary Wells's bunny Max, a chubby toddler-like character who bumbles through a series of board books targeted at young children. Enormously popular with the younger set, Max's struggles—whether the current struggle is to resist his bossy older sister, dress himself, or bake a cake for his grandmother's birthday—are comically evocative of the struggles that human toddlers endure on a daily basis.

More sophisticated children's literature uses rabbits to explore more sophisticated emotional themes. One of the most famous childlike rabbits in children's literature appears in Margaret Brown's *Goodnight Moon*. This short, lyrical book focuses on the experience of a little rabbit (in striped pajamas) who is getting ready to fall asleep. Brown's melodic litany contrasts the intimacy provided by familiar objects ("a comb and a brush and bowl full of mush") with the vast world beyond the windows ("good night stars, good night air, good night noises everywhere"), thus evoking both the comfort and mystery a child experiences while falling asleep.

The Runaway Bunny, another one of Brown's best-known books, focuses on an even deeper psychological theme of childhood—that of the conflicting needs to be both independent of the mother and assured that she will never be far away. In this book, a young rabbit asks his mother how she would respond if he were to leave her—for example, by running away or turning into a flower. Each time, she lovingly replies that she would turn into whatever object it would take to keep him with her. For very young children, the book provides the crucial message that a mother can both let her child explore and protect him.

Sam McBratney's *Guess How Much I Love You* features a personified lagomorph too—but this time, the character is a hare, not a rabbit. In Anita Jeram's lively illustrations, Little Nutbrown Hare and Big Nutbrown Hare look more wild than most storybook rabbits, and refreshingly, they wear no clothing. More important to this discussion, McBratney's story, again, plays on a common dynamic between child and parent—the game of exploring who loves the other more. Little Nutbrown Hare and Big Nutbrown Hare describe the boundaries of their love in very spatial terms—they claim to love each other as high as they can reach, as far as they can hop, "down the lane as far as the river," and "across the river and over the hills." Each time, the father, who is bigger and stronger, loves more. In the end, however, a sleepy Nutbrown Hare declares, "I love you right up to the MOON," and his father kindly concedes, "Oh that's far. That is very, very far." But then, as he snuggles his son into the shallow nest of dead leaves, Big Nutbrown Hare smiles and has the last word: "I love you right up to the moon—and back."

Not all storybook rabbits embody messages about an idealized parent-child bond or a perfect childhood world. Part of the astounding popularity of Beatrix Potter's 1903 *The Tale of Peter Rabbit,* for instance, stems from the fact that the tale has a disobedient main character and some truly tense action scenes. The whole story, in fact, revolves around the consequences that arise when Peter ignores his mother's command to avoid Mr. McGregor's garden ("Your father had an accident there," she warns; "he was put in a pie by Mrs. McGregor"). Peter raids Mr. McGregor's garden anyway, nearly gets caught by the angry farmer, and loses his jacket and shoes, but finally does manage to get home safely. Moral of the story: naughty little boys and bunnies get punished, in this case by getting a stomachache and missing dessert.

The chapter book *Rabbit Hill,* written in 1944 by American writer and illustrator Robert Lawson, was aimed at a slightly older children's market and has more complex themes. This delightful tale follows a family of very personified rabbits who wear clothes, cook soup, sit in rocking chairs, and get along quite amiably with foxes. But much of what takes place in the book portrays the lives of rabbits quite realistically. As such, it conveys animal welfare themes that were unusual for their time, even as it sounds a political note that very much resounds with the times in which Lawson wrote.

The story begins with the excitement provoked by the news that "New Folks" are coming to the "Big House" and the fact that no one knows whether those New Folks will be nice people who put in good gardens, mean people who do not put in gardens at all—or really mean people who would exterminate wild rabbits with dogs, ferrets, shotguns, or poison (all common ways of killing wild rabbits at the time).

As it turns out, the New Folks are exceedingly kind—they do not believe in fencing the vegetable garden *or* laying out poison or traps of any kind, much to the disbelief of the locals. The new residents save Willie the mouse when he falls into a rain barrel and Little Georgie when he is hit by a car. In fact, at the end of the book, the New Folks put up a statue of Saint Francis, with a sign that reads, "There is enough for all," and spread out plenty of vegetables and nuts and seeds for all the little animals. As a result, the animals end up protecting the New Folk's garden, even patrolling the property for "wandering marauders." This boggles the local handyman, who complains,

> Here's these new folks with their garden and not a sign of a fence around it, no traps, no poison, no nothing; and not a thing touched, not a thing . . . Now me, I've got all them things, fences, traps, poisons; even sat up some nights with a shotgun—and what happens? All my carrots gone and half my beets, cabbages et into, tomatoes tromp down, lawn all tore up with moles . . . I can't understand it.

Clearly *Rabbit Hill* is a morality tale, in which animals are kind to those humans who are also kind. And the tale tells readers, by the way, that it is OK not to exterminate the wild rabbits who live on one's land. But *Rabbit Hill* is also very much a product of its political times. Written at the end of World War II, the motto "There is enough for all" gives a hint of the optimism the world's people felt at the dawn of the peaceful postwar era.

Margery Williams's masterful *The Velveteen Rabbit: Or How Toys Become Real* (1922) uses a rabbit to illustrate even more complex themes in children's literature: that of the healing power of love. Also designed for older children, *The Velveteen Rabbit* features a stuffed rabbit, who, though "really splendid," feels himself "very insignificant and commonplace" because he is stuffed only with sawdust.

The little toy desires to become real, which as the Skin Horse explains, is not about "how you are made. . . . It's a thing that happens to you. When a child loves you for a long, long time, not just to play with, but REALLY loves you, then you become Real." In the course of this story, the stuffed toy actually becomes real twice. The first birth occurs after months of being adored by the little boy; he tells his nanny that the bunny "isn't a toy. He's REAL!'"—which brings the plush animal's heart and eyes alive with love and wisdom.

The second transformation occurs after the cherished rabbit is thrown away, along with all the other toys, because the boy has had scarlet fever. The rabbit is put into a sack and carried out into the garden to be burned. There he begins to reflect on his life, and a tear trickles down his "little shabby velvet nose" and falls to the ground. At that instant, a fairy appears and takes him to the woods, where he sees real, wild rabbits "dancing" on the grass. The velveteen rabbit discovers he, too, has real hind legs, and his joy is so great he goes "springing about the turf, jumping sideways and whirling round as the others did."

The fact that this rabbit is "born" first of human love and then reborn, or, one might say, resurrected, as a real rabbit makes *The Velveteen Rabbit* a clear analogy to the Easter story, itself long associated with rabbits. This Christian spring holiday originated from pre-Christian celebrations of the spring equinox, which involved worshipping the Teutonic moon goddess Ostara or the Anglo-Saxon moon goddess Oestre, both of whom were depicted as being part hare, being accompanied by hares, or shape-shifting into hares.

But what really makes the story resonate with children and their parents (at least on a more conscious level) is the fact that this bunny, who begins life feeling alienated, inferior, and unsure, is "brought to life" twice by the healing bonds of love and that in the end, even though he must leave the familial home, he finds his true place with truly kindred spirits. In other words, if *The Runaway Bunny* reassures toddlers that Mother will always be there, *The Velveteen Rabbit* reassures older children that even when it is time to leave the family home, the firm base established by familial love will allow them to come into their own true selves.

The Littlest Ministers

Some ancient cultures believed rabbits served as mediators between the secular and spiritual realms, the netherworld and the human world, the heavenly realms and the earthly ones. The ancient Egyptians, Romans, and Britons, for instance, believed that because rabbits slept with their eyes open, they had a gift for prophecy or divination. Less positively, some European cultures believed that rabbits were witches (or could turn into witches). The Celts believed that rabbits lived underground because they had deep connections to the netherworld; in this country, the indigenous Algonquins believed that rabbits held the keys to the afterlife and therefore refused to hunt them.

This idea of lagomorphs as mediators appears in some children's literature as well. The ever-late, always-hurrying White Rabbit from Lewis Carroll's *Alice's Adventures in Wonderland* ("Oh dear! Oh dear! I shall be late!"), for instance, is an endearing carica- ture of a stereotypically timid rabbit (as expressed in his fear of the Queen), but he is the only character, besides Alice, who readily moves between the real world and Alice's dream, or fantasy, world.

Uncle Wiggily, as depicted by Howard Garis, served as a more secular counselor. Wiggily is a kindly, pink-nosed, elderly rabbit gentleman, who lives, quite platonically, with Nurse Jane Fuzzy Wuzzy. The endearing character has a mild dilemma in each of his very short stories, but he always finds a way to help out a neighbor, child, or relative. In "Uncle Wiggily and the Apple Dumpling," for instance, he ends up giving a home- made apple dumpling to a starving squirrel family. In "Uncle Wiggily and the Wagon Sleds," the "nice old rabbit gentleman" helps two puppy dog boys shake a snow-bound bad mood by turning their wagons into sleds. As such, he plays a kindly, moralistic, even religious character, a source of wisdom and aide for the animals around him.

Bunnicula

Of course, the idea that rabbits have links to the underworld (or to heaven) is only one step away from the image of rabbits as beasts or monsters. This theme is best conveyed in the James Howe's *Bunnicula* series, in which a mild-mannered rabbit is mistaken for a vampire by the family's dog and cat. To modern Americans, much of the comedy in these books stems from the seemingly preposterous notion that a rabbit could be malicious. But to ancient peoples, this was no laughing matter. A number of artworks express the terror that some humans had of lagomorphs. One twelfth-century frieze on the collegiate church of Konigslutter, Germany, for instance, shows two hares tying up a prostrate man. A thirteenth-century carving on the Chartres Cathedral shows a demon in the form of a rabbit attacking a woman; another relief there shows a panicked soldier running away from a hare. *Bunnicula* draws on this theme of the monstrous rabbit to comic effect—but it is drawing on an ancient motif nonetheless.

Myth Connections

Why is it that rabbits so frequently appear in children's literature? On a very tangible level, rabbits' long ears, human-like eyes, silly big feet, and fluffy tails delight young human minds, while their soft fur delights young hands—which perhaps explains why *Pat The Bunny* is so popular with very young children. Yet something else seems to be at work here. Perhaps humans have for so long commingled and conflated images of humans and rabbits (i.e., with the worship of hare-headed goddesses and tales of shape-shifting witches) that rabbits have symbolically become a part of the human family in a way that, say, bears (who until the early twentieth century, when "teddy bears" were invented, were seen as predators rather than cuddly pets) have not. Sadly, however, the predominance of roly-poly, chubby-cheeked, suspender-wearing toddler-like rabbits has obscured the very complex intelligences and cultures of both wild and domestic rabbits—or the ancient idea that they are creatures worthy of worship. That is, for every carefully illustrated *Guess How Much I Love You* or subtle *Velveteen Rabbit*, there are dozens of storybooks that show rabbits as childish, cute, not very bright, and prone-to-mischief animals whom nobody cares all that much about. This devaluation, in turn, has led to a devaluation of rabbits themselves—as pets, as livestock, and as wild animals in need of study and protection.

Further Resources

Brown, M. W. (1947). *Goodnight moon*. New York: Harper Collins.

Carroll, L. (1865). *Alice's adventures in wonderland*.

Davis, S., & DeMello, M. (2003). *Stories rabbits tell: A natural and cultural history of a misunderstood creature*. New York: Lantern Books.

Lawson, R. (1944). *Rabbit hill*. New York: Viking Press.

Wells, R. (1996–). *Max* books. New York: Penguin Putnam Books for Young Readers.

Williams, M. (1981). *The velveteen rabbit*. Ontario: Running Mills Press.

Susan Davis

■ Literature
Human Communication's Effects on Relationships with Animals

To understand how human communication both informs and shapes human relationships with animals, one must first accept the idea that human languages and their associated communicative practices do not constitute a disinterested mirror that people use to simply reflect an objective reality. Rather, there is no absolutely neutral way of apprehending and representing the world. Because meaning is not fixed but fluid, humans use their symbolic systems—sometimes consciously, but generally unconsciously—to negotiate and construct understandings of reality. Thus, how people communicate about animals helps inform the way they think about animals and shape the way they experience animals.

Communication Carries an Action Plan

Every instance of communication serves to negotiate and construct meaning. If one refers to "the human relationship with *other* animals," the inclusion of the word "other" before the word "animals" emphasizes the state of humans as animals themselves. This is contrary to the popular usage of the term "animals," which has conventional meanings that generally exclude humans. A reader might take the next step and ask, How does this particular popular usage of the word "animals" serve to construct knowledge of the human relationship with animals—through its repeated usage does it help reproduce the idea that humans are different from animals or are not animals? Or a reader might ask, Why is this perception of humans as different or separate from animals ingrained in the way we communicate—do overarching cultural, scientific, economic, religious, or other social forces help maintain this understanding? These are precisely the types of questions asked by scholars who study discourse.

French philosopher and historian Michel Foucault (1926–84) developed the theory that discourses, or systems of language and representation, are productive, net-like organizations of communication that permeate society at every level with power relations. For Foucault, not only do discourses produce meaning, but this meaning also regulates the way people behave and constructs the way people view themselves and the world. Contemporary scholars have been profoundly influenced by this idea. This move toward discourse, or the cultural or discursive turn, is one of the most significant paradigm shifts in the social sciences to occur in recent years.

The level of words is a good place to start, looking at small-scale but important elements of discursive systems and their connection to the human relationship with animals. When a criminal is described on the evening news as "an animal," or a survivor of genocide exclaims, "We were treated like animals," certain culturally conventional meanings are associated with the word "animal." In the criminal's case, "animal" carries the connotation that one is violent and out of control; in the survivor's case, "animal" connotes that one is unworthy of respect or even life. Likewise, when someone calls another a "chicken" or a "pig," certain meanings are conventionally associated: "chicken" connotes cowardly, and "pig" connotes gluttonous or filthy. Although it may be obvious that the popular connotations in these instances are negative and often inaccurate (chickens bravely protect their chicks, and pigs avoid messing their living areas, tend not to overeat, and lack functional sweat glands to even "sweat like a pig"), scholars argue that these meanings serve to reinforce and are reinforced by larger-scale socially constructed understandings about animals.

So, how do these animal metaphors work to actually inform or shape knowledge? Without associated meanings, words are neutral symbols. It is through the sociocultural constructive process of communication that humans negotiate what these symbols signify. The words are generated within larger discourses and get their meaning by virtue of their relatedness to other words, grammars, and practices within their respective systems. In turn, the particular uses of words help to uphold a respective knowledge system, or ideology, giving hidden assumptions the appearance of being merely common sense, of being normal (as it should be), natural (as it is supposed to be), and neutral (neither bad nor good and having nothing to do with power).

The examples of "animal," "chicken," and "pig" are that of metaphor, and more extensive animal metaphors permeate discourse about the human relationship with animals. For example, one Western metaphor for animals is that of commodities or machines that generate resources for human consumption (e.g., "animal units" in agricultural talk). Metaphors, like other symbolic elements of communication, help to shape knowledge by privileging some options of conceptualizing and concealing others. George Lakoff and Mark Johnson argue that metaphors often serve as guides for future action that will fit the metaphor. Such action will, in turn, reinforce the power of the metaphor to make experience coherent. Metaphors, in this sense, can be self-fulfilling prophecies. If, in different discourses, an animal is seen as a dumb brute, a spirit guide, a majestic icon, a loved companion, a pest, a respected member of a shared ecosystem, a pet, or a powerful and sacred entity, that animal will accordingly be treated as such.

Scholars who study discourse assert that all communication is *interested*. By this, scholars mean that all communication contains an action plan of how to think about something or how to act with or toward something. Dominant ways of representing animals, therefore, favor certain ways of seeing and thinking about and relating with animals. At the same time, alternative ways of representing animals that might encourage different relations are often rendered difficult to select in part because taken-for-granted dominant representations preclude other such choices.

Texts of official discourses, in this case a municipal sign, symbolically represent animals in certain ways. This sign on a Miami boardwalk positions animals as equivalent to machines and vice, objects used for entertainment and exercise that can be a nuisance or dangerous to people taking a stroll. Symbolic representations help inform the way people perceive and experience relations with animals. Courtesy of Shutterstock.

Word and Grammatical Choice and Meaning

Some scholars argue that certain language choices help perpetuate a widespread phenomenon of human discrimination against other animal species, or speciesism. For instance, Joan Dunayer argues that in English, through the popular pronoun choice of "it" for most animals, humans erase not only animals' genders, but also their very uniqueness. By calling an animal "it," humans group that animal with lifeless objects, robbing the animal of sentience and the capacity to feel, think, and have consciousness.

More than mere word choice, scholars argue that grammar demands attention because communicators are generally less conscious about choices of grammar than about words. Grammar conveys latent ideology, or hidden cultural assumptions, and is powerful in shaping realities and reproducing "common sense," knowledge that appears natural and neutral but instead always is socially produced. For instance, some scholars argue that English grammar has been culturally constructed to privilege human agency—or the ability to consciously effect change—and conceal animal agency. Examples include choices about passive or active voice through verb arrangement and choices about transitivity through the order of a sentence that determines the "who does what to whom." English grammar use tends to set up the human as the active subject (the one who does things to others) and the nonhuman animal as the passive object (the one to whom things are done). For instance, in English and in some other languages, humans raise, breed, train, fatten, and control animals. Try to put these relations into grammatical arrangement where the animal has equal or more agency, and one will find it difficult with the English grammatical tools available.

On the other hand, certain languages, including many nonindustrial cultures' languages, offer more refined grammar that reproduces the human-animal relationship as interrelated rather than causal with human as agent. An alternative, for instance, is found in the case of certain animist cultures, in which animals are often seen as equal to humans, if not more powerful, and are grammatically represented as having agency used for or against humans.

Strategic Discourse

Although much of the discourse discussed so far has been communication that circulates among people in everyday interactions, in certain instances, communication is used more deliberately to legitimate certain relations with animals. Many such strategic discourses exist. An example to provide further illustration would be an institutional discourse such as meat-industry discourse.

Cathy Glenn examines how the meat industry uses two codependent discursive strategies to construct consumer support: "doublespeak" and speaking animals. The first strategy, "doublespeak," is the use of sterile language that is intentionally misleading by being ambiguous or disingenuous to hide violent processes internal to the industry. For example, in discussing internal practices, the industry uses the term "euthanasia" to describe the practice of workers killing piglets born too small (for industry uses) by holding their back legs and slamming their heads against the floor. The word "euthanasia" represents the practice as humane and conceals—and in the process condones—the details of the act and the inflicted violence and suffering. The second strategy, the use of "speaking" animals in advertisements to sell meat, involves smiling cartoon cows sitting on a grill or cartoon shrimp eating tiny "popcorn" shrimp. This strategy of showing happy animals cooking or eating themselves ironically works to construct ways of thinking that obscure the suffering of animals killed for their meat and endorse industry practices even in the face of serious concerns raised by environmental and animal advocates.

Arran Stibbe looks at how meat-industry publications use linguistic devices—from semantic classification schemes to pronoun usage—that work to reproduce ideological assumptions that make animal oppression seem both inevitable and benign and that encourage the disregarding of pain and suffering for the sake of market profit. One example is the industry's use of metonymy, or the symbolic use of a single characteristic or part to stand for a more complex whole. An excerpt from industry text provides an example: "There's not enough power to stun the *beef* . . . you'd end up cutting its head off while the *beef* was still alive." In this case, the more complex whole of a living cow is metonymically symbolized by the product the industry gets from killing the cow, creating the meaning of a cow as a meat resource for humans and concealing the meaning of killing a live, sentient being. One may imagine what the use of an alternative metonym, such as the cow's relational role (e.g., "There's not enough power to stun the calf's mother) would do to shift meaning.

Mastery View as Dominant Discourse

Discourse can be talked about at different levels, such as at the level of discourse in everyday communication (e.g., friends or strangers talking), at the level of strategic or institutional discourses (e.g., industry or scientific communication), or at the level of more widespread cultural discourses (e.g., overarching values and norms that infuse all scales of communication). Existing cultural discourses about the human relationship with animals include but are not limited to mastery discourse (humans having a relationship of dominion over animals), stewardship discourse (humans having a relationship of overseeing and taking care of animals), and mutuality discourse (humans having an interdependent relationship of reciprocity and respect with animals).

The dominant cultural discourse of human relations with animals in many Western settings is one of mastery. This mastery discourse is reproduced in economic, scientific, religious, governmental, and other institutional discourses and on an everyday interpersonal communication scale. The power humans exercise over other animals is both coercive (by force) and material (real and physical), and the coercion and its material results are both culturally justified and legitimated via this mastery discourse.

A core value assumption in mastery discourse is anthropocentrism, in which other animals are constructed as inferior to humans. Anthropocentrism shares traits with other oppressive discourses of racism and sexism. In anthropocentric discourse, nonhuman animals are in a similar role to that of the oppressed minority in racism or women in sexism. The animal is posited as the subordinated "Other" and the human is in the role of the dominating and oppressive "Center." Val Plumwood explores how the shared discursive traits of anthropocentrism, racism, and sexism include radical exclusion (through the Center seeing the Other as both inferior and radically separate); homogenization (in which the Center stereotypes the Other as interchangeable or replaceable); denial or backgrounding (in which the Center represents the Other as inessential and not worth noticing); incorporation (in which the Center defines the Other in relation to the Center, as lacking the Center's chief qualities, and devalued); and instrumentalism (in which the Center reduces the Other to a means to the Center's ends rather than according the Other value in its own right).

Much in the same way that discursive structures of racism and sexism set limits not only for the human objects of these discourses, but also for the perpetrators, anthropocentric discourse not only leads to the detriment of other animals but also distorts and limits the possibilities for those humans who use anthropocentric discourse. Mastery discourse constrains who people are and what they can become as humans relating to other animals—regulating people to hierarchical roles and indifference toward animals and denying alternative human-animal relationships.

Counterdiscourses

Nevertheless, although dominant meanings are reproduced, alternative meanings are also introduced and negotiated in communication, bringing with them different ways of understanding and practicing human-animal relations. Those who wish to challenge the dominant mastery discourse should keep the cultural ambivalence, or tension, between the dominant discourse and such counterdiscourses in mind. Counterdiscourses, in fact, are always in circulation and provide openings to resist dominant understandings.

Yet, although counterdiscourses provide choices of how to represent the human relationship with animals, the choice is still a strained one. The selection of a counterdiscourse and its respective ideology requires the choice of rejecting the dominant discourse and, with it, the decision of whether to represent or compromise one's own values, to oppose or agree with one's more or less powerful interlocutor's discourse, to be heard or not be heard, to be celebrated or to be retaliated against.

Discursive Struggles at the Zoo

Ethnographic fieldwork at an American zoo provides an illustration of the struggle between dominant discourse and counterdiscourses with an example of a schoolchildren's tour passing by the gorillas. The tour guide, in her role as zoo authority and lead adult among a group of mostly children, has extensive power to use communication to both establish and texture the dominant themes and meanings. In the observations that follow, the discursive constructions drawn on by the tour guide include dominant mastery and stewardship discourses, animals as performers, and anthropocentrism. A few children attempt to put forth counterdiscourses of captivity, connection, and freedom.

When the tour stops at the exhibit, a two-and-a-half-year-old gorilla just a little smaller than the children runs up to face them. She raises her arms above her head and begins to loudly pound her palms on the glass that separates her from them. The tour guide discursively frames the gorilla's actions as playful, fun, and performative, saying, "This baby's being really cute over here"; "She is going to entertain us here"; "This baby over here's just playing up a storm on the window"; "If we gave her a drum set, it might be really interesting to see what she'd do."

The power to represent reality through communication is often a site of struggle. Different people represented this young gorilla pounding on her glass cage as performing for humans, as communicating, or as wanting to get out. Courtesy of Tema Milstein.

A child, however, counters the tour guide's statements by saying, "Maybe he wants to be let out." The guide then does quick work reframing and reclaiming the authority to represent, saying, "Yeah, you *think* so. I think she's just playin'." Nonlinguistic elements of discourse also work in representation—see how the guide uses emphasis on the child's "think" and deemphasis on her own "think" to differentiate the weight and accuracy of each of their statements, subordinating the girl's "think" to her own.

Another child then says this of the young gorilla, "She's trying to get the lock undone." The guide uses her louder adult voice to speak over this resistant discourse and continue the work of reframing, again using her deemphasized, superordinated "think" to help do the work: "I think she's just showing off for you. Would you guys want to leave? It's a *beautiful* environment. They get fed every day." Another child is facing the gorilla, and the gorilla is looking at the child as she pounds the glass. The child says to the gorilla, "Hi."

The children here have introduced resistant counterdiscourses. Their discourses favor recognitions of captivity and desires for freedom as well as a connection with a sentient being. In her positioning as the authority, however, the guide has the final word, as well as the physical (coercive) control over the children in deciding when they are to stay at this discursive site and when they are to leave. The guide's final communication points one last time to the "fun" the baby gorilla is having, legitimizing the guide's dominant representation and then removing the children from the sight and sound of the gorilla:

"And she's just having a good ol' time here pounding away—all right, *all right*, my eagles, we're going to move along; *eagles*, this way."

Such discursive struggles take place among people of less obvious power differentials—for instance, with two friends in conversation: the first friend may engage a counterdiscourse (e.g., that animals have emotional lives and should have more legal protections) while the second friend engages a dominant discourse (e.g., that animals have only basic instinct and should not have more protections). In this case, neither the counterdiscourse nor the dominant discourse may be wholly fulfilling choices for the first friend. The animal emotions and rights counterdiscourse may be either devalued or dismissed by the second friend, who wields power by aligning with a dominant discourse. Conversely, if the first friend chooses the dominant discourse, this requires deferring the ability to represent an alternative human-animal relationship that she or he may believe in or practice. This theoretical framework of discourse-counterdiscourse struggle helps point to hurdles for individuals who have notions of the human-animal relationship alternative to the mastery discourse, as well as to societal-scale hurdles that stand in the way of transforming human-animal relations.

Alternatives and Resistance

The scholarly focus on discourse aims not only to raise awareness about dominant discourses, but also to specify emergent practices of resistance to dominant paradigms and to discern possibilities for change. Looking at counterdiscourses is a productive way to do this. Other fruitful studies include looking at alternative nonmastery discourses that exist as dominant discourses within different cultures and at deliberately formed counterdiscourses that are intended to strategically shift the human relationship with animals.

Dominant cultural discourses that do not reproduce a mastery worldview include those that reproduce worldviews of mutuality, connection, cooperation, and reciprocity among humans and animals. An example is the discourse of Peruvian Andean-Amazonian peasants in which the notion of "communication" is inseparable from the notion of mutual nurturance among animals, nature, and humans. Julio Valladolid and Frederique Apffel-Marglin explain how, in this culture, "for humans, to make *chacra*, that is to grow plants, animals, soils, waters, climates, is to converse with nature" (2001, p. 648). Notice in the explanation the grammatical shift from humans as the sole causative agent ("make," "grow") to the use of the verb "converse" to signify mutuality and interdependence. In this cultural discourse, all animals and nature, not only humans, converse with one another to make and nurture the *chacra*, including "the sun, the moon, the stars, the mountain, the birds, the rain, the wind . . . even the frost and the hail" (p. 648).

Valladolid and Apffel-Marglin, two indigenous Peruvian Andean-Amazonian scholars and development practitioners, "deprofessionalized" themselves to return to the villages of their people and dedicate their work to sustaining their people's cultural discourse and practice. Whereas scholars must take care to avoid misrepresenting, essentializing, or romanticizing indigenous cultural discourses about the human relationship with animals, indigenous and non-indigenous scholars alike argue it would be tragic to waste such accumulated knowledge and redundant for scholars to generate models of human relations with animals without learning from sustainable and reciprocal ways of life that have been practiced and in some cases remain vital.

Numerous dominant cultural discourses mediate the human relationship with animals in ways that differ from Western mastery discourse. Scholars, however, caution that in looking for alternative discourses, one must be careful not to misrepresent the discourse as an ideal alternative. An example can be found in primatology studies discourse in Japan. Japanese scientists approach their primate subjects differently from their Western counterparts, in that they do not bring to their studies Judeo-Christian–informed notions of separation or of human stewardship or dominion over animals. Japanese concepts of unity with animals are not, however, without hierarchy, albeit within a more horizontal, Buddhist-Confucian–informed hierarchical framework of karmic rebirth (e.g., for one's wrong deeds in this life, one may be reborn in the next as an animal). Nor is the Japanese discourse without culturally condoned cruelty (e.g., children might throw rocks at the primates without correction from adults). Care must also be taken to note that, as within any national borders, there is not only one Japanese discourse; rather, there are many different animal discourses.

In highlighting various dimensions of Japanese primatology discourse, Donna Haraway cautions against "the cannibalistic Western logic" that readily constructs other cultural possibilities as resources for Western needs and action. This caution is necessary. At the same time, cautiously learning from a spectrum of cultural discourses can help open new paths of thinking that resist oppressive discourses.

Alternatives to mastery discourse are also deliberately formed and practiced by animal advocacy groups in an effort to resist animal oppression and transform human-animal relations. People for the Ethical Treatment of Animals (PETA), for instance, uses intentional discursive approaches to bolster its social change campaigns. The group has linked discourses of the Holocaust to the meat industry with its "Holocaust on your Plate" campaign. By this discursive link, PETA relates the human treatment of animals to the way Nazis treated their victims and relates the suffering and overwhelming numbers of animals killed to the suffering and overwhelming numbers of Holocaust victims. With the title of the campaign, PETA also discursively implies that meat-eaters are complicit in this suffering, similarly to or even more so than those who remained silent during the Holocaust. In this case, PETA enlists rabbinical authorities to use their subject positioning to legitimate this discursive campaign. This strategy of introducing ideas and meanings that sharply contradict the mastery discourse elicits attention-getting rebukes, sometimes from Jewish groups and individual Jews; the media cover the rebukes and, as a result, deliver PETA's message to a wider audience, some of whom may for the first time consider the massive suffering of animals caused by a meat-centered diet. Similarly, PETA has also linked slavery discourses to the circus industry, enlisting African American civil rights spokespeople as authorities to use their subject positioning to legitimate comparisons of the oppression of animals in the circus to that experienced by Africans during slavery; again, this may be met by attention-getting refutations of the similarities between human and animal slavery.

PETA has also accessed the discursive genre of pornography to create humorous and attention-getting pro-vegetarian commercials that situate meat-eating men as impotent and vegetarian men as virile. More recently, to draw attention to animal cruelty and human health risks associated with dairy industry practices, PETA created a "Milk Gone Wild" campaign, simulating "Girls Gone Wild" videos with strategic image and word changes (women bare cow udders instead of human breasts; instead of "No rules, no parents, and, of course, no clothes!" PETA's commercial states, "No rules, no parents, and, of course, NO COWS!"). PETA produced television commercial spots for both the pornography and the "Milk Gone Wild" campaigns to air during Super Bowl football halftimes, but the television networks broadcasting the games refused PETA airtime. The action of censoring these commercials illustrates the discursive struggle. The network, in its role as a mechanism of representation (or a gatekeeper that filters which representations will circulate), was able to dismiss and largely repress the alternative discourse.

PETA, in turn, precisely because of this dismissal, was able to make a media event out of the censorship and to draw attention to its alternative message.

Language Change

Because discourse is systemic, the problem of dominant discourses such as mastery discourse creating a wide gulf between humans and animals cannot simply be fixed by erasing certain words from the vocabulary. These erasures do little if the meanings and associations of the new words that replace them reproduce similar configurations of meanings. Struggles over discourse, however, are a necessary and interrelated part of wider struggles for change. For example, the feminist movement, in addition to battles on economic, domestic, and public fronts, has waged a protracted and successful struggle over nonsexist discourse. Deborah Cameron writes that eliminating the use of "he" as a generic pronoun in the English language does serve to help change the repertoire of social meanings and choices available. Thus, change in linguistic practice itself can be social change if it coincides with and contributes to larger-scale societal transformations.

In ways similar to feminist language activism, scholars have begun to offer suggestions of word and grammar change for the human relationship with animals, including the use of narration to convey a sense of individual animals' lives, grammatical choices that make animals subjects if they are the primary actors or victims (e.g., the horse approached the girl), and verbs that imply animal emotional intention (e.g., the deer "fled" instead of "ran"). Dunayer also suggests avoiding expressions that elevate humans above other animals (e.g., "the sanctity of human life"); human-animal comparisons that patronize animals (e.g., "my dog is almost human"); terms that portray animals relatively free of human control and genetic manipulation as dangerous or inferior (e.g., "feral cat"); category labels that vilify animals (e.g., "vermin" or "pests"); and overqualified reference to animal thought and feeling (e.g., "the prairie dog *seemed* to recognize" or "it squealed *as if* it felt pain"). Besides the dismissive charge of "political correctness," other more substantive challenges to such attempts at change include the colossal task of coming up with a consistent and effective overall discourse for expressing anti-speciesist thought.

Changing Understandings through Conscious Communication

This essay began and will end with metaphors because they can be powerfully linked discursive structures in generating new knowledge of alternative human relations with other animals. One can use metaphors as tools for change by developing awareness about the metaphors one lives by and by having one's personal experiences with animals form the basis of alternative metaphors. In this way, one can develop what Lakoff and Johnson call an "experiential flexibility" to engage in an unending process of viewing one's life through new metaphors that open alternative ways of thinking.

Ethnographic research on whale watching in the Pacific Northwest provides an example. Many whale watch boat naturalists and captains frequently use the word "show" to describe the physically close and emotionally exciting experiences people have with the killer whales, or orcas. For example, a captain might tell another captain over the marine radio, or a naturalist might say to tourists on the boat, "That was quite a show today," after orcas swim close by the boat or engage in nearby boisterous activity, such as fin slapping or breaching out of the water.

Because each word gets its meaning by virtue of its relatedness to other words and discursive formations within its respective structure, one can interpret how the particular word choice of "show" relates to popular Western communication, informed by anthropocentric mastery representations of exciting or amusing animal behavior as performance or, more directly and ironically, representations by the marine entertainment industry of captive orcas trained to perform tricks for humans. Such mastery representations, however, are contradictory to the good-faith intentions of naturalists and captains to educate tourists about the behavior of whales in their natural habitats and to inform respectful understandings.

The wide and pervasive societal use of "show" to popularly describe positive human viewing experiences of animal behavior, however, makes a different choice of representation difficult. Well-intentioned naturalists and captains must select among existing discursive structures if they are to be understood. At the same time, the use of "show" can unwittingly serve to reproduce dominant discourse, directing a certain way of seeing the human relationship with the whales. Although this choice is metaphoric—indeed, the orcas' behaviors are not a show but are simply moments in their lives that humans happen to see—even the metaphoric nature of the word choice gets lost in the repeated use of "show," until "show" may become the very meaning of the representation.

When a couple of naturalists heard about this observation, they responded with a desire to come up with alternative ways of representing the experience, ways they felt more closely represented their actual experiences and feelings around the whales and the meanings they wanted to convey. One naturalist came up with "that was a really good

Tourists on a Pacific Northwest whale watch boat often communicate with each other to negotiate their understandings about whales. On-board naturalists also use communication to educate the tourists in particular ways about orcas. Courtesy of Shutterstock.

day," connoting that she felt fortunate and pleased to have been near the whales, and the tourists might also feel that way. Another naturalist came up with "that was a great encounter," connoting a valued and mutual interaction with the whales.

This type of reflection and the subsequent shifting of language to represent the experience itself can help shift thinking about human relations with animals. In this particular case, the naturalists were able to critically reflect on the whale watching industry's use of communication and then to consciously shift their use of language to fit both their actual perceptions of their experiences and their educational goals—to help tourists think about the orcas not as animals performing for human entertainment, but as animals who have their own value and agency, who make their own choices in their behavior and interactions, and whom humans were fortunate to encounter.

Scholars, therefore, argue that a crucial step to changing the human relationship with animals is in the act of deconstructing the use of language and associated communicative practices. One who wants to change human relations with animals must maintain a state of heightened discursive awareness and exercise a critical and self-reflexive sensibility. As such, one must refuse to take communication about animals at face value and must instead always question the status quo, or the preferred discursive "common sense" that circulates in communication about animals. A focus on discourse not only raises awareness about the discursive nature of human relationships with animals, but also allows one to begin to question the status quo of such relationships. For both scholars and everyday communicators, conscious communication can be both an emancipatory and a reconstructive undertaking, one of raising awareness about the social construction of human-animal relations and one of recognizing and creating compelling alternative visions of possible futures.

See also

Classification—*The* Scala Naturae
Ethics and Animal Protection—*Factory Farm Discourse*

Further Resources

Cameron, D. (1997). Demythologizing sociolinguistics. In N. Coupland & A. Jaworski (Eds.), *Sociolinguistics: A reader and coursebook.* Basingstoke: MacMillan. (Originally published 1990.)

Dunayer, J. (2001). *Animal equality: Language and liberation.* Derwood, MD: Ryce.

Glenn, C. B. (2004). Constructing consumables and consent: A critical analysis of factory farm industry discourse. *Journal of Communication Inquiry, 28*(1), 63–81.

Haraway, D. (1989). *Primate visions: Gender, race, and nature in the world of modern science.* New York: Routledge.

Lakoff, G., & Johnson, M. (1980). *Metaphors we live by.* Chicago and London: University of Chicago Press.

Plumwood, V. (1997). Androcentrism and anthropocentricism: Parallels and politics. In K. J. Warren (Ed.), *Ecofeminism: Women, culture, nature.* Bloomington: Indiana University Press.

Stibbe, A. (2001). Language, power, and the social construction of animals. *Society and animals, 9*(2), 145–61.

Valladolid, J., & Apffel-Marglin, F. (2001). Andean cosmovision and the nurturing of biodiversity. In J. A. Grim (Ed.), *Indigenous traditions and ecology: The interbeing of cosmology and community.* Cambridge, MA: Harvard University Press.

Tema Milstein

■ Literature
Melville, Herman

Herman Melville (1819–91) is best known for his novel *Moby Dick,* published in 1851 and now regarded as his masterpiece, as well as a masterpiece of American literature. Moby Dick, a white sperm whale, is obsessively hunted by Captain Ahab, who lost his leg in an encounter with this whale on a previous voyage. Ahab's search for revenge puts his whaling ship, the *Pequod,* and its crew, in peril. The novel, rich with detail about whales and whaling, can be viewed as a tale of human-animal interaction with multiple and complex layers of meaning and significance.

Melville went to sea at the age of seventeen and later worked on whaling ships in the Atlantic and Pacific Oceans during his early twenties. He had a keen eye for the ocean-dwelling animals, fish, and birds. He described the giant tortoises of the Galápagos Islands in *The Encantadas* (1854). The title refers to the Galápagos as the so-called "enchanted islands" because of the presence of these tortoises. Melville noted that they were brightly colored on the underside and dark on the upper shell; later, after hearing them rustle about the deck of his ship, he and his companions dined on them.

In several of Melville's lesser-known works, animals signal events about to happen or are symbols for larger themes. "The Haglets" is a poem about birds called haglets who fly above a doomed ship "like shuttles hurrying in the looms." Large flocks of seabirds signal land to sailors aboard a lifeboat in the novel *Mardi.* In the poem "The Maldive Shark," Melville observes that "sleek little pilot-fish" help out the larger, duller shark to find its prey. Both the pilot-fish and the shark need each other—the shark unwittingly protects the pilot-fish as they swim amidst the shark's wide teeth—and these two creatures work together symbiotically. In the short story "The Apple-Tree Table," Melville tells how several beautiful bugs emerge from a crack in a cloven-footed table, dazzling the little girl Julia and causing her to see the bug last to emerge as symbolic of the resurrection (Rollyson & Paddock, 2001, p. 5).

In *Moby Dick,* Melville provides lengthy descriptions of whales and those who hunt them for profit. Yet the novel is not merely an account of whales and whaling in the nineteenth century. Moby Dick, the white sperm whale, is in his wildness a symbol of human limitation. Moby Dick eludes capture and death at the hands of the sailors led by Captain Ahab, and all on the ship save one, the narrator Ishmael, go down in the attempt to kill the whale when he returns their attack. The whale here reminds us that nature, in spite of science, engineering, and other human creations, is ultimately beyond the control of humans. Moby Dick is absent of color and a specter for the crew of the *Pequod* and its monomaniacal captain, Ahab. Melville explores Moby Dick's appearance in chapter 42, "The Whiteness of the Whale"; the whale's whiteness is meant to signify the animal's difference, allure, and fearsomeness.

The novel's narrative of Captain Ahab's quest to seek revenge for the loss of his leg is interspersed with lengthy descriptions of whales and whaling. The beginning of the book presents a pedantic account of whales, where Melville satirizes the attempt of one he calls a "sub-sub-librarian" to think that by listing all the characteristics of whales, the pedant can capture and discern the animal's significance. By starting the novel in this fashion, Melville is able to then show that such categorization and description is simply a human creation. Chapter 32, "Cetology," catalogues the many kinds of whales. Even in this chapter, Melville reminds us that humans cannot fully comprehend whales through observation and description. The whale, as he says, has an "unwritten life."

Whales, being the largest mammals, inspire awe and fear as well as fascination. Melville's novel presents whales as mythic and mysterious creatures that nevertheless formed part of a highly profitable business. Whaling in the eighteen and nineteenth centuries was dangerous and demanding. Melville describes how these mammals were killed with harpoons and their huge bodies pulled to the side of the ship. The whale was butchered quickly, with all hands at work, so as to complete the task before sharks could attack. The blubber was then removed, along with the head, and the carcass was tossed back to sea. Oil was extracted from the blubber and the head through boiling them in large pots on deck of the ship. The oil, valued chiefly as an illuminant and lubricant, was then decanted and stowed in the hold.

The whale, though a mammal, is part of the sea and can exist there where man cannot. Whales thus fascinate human beings because they are like us, as mammals, and unlike us, as denizens of the oceans. Ishmael, the narrator of the novel, was drawn to whales and whaling to experience the remote, vast world beyond civilization. However, recent interpretations of the novel and whaling focus on specifically human-animal interaction themes. Philip Armstrong (2004) examines the discourse of whaling, specifically Melville's descriptions of the animals and the practice of hunting them. He sees Melville's work as prefiguring a cultural shift toward compassion for nonhuman animals.

Further Resources

Armstrong, P. (2004). *Moby-Dick* and compassion. *Society and Animals, 12*(1). Retrieved from http://www.psyeta.org/sa/abstract_12-1.shtml

Delbanco, A. (2005). *Melville: His world and work*. New York: Knopf.

Melville, H. (1851). *Moby-Dick; or, The Whale*. Retrieved from http://www.readprint.com/work-1207/Herman-Melville

Miller, J. (1993). *A reader's guide to Herman Melville*. New York: Octagon Books.

Rollyson, C., & Paddock, L. (2001). *Herman Melville A to Z: The essential reference to his life and work*. New York: Facts on File.

A. G. Rud

■ Literature
Metamorphosis

The word "metamorphosis" comes ultimately from the Greek and can be used to refer to any transformation of a living thing. Today, the word is most frequently used to mean a biologically determined change of structure as an organism passes from one stage of development to the next, for example, from tadpole to frog. This definition dates from the late seventeenth century, when naturalists and artists had begun to make more systematic and deliberate observations of living things. The older meaning of the term is a "magical" change of form, for example a witch who becomes an owl to fly about at night. A ritualistic metamorphosis was practiced in very early times by shamans of many cultures, who would go into a trance to assume the body of a bird or other animal and enter the spirit world. Such practices may be the ultimate origin of many tales of metamorphoses across the globe.

Nevertheless, mythological tales of shape-shifting, found in the folk literature of virtually all cultures, are an extension of patterns observed in the natural world. Insects and amphibians, especially, undergo a very radical change in appearance in transition from a larval to an adult state. Snakes shed their skins repeatedly, and insects cast off their exoskeletons; most plants can emerge from a seed, while birds and reptiles hatch from an egg. The most dramatic change of all is from a caterpillar to a butterfly, and that is the major model for stories of physical transformation. In a vast range of cultures from the Hopi Indians to the Greeks and the Japanese, the butterfly is a symbol of the soul. The Greek word "psyche" means both "butterfly" and "soul," as does the Latin word "anima."

Scholars of myth sometimes distinguish between "metamorphosis" and "metempsychosis," the transmigration of souls at death, but it is impossible to differentiate sharply between the two. In metempsychosis, the change tends to be irrevocable, whereas a metamorphosis is often simply a temporary disguise—for example, when Zeus changes into a bull to abduct the maiden Europa. Nevertheless, both concepts are represented symbolically in much the same way, and the butterfly emerging from the chrysalis is a nearly universal model for imagining the spirit leaving the body. Metempsychosis is found in many traditional religions, such as Hinduism and Buddhism, as well as in New Age beliefs.

The basic plots of folktales throughout the world are surprisingly constant, and their variety comes largely from the animals and plants that they feature. In Japan and China, for example, tales of transformations of foxes into people are particularly frequent. Fox women often try to charm men out of either malice or playfulness, but at times, they fall genuinely in love with the men. In a Chinese story recorded by Shen Chi-chi in the seventh century CE., a soldier marries a lovely lady named Jenshih, who later confesses to him that she is actually a fox, but she promises to remain in human form out of love for him. Fox maidens, however, can be recognized by other animals, and one day in the marketplace, some dogs catch the scent of a fox and begin to pursue her. Jenshih immediately changes back into a vixen but is soon caught and torn to pieces.

In Greco-Roman mythology, metamorphoses are very common, and they had a special fascination for writers, perhaps because shape-shifting was a fantastical motif in what many scholars consider a relatively rationalistic culture. In the *Odyssey*, for example, the enchantress Circe transforms the Odysseus's companions into pigs. Shape-shifting passes from myth into philosophy in the fifth century CE through the influence of the Pythagoreans, who believed in transmigration of souls. Socrates, in Plato's dialogue "Phaedo," suggests that people may become wolves in the next life if they were rapacious in this life, donkeys if they were foolish, or bees if they were good citizens.

The Roman poet Ovid recounted traditional tales of transformations in his long poem *Metamorphoses,* published in 7 CE, to dramatize his conception of a universe in constant flux. The stories of metamorphoses told by Ovid took place in a world that was highly structured, where there was little ambiguity in the distinctions between gods, men, women, plants, and animals. The transformations were similarly abrupt and unequivocal, and they signaled a temporary suspension of the cosmic order, which was quickly followed by a return to normalcy.

For Ovid, as for many sophisticated Romans and Greeks, metamorphoses are almost always a response to situations that are somehow extreme, whether it is in the love, grief, anger, fear, suffering, or boldness they inspire. Such situations challenged the cosmic order, and a metamorphosis was a way in which they might be controlled. Thus, in Greek and Roman mythology, when the maiden Arachne exceeds the limitations of the human condition both through her ability at weaving and through her pride, the goddess Minerva changes her into a spider. A metamorphosis here is a means to realign the material and spiritual dimensions of reality, when they are no longer in harmony.

The change in form may occasionally be a romantic apotheosis, as when Alcyone and her drowned lover Ceyx are reunited as birds. Far more often, however, the metamorphosis is deliberate violation of a person's bodily integrity. In many of Ovid's stories, a person is transformed into an animal as punishment for challenging a deity or sorcerer. The hunter Acteon is transformed into a stag for intruding on the goddess Diana in her bath, and the king Lykaon is enchanted into a wolf by Zeus for serving human flesh. The gods who perform transformations are a bit like managers in a firm who promote, demote, and transfer employees, so that operations can continue efficiently.

Considering how widespread stories of metamorphosis have been in the ancient world, it is remarkable that the Old Testament or Torah contains so few. The serpent in Eden undergoes a sort of metamorphosis after tempting Eve because it is compelled henceforth to crawl on its belly. There is also a metamorphosis when Moses, in order to demonstrate the power of Yahweh to Pharaoh, turns his staff into a snake. The magicians in the court of Pharaoh also change their staffs into snakes, but the snake produced by Moses devours the rest. In the New Testament, Jesus changes water into wine and fishes into loaves. The foremost mystery of Christianity is the Eucharist, wherein bread and wine become the body and blood of Jesus, an extraordinary event that demonstrates the power of God.

In the gospels, the physical form of Jesus rises from the dead, but there is no suggestion of a soul apart from the body. That idea entered Christianity from Greco-Roman culture, through the work of theologians such as St. Augustine. Imagery of metamorphosis then became very important in Christianity, especially in describing the process of conversion. Like the butterfly emerging from the chrysalis, a person who dedicates his or her life to Christ is said to be "born again."

The word "metamorphosis" literally means "beyond form," and the concept presupposes an identity that transcends the physical body of a creature. St. Augustine and other church fathers maintained that animals, to say nothing of plants, do not have souls. For much of the Christian era, in consequence, stories of metamorphosis have been largely confined to folk literature. In Dante's *Divine Comedy*, most of the dead retain their accustomed forms, though there are exceptions. The suicides, for example, are transformed into trees, which, however, begin to bleed and to speak when a branch is broken off. Because they scorned their bodily form in life, they remain deprived of it in death.

In the Renaissance, witches and warlocks were widely believed to be shape-shifters, roaming by night as creatures such as cats or ravens. Particularly, in the last decades of the sixteenth century and the start of the seventeenth, there was a wave of hysteria about werewolves, people that assumed the form of a wolf by night. Numerous tales of werewolves had been told in Norse, Greco-Roman, and many other civilizations, going back to remote antiquity, but diabolical associations made werewolves especially frightening in a Christian context. The images of werewolves and witches also focused the fears evoked by many political, religious, technological, and intellectual upheavals. In the story of one famous case reported in 1588, a hunter in Auvergne is attacked by a wolf at night, but manages to fend the animal off with his dagger and to cut off one of its paws. He later draws the paw from his sack to show it to a friend, to find that the appendage has turned into a lady's hand with a wedding ring. The friend immediately recognizes the hand of his wife; he returns home to find her missing one hand and trying to cauterize the wound with a candle.

In the high culture of the Renaissance, however, the idea of transformations became systematized in the discipline of alchemy. Metallurgy had provided inspiration for many ideas about shape-shifting since very early times because it can profoundly transform not only the shape but also the texture and color of metal ore. The great work of alchemists such as the legendary Nicholas Flamel and Robert Fludd was the transformation of common metals into gold in order to create the philosopher's stone, which could prolong human life indefinitely. This change became the guiding metaphor for other metamorphoses, both physical and spiritual, particularly the perfection of the soul. Paracelsus, a renowned sixteenth-century Swiss alchemist, shifted the focus of alchemy from metals to medicine. Many experiments of the alchemists involved attempts to create living things such as insects or even human life out of dirt and waste.

By familiarizing people to the idea that the forms of living things were not immutable, alchemists prepared them for the Darwin-Wallace Theory of Evolution. In popular culture for at least a century after the publication of Darwin's *Origin of Species* in 1859, evolution was portrayed as a metamorphosis, whereby creatures went through a series of fixed stages as they progressed to greater complexity. A fish was shown stepping out of the sea, to become a reptile, a mammal, an ape, and finally a human being. The late nineteenth-century biologist Ernst Haeckel developed a theory, now discredited, that the embryo goes through stages analogous to the "scale of evolution," as it develops from a fertilized egg to a human being.

Although philosophers such as René Descartes and John Locke had exalted human status as a condition of moral consideration, the theory of evolution made it more difficult to take this status for granted. Peoples such as Africans and Irish were often compared to apes, and individuals that seemed oddly proportioned could be stigmatized as "throwbacks" to an earlier evolutionary stage. In Franz Kafka's tale "The Metamorphosis" (published in 1915 in German as "Die Verwandlung"), a man wakes up one morning to find he has turned into a cockroach. He gradually comes to terms with his identity, until he is simply content to lie in his room all day looking at the ceiling, but his family tires of caring for him, and eventually, he dies of neglect. One theme of the story is the "dehumanization" that comes from numbing routines and formalized relationships.

In modern popular culture, shape-shifters have generally been replaced by human-animal composites. The werewolf in popular culture is no longer a person who is transformed into another species, but a hairy man with a lupine face. Among the numerous other composites in twentieth-century popular culture is the comic book hero Spiderman, a photographer bitten by a radioactive spider, who climbs buildings and swings on webs. There are also numerous ape-men, cat-women, bat-people, and so on. Perhaps the most popular monsters in movies, television shows, and novels are vampires, which usually appear as ordinary human beings until they open their mouths to reveal enormous fangs. If one of them sucks the blood of a person, the victim becomes a vampire as well.

The idea of a shape-shifter presupposes a clearly defined, if perhaps temporary, identity, and recent culture increasingly views the self as perpetually ambiguous. Just as Christianity once spiritualized metamorphoses, the secular culture of the modern world has psychologized them. The interest in the stories of composite creatures lies largely in the tension between the human and bestial aspects of their identity. Characters such as Spiderman and Batman are constantly struggling with their partial alienation from human society.

As we progress into the twenty-first century, not only individual identity but also human identity has become increasingly uncertain. People can radically change their appearance through plastic surgery and, if they have sufficient money, even their gender. Transplants of organs and genetic material from one species to another are starting to become routine. Human genetic material, for example, is sometimes placed in pigs, so that they will develop organs that can be transplanted safely into human beings. The pigs, then, become partly human, and the recipients of the organs become partly pig.

Genes from a firefly have been placed in a tobacco plant, to make the leaves glow in the dark. Researchers sometimes speak of individual plants and animals as if they were simply vessels of genetic material, to be recombined in endless possibilities. On the Internet, they can try out new personalities and even, in multi-user domains or MUDs, digital bodies. The very lack of stable identity, however, means that metamorphosis forfeits some of both its religious significance and its dramatic appeal.

The concept of a metamorphosis involves a balance between constancy and change. Scientists have argued that almost nothing is entirely static. Not only do living things evolve, but continents drift, and the very structure of space and time is warped by gravitation. This is flux far more profound than that postulated by Ovid, who seemed, for example, to consider the stars exempt from change. If, as many philosophers now maintain, the soul, for human beings as well as animals, is an illusion, a metamorphosis in the more traditional sense is impossible.

Nevertheless, though impossible to either verify or disprove, belief in metempsychosis remains widespread in contemporary culture throughout the world. The reason for the persistence of this belief throughout human history is probably that it addresses the basic human experience of identification with other creatures. It is one way to account for the way in which animals, from ants to whales, often seem uncannily "human."

Further Resources

Barkan, L. (1986). *The gods made flesh: Metamorphosis and the pursuit of paganism*. New Haven: Yale University Press.

Eliade, M. (1974). *Shamanism: Archaic techniques of ecstasy*. Princeton, NJ: Princeton University Press.

Kafka, F. (1983). The Metamorphosis (W. Muir & E. Muir, Trans.). In N. N. Glatzer (Ed.), *Franz Kafka: The complete stories* (89–139). New York: Schocken.

Newman, W. R. (2004). *Promethean ambitions: Alchemy and the quest to perfect nature*. Chicago: University of Chicago Press.

Ovid. (1955). *Metamorphoses* (R. Humphries, Trans.). Bloomington, IN: Indiana University Press.

Sax, B. (2001). *The mythical zoo: An encyclopedia of animals in world myth, legend and literature*. Santa Barbara, CA: ABC-CLIO.

Warner, M. (2002) *Fantastic metamorphoses, other worlds*. New York: Oxford University Press.

Boria Sax

■ Literature
Metaphors about Animals

Animals are referred to so often in human speech, including the English language, that we sometimes do not realize how certain figures of speech reflect animals, their characteristics, and their behavior. In some cases, however, the characteristics applied to the animal, such as greedy gluttony for a pig (as in "don't be such a pig"), may be misapplied, because, regarding this example, pigs do not necessarily overeat. The following animal metaphors are excerpted from *Speaking of Animals: A Dictionary of Animal Metaphors* (1995), by Robert A. Palmatier.

Batty. *To be—or go—batty.* To be or go crazy. IRCD: early-20th cent. Source: BAT. WNNCD: 14th cent. *Batty,* like *bats* (WNNCD: 1939), is probably derived from the expression *bats in your belfry* (WNNCD: 1907), although *batty* has been around since 1590 with the meaning "resembling a bat." The bat has been regarded as a crazy-looking, crazy-behaving animal throughout the 20th cent. *See also* Bat House; Bats; Bats in Your Belfry.

Beeline. *To make a beeline for someplace.* To proceed to a place rapidly and directly. WNNCD: 1830. Source: BEE. WNNCD: O.E. Forager bees proceed to flowering trees, bushes, and plants on the basis of instructions given to them in the hive by scouts. They learn approximate direction and distance but have to rely on scent once they approach the general area. However, on their way back to the hive they know where to go and make a direct and rapid flight. People make *beelines* on foot, in cars, on horses, in speedboats, etc., but their route is usually not as straight as that of the airborne bee. AID; ATWS; BDPF; CI; DEI; IRCD; SHM.

Cocky. *To be cocky.* To be conceited, arrogant, and overbearing. WNNCD: 1768. Source: COCK; WNNCD: O.E. On the farm, the cock, or rooster, *rules the roost* (q.v.) with a brash self-confidence—announcing the break of day, having the run of females, chasing away the other males, eating whatever and whenever he wants, and generally playing the role of *cock of the walk* (q.v.) to the hilt. He is the embodiment of *machismo.* ATWS; ID; IRCD; LCRH; ST.

Cool your heels. *To cool your heels.* To wait, or be forced to wait. Source: HORSE. A horse doesn't have "heels," but it does have fetlocks, which are comparable to human heels. (A horse doesn't have fingers or toes, either, but the hooves are comparable to combined fingernails/toenails.) After a long hot trip, a horse's metal shoes become overheated, in turn causing the hooves to become overheated. The remedy is to let the horse rest, esp. in the shade, or to let it stand in shallow water for a period of time. People "get hot" *because* they are made to wait, whether out of neglect or by design. A VIP sometimes lets a not very important person wait just to emphasize his/her superiority. MDWPO.

Dog (n). *A dog.* An unattractive woman (MDWPO: 1950): an ineffective racehorse; an unsuccessful literary or theatrical production. Source: DOG. WNNCD: O.E. The negative connotation of *dog* probably derives from the lowly status of the dog among domestic animals. In spite of the important work that dogs do for humans—hunting, herding, and guarding—they have long been regarded as dirty, lazy, good-for-nothing animals. Therefore anything undesirable, such as a homely woman, or ineffective, such as an over-the-hill racehorse, or unsuccessful, such as a flop play, can be called a *dog.* ATWS; NDAS.

Dog (v). *To dog someone's footsteps.* To follow someone closely and persistently. WNNCD: 1519. Source: DOG. WNNCD: O.E. A hunting dog is trained to follow the sight or scent of a quarry until it is found or killed. A bloodhound tracks an escaped criminal in the same way, as does a human bloodhound, or detective, although the track is more often a paper trail. Reporters and photographers sometimes *dog the heels* of celebrities; people sometimes *dog* other people until they get what they want; and projects are sometimes *dogged* (one syllable)—i.e., "plagued" or "hampered"—by bad weather or equipment problems: ATWS; IRCD; ST. *See also* Hound (v).

Dogged. *To be dogged* (two syllables). To be marked by persistence, obstinance, or stubborn determination. WNNCD: 1653. Source: DOG. WNNCD: O.E. Hunting dogs pursue their quarry relentlessly—i.e., with *dogged determination*—but the *bulldog* (q.v.) is probably the most *dogged* of all dogs. As the name implies, bulldogs were orig. bred and trained to bait bulls by clamping their jaws on the bull's nostrils and hanging *doggedly* until the bull gave up. People display *doggedness* by refusing, like the bulldog, to give up, even under the most trying circumstances. ATWS; ID; IRCD; MDWPO.

Dog and pony show. *A dog and pony show.* An elaborate sales pitch or public relations presentation. WNNCD: 1970. Source: DOG; HORSE. WNNCD: O.E. In show biz parlance, a dog and pony show is a *one-ring circus,* with nothing more than a trick dog and a *one-trick pony.* In PR parlance, a dog and pony show is a traveling sales pitch, with charts and diagrams and multimedia effects. The term is contemptuous when used about someone else's "circus," but is excusatory when used by presenters about their own presentation. LCRH; MDWPO; NDAS; ST.

Dyed-in-the-wool. *A dyed-in-the-wool conservative.* A staunch political conservative. WNNCD: 1579. Source: SHEEP. Wool that is dyed before it is processed (i.e., dyed *in the wool*) is more highly valued than wool that is dyed after it is woven or fashioned into a garment. Its color is fast and firm, like the views of a conservative. DOC.

Hightail it. *To hightail it out of here.* To get out of here as quickly as possible. WNNCD: 1925. Source: ANIMAL. When a deer or horse or rabbit is "spooked," it races away with its tail raised to the fullest height. In the case of the white-tailed deer, the raising of the "flag"—brown above, white below—is a signal to the other deer to take off. In the case of the horse, the raised tail can also be a sign of pleasure and pride, as when the horse is galloping around the lot or participating in dressage. For people, the expression probably derives from the cowboys of the Old West, who *hightailed it home,* along with their horses, when the work was done. The cottontail rabbit also raises its tail—brownish gray above, fluffy white below—when it becomes alarmed. AID; EWPO; IHAT; MDWPO; NDAS; ST.

Hole up. *To hole up.* To seek refuge or peace of mind in a safe or quiet place. WNNCD: 1875. Source: ANIMAL. Some mammals, such as the rabbit and the fox, hole up daily, as part of their normal routine; some mammals hole up seasonally, as with the "seven sleepers," who *go into hibernation* (q.v.) every winter; and some mammals take refuge in a hole, brushpile, or other cover when they are pursued. People hole up when they are pursued by the law, by bill collectors, by paparazzi, by fans, by relatives—or when they just want some peace and quiet.

Off the hook. *To be—or get—off the hook.* To be relieved of a problem or obligation: to get a reprieve. Source: FISH. If an angler fails to *set the hook* (q.v.) once the fish has *taken the bait* (q.v.), the fish may get free of the hook and even swim away with the bait. A person who is in trouble can be *let off the hook* by the person who put him/her there, just as an angler may release a fish that is too small to keep; but if that same person gets out of trouble on his/her own, he/she is said to *wiggle—or wriggle—off the hook.* AID; ATWS; BDPF; CI; ST.

Old goat. *An old goat.* A disagreeable, unpleasant, or stubborn old man; a lewd, lascivious, lecherous, or "dirty" old man. Source: GOAT. WNNCD: O.E. The goat has long been associated with the Devil because of its horns, its stubbornness, its lustfulness, and its dirty behavior: it will eat anything, including tin cans. An *old goat* is an established "sinner" who has no redeeming qualities, although he may think he is God's gift to the women. ATWS; CH; SHM. *Compare* Old Buzzard; Old Coot; Old Crow.

Recoil. *To recoil from something.* To draw back from something fearful or terrible. Source: SNAKE. Many poisonous snakes must strike from a coiled position, after which they must coil again, or *re-coil.* However, *recoil* is not based on the snake's behavior: it lit. means "to move backwards" (fr. Fr. *reculer,* fr. *cul* "bottom"). Since snakes have no "bottoms," they cannot, technically, *recoil* at all.

Root out. *To root out the cause of something.* To uncover the cause of a problem. Source: PIG. A pig—or hog—uses its snout to turn up the earth in search of roots, grubs, and truffles. If allowed to run free, it can turn a lawn into a plowed field. Humans use their brains to *root out* evil (etc.) wherever it may lurk, including underground. A variation of *root* is *rout,* which is applied to gouging out grooves in wood. ATWS.

Smart ass. *A smart ass.* A pretentious or sarcastic person. Source: ASS. WNNCD: O.E. *Smart ass* is an oxymoron. The ass, or donkey, has always been regarded as a stupid animal: a horse wannabe. A *smart ass* is a person who pretends to be a horse but is only as smart as a donkey. *See also* Ass; Make an Ass of Yourself. *Compare* Silly Ass.

Stick your nose into someone else's business. *To stick—or poke—your nose into someone else's business.* To intrude into other people's lives; to interfere with other people's affairs. Source: DOG. *Curiosity once killed a cat* (q.v.), but dogs do not hesitate to poke their noses in the "business," or spoor, of other animals. It is their way of determining whose territory they are invading, whether they are friend or foe, whether they are male or female, etc. Dogs also smell *each other*, for the same reason, but get in trouble, as people do, for *sticking their nose where it doesn't belong.* CI.

Still wet behind the ears. *To still be wet behind the ears.* To be inexperienced or unsophisticated. HOI: late-19th cent. Source: ANIMAL. When a foal, calf, or fawn is born, its mother begins to clean it up by licking the amniotic fluid from its body. Before the cleanup is finished, however, the newborn often tries to get to its feet and take its first steps, still having a little bit of fluid remaining in the recesses behind its ears. People who are *not yet dry behind the ears* are as naïve and innocent as a young animal struggling to find its land legs. AID; ATWS; BDPF; CI; DEI; DOC; EWPO; MDWPO.

Sudden death. The elimination of one of two competitors who are tied at the end of an election (by the toss of a coin, the drawing of straws or cards, or the roll of die) or at the end of a regulation game of professional football or hockey (the first to score wins the game) or at the end of a regulation golf math (the first to win a hole wins the match). Source: COCK. The sudden death of a gamecock in a cockfight results in an instant win by its opponent. The same is true in other illegal "sports," such as dogfights, in which killing the opponent is the ultimate goal. MDWPO; ST.

Worm (2). *You worm!* You lowlife! Source: WORM. WNNCD: O.E. The earthworm may not be the lowest form of life, but it is certainly one of the lowest to the ground—as low as a *serpent* (also known as a *wyrm* in O.E.). Worms are maligned because of their small size, slow movement, slimy appearance, and association with rotten food and rotting flesh. A human *worm* is a person of low moral character and despicable social behavior: a loathsome, contemptuous individual. ATWS; DEI. *See also* Varmint; Vermin; Worm (1); Worm (3).

Further Resources

Palmatier, R. A. (1995). *Speaking of animals: A dictionary of animal metaphors.* Westport, CT: Greenwood Press

Key to References

AID   Spears, R. A. (1987). *NTC's American idioms dictionary.* Lincolnwood, IL: National Textbook Company.

ATWS   Lyman, D. (1983). *The animal things we say.* Middle Village, NY: Johnathan David.

BDPF   Evans, I. H. (Ed.). (1981). *Brewer's dictionary of phrase and fable* (rev. centenary ed.). New York: Harper & Row.

CI   Kirkpatrick, E. M., & Schwarz, C. M. (Eds.). (1982). *Chambers idioms.* Edinburgh: W & R Chambers.

DEI   Gulland, D. M., & Hinds-Howell, D. G. (1986). *Dictionary of English idioms.* London: Penguin Books.

DOC   Rogers, J. (1985). *The dictionary of clichés.* New York: Facts on File.

EWPO   Hendrickson, R. (1987). *The Facts on File encyclopedia of word and phrase origins.* New York: Facts on File.

ID	Brock, S. (1988). *Idiom's delight*. New York: Times Books.
IHAT	Flexner, S. B. (1976). *I hear America talking*. New York: Simon & Schuster.
IRCD	Ammer, C. (1987). *It's raining cats and dogs*. New York: Dell.
LCRH	Claiborne, R. (1988). *Loose cannons and red herrings: A book of lost metaphors*. New York: Norton.
MDWPO	Morris, W., & Morris, M. (1988). *Morris dictionary of word and phrase origins* (2nd ed.). New York: Harper & Row.
NDAS	Chapman, R. L. (Ed.). (1986). *New dictionary of American slang*. New York: Harper & Row.
SHM	Degler, T. (1989). *Straight from the horse's mouth*. New York: Holt.
WNNCD	Mish, F. C. (Ed.). (1983). *Webster's ninth new collegiate dictionary*. Springfield, MA: Merriam-Webster.

Robert A. Palmatier

Caste Discrimination in Animals

Govindasamy Agoramoorthy

India has a lengthy and rich history of great empires, unique cultures, artistic ingenuity, numerous invasions, and legendary figures such as King Asoka, Gautama Buddha, and Mahatama Gandhi. No aspect of Indian history has provoked more controversy than the chronicles of social relations that include the sensitive caste divisions and untouchability. However, caste-like divisions are not uniquely Indian. Indeed, caste-like divisions were found in the history of most nations—whether in the American continent or in Africa, in Europe, or elsewhere in Asia. In some societies, such divisions were relatively simple, whereas in others, they were much more complex. In modern India, calling and treating other people as untouchables is now legally forbidden. However, many people are not aware that some animals carry caste-discriminating names even today.

Two fascinating and highly adaptable animals carry a touchy caste-discriminating name: "pariah." One is the pariah dog, and the other is the common pariah kite—both are commonly seen in villages, towns, and cities in India. The pariah dog is often used as a generic term for any domestic dog that is a stray or that lives in feral condition, without any geographical restriction. Although the origin of the domestic dog from wolf has been established, details about where and when the domestication occurred are not clearly known.

The Book of Indian Birds by Salim Ali states that the common pariah kite is the most common raptor in India. These large brown birds with forked tail, particularly in flight, are excellent acrobatic flyers; they can snatch away snacks from one's palm before the person's hand reaches his or her mouth. They feed on a variety of food items that include insects, spiders, worms, mice, lizards, frogs, small birds, and also leftovers from human kitchens. They scavenge gregariously near garbage dumps.

The word *pariah* was first recorded in English in 1613, and the *Encyclopedia Britannica* defines it as follows: "formerly known as untouchables but renamed by the Indian social reformer Mahatma Gandhi as 'Harijans' (children of the God Hari/Visnu, or, simply, children of God). The word pariah—originally derived from Tamil language word *paraiyar*, 'drummer'—once referred to the *Paraiyan*, a Tamil Nadu caste group."

■ Literature
Metaphors in Constructing Human-Animal Relationships

As anthropologists are fond of pointing out, the relationships a society forms with its animals are often indicative of the structure and nature of that society and of the patterns of thinking of the individuals in it. Perhaps no other aspect of the natural world is as rich and varied as the animal "kingdom"; as beings who are at once like us, yet different from us, our fellow animals provide a wellspring for our thought and imagination. A rich tapestry of real and mythical creatures run, fly, swim, and slither through our literature, art, taxonomy, religion, dreams, and legends. In our language, animals feature as adjectives, verbs, nouns, and metaphors: we can be *wolfish*, *sheepish*, *catty*, and *dog tired*; we might *badger*, *bug*, *horse around*, or be *foxed*; and our enemies might be denigrated as *pigs*, *dogs*, *weasels*, or *rats*.

The Metaphorical Gaze: Analogizing Our Relationships with Animals

Humans seem predisposed to engage in metaphorical expression. The categories that we use to carve the world up into discrete elements are rarely fixed; we blur boundaries, we explain aspects of our world in terms of how much they are like other aspects. Descartes compared animals to clockwork mechanisms, and the language of today's slaughterhouses and intensive livestock farms similarly speaks of animals as if they were objects or commodities. Before the nineteenth century, it was not uncommon for European courts to hold animals responsible for intentional "crimes" against humans and to sentence and execute them accordingly. We find it convenient to think of humans as sometimes animalistic but also to project our human societal concerns onto other species. As the psychologist James Hillman expresses it, "Animals are like walking metaphors of scratching, pecking, pushing, snarling, domination and anxiety."

The anthropologist Claude Lévi-Strauss described animals as "good to think." Engagement with animals seems crucial to the development of imagination and self-identity; the dreams and stories of children abound with animal images, and real and toy animals form the basis of children's early attempts at categorization. Modern psychology has tended to view categories as abstract representations of real divisions in the world. More recent approaches in cognitive science have shown, however, that our thinking seems to be very much rooted in feeling, sensation, and perception and that human categories do not tend to be fixed with clear boundaries. The linguist George Lakoff and philosopher Mark Johnson have suggested that our thinking is primarily metaphorical and that even our abstract concepts and categories are structured metaphorically. For example, when we talk about mood states, we often resort to orientational metaphors, as in "I'm on a *high*," "She's feeling *down*," and "He cheered *up*." In a very real sense, when we are feeling upset or sad, our physical posture tends to slump; we have neither the physical nor the mental energy to hold ourselves upright. Such metaphors are not simply figures of speech; they help us to make sense of the world by mapping similar experiences on to one another. So *up* is healthy and good, and *down* is unhealthy and bad.

It is perhaps not surprising that our philosophy and science are also replete with metaphors; they seem fundamental to our thinking. A similar orientational metaphor can be detected in early Western notions of nature as a linear hierarchy, such as the great chain of being. In this medieval Judeo-Christian model of the world, humans are seen as "higher" on the scale (and closer to God) than animals, who are seen as inferior and closer to the Earth (and to the Devil). Here too, then, we find the metaphor that *up* = *good* and *down* = *bad*. This metaphor underpins many traditional animal classifications in the West. For example, according to Elizabeth Lawrence, the honeybee has long been seen as embodying goodly (and godly) characteristics, such as capacity for toil, perceived virginity and purity, and the ability to communicate in a sophisticated way. Honeybee communities were likened to monastic cloisters, whose inhabitants labored solely for the service of God and others. The pig, by contrast, could not be further away from the dizzy heights of the natural hierarchy, dwelling as it does in the muddy confines of domesticity. Unlike the bee, the pig has been the subject of rather less complimentary descriptions, variously considered as a filthy, gluttonous, lazy, and stupid animal.

If animals can be considered in terms of their human-like qualities, so too can humans be described as animals. It is from the animal realm that we choose the most abusive terms with which to describe our fellow humans. In his classic essay "Animal Categories and Verbal Abuse," the anthropologist Edmund Leach suggests that there are fundamental similarities between social and natural categories. Cross-culturally, people tend to draw boundary lines between close kin, distant kin, friends and neighbors, and strangers. We similarly classify those animals that fall within our circle of close associates as companion animals who enjoy the same status as human family members. Occupying the middle ground are domesticated animals that may be eaten, such as livestock and farm animals. Inhabiting more distant realms are wild animals who may be frightening or mysterious. Particularly problematic are those animals that fall between categories, that cannot be easily classed as edible or non-edible, wild or tame. Such animals often invite specific taboos, for example the biblical taboo against touching and eating pigs. Although the pig is a domesticated animal, it is unusual because it is one of only a few animals that are reared solely for food and then slaughtered. Until relatively recent times, pigs were kept in close proximity to domestic households, enjoying close relationships with humans. Perhaps it is the ambivalence of the human-pig relationship, or even, as Leach suggests, a sense of shame about it, that has ensured its status as the origin of a number of insults applied to those whom we consider undesirable or otherwise marginalized. It is interesting that the English language features a number of euphemistic terms when applied to the large, often domesticated animals that humans live closely to but also eat; we do not speak of eating sheep, pigs, cows, and deer, but rather of consuming mutton, pork, beef, and venison. Such euphemisms do not seem to be needed when we discuss the meat of smaller wild animals that we do not have close relationships with; we seem happier to discuss eating hare, pigeon, or grouse, perhaps because these smaller animals inhabit wild, rather than domesticated, settings.

Our discourse about animals is often ambivalent; as companions, pests, archetypal symbols, or meat, animals fulfill multiple roles in human society and almost invite contradictory categories and mixed metaphors. For example, that most familiar of human associates, the dog, has been classed as a cherished companion, as a working animal, as a subject for laboratory research, and as food, depending on particular cultural perspectives. Our discourse about animals is often structured according to the concerns of a particular society at a particular point in time. For instance, as Harriet Ritvo has convincingly argued, relationships with wild and domestic animals in the Victorian era reflected the imperialistic and doministic concerns of a colonial society. The animals that were held in highest regard (and considered most intelligent) were the most "subservient": the sheep, ox, horse, and particularly the dog, who was considered to be the most willing and malleable of servants. Cats, however, did not enjoy such high regard. Despite being useful in catching rats and mice in domestic households, the cat was derogated as being deceitful and villainous; its nocturnal habits and carnivorous nature reflected its independent and untamable nature and probably assured its status as an ambiguous animal, one that inhabited a marginal realm between domesticity and wildness. Cats have long been metaphorically associated with femininity and were traditionally the companions of marginalized women (e.g., those accused of witchcraft and other antisocial vices). Their status as supernatural or sacred animals seems more positive in some other cultures; in Buddhist temples in the East, cats are considered valuable guardians.

Metaphors help us to negotiate the shared space between ourselves and other species, yet the adoption of a particular metaphor always entails a partial view of reality, a vision of the animal that can at once be symbolic and real, meaningful and confused. The snake is a good example of how our attitudes toward animals are infused with the weight of cultural tradition. Snakes inspire ambivalent responses, ranging from fascination to extreme fear. Some authors have argued that humans seem preferentially phobic toward animals such as snakes because they are potentially dangerous features of the environment. Snake phobias, however, seem more culturally than biologically determined; as Jared Diamond has pointed out, the indigenous people of New Guinea, although living in a landscape where over one-third of the snake species are poisonous, show no sign of a phobia toward snakes. It seems likely that the snake's slow, silent movement, lack of detectable expression, and unblinking gaze make it difficult to interpret and interact with. Indeed, the snake's gaze has acquired a multitude of metaphorical and symbolic meanings, inspiring legends of snakes "charming" or hypnotizing their human or animal victims prior to devouring them. In an examination of traditional attitudes toward rattlesnakes, Boria Sax has argued that there are clear parallels between early accounts from the New World of the rattlesnake's ability to kill with a glance and European legends of the mythical basilisk, very often represented as a dragon or serpent with a terrifying and fatal gaze. Because our metaphors are grounded in the meanings a particular culture assigns to animals, it is often difficult to detect them; a metaphorical comparison that is used often enough eventually ceases to sound like a metaphor—it becomes judged as literal. In today's rational age, rattlesnakes are no longer thought capable of unnatural hypnotic powers, yet popular Western attitudes toward them are still characterized by loathing and disproportionate fear.

Our relationships with animals anchor our perception and understanding and provide the language by which we describe human social relationships and experiences. This kind of understanding is not just a Western phenomenon. In ancient China, animals were not classified according to modern Western taxonomic principles, or in terms of the kind of linear hierarchy that characterized Greek accounts of the biological scale. Rather, their classification reflected sociopolitical or moral concerns. For example, Confucius observed that "a swift horse is not praised for its physical strength but for its virtue." In Chinese society, real and mythical animals were not distinguished, and species classifications were malleable, such that the categories of serpent and dragon were considered interchangeable. Yet such classifications embodied more than a simple usage of animals as descriptors for human social affairs, as convenient analogies to link human and animal worlds. Rather, they were a means by which, through metaphorical projection, humans and animals might be seen as inhabiting a similar cosmos, as embodying the same natural principles. The category of "numinous" animals, for instance, included snakes, unicorns, turtles, and human sages. Numinous animals were those thought able to transform their physical shape, size, or color or able to inhabit different realms, moving easily between air, earth, and water. The sage who had perfected his spiritual and energetic powers became able to transform himself, like other sacred animals.

A study of the ways in which we conceptualize other species illustrates the metaphorical and symbolic bases of our thought and language. At the same time, it provides a window on how we construct our positive and negative relationships with other animals.

Metaphors are always partial and culturally bound, yet because they are based on experience, they can illustrate common human ways of responding to particular species. The ambivalence toward snakes is apparent in both American fears about rattlesnakes and ancient Chinese awe at the snake's ability to metamorphose by sloughing off its skin. The marginal status of the fox is evident in English folktales of the cunning, wily Reynard and in Japanese accounts of possession by fox spirits. Elizabeth Lawrence suggests that the paucity of animal encounters that characterizes modern urban existence may potentially result in the development of a richer imaginative life, inhabited by the wild animals we no longer live close to. Yet there is a bleaker alternative to her prediction. If our thinking is grounded in physical experience, in meaningful encounters with the natural world, the increasing urbanization of nature may leave us wondering, as did Paul Shepard, what fate awaits us, as inhabitants of a world no longer "peopled" by animals:

> During nearly all the history of our species man has lived in association with large, often terrifying, but often exciting animals. Models of the survivors, toy elephants, giraffes and pandas, are an integral part of contemporary childhood. If all these animals become extinct, as is quite possible, are we sure that some irreparable harm to our psychological development would not be done?

See also

Ethics and Animal Protection—*Factory Farm Industry Discourse*
Literature—*Human Communication's Effects on Relationships with Other Animals*
Literature—*Metaphors about Animals*

Further Resources

Belk, R. (1996). Metaphoric relationships with pets. *Society and Animals, 4*(2), 121–46.
Diamond, J. (1993). New Guineans and their natural world. In S. R. Kellert & E. O. Wilson (Eds.), *The biophilia hypothesis* (pp. 251–71).
Hillman, J. (1997, January). Going bugs with James Hillman. The *Satya* interview. *Satya*. Retrieved from http://www.satyamag.com/jan97/going.html
Lakoff, G., & Johnson, M. (1980). *Metaphors we live by.* Chicago: University of Chicago Press.
Lawrence, E. (1993). The sacred bee, the filthy pig, and the bat out of hell: Animal symbolism as cognitive biophilia. In S. R. Kellert & E. O. Wilson (Eds.), *The biophilia hypothesis* (pp. 301–41).
Leach, E. (1964). Anthropological aspects of language: Animal categories and verbal abuse. In E. H. Lenneberg (Ed.), *New directions in the study of language* (pp. 23–63).
Ritvo, H. (1987). *The animal estate: The English and other creatures in the Victorian age.* Cambridge, MA: Harvard University Press.
Sax, B. (1994). The basilisk and rattlesnake, or a European monster comes to America. *Society and Animals, 2*(1), 3–15.
Shepard, P. (1978). *Thinking animals: Animals and the development of human intelligence.* Athens & London: University of Georgia Press.
Sterckx, R. (2002). *The animal and the daemon in early China.* New York: State University of New York Press.

Diane Dutton

Metaphors and the Treatment of Animals

Traci Warkentin

Animal metaphors are everywhere, in classrooms, on television, in textbooks and magazines, even in casual conversations. Many are so common that we do not even notice them, and yet they may have dramatic influences on our human relationships with other animals. Let us take cows for example and look at the way they are talked about in the dairy industry. A Web site promoting the use of growth hormones for dairy cows makes the following general statements: "milk production can be made more efficient so that milk yield increases with feed intake, and nutrient intake replenishes the cow's body reserves." This is probably familiar language, which, upon closer inspection, reveals an underlying metaphor that a cow is a milk-making machine.

It may seem like common sense that cows produce milk, but to think of them in terms of milk-producing *mechanisms* may have powerful implications for their everyday lives. In fact, this metaphor reflects the treatment of many cows in industrialized agriculture, particularly in high-density factory farming. Food is the input, and milk is the output of the generic cow machine. When thought of in abstract, mechanical terms of productivity, the nature of what it means to be a cow may be obscured or forgotten. In traditional forms of pastoral agriculture, cows roamed freely in the fields, grazing grass and breathing fresh air. Many dairy cows nowadays typically spend their entire lives inside factory walls.

Such animal metaphors show us the reciprocal nature of how the way we talk about animals influences how we think about and treat them, and in turn how the current relationships we have with animals reflects the way we talk about them. Can you think of other common animal metaphors? How about companion animals, such as cats and dogs? They are often referred to as human children: "This is Fluffy, my little baby" and "I'm Rover's mom." Although these expressions convey great love and personal affection, as if the dog or cat is a genuine member of the human family, it can also be demeaning by infantilizing the adult animals. It can obscure the significant differences that exist between adult canines, adult felines, and human children and therefore may not lead to the best ways for adult humans to interact and develop relationships with them. This metaphor of animals as honorary human citizens is often applied in animal rights arguments and can be both effective and detrimental to the cause. Perhaps one of the most disturbing reversals of metaphor occurs when someone who is a victim of sexual abuse describes herself as being treated as "a piece of meat." In this case, the woman is thinking metaphorically that her body is a dead food animal, suggesting many complex implications of the language for both people and animals.

There are other kinds of animal metaphors too, such as those we use to describe other people. For instance, in a disagreement, you might claim that your friend has her "claws out," which implies that she is a cat. What does this say about her? What does this say about cats? In a very different sense of the metaphor, a person practicing yoga may embody a cat pose, a downward dog, and the cobra. For centuries, animal metaphors have been an integral part of Hatha yoga, a spiritual and physical practice originating in India.

Although not all metaphors necessarily have negative effects on human-animal relationships, it is constructive to become aware of them and think critically about what they mean in practical terms. Ultimately, as we explore relationships between humans and animals, it is helpful to examine the language we use to talk about them.

(continues)

■

Metaphors and the Treatment of Animals (continued)

Further Resources

Adams, C. (2000). *The sexual politics of meat: A feminist-vegetarian critical theory* (2nd ed.). New York: Continuum.

Lakoff, G., & Johnson, M. (1980). *Metaphors we live by.* Chicago: University of Chicago Press.

Midgley, M. (2001). *Science and poetry.* New York: Routledge.

Radha, S. (1995). *Hatha yoga, the hidden language: Symbols, secrets and metaphor.* Spokane: Timeless Books.

Sabloff, A. (2001). *Reordering the natural world: Humans and animals in the city.* Toronto: University of Toronto Press.

Warkentin, T. (2002). It's not just what you say, but how you say it: An exploration of the moral dimensions of metaphor and the phenomenology of narrative. *Canadian Journal of Environmental Education, 7*(2), 241–55.

Wayne, K. (1995). *Redefining moral education: Life, Le Guin and language.* Bethesda: Austin & Winfield.

■ Literature
Nature Fakers

It seems odd that the president of the United States would publicly attack a man who wrote widely respected children's books. Odder still, the focus of his criticism was, he argued, the author's misunderstanding of natural history and how animals think. Still, this is just what happened back in 1907. The president was Theodore Roosevelt, and the author was William J. Long.

Roosevelt's attack was the culmination of a four-year-long public debate called the Nature Fakers controversy. It began when John Burroughs, the widely respected dean of American nature writers, published "Real and Sham Natural History" in the March 1903 *Atlantic Monthly*. He was upset by a new approach to nature writing that, he felt, overly sentimentalized animal life in the wild. These authors also claimed that wild animals think much like humans and even school their young in the ways of the wild. Criticizing Long's then recently published book *School of the Woods*, Burroughs wrote,

> There is a school of the woods, as I have said, just as much as there is a church of the woods, or a parliament of the woods, or a society of united charities of the woods, and no more; there is nothing in the dealings of animals with their young that in the remotest way suggests human instruction and discipline. The young of all the wild creatures do instinctively what their parents do and did. They do not have to be taught; they are taught by nature from the start. The bird sings at the proper age, and builds its nest, and takes its appropriate food, without any hint at all from its parents.

Long and other writers of his sort were, Burroughs argued, sham naturalists who wrote fiction, but represented their work as accurate natural history. Ernest Thompson Seton's foxes did not really ride across fields on the back of sheep in order to escape the hounds, wrote Burroughs, nor did they lure dogs onto railroad trestles in time for them to be killed by passing trains. Yet Seton proclaimed in the prefaces to his books that his stories were truthful.

Burroughs would not have been as upset about these books if they did not sell. At the beginning of the twentieth century, the United States was in the midst of its first environmental movement. The nation had discovered that the frontier was closed and that natural resources were dwindling. There was already a movement to protect declining species of birds. Efforts to protect wilderness had begun, and the natural resources conservation movement was about to be born. Nature study was becoming the cutting-edge pedagogy of the day, and there was a hot market for books about nature. The growing population of people living in urban and suburban regions was eager for tales that provided escape from their pressured lives and retreat to the imagined comforts of nature.

Long and Seton helped to create a new literary genre, the realistic wild animal story. They told stories from the perspective of the animals themselves. Their readers entered the animals' minds (or thought they did) and empathetically experienced their lives from the animals' points of view. British writer Anna Sewell had pioneered this approach in 1877 with her book *Black Beauty*, which told the story of horrifying abuse to a horse from that horse's perspective. Sewell's book was so effective in mobilizing public sentiment to protect horses that it was dubbed "The Uncle Tom's Cabin of the Horse." A year later, the American author Charles Dudley Warner did for wild animals what Sewell did for horses and other domesticated animals. He published a story, "A-Hunting of the Deer," that described a deer hunt from the point of view of the deer.

The wild animal story gained wide popularity by the late 1890s, particularly as a result of the craft of Canadian authors Charles G. D. Roberts and Ernest Thompson Seton. Their popularity may have been influenced by the great success of the fictional animal stories in Rudolph Kipling's *Jungle Book*. Seton's *Wild Animals I Have Known* was particularly successful. First published in 1898, it went through at least sixteen printings by 1902. Such publishing success attracted many imitators who were eager to sell books to a burgeoning market. Authors of varying talents, from the incompetent to highly skilled, climbed aboard the bandwagon. One author reported seeing a tree full of birds all in commotion. Then a bird fell dead. He concluded that he had witnessed a trial and execution. On the other hand, Jack London's widely respected novels *Call of the Wild* and *White Fang* fit into this genre.

The first story in Seton's *Wild Animals I Have Known* is perhaps his best-known animal story. (Burroughs argued that this book should have been titled *Wild Animals I Alone Have Known*.) It is a tale from Seton's days as a wolf hunter in New Mexico about his effort to kill a wolf known as Lobo. Lobo was legendary for outsmarting every effort to shoot or trap him. Seton finally succeeded by using the body of Lobo's mate, Blanca, to lure him into an array of traps. "Poor old hero," Seton wrote, "he had never ceased to search for his darling, and when he found the trail her body had made he followed it recklessly, and so fell into the snare prepared for him." Although he tried to keep Lobo alive, the wolf died a few days later, apparently of a broken heart. Lobo's body was placed beside Blanca's, and a cattleman commented, "There, you *would* come to her, now you are together again." The tragic tale of Lobo, a symbol of the wild independence and craftiness, as well as of faithfulness to his mate, brought tears to the eyes of Seton's readers.

Seton's animals were the embodiment of wild intelligence and goodness. He argued that they followed a code of conduct akin to the Ten Commandments. However, Seton pointed out in the preface to this book, "The life of a wild animal *always has a tragic end.*" Charles G. D. Robert's animals embodied a more overtly Darwinian struggle for survival. Typical of his tales, the first story in his 1896 book *Earth's Enigmas* tells of a pair of hungry mountain lions who set out to find meat for themselves and their cubs. They find a child in a cabin and are about to kill him. Just as they are about to pounce on him, the boy's father shoots and kills them. There is, however, a twist to the story. A few weeks later, that father chances upon the lions' den and finds the rotting corpses of their cubs who starved to death. With this shift of perspective, Roberts gave the lions and the father moral parity. All were trying to protect the lives of their young. All were caught in a tragic struggle for survival.

William J. Long, on the other hand, viewed the lives of wild animals as "gladsome." Although animals died, they did not suffer, and there was even the possibility of an after-life. Parents taught their young in much the same way as do human parents. Animals may communicate telepathically with each other, he argued. He even claimed that wild animals administer medical attention to themselves and reported seeing a woodcock apply a mud cast to its broken leg. Wolves may be fierce, but they do not kill needlessly. In one book he even had a wolf lead children to safety. In Long's world, nature was guided by a loving God, rather than by a Darwinian struggle. His *A Brier-Patch Philosophy* is an unsung classic in animal rights literature. Long's books were appealing and widely read; many of his stories were published in a series of classroom readers, which expanded his readership and reputation.

The wild animal story was initially a Canadian creation, but it quickly spread to the United States. Roberts and Seton eventually settled in the United States. Some literary critics have argued that the theme of survival is especially prevalent in Canadian litera-ture and is reflected in the Darwinian character of Roberts's and Seton's stories. Long's tales, on the other hand, do not embody such conflict.

The controversy that John Burroughs began focused on a number of issues. Princi-ple among them was whether or not animals can think and whether some of the authors were honest in what they reportedly witnessed. In general, the critics of those who came to be called Nature Fakers claimed that these writers were simply making money by pro-ducing books that overly dramatized animals in order to satisfy a market of eager, and gullible, readers.

Things were more complicated than this, however. The discussion of animal rea-soning was hampered by the options that psychology made available at that time. Ani-mals were either instinctive machines, or they could reason. Burroughs was a member of the first camp. He believed that animals were incapable of reason. Theodore Roosevelt was more complicated in his understanding and cautioned Burroughs to consider that, for example, a salamander and a chimpanzee may differ in the extent of their reasoning abilities. Others wondered, if animals can think in ways akin to people, then can we not understand them empathetically? Long, a former Methodist minister, argued that human and animal minds reflect the mind of God and that thus, we are capable of understand-ing animal minds through empathy.

The issue of whether authors were faking nature is also complicated. On one hand, it was clear that some were fabricating tall tales, which they misrepresented as truthful. On the other hand, different people bring different understandings to their interpretation of animals. Early in the twentieth century, for example, most people (including Roo-sevelt) believed that wolves were savage creatures eager to kill both humans and other creatures. Many wild animal story writers, on the other hand, presented them as socia-ble creatures who killed in order to survive. Our present understanding of wolves and other predators tends to be closer to that of those called Nature Fakers.

In general, though, those who wrote wild animal stories viewed these animals as autonomous beings living lives largely independent of humans. They represented a new nonutilitarian view of wild animals that increasingly spread through the population as the century progressed.

Theodore Roosevelt contacted Burroughs soon after his *Atlantic Monthly* article appeared. This began a close friendship between the two men. Roosevelt was a skilled naturalist and encouraged Burroughs to moderate his views regarding animal psychology. However, Roosevelt was angry about the way some authors, particularly Long, appeared to misrepresent nature. He was especially upset that their books were required reading in classrooms. Burroughs urged him to speak out publicly against them, but the president refused, saying that it was not proper for someone in his office to attack private individuals.

When the controversy erupted, few of the so-called sham naturalists responded, fearing that doing so would fan the flames. Roberts ignored the debate. London decided, as he later wrote, "to climb a tree and let the cataclysm go by," and did not defend himself until the controversy was long over. Seton invited Burroughs to visit his home in Connecticut and proved his skills as a naturalist. Long, however, launched a vigorous defense of himself and his stories. This made him the lightning rod of the controversy.

Finally, after the debate dragged on in newspapers and magazines for four years, Roosevelt wrote to Burroughs that he needed a diversion and decided to speak out against Long in an interview. It was published in June 1907. "I don't believe for a minute," Roosevelt said, "that some of these men who are writing nature stories and putting the word 'truth' prominently in their prefaces know the heart of wild things." He continued,

> As for the matter of giving these books to the children for the purpose of teaching them the facts of natural history—Why, it's an outrage. If these stories were written as fables, published as fables, and put into the children's hands as fables, all would be well and good. As it is, they are read and believed because the writer not only says they are true but lays stress upon this pledge. There is no more reason why the children of the country should be taught a false natural history than why they should be taught a false physical geography.

Needless to say, the press was delighted with the controversy. It generated lots of copy and sold lots of newspapers. Long was incensed and rushed a response to newspapers around the nation. The *New York Times* gave it an entire page in its editorial section under the title "'I Proposed to Smoke Roosevelt Out'—Long." In addition to defending himself and his stories, he attacked Roosevelt for being a hunter who, he argued, only studied wildlife while sighting along the barrel of a gun.

> The idea of Mr. Roosevelt assuming the part of a naturalist is absurd. He is a hunter. . . . Who is he to write, "I don't believe for a minute that some of these nature writers know the heart of the wild things." As to that, I find after carefully reading two of his big books that every time Mr. Roosevelt gets near the heart of a wild thing he invariably puts a bullet through it. From his own records I have reckoned a full thousand hearts which he has known thus intimately.

Roosevelt responded with a magazine article from his own pen. It was published in September, together with a collection of statements in support of the president from prominent naturalists and scientists throughout the nation. "Roosevelt Whacks Dr. Long Once More," proclaimed the *New York Times*.

The combination of Roosevelt's article and the scientists' statements provided a killing blow. Long's next book was a history of English literature. He did not publish another nature book until after Roosevelt's death. Nevertheless, his publisher kept many of his books in print. One remained in print until 1940, a demonstration of the enduring appeal of his writing.

The Nature Fakers controversy marked the presence of nonutilitarian fellow-feeling toward wildlife on the part of a growing number of Americans. Some began to think of animals as persons with an innate right to live their lives without human disturbance. At the same time, the controversy marked the growth of scientists as the arbiters of knowledge about wildlife. The Nature Fakers gained their understanding of animals through personal observation and anecdotal knowledge. The scientists who supported Roosevelt argued that such knowledge is not sufficient; one must rely on systematic scientific study.

Robert MacDonald has argued that the wild animal stories represented a "revolt against instinct" and the Darwinian vision of nature. On the other hand, Thomas Dunlap suggested that they provided a way for the public to assimilate Darwinism in an acceptable way by showing that we and our other animal companions on this planet share common emotional, mental, and moral ground.

The major writers who were dubbed Nature Fakers by Roosevelt and Burroughs believed what they wrote. They may have been mistaken, but they did not intend to deceive their readers—not even Long. However, the controversy promoted a higher level of professionalism within their field. As one of the long-forgotten writers of that time wrote about Roosevelt's attack, "I now realized that if I ever make a bad break in regard to my natural history statements that I was doomed."

Ernest Thompson Seton was a skilled artist and naturalist. He wrote *Art Anatomy for Animals*, the first such anatomy book for artists. His wealth of notes and specimens was sufficient to convince Burroughs of his qualifications to write about animals. After the controversy, he wrote a four-volume natural history of North American mammals to prove his expertise. He also became a founder of the Boy Scouts of America and its first Chief Scout.

Charles G. D. Roberts also had a good foundation in natural history based on his youthful observations of nature in Canada and subsequent informal studies. His career was primarily of a literary nature, though. He was one of the Confederation Poets and has been called the father of Canadian literature. He was knighted for his literary contributions.

William J. Long began his career as a teacher, but then shifted to the ministry. He studied at Andover Theological Seminary and then studied philosophy, theology, and history at the University of Heidelberg, which awarded him a PhD. He was a lifelong outdoorsman and observer of wildlife, but his theology shaped his interpretations of what he saw. Following the controversy, he began a new career writing highly respected books on the history of American and English literature and later published additional wild animal stories.

The genre fell out of favor following the controversy. William Magee suggested that this happened because there are few plots available for stories about animals who live for themselves, rather than humans. However, the genre was reborn in the 1940s with Rachel Carson's *Under the Sea Wind* and Sally Carrighar's *One Day on Teton Rock*. These books were enlivened by new plots that reflected a new ecological understanding of the world. The genre continues today through just about every wild animal film that we see on television and in the theater. And debates over their authenticity and interpretations continue.

See also

Literature—*Animal Stories across Cultures*

Further Resources

Burroughs, J. (1903, March). Real and sham natural history. *Atlantic Monthly, 91*, 298–309.

Clark, E. B. (1907, June). Roosevelt on the nature fakers. *Everybody's Magazine, 16*, 770–74.

———. (1907, September). Real naturalists on nature faking. *Everybody's Magazine, 17*, 423–27.

Dunlap, T. R. (1992, April). The realistic animal story: Ernest Thompson Seton, Charles Rogers, and Darwinism. *Forest & Conservation History, 36*(2), 56–62.

Long, W. J. (1906, June 2). "I propose to smoke Roosevelt out"—Long. *New York Times,* part 5.

Lutts, R. H. (1990). *The nature fakers: Wildlife, science & sentiment.* Golden, CO: Fulcrum.

———. (1992, October). The trouble with Bambi: Walt Disney's *Bambi* and the American vision of nature. *Forest & Conservation History, 36*(4), 160–71.

———. (1996, Fall). John Burroughs and the honey bee: Bridging science and emotion in environmental writing. *ISLE: Interdisciplinary Studies in Literature and Environment, 3*(2), 85–100.

———. (Ed.). (1998). *The wild animal story.* Philadelphia: Temple University Press.

MacDonald, R. (1980, Spring). The revolt against instinct: The animal stories of Seton and Roberts." *Canadian Literature, 84*, 18–29.

Magee, W. H. (1964, Summer). The animal story: A challenge in technique. *Dalhousie Review, 44*, 156–64.

Roosevelt, T. (1907, September). Nature fakers. *Everybody's Magazine, 17*, 427–30.

Ralph H. Lutts

■ Literature
Rabbits' Roles in Watership Down

When Rex Collings Ltd., a small publishing firm in London, released Richard Adams's *Watership Down* in late 1972, it was an instant hit. True, the first printing was only 2,500 copies. But that run quickly sold out, and Adams notes in the introduction to the 2005 Scribner edition that he was "staggered" by the number of favorable reviews the book received.

The book's success hardly stopped there. Since being a best-seller for ten months in 1974 and 1975, *Watership Down* has never been out of print, has been translated into a dozen languages, and has been made into both a highly acclaimed animated film and a somewhat less successful television series. George Lucas has said that *Watership Down* was the inspiration for his *Star Wars* series; in a 2003 survey of its viewers, the BBC found that *Watership Down* came in forty-second of the 100 greatest books of all time. *Watership Down* "plainly has a wide appeal," Adams writes modestly, adding, "although the reason has never been altogether clear to me."

Certainly, the fact that the book features wild rabbits—rather than the roly-poly, child-like "bunnies" that hop through the pages of much juvenile fiction—was one reason the book had such wide appeal, especially to adults. Childlike rabbits indubitably attract children (just think of the characters Peter Rabbit, Benjamin Bunny, and Max). But the depth of Adams's adult rabbit characters, as well as the accuracy with which he described their natural behaviors, clearly engaged more sophisticated readers. So, too, did the complexity of the rabbits' civilizations and the dramatic adventures through which Adams put his long-eared characters.

Not all the reviews were positive. Some reviewers complained about the book's anthropomorphism, whereas others lamented the basic fact that the novel featured rabbits. But several reviewers went so far as to compare the novel to epic works such as the Odyssey and the Aneid, and many waxed enthusiastic at the book's tautly structured drama. Adams "has bravely and successfully resurrected the big picaresque adventure story," Nicholas Tucker wrote in the *New Statesman* in 1972, "with moments of such tension that the helplessly involved reader finds himself checking whether things are going to work out all right on the next page before daring to finish the preceding one."

Whether enthused or not, few reviewers identified the very pivotal role that humans play in the book (perhaps because so few literary critics would be familiar with the history of rabbit-human relationships). But actually, Adams very deftly portrays the conflicting types of relationships that humans have had with rabbits historically and those relationships' very deleterious effects (on the rabbits, not the humans). Further, those relationships provide the impetus for much of the major action in the book.

Rabbits as Prophets

At the beginning of the novel, the rabbits leave their home warren in the opening chapters because Fiver, a rabbit who is a prophet, senses danger. This is a historically accurate characterization: many ancient peoples, including the Egyptians, Romans, and Britons, believed that rabbits had supernatural powers of divination (or fortune telling). Similarly, the Native American Algonquins refused to hunt rabbits because they were seen as the keepers of life's secrets and a key to the afterlife.

Other rabbits in *Watership Down* also have what we might call psychic powers: Hyzenthlay, a doe in the Efrafa warren, foresees the bloody conflict in the book's conclusion, and Hazel, the leader of the Watership Down gang, receives a vision from El-ahrairah (the rabbit god) that shows him how to beat the Efrafan warriors who attack his warren. Although some of the characters in the book are initially skeptical that such visions can be accurate, the fact that rabbits can act as conduits to other worlds and other sources of knowledge comes to be a given by the end of the book, just as it was a given in many ancient cultures.

Rabbits as Pests

Fiver's original vision is of "some terrible thing—coming closer and closer" and of a field that is covered in blood. Later in the book, we learn that men did indeed go to the warren, block up the holes, and fill them with poisonous gas to get rid of the rabbits. This, too, is historically accurate. Over the centuries, humans have transported wild European rabbits (*Oryctolagus cuniculus*—the same species, by the way, as our domestic rabbits) from their native Spain to other parts of Europe, South America, Australia, and New Zealand. In the absence of native predators, the rabbit populations have exploded and ended up devouring crops, pasture lands, shrubs, and trees, making them some of the most destructive exotic pests known to humankind.

In response, humans have used all sorts of weapons—including guns, ferrets, dogs, and the aforementioned poison—to wipe out the rabbits. Most of these are mentioned in the book. At Cowslip, the first warren that Hazel and his gang join, for instance, the rabbits are regularly snared. And Efrafa, the second warren the Watership Down bucks infiltrate, is so tightly (and unnaturally) controlled by the tyrannical General Woundwort precisely because his own mother had died of "the white sickness," or myxoma virus. Again, Adam hits the historical nail on the head. First identified among domestic

European rabbits in Brazil in 1896, the myxoma virus was deliberately released in France by an enterprising French pediatrician who had gotten the virus from a friend at a local laboratory. Unfortunately, the virus causes incredible suffering in rabbits and, because it is spread by mosquitoes, can infect domestic populations. Within three years, 98 percent of the wild rabbits in France alone were dead, as were 40 percent of the domestic rabbits.

A year later, the virus was brought to England by farmers also eager to rid their land of rabbits, with similar effects. Both the New Zealand and Australian governments have also used the virus, with limited success. In Europe, however, the virus is considered the worst wildlife disease of the twentieth century.

Rabbits as Products

One of the ironies of the human relationship to wild rabbits is that even as they have been considered pests, humans have cultivated them as products. Certainly, this is the case with the Cowslip Warren, where the farmer actually fed the rabbits, amply, in order to fatten them up for snaring and, presumably, meat and fur. That warren is an allusion to the Medieval "leporaria" or rabbit gardens, which were simply enclosed pastures in which monks or lords kept rabbits and hares to be harvested at will. Today, although rabbit gardens no longer exist, wild rabbits are still a valued game species in a number of countries, even as they are reviled as pests.

Domestic rabbits also provide products to humans—including meat, fur, "lucky" feet, and laboratory experimentation. In fact, this is by far the dominant use of domestic rabbits in our world today: in the United States alone, about one billion rabbits are raised for commercial purposes, versus perhaps five million who are kept as pets.

Rabbits as Pets

Rabbits were domesticated somewhere between 500 and 1,000 years ago, much later than other popular pets, such as dogs (10,000 years ago) or cats (perhaps 5,000 years ago). Fittingly, rabbits appear as pets only briefly in *Watership Down*. Indeed, the only domesticated rabbits that appear in the book are the pets of Lucy, a local farmer's daughter, and even they seize the opportunity to flee their hutch life when given a chance.

Lucy shows up again at the end, when she rescues the wounded Hazel from her cat and then transports him—via the family doctor's car—back to Watership Down. It is a nice ending, and a convenient one. But to say that Adams looked to humans to solve the problems that humans had created for rabbits would be too much of a stretch. Why? Adams based his book largely on Ronald Lockley's *The Private Life of Rabbits,* the first popular account of wild rabbit habits published. Lockley was without question fascinated with rabbits: he built artificial warrens with observatory windows attached on his beloved Skokholm Island, so that he could watch them, and his book is littered with sentimental (even erotic) references to rabbit personalities. But his original aim had been to raise domestic, fur-bearing Chinchilla rabbits on his island. In the course of trying to clear the island of wild rabbits, he tried nearly every extermination method in the book—including trapping, netting, poisoning, and ferreting, all to no avail.

Adams, who became close friends with Lockley, was not much more of a rabbit fan. He told the *Ottawa Citizen* in 1998 that he supported a massive cyanide "cull" of wild rabbits in England because it was a "regrettable necessity" and once told a London newspaper, "I've never been one of these sentimentalists. I'm not a fluffy bunny sort of person at all. If I saw a rabbit in my garden, I'd shoot it."

Despite the author's pragmatism, the very fact that the book has been such a smash hit speaks to one of the primary relationships that humans have to rabbits: reverence. This is the story, after all, of a ragtag bunch of wild rabbits that flee their warren, seek a new home, and are victimized, mercilessly, by humans. Most people who do not have rabbits as pets scorn them as "dumb" or "boring" animals (no doubt because most pet rabbits live in cages, where they become so depressed they *seem* boring). Few people are interested in the behavior or ecology of wild rabbits (perhaps because they are so common). And few animal welfare organizations have tackled the very bleak conditions under which most commercial rabbits are raised and slaughtered. Yet *Watership Down* was on the best-seller list for ten months, has sold millions of copies, was made into a smash movie, and has spawned dozens of fawning Web sites. The very popularity of the book suggests that something in us is still powerfully drawn to the rabbit—that we still believe, on some level, that they have some haunting connection to other worlds; that we recognize, on some level, their cunning adaptation to their prey status; that we understand, even remotely, the very shabby way our species has treated their species. As Lockley once noted, "Rabbits are so human, or is it the other way round—humans are so rabbit?"

Further Reading

Adams, R. (2005). *Watership Down*. New York: Scribner.

Davis, S. E., & DeMello, M. (2003). *Stories rabbits tell: A natural and cultural history of a misunderstood creature*. New York: Lantern Books.

Lockley, R. M. (1954). *The private life of rabbits*. London: Corgi Books.

Susan Davis

■ Literature
Thoreau and the Human-Animal Relationship

"Wonderful, wonderful is our life and that of our companions! That there should be such a thing as a brute animal, not human! And that it should attain to a sort of society with our race!" So wrote Henry David Thoreau (1817–62), American writer and naturalist, in his journal for December 12, 1856. It is one of many passages in Thoreau's writings that show an enduring interest in animals, an interest that often broadened out into loving appreciation or wonder at his good fortune at living among them.

"Think of cats, for instance," the passage continues. "They are neither Chinese nor Tartars. They do not go to school, nor read the Testament." Yet they are another tribe of animate beings with their own history, manners, and experience of the world. "What sort of philosophers are we, who know absolutely nothing of the origin and destiny of cats?" he asks. "At length, without having solved any of these problems, we fatten and kill and eat some of our cousins!"

Thoreau observed and described animals with a scientist's precision, a poet's sensibility, and a lover's heart. He was intrigued by animals of all sorts—large and small, wild and domestic, common and rare—devoting thousands of journal pages to chronicling their appearance, behavior, migrations, and more. Some of his favorite experiences and best writing focused on animals.

A Complete Science

Although always a careful observer, Thoreau's early writings sometimes show him looking through animals to their poetic, moral, or spiritual meaning. The song of the wood thrush, he wrote, "declares the immortal wealth and vigor that is in the forest . . . Whenever a man hears it, he is young, and Nature is in her Spring." The thrush "sings to make men take higher and truer views of things. He sings to amend their institutions; to relieve the slave on the plantation and the prisoner in his dungeon, the slave in the house of luxury and the prisoner of his own low thoughts" (*Journal*, July 5, 1852).

As he grew older, Thoreau's descriptions of animals became more scientific and ecologically informed. Although still awake to nature's beauty, he became more likely to find that beauty in a detailed description of the animal itself, rather than in its symbolic associations. He also became more alive to animals' ecological roles. Thoreau's late manuscripts *The Dispersion of Seeds* and *Wild Fruits* are filled with detailed accounts of how insects, birds, and small mammals further seed dispersal and thus influence forest succession.

As Thoreau's observations became more scientifically informed, he struggled to specify an approach to animals that would marry accuracy and appreciation. In 1858, four years after the publication of his masterpiece *Walden*, he discovered a fish he had never seen before in Walden Pond, shaped like a bream but with markings like a perch. He collected several dozen of the little fish and made a minute description in his journal:

> They are one and one sixth inches long by two fifths of an inch wide . . . dorsal fin-rays 9–10 (Girard says 9–11), caudal 17, anal 3–11, pectoral 11, ventral 1–5. They have about seven transverse dark bars, a vertical dark mark under eye, and a dark spot on edge of operculum . . . They are exceedingly pretty seen floating dead on their sides in a bowl of water, with all their fins spread out. (*Journal*, November 26 and 27, 1858)

"Are they not a new species?" Thoreau wondered excitedly, presenting them at the next meeting of the Boston Natural History Society. Opinion was divided at the meeting, but subsequently, the specimens were identified as the previously described *Pomotis obesus*.

Thoreau, nonetheless, had known the thrill of discovery, filled in a detail of the natural history of Massachusetts, and seen new beauty in nature. All this was a function of careful observation—and skill in field collecting and dissection. Not for the first time, Thoreau wondered whether such killing in the service of knowledge was justified.

Thoreau's journal for the next week is filled with pleasure over his discovery. Significantly, as he considered its meaning, he imaginatively placed the fish back in the pond:

> I cannot but see still in my mind's eye those little striped breams poised in Walden's glaucous water. They balance all the rest of the world in my estimation at present, for this is the bream that I have just found . . . But in my account of this bream I cannot go a hair's breath beyond the mere statement that it exists,—the miracle of its existence, my contemporary and neighbor, yet so different from me! I can only think of precious jewels, of music, poetry, beauty, and the mystery of life . . . I want you to perceive the mystery of the bream . . . I have a friend among the fishes, at least a new acquaintance. Its character will interest me, I trust, not its clothes and anatomy.

Thoreau's pleasure is a function of the novelty of what he has uncovered. Walden Pond has surprised him with something new after all these years. Yet nature is filled with such wonders. Science is valuable, in part, because it shows us new diversity in a too familiar nature, awakening us to the beauty, sometimes to the very existence, of our "neighbors."

A scientific curiosity motivated Thoreau's interest in the bream. Yet considering what science would make of such a discovery, he was unhappy:

> A new species of fish signifies hardly more than a new name. See what is contributed in the scientific reports. One counts the fin-rays, another measures the intestines, a third daguerreotypes a scale, etc. etc.; otherwise there's nothing to be said . . . A dead specimen of an animal, if it is only well preserved in alcohol, is just as good for science as a living one preserved in its native element.

Those like Thoreau who are committed to a scientific understanding of animals cannot completely jettison collection, dissection, and rational analysis—despite their limitations and inherent violence. Still, the zoology of Thoreau's day slighted ethology; in his words, it left the *anima* out of animals. Anatomy and physiology are important—even beautiful—but so is "character," if by character we mean behavior and an animal's varied relations to its environment. The holism advocated by Thoreau has been incorporated into modern biology, which seeks to explain community and ecosystem interactions, the mechanisms of evolution, and the evolutionary histories of particular species. For ecology, ethology, and evolutionary biology, a dead specimen is *not* as good as a live one "preserved in its native element."

But even this more complete science must be completed, Thoreau believes—first, by personal acquaintance with our own neighbors: the plants and animals, forests, and fields around us; second, by "friendship," or, if that seems too strong a word for any possible relationship to something as cold-blooded as a fish, then by "appreciation"; third, by poetry and celebration of these new (to us) forms of life; and finally, and most importantly, by protection. Thoreau was one of the first Americans to call for the preservation of the wild lands necessary to protect wild species. In his first book, *A Week on the Concord and Merrimack Rivers,* he even advocated sabotaging dams to protect salmon and other endangered anadromous fish species.

Threats and Challenges

Thoreau sees three main threats to healthy human relationships with animals. First, there is simply indifference. "First notice the ring of the toad, as I am crossing the Common in front of the meeting-house," Thoreau wrote one May Day. "The bell was ringing for town meeting, and every one heard it, but none heard this older and more universal bell, rung by more native Americans all the land over" (*Journal*, May 1, 1857). Five springs earlier, noting the spring bird migrations, Thoreau had imagined himself "sharing every creatures suffering for the sake of its experience & joy." He asked himself, "The song-sparrow & the transient fox-colored sparrow, have they brought me no message this year? . . . I reproach myself because I have regarded with indifference the passage of the birds" (*Journal*, March 31, 1852).

Second, for wild animals, there is the danger of extinction. Thoreau loved the fields, forests, and streams of his native Concord. "But when I consider that the nobler animals have been exterminated here," he wrote, "the cougar, panther, lynx, wolverene, wolf, bear, moose, deer, the beaver, the turkey, etc., etc., I cannot but feel as if I lived in a tamed, and, as it were, emasculated country." "I take infinite pains to know the phenomena of the spring, thinking that I have here the entire poem, and then, to my chagrin, I hear that it is but an imperfect copy that I possess and have read, that my ancestors have torn out many of the first leaves and grandest passages, and mutilated it in many places . . . I wish to know an entire heaven and an entire earth" (*Journal*, March 23, 1856).

Third, there is cruelty and mistreatment of animals, particularly domestic animals. *Walden* includes a detailed argument against hunting and for vegetarianism. In the chapter "Higher Laws," Thoreau claims that "no humane being, past the thoughtless age of boyhood, will wantonly murder any creature, which holds its life by the same tenure that he does. The hare in its extremity cries like a child" (212). Elsewhere he writes, "It would be worth the while to ask ourselves weekly, Is our life innocent enough? Do we live inhumanely, toward man or beast, in thought or act? To be serene and successful we must be at one with the universe. The least conscious and needless injury inflicted on any creature is to its extent a suicide" (*Journal*, May 28, 1854).

These quotations show that Thoreau saw these three problems as threatening losses both to animals and to human beings. Our indifference to animals leads us to miss wonderful opportunities to know and appreciate these "brute neighbors." Our failure to preserve wildness on the landscape tames and dulls us, too, because we miss chances for poetry, scientific and historical knowledge, and beauty. Cruelty to animals, even when careless or indirect, cannot fail to make us more callous in all our dealings with the world.

Thoreau gives a personal and fairly comprehensive answer to these three threats in *Walden*. There he advocates vegetarianism (to spare farm animals) and simple, low-impact living (to spare habitat for the wild animals with whom he shares the forests around the pond). In later works such as *The Maine Woods* and *Wild Fruits*, Thoreau proposes town forests and greenbelts along rivers, to preserve local wildlife, as well as national parks or "preserves" that would protect the full complement of native wild animals in northern Maine and other still-wild areas of the country.

Above all, Thoreau asks us to keep our eyes and hearts open to the beauty, complexity, strange otherness, and striking similarities to be found among our animal kin. We should do this for the animals' sakes, he says. And for our own.

Further Reading

Cafaro, P. (2004). *Thoreau's living ethics:* Walden *and the pursuit of virtue*. Athens: University of Georgia Press.

Thoreau, H. (1962). *The journal of Henry D. Thoreau* (14 vols.). New York: Dover. (Originally published 1906)

———. (1971). *Walden*. Princeton: Princeton University Press. (Originally published 1854)

———. (1972). *The Maine woods*. Princeton: Princeton University Press.

———. (1980). *A week on the Concord and Merrimack Rivers*. Princeton: Princeton University Press. (Originally published 1849)

———. (1981–). *Journal*. Princeton: Princeton University Press. [New edition, seven volumes so far].

———. (1993). *Faith in a seed:* The Dispersion of Seeds *and other late natural history writings* (B. Dean, Ed.). Washington, DC: Island Press.

———. (1993). *Thoreau on birds: Notes on New England birds from the journals of Henry David Thoreau* (F. Allen, Ed.). Boston: Beacon Press.

———. (2000). *Wild fruits: Thoreau's rediscovered last manuscript* (B. Dean, Ed.). New York: Norton.

Walls, L. (1995). *Seeing new worlds: Henry David Thoreau and nineteenth-century natural science*. Madison: University of Wisconsin Press.

Philip Cafaro

■ Living with Animals
Bats and People

The acute lack of rapport between humans and bats is legendary and is perhaps best explained using the framework put forth in *Landscapes of Fear* by Yi-Fu Tuan. Fear drives the manifestations of chaos, both natural and human, he states. Human fear of the dark occurs worldwide and, although not present in the first year of life, is firmly developed by the age of ten months. Fear of certain animals has similar development in the human child; it grows with age and continues into adulthood with little or no reinforcement. These two basic universal and apparently innate fears collide typically when humans and bats encounter one another. Indeed, historical and contemporary contacts with bats are best described as nocturnal chaos that flows unbounded through our landscapes of fear.

Human distain and ignorance of an animal poorly understood is centuries old and almost surely predates human literacy. Even today, most of what people know of bats is based on mythology. Many of the myths are hard to originate or track, but the linking worldwide of bats to witchcraft and magic has given rise to many of the irrational reactions people manifest at the very mention of the word "bat," and in some cases the response is deadly. One of the more famous historical accounts occurred in 1332, when Lady Jacaume of Bayonne in France was publicly burned because "crowds of bats" were seen about her house and garden.

Shakespeare invoked bats and witches in several of his plays. For example, the "wool of bat" in the brew of *Macbeth*'s three witches. In ancient Greece and Rome, it was believed that one could prevent sleep by placing the engraved figure of a bat under one's pillow, or if one tied the head of a bat in a black bag and laid it near one's left arm. In Roman antiquity, Pliny maintained that a man could stimulate a woman's desire by placing a clot of bat blood under her pillow. In many parts of Europe, a practice said not only to ensure wakefulness, but also to protect livestock and prevent misfortune, was to nail a live bat's head down above a doorway. Canadian Indians relate that placing the head or dried intestines of a bat in an infant's cradle will cause the baby to sleep all day. In a similar vein, Mescalero Apaches believe that the skin of a bat attached to the head of a cradle will protect a baby from becoming frightened.

In Texas, one lovesick suitor was told to place a bat on an anthill until all its flesh was removed, wear its "wishbone" around his neck, pulverize the remaining bones, mix them with vodka, and give the drink to his beloved. A similar love potion from Europe recommends mixing dried, powdered bat into a woman's beer. Wrapping a bat's heart in a silk handkerchief or red ribbon and keeping it in a wallet or pocket, or tying it to the hand used for dealing cards is supposed to elicit good luck. Some also believe that tying a silk string around a bat's heart will bring money. Many beliefs in Europe and the United States relate the value of bats' blood, or their excrement, as a depilatory (hair remover). But in England and North Carolina, the use of bats' blood has been advocated to prevent baldness. In India, using a hair wash of crushed bat wings in coconut oil is said to prevent both baldness and graying of hair.

As recently as 1957, a California taxidermist sold bat blood, presumably for witchcraft, and a report from Ohio claims that bat blood can be used to summon evil spirits. Some Illinoisans assert that bat blood gives witches "the power to do anything." There are also reports of bats used for witchcraft in Mexico's Yucatan, and bat wings are often found in the conjure bags of African Americans in Georgia.

Vampire myths go back thousands of years and occur in almost every culture around the world. Their variety is almost endless—from red-eyed monsters with green or pink hair in China; to the Greek Lamia, which has the upper body of a woman and the lower body of a winged serpent; to vampire foxes in Japan; to a head with trailing entrails known as the Penanggalang in Malaysia. On exception, the Chinese revere bats as a sign of good luck and happiness; the Chinese word for bat is "Fu," meaning happiness.

Factually, human exploitations of bats have been many, and some are quite bizarre. Naturally, bats have historically been a source of food to many peoples in Africa, Asia, and Indonesia and also to aboriginal Australians for as long as the last 10,000 years. The droppings of bats have been commonly used as fertilizer. So-called guano caves, such as Carlsbad Caverns in New Mexico, were mined in the 1950s for making gunpowder for the military. One of the more bizarre and perhaps disturbing human attempts to exploit bats was a plan by the U.S. Government called Project X-Ray, which proposed using large numbers of Brazilian free-tailed bats (*Tadarida brasiliensis*) to carry incendiary devices (bombs), which were attached to their bellies by a surgical clip and a string. The idea was to release bats placed in cages with parachutes over enemy territory from airplanes; the cages would open at a specific altitude, releasing the bats to fly into enemy buildings to roost. The bats would predictably chew the string, releasing the "bomb," which would explode and burn down the building. However, controlling where exactly the bats would roost after release turned out to be a problem, as was keeping them alive in captivity. In one case, bats escaped a desert test range and took up residence under an elevated gasoline tank in a nearby town. The project was cancelled shortly after this incident.

Taking advantage of the natural ability of bats to eat large numbers of insects, a medical doctor, Louis Campbell, decided to use bats to combat malaria in Texas in 1911. He constructed a large tower-style bat house and collected large numbers of Brazilian free-tailed bats and placed them in his structure, which he named Dr. Campbell's Malaria Eradicating, Guano Producing Bat Roost. For years, his attempts to attract bats to his tower were unsuccessful because he knew little about their biology. After taking a year off from his practice and studying bats in caves, he reattempted his bat tower idea with success, and even as late as the 1948, his last heir received a royalty check in the sum of $500 from the sale of guano fertilizer.

Early attempts by scientists to study bats were also confusing and misleading, fueling further human misunderstanding. The bats were for a long time difficult to even classify correctly. Originally, they were seen as strange birds. In the seventeenth century, John Ray placed bats among "the four-feet animals." In the eighteenth century, the great taxonomist Carl Linnaeus observed that bats differed from other birds by giving birth to living suckling young and that this qualified bats to belong to the group he called Mammalia. Within mammals, Linnaeus struggled with categorizing bats into his taxonomic scheme and decided to place the then seven known species into the order Primates, based on characters he saw as consistent with human anatomy, in particular the musculoskeletal and nervous systems.

Lazaro Spallanzani, a contemporary of Linnaeus, set out to understand how bats could fly in complete darkness while avoiding collisions with inanimate objects and each other (one of the apparently supernatural abilities of bats). Blindfolds that he placed on bats had no affect on their nocturnal abilities. A colleague, Louis Jurine, elaborated on Spallanzani's experiments by plugging the bats' ears, which caused them to become confused and disoriented in flight. Spallanzani and Jurine hypothesized that bats emit ultrasonic pulses to navigate their flight, and with their ears plugged, the bats could not hear themselves and thus became disoriented. The Spallanzani/Jurine hypothesis went unexplored further for almost 300 years. In fact, other biologists even mocked their work, making statements such as, "Since bats see with their ears, I suppose they hear with their eyes." It was not until the 1930s that Dr. Donald Griffin, intrigued by the Spallanzani/Jurine hypothesis, brought a crate of big brown bats (*Eptesicus fuscus*) to a physicist's lab, where one of the first developed ultrasonic-detection devices allowed the hidden world of bat sonar to be revealed for the first time in human history in a cacophony of song later coined by Griffin as "echolocation." Thus, 300 years after the Spallanzani/Jurine experiments, one of the great misconceptions of bats as supernatural or mythological creatures was dispelled and replaced with scientific reality and reason.

Such excellent science, however, had little effect on public impressions of bats that remain deeply and negatively rooted in the human psyche. But the establishment of international conservation societies, such as Bat Conservation International in Austin, Texas, and many regional and local conservation groups, has begun to change the unconstructive and damaging tide of human consternation about bats. Many people now see value in bats as a part of our diverse natural heritage of life on Earth. Bat species make up almost one-quarter of all living mammal species and occur in almost every habitat. They are voracious insect eaters, seed dispersers, pollinators, and carnivores. Many species are critical to ecosystem health and vitality through their interactions as predators and reproductive agents of plants, especially in tropical ecosystems. Some studies in Europe have shown that simply playing recordings of bat sonar in agricultural fields significantly dropped the number of agricultural insect pests. The amazing array of niches occupied by bats underscores their import as keystone species in many regions. Even those who view nonhuman animals in purely human-centric terms can find value in bats, whose biology has impacted technological advances in the areas of aircraft design, the development of sonogram technology used by pediatricians and by submarines, the advance of sonar devices to help the blind, and anticoagulants used as blood thinners.

The association of bats and rabies has had detrimental effects on human impressions of bats. Although bats can and do carry rabies, typically the incidence is quite low in natural populations, and some have estimated that worldwide, bats rank below feral pigs in incidence of this deadly disease. However, because most human encounters are with sick or injured bats that have become grounded, many of these individuals are rabid, and thus, a human health concern is warranted under these condition. Typically, healthy bats avoid human contact.

Although many have begun to acknowledge the importance of bats, human impacts on populations, whether intentional or unwitting, remain high, and many populations show serious declines. The direct loss of, or human-centered changes in, natural habitat remains a large reason for declines in bats, as is human disturbance at nursery colonies where females raise newborn young and at cave hibernation sites. Bats commonly use caves and abandoned mines, and human incursions (i.e., spelunking) into these sites at certain times of year can cause severe loss of bats. Removal of tree snags by foresters and disturbance at rock crevices by rock climbers can also cause disruption of colonies and reproduction. Forest-thinning practices typically remove large snags and large old-growth trees that are roosting sites for many bat species. Abandoned mines and bridges used as bat roosts are examples of ways that humans have unwittingly helped bats, as have human-made structures such as barns, outbuildings, and houses.

Humans have also intentionally helped bats. With the advent of the bat house brought forward in the 1980s by Bat Conservation International, humans intentionally provide alternative roost sites for bats that are evicted from houses and barns. Also on the horizon today, however, is an increase in windmill turbines to catch wind energy in a relatively cheap and environmentally healthy way. Unfortunately, in some cases, wind turbines have killed large numbers of bats when placed in migration routes. In one case, more than 400 bats were killed in one weekend by three wind turbines positioned on a mountain ridge top in West Virginia.

Further Resources

Adams, R. (2003). *Bats of the Rocky Mountain West*. Boulder: University Press of Colorado.

Allen, G. M. (1939). *Bats*. Boston: Harvard University Press.

Griffin, D. R. (1958). *Listening in the dark*. New Haven: Yale University Press.

McCracken, G. F. (1993). Bats and the netherworld. *Bats, 11,* 16–17.

Tuan, Y. (1979). *Landscapes of fear*. New York: Pantheon Press.

Rick A. Adams

■ Living with Animals
Birds and Problems from Humans

> Dark as thunderclouds, each morning a roar of wings filled the sky as the males returned to the eggs, to the females who sprang hungrily into the air, so that spiralling funnels rose and fell on every hand. It remained magnificent and terrifying; a marvel they never collided; a mystery that in those roiling millions, each found its own nest, its own mate.

Thus Graeme Gibson in *Perpetual Motion* writes of the passenger pigeon (*Ectopistes migratorius*). Once, it was probably the most common bird in North America, perhaps even in the world, but a century of overhunting drove it to extinction. Any essay on what humans do to birds must unfortunately be dominated by such stories. Our effects on birds tend to be detrimental, and even our positive effects generally represent attempts to reverse or dampen the negative impacts.

Besides the passenger pigeon, humans have been instrumental in the extinction of about ninety other species of birds in the last four hundred years, mostly from islands. Data from before 1600 is scarce, but evidence suggests that we have been involved in bird extinctions for thousands of years. For instance, at least thirty-four land birds on New Zealand went extinct between when the Maori people settled there (around 1300 AD) and European arrival in 1769, and a recent study links Maori hunting practices to the extinctions. Our impact has grown over time. Over 1,000 species of birds, about 10 percent of the total, are threatened with extinction today. Most at risk are island species, Neotropical migrants, grassland species, migrant shorebirds, seabirds of the Southern Hemisphere, and rainforest species with restricted ranges.

Of the threats to bird populations today, the introduction of alien competitors and predators is considered the most serious. Habitat destruction and conversion is in second place. Sometimes the detrimental effect of land conversion is masked by the presence of a large overall number of birds because species that prefer edges or that are tolerant of human proximity often do well in suburban or even urban areas. Diversity, on the other hand, or the number of bird species present, can decline drastically when land is converted for intensive human use, resulting in an ever greater area dominated by the same few species. Coffee plantations provide an important example of the effect of land conversion. Coffee naturally grows in shade, and such traditional plantations preserve a bird diversity second only to undisturbed forest. Most coffee sold today, however, has been altered to grow in the sun, increasing yields, but cutting bird diversity by 75 to 95 percent. Habitat fragmentation is a landscape-level effect of land conversion that is associated with declines in bird populations, especially in species with particular habitat requirements such as forest interior. Forest birds in fragmented habitat can be more susceptible to predators and brood parasites (birds that lay eggs in other birds' nests) that are excluded from large forests but penetrate smaller patches. Another significant impact on birds in many areas of the world is overhunting or capture. For instance, eggs of wading birds are still taken unsustainably in the Peruvian Amazon; parrots are the most endangered group of birds on account of the pet trade; and tens of thousands of petrels and albatrosses are by-caught each season in illegal fisheries.

Aside from factors stemming from habitat conversion and introduced birds, the following are some direct or indirect effects humans have on birds today in North America, in approximate descending order of concern:

- Predation by domestic cats (about 4 million a *day* in the United States, or a billion a year)
- Deaths by pesticide poisoning (about 67 million per year, or 10 percent of those exposed, are killed immediately, 7 million of which cases are a result of homeowner use of pesticides; an unknown additional number of birds die after illness or from eating contaminated prey)
- Deaths by collision with windows (about 80 million per year in the United States; office building lights should be turned off in cities at night, especially during migration)
- Deaths by collision with motor vehicles (about 57 million per year in the United States)
- Deaths by collision with communication towers (4–10 million per year in the United States; researchers suggest that towers under 300 feet in height are not a serious problem)
- Physiological changes from electromagnetic fields, especially from power lines (the detriment to the birds is not yet clear)

The most succinct encapsulation of the history of our effects on wild birds is that, largely by overlooking them, we have made life difficult for many of them and impossible for some. However, some of us have awakened to this problem and have worked to ameliorate our effects. In recent decades, many have taken action to mitigate or slow habitat destruction, reserve crucial habitat, guard against invasive predators and competitors, and protect birds with legislation or service by private groups. But it is the growing realization by individuals of what we have to lose that is probably the most hopeful of starting points toward recovery of our bird life. Returning to the passenger pigeon, this time from *A Sand County Almanac*, Aldo Leopold writes,

> For one species to mourn the death of another is a new thing under the sun. The Cro-Magnon who slew the last mammoth thought only of steaks. The sportsman who shot the last pigeon thought only of his prowess. The sailor who clubbed the last auk thought of nothing at all. But we, who have lost our pigeons, mourn the loss.

Further Resources

Chace, J. F., & Walsh, J. J. (2006). Urban effects on native avifauna: A review. *Landscape and Urban Planning, 74*, 46–69.

Duncan, R. P., Blackburn, T. M., & Worthy, T. H. (2002). Prehistoric bird extinctions and human hunting. *Proceedings of the Royal Society of London B, 269*, 517–21.

Faaborg, J. (2002). *Saving migrant birds: Developing strategies for the future.* Austin: University of Texas Press.

Gill, F. B. (2006). Conservation of endangered species. In *Ornithology* (3rd ed., ch. 24). New York: Freeman.

Norris, K., & Pain, D. J. 2002. *Conserving bird biodiversity: General principles and their application.* New York: Cambridge University Press.

Votier, S. C., Hatchwell, B. J., Beckerman, A., McCleery, R. H., Hunter, F., Pellatt, J., Trinder, M., & Birkhead, T. R. (2005). Oil pollution and climate have wide-scale impacts on seabird demographics. *Ecology Letters, 8,* 1157–64.

Youth, H. (2003). *Winged messengers: The decline of birds.* Washington, DC: Worldwatch Institute. Available at http://www.worldwatch.org/pubs/paper/165/

David C. Lahti

■ Living with Animals
Cat and Human Relationships

Over 200 million cats were estimated to be living as companions to humans worldwide as of 2005, and the numbers are growing. Cats now outrank dogs as household pets in the top fifteen pet-keeping countries of the world. In the United States alone, there were over 76 million pet cats in 2005, versus 60 million dogs. Cats are often considered members of the family, may provide health and psychological benefits, and are well-loved by many. However, an estimated 20 to 50 million cats are believed to be living as feral animals in the United States alone, abandoned or born outdoors and surviving on their own. Many others live in shelters or in the homes of "hoarders," who harbor abnormally large numbers of animals even though they are unable to care for them. And sadly, many are physically abused, targeted as human surrogates, or mistreated simply because they are small and accessible; millions more are euthanized each year for lack of food, space, or care. Human-cat interactions, then, can be both extremely positive and upsettingly negative.

Cat-Human Overview

Domestic cats (*Felis catus*) are part of a lineage of small cats that is thought to have emerged approximately 6 million years ago. Genetically, they are apparently most closely related to African wild cats (*Felis lybica*), although they also share morphology and genetics with the European wild cat (*Felis sylvestris*), and some researchers consider all three related subspecies. Their relationship with humans has been documented primarily through artifacts and grave remains of ancient cultures. Although the ancient Egyptians are often given credit for first domesticating African wildcats over 4,000 years ago, archeological discoveries in the Mediterranean suggest that cats and humans may have had close bonds for over 9,500 years. Relationships between humans and most domestic animals seem to have begun primarily as utilitarian interactions, with the animal supplying something humans needed. For cats, this most likely was pest control, especially in areas where agriculture—planting crops and storing grains and other plant foods for later use—played an increasingly important role in human survival. Control of rodent and snake populations around crops may have resulted in cats being valued by humans and being fed and cared for. In some ancient societies, cats became so highly valued that they were seen as companions and guardians of gods or even as gods themselves.

Although it is clear how a relationship with humans might benefit cats, providing them with a steady food supply (either directly or indirectly), shelter, and more recently, medical aid, it is less clear what benefits humans might derive from cats beyond help with pest control. Although studies since the 1970s have shown that pets can provide health benefits to humans, by lowering blood pressure and other stress responses and providing a social anchor, most of the research involved dogs, not cats. Studies specifically focused on cats have shown more subtle, but perhaps just as valuable, effects. People state that they like the ease of care, affection, and companionship cats provide and that their personalities, appearance, and behavior, including such endearing actions as "purring on your lap," make them a "comforting presence." And even though research shows cats are not as overt in their displays of affection as dogs, they do seem to provide emotional support and play a real role in people's social networks, even taking on the role of a significant partner for many older people living alone.

Cats are also very flexible in their needs and their ability to adapt to social situations. They can be socially self-sufficient, can be left alone for longer periods than dogs, and are more likely to accept care from strangers (such as neighbors or friends) if their human companions go away or become ill or hospitalized. They require less effort, are less demanding of attention than dogs, and require less vigorous interaction, a positive aspect for humans who prefer a more low-key interaction than that usually offered by dogs or who are older, disabled, ill, or caring for others who are ill. Cats are also able to live in a variety of situations, from tiny apartments to large mansions, with one person or many, one cat or several, without other animals to a large menagerie. Although trouble can arise, many cats live in these situations with few overt problems.

Cats can also provide therapy to those who need companionship, for example serving Animal Assisted Therapy programs (AAT) in hospitals or long-term care facilities. Very friendly cats who seek and respond well to petting and being held on a person's lap can provide social warmth and serve as an external focus for people who would otherwise be withdrawn; they are especially helpful to those individuals who are afraid of or allergic to dogs.

Unlike dogs, which have been intensively bred to help humans with a variety of tasks, cats have not been subject to a long history of selective breeding. There are currently thirty-seven pedigreed and four "miscellaneous" breeds recognized by the Cat Fanciers' Association in the United States (www.cfainc.org), with similar numbers listed by the Governing Council of the Cat Fancy in the United Kingdom (www.gccfcats.org). This is in comparison with 150 dog breeds recognized by the American Kennel Club (www.akc.org; with an additional fifty-seven breeds under consideration). Although both cat organizations and many popular cat books describe physical and behavioral characteristics of different breeds, for example the "intelligence, inquisitive personality, and loving nature" of the Siamese or the "playful but not demanding and tremendously responsive nature" of Persians, only a few research studies exist that directly examine the behavior of pet cats, pedigreed or not, as they interact with other cats, other animals, and people in the home.

Cat-Human Interactions in the Home

The relatively few studies of cats in the home setting reveal patterns of behavior that are familiar to humans who live with cats and that are similar in many ways, although not all, to those of cats in outdoor colonies. They indicate that cats are extremely adaptable, acutely aware of one another, and adept socially, able to deal with each other in subtle as well as obvious ways within the constraints posed by their human caretakers. Long gone is the notion that cats are solitary animals.

Extensive studies of outdoor cat colonies have provided a picture of cat social behavior independent of the human home. It is presumed that in this setting, domestic cats, left to themselves, are demonstrating "natural" behavior. However, considering that they have been domesticated for several thousand years, it is not clear which setting is more natural for cats, being with humans or being alone.

Barn cats, temple cats, and cats in feral colonies demonstrate patterns of behavior that are familiar to most humans. The cats tend to form groups when food is concentrated and are more solitary when food is scarce. In larger colonies, females and their kittens tend to form the most stable social groups, with female kittens then tending to stay near their mothers as they become adults and males tending to disperse. Individual males and females form long-term associations ("friendships") with some while avoiding others, and a variety of social behaviors occur, including mutual grooming (allogrooming), rubbing against each other (head butting and side or cheek rubbing), and lying in physical contact or near one another. Individuals tend to stay within home ranges, areas of normal use, rather than in defended territories. Fighting does occur, but at fairly low levels, even when females are in heat and males might compete for opportunities to mate. Strangers are not well tolerated, and a long period of time may pass before a new cat is accepted into a group.

In the home, cats show somewhat similar behaviors. They are able to cope with a variety of living arrangements, from one cat in a large house to several cats in a small apartment to many cats in a medium-sized house; cats in the home have been documented living in stable social situations at densities greater than fifty times the highest densities found in groups of cats outdoors. Cats divide up the house into home ranges, and although some areas may be unique to some individuals, most are overlapping, cats adjusting where they are and who they might interact with through a constantly shifting set of movements throughout the house with respect to one another. Individual relationships also clearly play a role, as they do in wild cat colonies, with particular cats sharing specific rooms or particular favored spots with one another (either by physically sharing or time-sharing, using the same spot but at different times) and avoiding specific other individuals. Gender can affect relationships as well, and studies of two-cat households indicate that male cat pairs in two-cat households interact with less aggression and more affiliative behaviors than female cat pairs or male-female pairs. This is somewhat different from what is commonly seen in outdoor colonies, where related females tend to form central groupings and spend time together.

However, cats in the home also have the additional task of having to cope with humans and other non-cat animals. Studies that focus on how humans and cats interact in the home find there is generally a low level of interaction, and most interactions are of fairly short duration (one minute or less). In one of the few in-home studies, humans tended to approach within one meter of the cat more often than the reverse, but when the cat did the approaching, the human and cat stayed within a meter of each other for a longer time. Women, men, boys, and girls interacted differently with cats; for example, adults vocalized toward the cat earlier in an interaction and for longer than did children. Women spent more time at home and therefore had more interaction with cats than did men; juvenile humans (11 to 15 years of age) were least likely to be within one meter of the cat and had the least amount of interaction, although it was not clear why. Single cats stayed closer to owners for longer and had more social play and more interactions in general with owners than did multiple cats. Interaction, proximity, and rubbing by the cat were moderately more frequent in smaller than in larger families. Studies of pedigreed cats showed they tended to be more predictable in their behavior than non-pedigreed cats, which was highly valued by their human companions. In view of the increasing popularity of cats as companion animals, follow-up studies are sorely needed in this area.

What is most surprising is how little research has focused on many of the most familiar cat-human interactions. These include petting, feeding, tail signaling, vocalizing (either cat to human or human to cat), marking, and contact behaviors such as rubbing. Attempts have been made to parse cat vocalizations by context and to assess human perceptions of the calls, and researchers using experimental situations have found we tend to address cats the same way we address small children, using short sentences, different tone of voice, and so on. But there is little detailed study of vocalizations in naturally occurring situations. There is a large set of veterinary literature on food products for pet cats, and research is ongoing to explore what foods best help cats develop good health, maintain it, and avoid potentially fatal problems. But few studies have targeted the actual feeding interaction itself (initiation, coordination, ending), even though this is one of the most common, frequent, and important interspecific interactions in which humans and cats engage and where communication and manipulation by one or both parties may play important roles. Anecdotal information about litter-box placement, litter type, and number of boxes per social group is slowly being updated with formal studies. Contact-seeking behaviors, such as petting, rubbing, and sleeping on human laps or on beds with humans, are currently under investigation. There is some evidence that cats may suffer when their human companions leave them alone in the house, developing clinical signs of separation anxiety, a phenomenon more typically associated with dogs.

Another area that has not been well studied involves interactions that occur between cats and other non-cat animals in the human setting, where the animals interact as if they were companions—in other words, cats having other animals as friends. For example, some cats and dogs in the same household regularly play and sleep together, and some cats regularly stay with and interact with horses in barns. Are these cases of pets keeping their own pets, of companions having their own companions? Although there is much anecdotal information about this phenomenon, research has not directly addressed this aspect of cat relationships.

Perhaps the most research concerning the cat-human bond has been conducted on cat socialization—that is, the ability of cats to relate well to other cats and to humans. Because the way cats respond to people is key in whether they receive care, understanding socialization is important. Research indicates that socialization depends on both early experiences and genetics. In general, kittens who spend at least the first few weeks of their lives with their mother, who are handled by humans during that time, who have their mother or other adult cat present when interacting with humans, and whose genetic father was friendly with humans tend to be more comfortable with humans as adults. However, these are not hard and fast rules, and there are many exceptions: there are cats who seek human companionship despite never having been handled, cats who were handled as kittens who always remain aloof, and cats who change in their social behavior depending on situation and time. Many anecdotes exist about feral cats who have never dealt with humans but who slowly become good companions after being taken in, or cats who become more or less social as their situations change (i.e., as humans, cats, and other animals enter or leave the household). Little research has directly addressed the issue of cat personality, beyond finding that some seem investigative and bold as kittens and in the early years of adulthood, whereas others seem more aloof or fearful. Some researchers have developed a cat temperament test that would help shelter staff, veterinarians, and others better assess "cat sociability, aggressiveness, and adaptability." However, it is clear that cats, like people, can and do change.

There are obvious responsibilities for the human in the relationship. Humans should be providing adequate and appropriate nutrition, shelter, vaccinations, and other preventative health care and a means of identification to prevent loss. However, humans with cats have traditionally been less likely than humans with dogs to take their pets to veterinarians. The American Veterinary Medical Association in the United States surveys humans every two to three years. Based on that information, humans caring for cats make an astounding 70 million visits to veterinarians, but humans with dogs make over 117 million visits, despite more cats being kept as pets. This may reflect less need for cats to see veterinarians or a difference in perception by owners of need for medical care.

Humans are concerned about cats and their possible effects on human health. Concerns include zoonoses (diseases naturally transmitted between animals and humans), as well as allergies, asthma, bite injuries and infections, flea and parasite transmission, and other hazards. Most studies focused on these concerns emphasize the low risk of disease associated with cat ownership. Even those humans with compromised or depressed immune systems, including the sick and very young, may benefit from and be able to continue pet ownership with precautions. The introduction of long-lasting rabies vaccines; vaccines for other diseases (e.g., feline leukemia, feline immunodeficiency virus, and so on); the finding that toxoplasmosis can be introduced from a variety of sources, not just from cats; and other factors (such as owners keeping cats indoors so that they are less likely to be exposed to parasites or to transmissible diseases) seem to have decreased human concerns about cat-caused illness. Some studies have even reported surprising results—for example, that being exposed at home to pet cats and dogs early in life may have a protective effect against allergy or even asthma rather than exacerbating these conditions.

Cat-Human Problems

Despite the obvious positives in the cat-human relationship, cats can also be a source of fear and problems. "Ailurophobia," or fear of cats (also known as "felinophobia"), is an ancient dread. Throughout ancient Asia and medieval Europe, cats were considered "malevolent demons, agents of the Devil." In many countries throughout history, including the United States, cats have been burned and shot as agents and familiars of witches or simply as harbingers of "bad luck." Even in the twenty-first century, humans in many countries continue to voice concerns that cats might sleep on children's faces, smothering them or "stealing their breath." Humans voice similar concerns in a more modern form when they worry about cats causing allergies or asthma in children, this despite studies that show growing up with cats does not necessarily promote either condition.

Cats, in part because of their small size in relation to dogs and in part because of human fears, are often targets of physical and emotional abuse. Case reports of cruelty and chilling first-person accounts vividly demonstrate a connection between animal abuse and violence among humans, showing that spousal and child abuse are often coupled with pet abuse. Although there is clearly a connection between abuse of animals and abuse of people, one of the more disconcerting findings of research in this area is that many instances of animal abuse are not performed by individuals who go on to hurt humans, but rather by "regular" individuals who do not see dogs and cats as deserving of care or respect and do not see anything wrong with harming them. Cats also become targets of hoarding behavior, in which individuals live with dozens to hundreds of living and dead animals. These individuals often show signs of pathological self-neglect, a variety of psychological conditions, and a lack of ability to care for themselves or their animals. There is a lack of consensus about how to deal with hoarding and animal abuse, and written laws and the handling of court cases show a range of responses, from extremely lenient to strongly punishing. Housing policies that exclude animals are also a problem, forcing humans to choose between affordable housing and keeping a loved cat; this issue is also being contested in the courts.

Even the normal human-animal bond in the home can go wrong in serious ways. Despite cats being socially flexible, stressful situations can and do occur. Problems may result from crowding, incompatibility among individuals, lack of owner attention, lack of owner knowledge about what constitutes normal cat behavior, relationship problems of cats with particular people or other pets, and the owner having specific expectations about the cat's role in the household that are not being met (such as wanting a close companion when a particular cat is more aloof in personality). When behavior problems occur, cats are at risk for relinquishment to shelters and animal control facilities or abandonment to the outdoors. Research has demonstrated that risk factors for relinquishment are often ones that can be modified with proper intervention and education. Owners that had read a book or other educational material about cats were less likely to relinquish their cat.

Several trends have developed in the last decade in response to the growth of cats as pets and the problems that have resulted. First, there has been explosive growth in research dealing with pet behavior problems. Veterinary clinicians; researchers in ethology or animal behavior, psychology, sociology, and nursing; and a variety of other individuals with animal experience (e.g., trainers, breeders) have developed private practices dedicated to helping people deal with pet problems. Specialties in behavior are now recognized in veterinary medicine, and boards of experts have been formed to certify practitioners (for example, see the American Veterinary Society of Animal Behavior Web site at http://www.avsabonline.org/, the Animal Behavior Society Web site at http://www.animalbehavior.org, and the Web site of the Association of Pet Dog Trainers at http://www.apdt.com.

Second, perhaps not surprisingly, animal law has become a rapidly developing field with a burgeoning number of courses, books, conferences, and at least one law review dedicated to the topic. Legal debate swirls around the use of the terms owner, companion, and guardian, and some states are considering requiring malpractice insurance for veterinarians. Government agencies that fund medical, veterinary, and other research that involves animal testing have enacted strict guidelines for animal housing, care, and use. And national humane societies (e.g., Humane Society of the United States, American Society for the Prevention of Cruelty to Animals, and the American Humane Association) have developed guidelines for their affiliated shelters. However, there are still few laws, regulations, or state and national guidelines regarding feral cat populations, animal control facilities, or nonaffiliated shelter programs, especially at the local level.

Third, despite problems that can occur between cats and humans in the home, the last decade has seen an increase in the United States in the number of cats being kept completely indoors. Traditionally in the United States, cats were allowed to roam free at their will. Yet studies since the early 1990s have shown that 50 to 60 percent of pet cats in the United States are now kept indoors at all times, with an additional 10 to 20 percent having restricted outdoor access (e.g., walked on leashes, sitting with owners on a deck, kept on a lead in the yard). This is in contrast to cats in the United Kingdom, where nearly 75 percent are allowed outdoors. Restrictions in the United States are apparently in response to a number of pressures: concerns that domestic cats are essentially an introduced and invasive species that could threaten local wildlife, especially birds; increased threats to the cats themselves from disease, from their own predators (primarily coyotes and raptors), and from increasing vehicular traffic; and human fears that they or their cats could be harmed by cat-transmitted zoonoses. Although research does not support most of these concerns, they seem to have affected cat-keeping in the United States

Fourth, although it is not clear whether the increased time cats now spend indoors in the United States has resulted in an increase in behavior problems, there has been an alarming increase in the number of cats forced to live completely outdoors as abandoned, stray, and feral individuals. Major animal welfare groups (e.g., the Humane Society of the United States, the American Society for the Prevention of Cruelty to Animals, Alley Cat Allies), as well as the American Veterinary Medical Association, provide information, education, and guidelines about this growing problem. In many communities, volunteer "caretakers" feed and care for cats in feral colonies and engage in TNR programs (trap, neuter, return), which are designed to reduce reproduction and disease by neutering and vaccinating colony members. Studies have found that TNR programs create healthier colonies and ultimately decrease colony size or even eradicate colonies (as animals naturally age and die without reproducing). These programs also seem more successful than those in which cats are killed (new ones take their place) or trapped and removed to animal control facilities or shelters (where most are euthanized). But again, there are few laws or policies available to guide local communities. Increased education about successful programs is an important need at this time.

Once abandoned, it is difficult for a cat to be adopted back into a home. Many more cats are euthanized each year (an estimated 4 to 10 million in the United States) than are adopted, and kittens, although most likely to be adopted, are also most likely to be relinquished. Relinquishment rates are lower and adoption rates higher when veterinarians and others provide guidance, support, and education to owners in an effort to modify owner perceptions and understanding of cat behavior.

> In the end, cats and humans continue to have a double-edged relationship; cats are seen as "cute and endearing" members of the family and are clearly a pet of choice on the one hand, and yet they also continue to be an object of fear and loathing on the other. Research and education are helping humans become more aware of cat social behavior and social needs, which in turn should help both of the animals in the relationship, cat and human, live better lives.

See also

Animals as Food—*Dogs and Cats as Food in Asia*
Bonding—*Companion Animals*
Ethics and Animal Protection—*Hoarding Animals*

Further Resources

Arluke, A. (2006). *Just a dog: Understanding animal cruelty and ourselves.* Animals, Culture, and Society series. Philadelphia: Temple University Press.

Arluke, A., & Sanders, C. R. (1996). *Regarding animals.* Philadelphia: Temple University Press.

August, J. R. (2006). *Consultations in feline internal medicine.* St. Louis, MO: Elsevier Saunders.

Rochlitz, I. (2005). *The welfare of cats.* The Netherlands: Springer.

Slater, M. R. (2002). *Community approaches to feral cats: Problems, alternatives & recommendations.* Washington, DC: Humane Society Press.

Turner, D. C., & Bateson, P. (Eds.) (2000). *The domestic cat: The biology of its behaviour* (2nd ed.). Cambridge: Cambridge University Press.

Penny L. Bernstein

■ Living with Animals
Corals and Humans

Learning how to be a part of the world means recognizing that "we" are not alone in being *Homo sapiens* and that our futures are predicated on the relationships we build with the world as part of the world. Corals are instructive in this way. They exist only through their contact with others. Relationships are intrinsic to corals, at once their substance and their limits.

How Corals Live

Reefs are large underwater structures of coral skeletons, made from calcium carbonate secreted by generation after generation of tiny coral polyps over sometimes millions of years of coral growth in the same location. Coral reefs are extremely sensitive. Slight changes in the reef environment may have detrimental effects on the health of entire coral colonies. These changes may be a result of a variety of factors, but they generally fall within two categories: natural disturbances and anthropogenic disturbances. Although natural disturbances may cause severe changes in coral communities, anthropogenic disturbances have been linked to the vast majority of decreases in coral cover and general colony health when coral reefs and humans occur together.

Primary formers of reefs are colonial scleractinians (*scler* = hard, *actinia* = ray), or stony star corals. Scleractinia are to a greater or lesser extent carnivorous, catching food with their tentacles that are lined with stinging cells (nematocysts). However, the majority of scleractinian corals have also formed a symbiotic relationship (beneficial to both parties) with zooxanthellae, single-celled algae that live within their tissues. The algae carry out photosynthesis, sharing the sugars and oxygen produced with the coral polyp. The coral in return provides protection, as well as useful waste products such as carbon dioxide and nitrate plus simple minerals. These corals with a symbiotic relationship make up the majority of hard corals and live close to the surface, where there is plenty of sunlight.

Each polyp secretes a calcium carbonate exoskeleton (corallite) that protects the soft, sack-like body inside. A polyp's tubular body has only one opening, which is surrounded by tentacles. The closed end of the tube forms the base of the polyp, where an exoskeleton structure is created, called a basal plate, which has an annular thickening around the edge. In addition, the basal plate is reinforced at six points by separate radial calcareous ribs. These ribs grow upward, so that they and the basal plate that surrounds them finally extend as the so-called septa into the stomach cavity of the polyp. By means of this exoskeleton structure, the polyp creates a shelter into which it can retire if danger threatens. Finally, once the septa have reached a given height, the polyp creates a new basal plate above them. When this occurs, the lower part of the polyp is ligatured, and over a continuous series of new levels is created and the stem of coral grows in height. The coral polyp secretes the lime (calcium carbonate) on the external surfaces of its basal plates. Corals generally use aragonite, a carbonate material, to make the calcium carbonate.

In a recent study published in *Geology*, Justin Ries, Steven Stanley, and Lawrence Hardie have suggested that scleractinian corals may have the capacity to alter their skeleton to match the changing chemistry of seawater, making them the only known animals to achieve such a feat. The study shows that when there is a decrease in the ratio of magnesium to calcium in the seawater, corals can switch to calcite for producing calcium carbonate. "This is intriguing because, until now, it was generally believed that the skeletal composition of corals was fixed," said coauthor Justin Ries, a postdoctoral fellow at Johns Hopkins University. These researchers formulated six different magnesium-to-calcium ratios that existed throughout the 480-million-year history of corals and then added three species of corals. They examined the mineral composition of the coral skeletons and found that each kind of coral had produced its skeleton based on the kind of water it was in. Ries postulates that this "mineralogical flexibility" provides corals with an "evolutionary advantage" because it would take more energy for corals to produce skeletons that are not favored by the chemistry of the seawater surrounding them. "This is particularly significant given recently observed and predicted future changes in the temperature and acidity of our oceans via global warming and rising atmospheric [carbon dioxide], respectively—that will presumably have a significant impact on corals' ability to build their skeletons and construct their magnificent reefs," Ries says.

What Corals Teach Us

In the "thick of things," scleractinian corals teach us that inside and outside, body and place are not exclusive domains. Their bodily structure is part of their dynamic engagement with water. Their morphology—their very skeletal structure—entails the chemistry of their ecosystem. The boundary-making practices by which scleractinians differentiate between "themselves" and their "environment" are constitutive—they incorporate elements of their environment within themselves in order to produce conditions that will support a reef ecosystem, their own home. They are engaged in making changes in the ocean as part of the ocean. They are not just *in* the water; they are of the water. The relationship between their physiology and their ecology is fluid and intimate. The ongoing materialization of their bodily boundaries is contiguous with their surrounding environment. Being a scleractinian coral is a matter of responding to marine conditions that the scleractinians participate in materializing and that constitutively help to materialize them. Indeed, coral survival depends on their ability to adapt to their changing environment. They show us that perhaps nothing stands independently constituted and positioned. Ontology (being-ness) is never an issue of existing in the world, but of being of the world.

Enmeshed in eco-sensitive formations concerning their embodiment, corals are, Justin Ries suggests, evolutionarily sensitive to processes of differentiation. Coral skeletons mark the limits of the exclusivity of boundaries. This is not about dissolving boundaries; this is about how boundaries, as very real material things, are activated, in process. As changes in oceans become more pressing for us all, corals cannot afford to ignore the chemistry and temperature of their habitat. The coral's ontology is constrained by interchange. Thus, corals are beings dynamically produced, sustained, and entangled with other phenomena. Corals are always interchanging with their marine environment and respond to differential input produced through these same enmeshments, reworking their bodies in order to survive. That is, corals must recognize and adapt to changes in their environment to elude death. These are life and death stakes.

Let us not forget that in some kinds of scleractinian corals, the colony of polyps functions as a single organism by sharing nutrients through a gastrovascular network. Genetically, the polyps are clones, each having exactly the same genome. Each polyp generation grows on the skeletal remains of previous generations, forming a structure that has a shape characteristic of the species, but also subject to environmental influences. Corals are entangled; they are both individuals and colonies. Corals are differentiated, while challenging anthropocentric ideas of individuation. Difference between one polyp and another is always constitutive and relational. Corals are good teachers for understanding how ontology literally matters, the materiality of being. Coral embodiment is always ongoing, dynamic, and situated. Corals are always becoming themselves through entanglements with others. Never only themselves, corals are ensembles; they are made up of relationships. To be coral is to cohere.

Corals are living, metabolizing, reproducing, interchanging beings that exist by crossing divides between organic and inorganic, ocean and flesh, aragonite and calcite. They show us that being-ness is not an individuated process that preexists interchange, but rather ontology is constituted through relationship. The being so constituted exists across and beyond its calcium carbonate exoskeleton. Being, in this way, is a distributed practice. Being is an enmeshment, enmeshing "self" and "other." Being is an *engagement with the world as part of the world.* Crucially, ontology is not about *things,* but about relationships. *To be* is not to be contained, grounded, or bounded, or not this alone. It is to be able to live multiplicity. The multiply-lived is not a clear, coherent, and positive integrity that can be seen and held together all at once. Being-ness is not integrity; it is resource, reach, and possibility. And as corals produce reefs, "we" (for example) participate in materializing what constitutively helps to materialize "us." Our future as humans depends on the relationships we build with the world as part of the world. Like corals, we create our world with our bodies, and by doing so recreate our bodies as part of that world. We are always in transition, are always work in progress.

Further Resources

Pearse, V., Pearse, J., Buchsbaum, M., & Buchsbaum, R. (1987). *Living invertebrates.* Palo Alto, CA: Blackwell Scientific Publications.

Reaser, J. K., Pomerance, R., & Thomas, P. O. (2000). Coral bleaching and global climate change: Scientific findings and policy recommendations. *Conservation Biology, 14*(5), 1500–11.

Ries, J. B., Stanley, S. M., & Hardie, L. A. (2006). Scleractinian corals produce calcite, and grow more slowly, in artificial Cretaceous seawater. *Geology, 34*(7), 525–28.

Roessler, C. (1990). *Coral kingdoms.* New York: Abradale Press.

Smith, S. V., & Buddemeier, R. W. (1992). Global change and coral reef ecosystems. *Annual review of ecology and systematics, 23,* 89–118.

Willmer, P. (1990). *Invertebrate relationships: Patterns in animal evolution.* Cambridge: Cambridge University Press.

Eva Hayward

■ Living with Animals
Dolphins and Humans: From the Sea to the Swimming Pool

I have met with a story, which, although authenticated by undoubted evidence, looks very much like a fable.

—Pliny the Younger

The extensive history of dolphin-human interactions represents a complex and unique quality of interspecies relationship that has been both revered and abused by humans. Dolphins are distinctive among wild animals in that it is not unusual for them to initiate sociable contact with humans in the absence of humans provisioning them with food.

And few wild animals are featured so prominently—and often so positively—as dolphins in human culture, mythology, art, architecture, literature, anthropology, history, and the media. In fact, the degree to which they have formed close social bonds with humans has remained somewhat of an "enigma," as long-time researcher Christina Lockyer noted in 1990. However, dolphins are also considered significant economic resources to humans—for recreation, education, tourism, and entertainment and sometimes as food (Frohoff, in press; Twiss & Reeves, 1999).

The term "dolphin" is used here loosely to refer to cetaceans within the suborder Odontoceti: the toothed cetaceans (as contrasted with the other cetacean suborder, Mysticeti, which includes whales having baleen instead of teeth). Odontocetes make up the vast majority of cetaceans—approximately seventy-one of the approximately eighty-six species. They are represented by diverse species within six families: Delphinidae ("true" dolphins), ranging from the tiny Commerson's dolphin to the majestic orca; Monodontidae, consisting of beluga whales and narwhals; Phocoenidea ("true" porpoises); Physeteridae, including the largest of odontocetes, the sperm whales; Platistidae, the river dolphins; and the somewhat elusive beaked whales of the family, Ziphiidae. Some baleen whales are also interactive, such as gray whales in Mexican lagoons, but are outside the scope of this review.

History

The history of dolphin-human interactions in many cultures around the world is replete with mythology, folklore, and accounts of actual events. Petroglyphs of dolphins date back as far back as 9,000 years. Other cultures, such as the indigenous New Zealand Maori, Australian Aborigines, Amazonians, and Northwest Coast tribes in North America, also exhibited numerous representations of dolphins in their history and myth—some of which still continue. And dolphins were prominently venerated in the art, coins, and buildings (such as the ancient Greek temple of Delphi) of various other civilizations, such as those of the ancient Greeks and Romans. One of the earliest confirmed artistic representations of dolphins in Crete dates as far back as 3500 BCE. The most established literary reference to dolphins from classical antiquity is Aesop's 600 BCE fable "The Monkey and the Dolphin." The intelligence of dolphins was highly regarded by Aristotle and others at least as early as 384 BCE.

The importance of protecting dolphins was proclaimed by Oppian around 200 CE when he wrote, "The hunting of dolphins is immoral, and that man can no more draw night to the gods as a welcome sacrificer nor touch their alters with clean hands, but pollutes those who share the same roof with him, who willingly devises destruction for the dolphins. For equally with human slaughter the gods abhor the deathly doom of the monarchs of the deep." Yet such reverence for dolphins was not so clearly expressed by other civilizations that did hunt dolphins at that time. Today, a dichotomy still exists in how different cultures relate to dolphins. Some view dolphins as little more than food, fertilizer, or a form of commerce, whereas others demonstrate a high degree of respect for these animals. Beginning around the 1960s, television shows such as *Flipper*, scientists such as Drs. John Lilly and Ken Norris, and writers such as Jim Nollman and Wade Doak both contributed to and reflected an elevated status of dolphins in the popular culture of some societies.

"Friendly" Dolphins

The line between fact and fable is blurred in multiple tales about dolphin-human interactions from classical antiquity to the present. Ancient stories of free-ranging dolphins "befriending" humans and giving them rides at sea, saving the lives of humans, and cooperatively fishing with them continue into modern times, with a good number of such accounts being documented by scientists and the media over the last century. Associations between free-ranging dolphins and humans typically develop when both experience repeated mutual reinforcement of positive sociable behavior. With a few exceptions, provisioning the dolphins with food has not been a typical component of sociable interactions between dolphins and people (Frohoff & Peterson, 2003; Gales et al., 2003).

Some solitary dolphins who are rarely, if ever, observed in the company of other dolphins initiate social interaction with people, sometimes even forming close bonds with individuals over time. Pliny the Elder (23–79 CE) told one of the first stories about friendly solitaries, in which a dolphin named "Simo" in the Mediterranean Sea formed a close relationship with a boy and even gave the boy "rides" to school and back. Then in 1955, news appeared of a solitary, sociable bottlenose dolphin, named "Opo" by his admirers, in the town of Opononi, New Zealand. Since this time, over seventy solitary, sociable dolphins have been documented in coastal locations around the world (Gales et al., 2003). Primarily, these have been bottlenose dolphins, but within the past decade, numerous solitary sociable beluga whales have been observed by Cathy Kinsman and other researchers in eastern North America, and two orcas in western North America have also been documented (Frohoff & Peterson, 2003; Gales et al., 2003). Dolphins in groups also interact with people who are in boats, on shore, or wading or swimming with a varying degree of contact and habituation.

Stories of dolphins assisting swimmers and boaters in distress have been reported in ancient as well as in recent times. Plutarch wrote, in "On the Intelligence of Animals" (66 AD), about the dolphin, "It has no need of any man, yet is the friend of all men and has often given them great aid." People of the Amazon have recounted incidences in which river dolphins have saved the lives of people whose boats had capsized. In modern times, dolphins have been reported, often by reputable sources (such as scientists and government personnel), to have assisted or saved the lives of people who may have otherwise drowned by holding them at the surface or by pushing them toward shore. Dolphins have also protected people from shark attacks. This is not entirely surprising, considering the numerous accounts of dolphins defending each other from danger. Yet it is unknown why dolphins would risk themselves to benefit those of another species by exhibiting what may be pure altruism. But it is also evident that people in danger have not always been assisted by dolphins when in their presence.

An example of mutually beneficial dolphin-human interactions is seen where dolphins cooperatively hunt for fish with humans (Gales et al., 2003). Generally, both dolphins and fishermen benefit by working together to herd and catch fish, and various species of river and oceanic dolphins have been involved. Pliny the Elder (23–79) wrote that dolphins worked with humans to drive mullet into the nets in the Rhone River in France. Later, Oppian reported cooperative fishing at Emboea, and more recently, this practice has been observed and scientifically documented in the Mediterranean, North Africa, New Zealand, and South America. The Australian aborigines of Moreton Bay and Amazonian fishermen have also fished with the regular assistance of dolphins. In Laguna, Brazil, fishermen continue to fish cooperatively with bottlenose dolphins as they have done for several generations. Apparently, these cooperative and often highly sophisticated relationships are initiated and controlled by the dolphins rather than the fishermen. A well-known group of orcas in New South Wales, Australia, engaged in cooperative hunting of whales with humans in the mid-1800s, which lasted for at least five decades.

Dolphins as Living "Resources"

It is encouraging that in the past few decades alone, it has become far more popular and profitable to watch cetaceans than to kill them in many parts of the world. But there is a wide continuum of activities in which humans can view and interact with dolphins, ranging from nonconsumptive (e.g., watching them from land) to minimally consumptive (watching them from boats that are operated knowledgably and cautiously) to highly consumptive (e.g., inadvertent harassment or intentional capture from the wild) (Frohoff, in press, 2000; Gales et al., 2003). Increasingly, the appeal of swimming, and otherwise interacting closely, with dolphins in both captivity and in the wild has proliferated and has been commercialized into an international industry in which dolphins are often exploited. Dolphins are increasingly sought-after as lucrative sources of entertainment, recreation, and to a lesser degree, educational and therapeutic purposes. Yet responsible research, management, and public awareness of dolphin interactions in both captivity and in the wild have lagged sorely behind this expansion, resulting in not only dangerous, but occasionally fatal consequences for the dolphins.

Incidents in which dolphins harm humans are typically limited to abnormal or stressful conditions for the dolphins; such as in the absence of companionship with other dolphins, in situations where the public regularly provisions them with food, when intentionally or unintentionally disturbed or harassed, or in captivity (Frohoff, 2000; Gales et al., 2003). This is occurring more frequently with the increasing public demand for close interactions with these animals who are wild, even when well-trained. The only confirmed account of a human fatality resulting from a bottlenose dolphin occurred in Brazil when a solitary, sociable dolphin was physically restrained and abused by two men—an obvious case of self-defense as documented by biologist Marcos Santos. Two other human fatalities (one person being a trainer) were attributed to captive orcas (Rose et al., 2006). It is not uncommon for members of the public to become injured from swimming with captive dolphins or even from interacting with them at the poolside—injuries have included broken bones, internal injuries, bruises, and lacerations, some requiring hospitalization. There is also a very real potential for disease transmission between cetaceans and humans.

Dolphins in the Wild

As Erich Hoyt has noted, dolphin and whale watching in the wild has become one of the fastest-growing segments of the tourism market over the past few decades. In Shark Bay, Australia, a group of bottlenose dolphins has allowed close contact with waders who have regularly fed them for approximately forty years, which has evolved into an important source of local economic revenue and which has inspired other dolphin-feeding enterprises (Frohoff & Peterson, 2003; Gales et al., 2003). Other dolphin groups have been targeted by large numbers of humans seeking close contact in the absence of food provisioning. For example, tens of thousands of people attempt to swim with bottlenose and dusky dolphins from commercial boats every year in New Zealand. Spotted and bottlenose dolphins have been interacting with swimmers and divers in offshore Bahamian waters fairly consistently for over decades, and in Hawaii, spinner dolphins are regularly sought-after by swimmers and kayakers. Even in Japan, people flock to swim with the same free-ranging bottlenose dolphins who, ironically, may also be killed by local fishermen. These situations often begin noncommercially, involving only a few people and dolphins, and then become the focus of businesses attracting large numbers of people, many of whom are tourists.

Severe and lethal injuries from boats, cumulative and long-term changes related to breeding behavior, disruption of rest, displacement from habitat, and long-term responses to human activity (i.e., sensitization and habituation) have been associated with vessel and swimmer impacts (Frohoff, 2000; Gales et al., 2003). In particular, increased mortality was documented for solitary, sociable dolphins and for free-ranging dolphins who were regularly fed by humans. Lars Bejder and others observed that in some areas, the impact of numerous swimmers was minimal, whereas the impact of even a single swimmer in other habitats was severe enough to result in repeated disruption of rest, feeding, and other important dolphin activities. For instance, Rochelle Constantine has found that over time, the bottlenose dolphins in the Bay of Islands in New Zealand have been interacting with swimmers less and avoiding them more. David Lusseau recently found that an entire population of dolphins is being threatened by the numerous tourist boats that come to see the dolphins in Milford Sound, also in New Zealand. He documented that dolphins are being injured and killed by the boats, and up to 7 percent of them bear visible scars from boat collisions. Anna Forest observed that spinner dolphins in Kealakekua Bay, Hawaii, may be avoiding important areas for resting, nursing, and mating because of the increased presence of swimmers and boaters. Intensive whale-watching activity from boats appears to regularly disrupt important activities such as resting, feeding, and mating in seriously depleted populations of orcas in North America. Even when dolphins do not exhibit recognizable signs of stress in response to human interaction, excessive proximity to humans places them at higher risk of being deliberately or accidentally harmed by them.

Dolphins in Captivity

The first dolphin kept in captivity might have been in the first century CE, when an orca that had stranded was kept by Roman guards for sport (Twiss & Reeves, 1999). Private collections of dolphins held in tubs and pools were attempted as early as the 1400s. In the mid-1800s, P. T. Barnum publicly exhibited captive bottlenose dolphins and belugas in a New York museum. Facilities dedicated to this purpose began to burgeon in the late 1930s. Although accurate numbers of captive dolphins and facilities are not available, reports have estimated that at least 200 captive dolphin exhibits are currently in existence in at least sixty countries (e.g., Couquiaud, 2005; Rose et al., 2006). Captive marine mammals have traditionally been simply on display or have been used to perform trained behaviors for public performances. But over the past few decades, the popularity of interactive programs allowing the public to swim with, wade with, touch, and feed captive dolphins has created a tremendous increase in the international market for captures of these animals from the wild.

Capture from the wild and transport are clearly stressful and dangerous for dolphins (Frohoff & Peterson, 2003). A study by R. J. Small and D. P. DeMaster in 1995 revealed that mortality rates of bottlenose dolphins increase six-fold immediately after capture from the wild for approximately five days and do not resume to base captive mortality rates for up to thirty-five to forty-five days. This and other studies found that survivorship rates in bottlenose dolphins and orcas through the mid-1990s remained persistently lower than those of these species in the wild (although the differences ceased to be statistically significant for bottlenose dolphins) (see Rose et al., 2006). Although this indicates that dolphin husbandry has improved over the years, at least in some of the selected facilities, it has not done so to the extent that cetaceans live longer in captivity. And most of the facilities examined in these analyses were in North America, where captive standards have generally been superior, where staff are more

A dolphin in a feeding/petting pool who may only be seeking food yet must tolerate a multi-tude of hands touching him to obtain it. Courtesy of Toni Frohoff.

experienced, and where regulations have been stricter than in most other countries. Naomi Rose et al. (2006) noted research revealing that the overall mortality rate of captive orcas is at least two and a half times (and up to six times as high) as that of orcas in the wild.

These data for dolphin longevity are contradictory to what might be predicted, given that captive dolphins often have access to veterinary care, consistent food availability, and protection from natural predators and other threats encountered in the wild (Rose et al., 2006). Psychological and physical stress associated with captivity is probably the primary reason that cetaceans do not live as long or longer than their wild counterparts (Frohoff, in press; Frohoff & Peterson, 2003). Numerous peer-reviewed publications have related capture and captivity with stress, as evidenced by behavioral and physio-logical hormonal changes (e.g., elevated adrenocortical hormones) and diminished immunological response associated with illness and death (e.g., Couquiaud, 2005; Rose et al., 2006; Sweeney, 1990). Behavioral abnormalities include stereotyped behavior; unresponsiveness; excessive submissiveness, excessive aggressiveness, and excessive sex-ual behavior; self-inflicted trauma; and stress-induced vomiting. Other causes of death associated with captivity include physical injury, shock, ingestion of foreign objects, ulcers, heat stroke, exposure to chemicals, lack of escape from environmental hazards such as hurricanes, poor veterinary care, drowning as a result of entanglement, and injury or death from other dolphins or objects in the enclosure. Zoonotic disease trans-mission is also a serious concern, especially in programs that permit the public physical contact with the dolphins.

Captive enclosures, regardless of how natural they look to human tourists, are often designed primarily with the appeal to human visitors in mind (Couquiaud, 2005) and are insufficient to allow marine mammals to exhibit a normal range and quality of behaviors and social groupings. Chemical sanitation of the water in the pool can cause physical irritation to the animals' skin and eyes (Couquiaud, 2005; Frohoff & Peterson, 2003). Although some facilities are making attempts to improve their dolphin enclosures and husbandry techniques, a proliferation of new and expanded facilities often reflects a lack of experience and low level of professionalism in the maintenance of these animals. Examples include dolphins confined to small swimming pools and even shallow holes dug into dirt and lined with plastic without filtration systems; dolphins used in traveling road "shows"; dolphins housed alone in pools in buildings without natural light or appropriate social companions; and arctic beluga whales placed in abnormally hot climates (Rose et al., 2006). Sea pens and cages in which dolphins are fenced in natural water facilities might appear to be better for the dolphins in the public eye, yet they are often completely inadequate for the maintenance of dolphins, largely because of a lack of environmental controls (Couquiaud, 2005; Frohoff & Peterson, 2003).

Interactive programs are proliferating because of large tourist revenues, many of which are often generated from the cruise ship industry, which brings tourists to interact with captive dolphins as part of their recreational itinerary. Bottlenose dolphins are the odontocetes most frequently used in these facilities, but beluga whales, orcas, and other odontocetes are also used in these types of programs. The animals are typically afforded little or no control over the presence of human "visitors" in their enclosures. Increased noise, environmental stimuli, and disruption of rest, as well as greater risk of disease, harassment, and physical injury from the public, are sources of stress in addition to those already encountered in the captive environment (Frohoff, in press). The three studies that focused on behavioral indicators of stress in captive dolphin swim programs and the one study on touching and feeding encounters used in these programs all reported stress-related and avoidance-related behaviors in the presence of swimmers—some indicative of long-term negative physiological impacts. According to various researchers including Betsy Smith (the founder of dolphin-assisted therapy using captive dolphins), there is no research substantiating that using dolphins is any more effective than therapy involving domestic animals (Frohoff & Peterson, 2003). Although dolphin-assisted therapy has become increasingly popularized and controversial, it is responsible for the acquisition of numerous dolphins and the creation of new facilities, many of which are poorly suited to maintain dolphins.

Many countries have few, if any, legal requirements governing the welfare and unique needs of captive dolphins. This often results in diminished welfare and even fatal consequences for the animals (Couquiaud, 2005; Frohoff & Peterson, 2003). However, some countries, such as Chile, prohibit the capture or confinement of these cetaceans, and others such as Italy and Brazil prohibit interactive programs involving physical contact and public feeding. Increasing numbers of countries are phasing out captive public display facilities, and some have denied permits to capture, import, or export dolphins (including Australia and Mexico, the latter of which has most recently banned these practices).

Implications of Captures on Populations

Capturing marine mammals from the wild can do far more than harm the targeted animals, so it is not a concern restricted to those concerned about animal welfare. As has been noted by the United States National Marine Fisheries Service (NMFS) and the World Conservation Union (IUCN), live captures can threaten the welfare and survival of remaining individuals in pods and even entire populations. This is largely because few, if any, appropriate scientific assessments of the population-level effects are conducted on targeted populations prior to approving captures. For example,

A drive fishery, one of many slaughters of dolphins and other cetaceans in Japan, which still continue. Courtesy of Toni Frohoff.

approximately 100 Indio-Pacific bottlenose dolphins in the Solomon Islands were captured for public display and export despite a lack of scientific surveys of the local population. Perhaps the most inhumane and depleting method of obtaining cetaceans for public display is the drive fishery in Japan, which receives large sums of money from some dolphinaria that make requests for live dolphins and which may actually be encouraging a practice that might otherwise be discontinued.

There do not appear to be any peer-reviewed studies documenting significant educational benefits of captive facilities or interactive programs (Frohoff & Peterson, 2003; Rose et al., 2006). Some professionals, such as members of the International Marine Animal Trainer's Association, maintain that captive facilities provide a unique opportunity for viewing and learning about marine mammals. Others argue that such facilities are little more than aquatic circuses or petting zoos and that the quality of information imparted to people who visit them is misleading or even dangerous to dolphins in the wild. For example, a U.S. government biologist stated, "There is growing concern that feeding pools, swim programs, and other types of interactive experiences with marine mammals in captive display facilities may perpetuate the problem of the public feeding and harassment of marine mammals in the wild" (Frohoff & Peterson, 2003, p. 67). The only comparison of viewing cetaceans in captivity with viewing those in the wild appears to be the study recently conducted by Carole Carlson and Toni Frohoff, who concluded, that when conducted in a responsible and precautionary manner, viewing cetaceans in the wild seems to offer more benefits and fewer negative impacts—to both cetaceans and people—than viewing them in captivity and can also provide a uniquely important form of tourism and income to local communities.

Other Threats to Dolphins

In addition to the exploitation of captive and free-ranging dolphins for recreation, directed dolphin hunting still continues in some areas, despite international objections from scientists, international regulatory bodies, conservationists, and the general public (Gales et al., 2003; Twiss & Reeves, 1999). Dolphins killed are sometimes used for human consumption, despite the high levels of mercury and other contaminants in these animals that exceed health safety standards. Most notably, dolphins in Japan are killed in what are referred to as "drive" fisheries, in which dolphins are driven into shallow areas and then killed with harpoons, knives, or other sharp objects. Despite depleted populations, hundreds to thousands of various species of dolphins and porpoises are slaughtered annually and used for human food, pet food, and fertilizer. It has been estimated that approximately 400,000 Dall's porpoises have been killed by Japan in the past twenty-five years. In the recent 2003–2004 drive hunt season, in only one of the hunting towns, over 1,000 dolphins were reportedly killed, and at least seventy-eight were captured alive for the captivity industry. In Denmark's Faeroe islands, approximately 1,000 pilot whales and sometimes other dolphins are killed annually in a similar manner. In some regions in South America and Asia, dolphins are hunted for food or used as bait.

Subsistence hunting of beluga whales by aboriginals, their ancestors, and others has occurred in regions of Chukotka, Alaska, Canada, and Greenland for thousands of years—narwhals have been hunted in Greenland and eastern Canada for centuries (Twiss & Reeves, 1999). However, in recent centuries, Canadian Inuit were hired by commercial whalers, hunting in Greenland became commercial, and modern boats and equipment are now frequently used in these hunts. Other species such as bottlenose dolphins and orcas are also occasionally hunted in Eastern Canada and Greenland. In 2000, Cook Inlet beluga whales in Alaska were recognized as "depleted" under the Marine Mammal Protection Act (MMPA) when the population was estimated to have decreased to roughly 350 animals. Prior to the reduction in the annual harvest, the government estimated that roughly eighty belugas were hunted in 1998 alone. Although the allowable Alaska Native harvest has been substantially reduced, the population is now highly susceptible to adverse impacts.

Dolphins face numerous and increasingly challenging threats to their habitats, including ship strikes; prey reduction resulting from human overfishing and climatic changes caused by human industrial activity (such as the acidification and warming of the Pacific Ocean because of carbon dioxide emissions); and chemical pollutants such as heavy metals and persistent organochlorine chemicals such as DDT and PCBs (Gales et al., 2003; Twiss & Reeves, 1999). Acoustic threats that dolphins have encountered for decades include vessel noise, seismic research, acoustic devices used in fishing, explosives, and drilling noises associated with obtaining gas and oil and other commercial activities. Although these continue to increase, a newer and perhaps more dangerous acoustic threat from military sonar emissions has been associated with extreme behavioral events and disturbing incidents of deaths of dolphins around the world (Frohoff & Peterson, 2003; Twiss & Reeves, 1999).

Many dolphin deaths related to human fishing activities (entanglement, bycatch in high-seas driftnets, gillnet, and other fisheries) are incidental, yet continue to pose serious threat to dolphins. For example, thousands of dolphins in the United Kingdom are estimated to be dying each year because of fishing nets, despite European laws designed to protect them. Perhaps the most famous of these situations is that of the mortality associated with tuna fishing fleets in the eastern tropical Pacific that have deliberately targeted dolphin schools because of their association with yellowfin tuna. But the efforts of nongovernmental organizations such as the Earth Island Institute and legislation such as the U.S. Marine Mammal Protection Act have reduced the mortality of Eastern spinner dolphins and spotted dolphins in tuna nets by more than 97 percent since 1990, from an annual mortality of at least 100,000 per year to less than approximately 3,000 annually (primarily from tuna caught in Mexico and Venezuela). "Dolphin-safe" labeling on cans of tuna represents one of the greatest impacts of conservation in the commercial sector.

Numerous international agreements, of which many but not all governments are a part, have been established in addition to regional, national, and local forms of regulation and legislation designed to regulate hunting, trade, and protection of cetaceans (Twiss & Reeves, 1999). As Erich Hoyt has demonstrated, marine protected areas in which entire ecosystems are protected, as opposed to the protection of a single species within an ecosystem, offer one of the most important approaches to conservation. Such efforts are particularly important for cetaceans because recovery from depletion may be slow given that they typically have a low population growth rate and often are far-ranging.

Dolphin-Human Communication

Perhaps because of the charisma, impressive cognitive abilities, and sociability of dolphins, efforts to develop methods in which people can communicate with them have been made by scientists as well as numerous others. Researchers are increasingly observing sophisticated social learning abilities, including vocal and motor imitation of humans. Compelling evidence of culture in odontocetes has become well established, as evidenced by tool use and the sharing of vocal dialects through generations. In a 2001 article in the *Journal of Behavioral and Brain Sciences*, Luke Rendell and Hal Whitehead noted, "The complex and stable vocal and behavioural cultures of sympatric groups of killer whales (*Orcinus orca*) appear to have no parallel outside humans, and represent an independent evolution of cultural faculties." And as Lori Marino et al. (1994) have observed, dolphins "share those neurological, cognitive, and social characteristics with great apes and humans that are generally regarded as having been important for the development of self-awareness in primates."

Over the years, many attempts have been made to communicate with dolphins with a variety of means, scientifically and non-scientifically, through computers, music, and other media. The most well-known of these efforts were probably those of dolphin-research pioneer Dr. John Lilly, whose initially conventional, then subsequently unconventional and controversial, methods popularized the idea of human-dolphin communication beginning in the 1960s—to the dismay of much of the scientific community. Thereafter, the study of dolphin-human communication was generally academically dismissed until recent application of rigorous and systematic scientific techniques were applied to the study of interspecies communication within the last few decades (Frohoff & Peterson, 2003). Research on dolphin-human communication can provide an invaluable window into dolphin-dolphin communication as well as dolphin cognition, emotion, and conservation.

In captivity, reinforcement training may be considered one form of interspecies communication—albeit one in which humans are in a dominant position. Numerous controlled experiments by Louis Herman, Adam Pack, and many others have been conducted to assess language comprehension in captive bottlenose dolphins and have demonstrated that dolphins can understand syntax and semantics; generalize known words into actions and concepts, regardless of color, shape, and size; and demonstrate remarkable learning capabilities. In the wild, interactions between humans and sociable dolphins have indicated a sophisticated level of communication (e.g., Frohoff & Peterson, 2003).

These studies indicate that dolphins have a highly sophisticated communication system, even if it is very different from that of humans. Interspecies *mis*-communication also may occur, such as when humans in boats misinterpret indications of disturbance in dolphins as signs that the dolphins are excited to be in their company.

In conclusion, an interspecies form of cultural exchange and coevolution appears to have been occurring between some dolphins and humans for millennia. However, the beauty of both the myth and the reality is marred by a tragic irony—that the more closely and frequently dolphins and human interact, the more vulnerable dolphins are to being injured or killed by humans and human activity. Consequently, many researchers have concluded that the precautionary principle should be applied to such interactions, so that people watch and learn about dolphins in a manner that is not at the dolphins' expense.

Further Resources

Couquiaud, L. (2005). A survey of the environments of cetaceans in human care. *Aquatic Mammals, 31*(3), 279–80.

Frohoff, T. G. (2000). *Behavioral indicators of stress in odontocetes during interactions with humans: A preliminary review and discussion.* International Whaling Commission Scientific Committee, SC/52/WW2.

———. (in press). Marine animal welfare in captivity. In M. Lück (Ed.), *Encyclopedia of tourism and recreation in marine environments.* Wallingford, UK: CABI Press.

Frohoff, T. G., & Peterson, B. (Eds.). (2003). *Between species: Celebrating the dolphin-human bond.* San Francisco: Sierra Club Books.

Gales, N., Hindell, M., & Kirkwood, R. (Eds.). (2003). *Marine mammals: Fisheries, tourism and management issues.* Collingwood, Australia: CISRO.

Marino, L., Reiss, D., & Gallup, G. G., Jr. (1994). Mirror self-recognition in bottlenose dolphins: Implications for comparative investigations of highly dissimilar species. In S. T. Parker, R. W. Mitchell, & M. L. Boccia (Eds.), *Self-awareness in animals and humans: Developmental perspectives* (pp. 380–91). New York: Cambridge University Press.

Rose, N. A., Farinato, R. H., & Sherwin, S. (Eds.). (2006). *The case against marine mammals in captivity.* Washington, DC: The Humane Society of the United States and the World Society for the Protection of Animals.

Sweeney, J. C. (1990). Marine mammal behavioral diagnostics. In L. A. Dierauf (Ed.), *CRC handbook of marine mammal medicine: Health, disease, and rehabilitation* (pp. 53–72). Boston: CRC Press.

Twiss, J. R., II, & Reeves, R. R. (Eds.). (1999). *Conservation and management of marine mammals.* Washington and London: Smithsonian Institution Press.

Toni Frohoff

■ Living with Animals
Elephants and Humans

Just as there is a diversity of human cultures, there is a diversity of human-elephant relationships. Every culture, and indeed every individual within a culture, crafts a unique way of being in relationship with members of other species. Subsequently, when we wish to understand the relationship between elephants and humans, it is necessary to first consider the specific culture and period in history. How elephants live and elephant-human relationships are tied to human history.

Up until a century or more ago, elephants roamed most of Asia and Africa. In Africa, some tribes hunted elephants, but the tribes killed relatively few elephants compared to the decimation brought by European settlement. For example, for the Acholi people of Uganda, the elephant—or *lyec*—formed an integral part of the culture, symbolizing strength and virtue. The two cultures—elephant and human—lived in a shared system of values and law, as revealed in the Acholi story of Gipir and the Elephant. Here we get a glimpse of the Acholi-elephant relationship long before the colonialization of Africa:

> One day, an elephant raided the field of the farmer Gipir. Gipir quickly took the first spear he could lay hands on to frighten off the elephant, who ran into the forest wilderness. However, Gipir forgot and left the spear in the forest. The spear belonged to Labongo, Gipir's brother, who wanted the precious spear back. Gipir searched and searched in the forest wilderness but was unable to find the spear. Just as he was about to give up, he suddenly encountered *Min Lyec,* the Elephant Matriarch—the leader of all elephants. *Min Lyec* did not charge Gipir even though the spear had been used to threaten the other elephant. Instead, she gave over the spear to Gipir because she knew he had meant no harm but was only telling the other elephant to not take his families' food—just as an elephant might charge a human who took succulent greens away from an elephant family. *Min Lyec* and the village Elders both looked after each other's people: elephant and human alike.

Like the American bison and Sioux Indians of North America, elephants and many African indigenous cultures reflected each other's ways and customs. As we saw in the Acholi story of *Min Lyec*, people and elephants shared many customs and cultural patterns. This is not surprising, because as different as elephants may physically appear from humans, they have much in common.

Both species are social species that form lifelong friendships and live in families. Elephants spend their entire lives eating, sleeping, birthing, dying, and loving together. Similar to many human cultures, elephant natal families are composed of many generations. These constellations of female caretakers responsible for the raising of young elephants are called "allomothers." Elephants and humans both show complex emotion: grief, joy, excitement, anxiety, and fear. They even share similar cultural rituals. When one of their family or friends dies, the older members bring the younger elephants to visit the bones and graves. The elephants touch the whitened bones and skulls in the same ways that they greet each other when alive.

The fact that elephant and human cultures share so much in common is not surprising. In addition to the way they behave, elephants and human brains have much the same basic structures and mechanisms that give us personality, culture, language, and emotions. Research in the neurosciences has shown that all mammals share the same basic brain structures. Associated with these structures are specific physiological (e.g., autonomic, cardiovascular, immunological, nervous) and psychological and behavioral (e.g., attachment and social bonding, maternal behavior, facial recognition, learning, fear conditioning, pain, aggression, and the regulation of and experience of emotion) systems. Like humans, elephant brains at birth are only a fraction of the volume they attain as adults. At birth, elephant brains are approximately 35 percent adult size and are characterized by human-like features that account for high intelligence, emotionality, and a capacity for complex social interactions. And like humans, elephant young spend extended periods of time in the care and protection of adults from birth through adolescence.

It is perhaps this feeling of relatedness that makes elephants such an important figure in the mythology, religion, literature, and arts of many human cultures. In India, the elephant is worshipped as the god Ganesha. Unlike the situation in most of Africa, however, the Hindu religion prohibits the killing of elephants. Yet despite their status as gods, many Asian elephants lead a very ungodlike existence, and work in religious parades, zoos, circuses, and logging operations. In India, the process of domestication is called *Mela Shikar*. Young elephants who are chosen to work for humans are taken from their families and may be chained, starved, tortured, and beaten into submission. Riding already captive broken elephants (*kunkis*), a wild elephant is captured by slipping a rope noose around his or her neck. To avoid strangulation, the elephant finally stops fighting to get loose. More knotted ropes bind the neck, the back legs, and the forelegs. In cases in which the animal is more difficult to break (which means it complies with whatever the *mahout*, or elephant owner wishes), their head is tied to the ground, preventing them from rising or moving without strangling. They are also refused food and water.

Elke Riesterer, a massage therapist who works in the rehabilitation of injured, captive elephants, recounts how the mahouts treat Asian bull elephants. The following passage vividly illustrates the incongruities that characterize many human-animal relationships— worship and admiration mixed with tremendous cruelty and the desire to conquer and subordinate:

> One day, in visiting a mahout camp, I came upon a bull chained at the end of a shallow pond next to a hillside. He was pulling desperately at his chain and rocking back and forth. An employed elephant in *musth*—the period when a male elephant experiences a hormonal shift and begins to search for a mate—will lose much money for its owner since the bull is unable to work. In order to shorten the length of musth, which can last anywhere from a few days to months, the owner or mahout tied the bull to a tree on a very short length of chain. The mahout had stopped feeding the bull and provided very little water to reduce the elephant's strength and therefore control him more. The veterinarian students watching this spectacle clamored around closing in as closely as they dared to witness the spectacle, curious yet detached from the bull's tortured experience. Agitated, the bull yanked at his chain, trying to pull farther away from the crowd. The fluid still streamed from his temples. Tears welled up in my eyes.

Life has become extremely stressful for elephants in all countries since the era of widespread European colonialization. Overall, the relationship between humans and elephants today can be characterized as violent and hostile. In both Asia and Africa, elephants have little habitat left and are confined mostly to small reserves. Whereas elephants and humans in places like Kenya, Tanzania, South Africa, and Uganda once shared cultures and land, elephants are now besieged by a burgeoning human population. This recent change in the human-elephant relationship has been termed *Human-Elephant Conflict*, or HEC.

In general, Europeans who came to elephant lands came to hunt and kill for ivory and trophies, a practice that resulted in the near extinction of the species. The European presences also perturbed the traditional relationships between local elephant and human cultures. For this reason, parks and reserves had to be established to protect the few remaining herds, and the African elephant remains in serious danger because of widespread genocide from human poaching, legal culls (the systematic killing of elephants for meat, ivory, and population control), and encroachment by humans into their shrinking habitat. As human populations continue their inexorable growth in both Asia and Africa, a struggle with elephants is being waged for the few remaining fragments of undeveloped lands. Elephants have few refuges remaining that are not inhabited by humans, and as a consequence many are starving.

Unexpectedly, human violence against elephants coupled with habitat degradation has also caused elephants to behave differently. Here, sadly, we see yet again another similarity between elephants and humans. Like humans, elephants are vulnerable to trauma and stress and may exhibit symptoms of posttraumatic stress disorder (PTSD). Elephants are highly stressed and "on edge." Today in Uganda, both Acholi and elephant societies have been shattered by violence. Civil war has left a genocide in it its wake. In parallel, elephant communities have been broken and decimated from massive poaching to obtain ivory to sell for guns and food. Like the older elephants who are poached for their ivory, Ugandan elders have been systematically killed. The results have been similar, with bands of young elephants and Acholi children roaming throughout the countryside without family or homes. Elephants and humans are cultures in mutual crisis, and the traditional bond of respect between Acholi and elephant is gone.

Elsewhere in the world, the elephant-human relationship is also often based in violence—and inconsistencies. For example, elephants may be viewed as fantastic, loveable creatures like Disney's Dumbo, yet still they are confined to zoos and circuses, where their existence differs little from the captive elephants of Asia. They often live in depauperate environments of concrete stalls, poor air, and limited space. As a result, captive zoo and circus elephants can often develop terrible foot problems, often requiring partial amputation. They are frequently kept in isolation or have limited contact with other elephants and may be moved to different zoos or parks without consideration of their social relationships.

Men groom Boon Khum, one of the adult Asian elephants at a sanctuary. Location: Elephant Nature Park, Chiang Mai, Thailand. ©William Albert Allard/National Geographic Image Collection.

Amidst this violence, there are individuals trying to help elephants. These individuals provide us with an alternative way in which humans can relate to elephants. Instead of wanting to have power over an elephant, these people have learned how to live with elephants as equals. For example, just outside Nairobi (a hub for much illegal poaching and ivory trade), the David Sheldrick Wildlife Trust cares for young elephants that have been orphaned by poaching, culls, and other human activities. There, founder Daphne Sheldrick and a team of Kenyan Elephant Keepers not only tend to the physical needs of the orphans in shock from witnessing the deaths of their families, but also help heal the minds and hearts of these young elephants. The Keepers recreate the constellation of allomothers by nursing, caring, playing, feeding, and sleeping with the orphans twenty-four hours a day. They also teach the young elephants what it means to be an elephant, so that when the young elephants are released back into the wild they are able to live successfully with the herd. Often these youngsters remember their early healing relationship with the Keepers and return with their children and grandchildren to visit their human family at the Trust. The Keepers of the Trust are guardians of elephant culture.

Elephant sanctuaries operate elsewhere. In the United States, at the Elephant Sanctuary in Tennessee (www.elephants.com), Carol Buckley and Scott Blais have established a haven covering thousands of acres where circus and zoo elephants can live together in freedom and care. Here, the human-elephant relationship is one of parity and equality, and each species lives in mutual respect. By supporting the individual's path of life, humans and elephants enrich each other's lives and enhance the possibilities for living in peace. In India, Suparna Ganguly is working to secure basic rights for temple elephants and elephants who are used for labor in the forests. In all these examples, we see models of new relationships with elephants that are based in parity and mutual respect instead of domination. The future of elephants is difficult but the choice is clear. And as we saw with the Acholi and elephants of Uganda, the lives and fates of humans and elephants are intertwined. We as humans can have the joy and honor of living alongside the wonder of elephant society if we are willing to change ourselves.

Further Resources

Abe, L. E. (1994). The behavioural ecology of elephant survivors in Queen Elizabeth National Park, Uganda. Ph.D. Dissertation. Cambridge University.

Bradshaw, G. A., Schore, A. N., Poole, J. H., Moss, C. J., & Brown, J. L. (2005). Elephant breakdown. *Nature, 433*, 807.

The David Sheldrick Wildlife Trust. http://www.sheldrickwildlifetrust.org/

Douglas-Hamilton, I., & Douglas-Hamilton, O. (1975). *Among the elephants.* New York: Viking Press.

Fuentes, A. (2006). The humanity of animals and the animality of humans: A view from biological anthropology inspired by J. M. Coetzee's *Elizabeth Costello. American Anthropologist, 108,* 124–32.

Ghosh, R. (2005). *Gods in chains.* Bangalore, India: Foundation Books.

Kurt, F., Mar, K. U., & Garaï, M. E. (2005). *Giants in chains: History, biology and preservation of Asian elephants in captivity. Ethics and Elephants.* Baltimore, MD: Johns Hopkins University Press.

Moss, C. J. (1988). *Elephant memories: Thirteen years in the life of an elephant family.* New York: William Morrow.

Williams, C. (2006, February 18). Elephants on edge. *New Scientist,* 39–41.

Gay A. Bradshaw and Evelyn Lawino Abe